THE BIOLOGY OF GALL-INDUCING ARTHROPODS

Editors:

Gyuri Csóka
Hungarian Forest Research Institute
Mátrafüred Research Station
3232 Mátrafüred, P.O. Box 2, HUNGARY

William J. Mattson
Forestry Sciences Laboratory, U.S.D.A. Forest Service
5985 Highway K
Rhinelander, WI 54501-0898

Graham N. Stone
Institute of Cell, Animal, and Population Biology
University of Edinburgh, Ashworth Laboratories
King's Buildings, West Mains Road
Edinburgh EH9 3JT

Peter W. Price
Department of Biological Sciences
Northern Arizona University
Flagstaff, AZ 86011-6540

PREFACE

This proceedings is the result of an international symposium that was held August 14-19, 1997 in Mátrafüred, Hungary. It was organized under the guidlines of the International Union of Forestry Research Organizations (IUFRO) by the Hungarian Forest Research Institute, Department of Forest Protection, and the North Central Research Station of the U.S. Forest Service. IUFRO Working Party S7.03-02 (Gall-forming Insects) and Subject Group S7.01-00 (Physiology and Genetics of Plant-herbivore Interaction) convened for this conference.

The proceedings is published and distributed by the USDA Forest Service, North Central Research Station in recognition of and unstinting endorsement of the IUFRO goals of promoting strong world partnerships, particularly in forest science, for the sustained management, conservation, and betterment of the World's forests and woodland ecosystems.

ACKNOWLEDGEMENTS

The editors gratefully acknowledge the crucial support of the Hungarian Forest Research Institute, the Forest Department of the Hungarian Ministry of Agriculture, and the private companies Agroinform Publishing House, InterDnet, and Matra Forest and Wood Company for underwriting the costs of organizing the symposium, and the North Central Research Station for assembling, editing, printing, and distributing this publication. Finally, the editors also wish to thank all of the participants whose spirited contributions helped make this scientific exchange so worthwhile and extraordinary.

TABLE OF CONTENTS

BIOGEOGRAPHY AND BIODIVERSITY

Gall-inducing and other gall midge species (Diptera: Cecidoymiidae) associated with oaks (*Quercus* spp.) (Fagaceae) in the Palaearctic region
M. Skuhravá, V. Skuhraváy, and K. Dengler 1

The zoogeographic significance of European and Asian gall midge faunas (Diptera: Cecidomyiidae)
M. Skuhravá and V. Skuhraváy 12

Gall midges infesting Chenopodiaceae: their biology and taxonomy
N. Dorchin 18

Actilasioptera (Diptera: Cecidomyiidae), a new genus for Australasian and Asian gall midges of grey mangroves, *Avicennia* spp. (Avicenniaceae)
R.J. Gagné and L.J. Law 22

Tephritid galls and gall Tephritidae revisited, with special emphasis on myopitine galls
A. Freidberg 36

The diversity of insect-induced galls on vascular plants in Taiwan: a preliminary report
M.-M. Yang and G.S. Tung 44

Gall-forming and seed-eating insects, and their associates on joint pines, *Ephedra* spp. in Spain, a rich and newly discovered fauna
R. Askew and J. Blasco-Zumeta 54

Is there a cool-wet to hot-dry gradient of increasing gall-forming insect diversity?
R. Blanche and J. Ludwig 57

ECOLOGY AND POPULATION BIOLOGY

Role of *Periclistis* (Hymenoptera: Cynipidae) inquilines in leaf galls of *Diplolepis* (Hymenoptera: Cynipidae) on wild roses in Canada
J.D. Shorthouse 61

Sexual selection via female choice in the gall-making fly, *Eurosta solidaginis* Fitch (Diptera: Tephritidae)
L.J. Faust and J.M. Brown 82

The yew gall midge, *Taxomyia taxi*: 28 years of interaction with its chalcid parasitoids
M. Redfern and R.A.D. Cameron 90

On the beginnings of cecidology in 19th century North America: Bassett's discovery of heterogony in oak gall wasps (Hymenoptera: Cynipidae)
A. Wehrmaker 106

Wolbachia-induced thelytoky in cynipids
O. Plantard and M. Solignac 111

The gall-inhabiting weevil (Coleoptera) community on *Galenia africana* (Aizoaceae): co-existence or competition?
S. Louw 122

Gall-forming aphids on *Pistacia*: a first look at the subterranean part of their life cycle
D. Wool 127

The life-cyle of the blackcurrant gall mite, *Cecidophyopsis ribis* Westw.(Acari: Eriophyidae)
D. Gajek and J. Boczek 131

PLANT RESPONSES TO GALL-INDUCERS

Can inducing resistance cost as much as being susceptible?
C. Glynn, S. Larsson, and U. Gullberg 136

Plant hypersensitivity against tissue invasive insects: *Bauhinia brevipes* and *Contarinia* sp. interaction
T.G. Cornelissen and G.W. Fernandes 144

Reprogramming plant development: two approaches to study the molecular mechanism of gall formation
K. Schönrogge, L.J. Harper, S.E. Brooks, J.D. Shorthouse, and C.P. Lichtenstein 153

Plant hormones and gall formation by *Eurosta solidaginis* on *Solidago altissima*
C.C. Mapes and P.G. Davies 161

The effects of gall formation by *Lipara lucens* (Diptera: Chloropidae) on its host *Phragmites australis* (Poaceae)
L. De Bruyn, I. Vandevyvere, D. Jaminé, and E. Prinsen 173

Analysis of pigment-protein complexes in two cecidomyiid galls
C.-M. Yang, M.-H. Yin, K.-W. Chang, C.-J. Tsai, and S.-M. Huang 188

The development of an oval-shaped psyllid gall on *Cinnamomum osmophloeum* Kaneh. (Lauraceae)
G.S. Tung, M.-M. Yang, and P.-S. Yang 193

EVOLUTIONARY PERSPECTIVES ON PLANT-GALLER INTERACTIONS

Adaptive radiation of gall-inducing sawflies in relation to architecture and geographic range of willow host plants
P. Price, H. Roininen, and A.G. Zinovjev 196

Palearctic sawflies of the genus *Pontania* Costa (Hymenoptera: Tenthredinidae) and their
host-plant specificity
A.G. Zinovjev 204

Host-plant associations and specificity among cynipid gall-inducing wasps of eastern USA.
W.G. Abrahamson, G. Melika, R. Scrafford, and G. Csóka 226

Molecular phylogeny of North American oak-galling Cynipini (Hymenoptera: Cynipidae) supports need for generic revision
D.M. Drown and J.M Brown 241

Molecular phylogeny of the genus *Diplolepis* (Hymenoptera: Cynipidae)
O. Plantard, J.D. Shorthouse, and J.-Y. Rasplus 247

Patterns in the evolution of gall structure and life cycles in oak gall wasps (Hymenoptera: Cynipidae)
J.M. Cook, G.N. Stone, and A. Rowe 261

The population genetics of postglacial invasions of northern Europe by cynipid gall wasps
(Hymenoptera: Cynipidae)
G. Csóka, G. Stone, R. Atkinson, and K. Schönrogge 280

Tests of hypotheses regarding hybrid resistance to insects in the *Quercus coccolobifolia* x *Q. viminea* compex
W.J. Boecklen and R. Spellenberg 295

Genetic and environmental contributions to variation in the resistance of *Picea abies* to the gall-forming adelgid, *Adelges abietis* (Homoptera: Adelgidae)
W.J. Mattson, J. Levieux, and D. Piou. 304

POSTER ABSTRACTS

Recolonization of northern Europe by gall wasps: replicate analyses of the genetic consequences of range expansion using allozyme electrophoresis of four gall wasps in the genus *Andricus* (Hymenoptera: Cynipidae)
R. Atkinson and G. N. Stone 316

Insect-induced galls in the Guandaushi forest ecosystem of central Taiwan
S.-Y. Yang, J.-T. Yang and M.-Y. Chen 318

Cecidomyiid-induced galls of *Machilus thunbergii* Hay. in the Guandaushi forest ecosystem of central Taiwan
S.-Y. Yang, J.-T. Yang and M.-Y. Chen 319

Chalcid parasitoids in oak cynipid galls (Hymenoptera: Cynipidae) in the Carpathian Basin
C. Thuróczy, G. Melika, and G. Csóka 320

Gall midge faunas (Diptera: Cecidomyiidae) of four Mediterranean islands
M. Skuhravá and V. Skuhravy 321

Review of Giraud's types of the species of *Synergus* Hartig, 1840 (Hymenoptera: Cynipidae)
J. P. i Villar and P. Ros-Farré 322

Gall-inducer and gall-inquiline cynipids (Hymenoptera: Cynipidae) collected in Andorra with a malaise trap
P. Ros-Farré and J.P. i Villar 323

Gall midge-parasitoid interactions in a patchy landscape
O. Widenfalk 324

Induced resistance in willow against a gall-former: active defense or lack of resistance?
P. Saarikoski, S. Höglund, and S. Larsson 325

Expression of gall midge resistance in a willow genotype: effects of shading
S. Höglund and S. Larsson 326

Broad overview of a gall-system: from plant anatomy to ecological interactions
M. Mendonca, H. P. Romanowski, and J.E. Kraus 327

Two species of eurytomid gall formers with potential as biological control agents of strawberry guava, *Psidium cattleianum*, in Hawaii
C. Wikler and J.H. Pedrosa-Macedo 329

GALL INDUCING AND OTHER GALL MIDGE SPECIES (DIPTERA: CECIDOMYIIDAE) ASSOCIATED WITH OAKS (*QUERCUS* SPP.) (FAGACEAE) IN THE PALAEARCTIC REGION

Marcela Skuhravá[1], Vaclav Skuhravy[2] and Klaus Dengler[3]

[1] Czech Zoological Society, CZ-128 00 Praha 2, Vinicná 7, Czech Republic
[2] Institute of Entomology, Czech Academy of Sciences, CZ-370 05 Ceske Budejovice, Branisovská 31, Czech Republic
[3] Fachhochschule Rottenburg, Hochschule fur Forstwirtschaft, Schadenweilerhof, D-72108 Rottenburg am Neckar, Germany

Abstract.—In the Palaearctic region, 56 gall midge species are associated with 14 species of *Quercus*. Thirty-four species are gall makers whose larvae cause galls of various shapes on organs of its host plants. Fourteen species are inquilines, of which eight spp. develop in galls caused by gall midges and six spp. develop in galls of gall wasps (Cynipidae, Hymenoptera); larvae of *Resseliella quercivora* feed with cambium sap; *Xylodiplosis nigritarsis*, *X. praecox* and *X. aestivalis* are xylophilous and their larvae develop inside the xylem vessels of fresh cut oak wood; *Clinodiplosis cilicrus*, *Coquillettomyia umida*, *Trisopsis abdominalis* and *Winnertzia maxima* are mycophagous. Because of a concentration of research effort in Europe, the majority of gall midges currently known to be associated with oaks occur in the western Palaearctic. The gall midge fauna associated with oaks in the eastern Palaearctic is poorly known.

INTRODUCTION

Oaks (*Quercus* L.) and their close relatives (*Fagus* L., *Castanea* Mill., and *Nothofagus* L.) belong to the family Fagaceae. The genus *Quercus* is rich in species including more than 600 recent and many fossil species. Oaks are usually stout trees, rarely shrubs, and have deciduous or evergreen leaves. Several oak species have wide geographic distributions and are important (sometimes dominant) components of the vegetation cover characteristic of particular biogeographical regions or altitudinal zones. Oaks are most species-rich in North America and eastern Asia, with fewer species in Europe. Below we use the *Quercus* taxonomy proposed by Tutin *et al.* (1964) and the gall midge taxonomy used by Skuhravá (1986).

In the Palaearctic region, 14 oak species host a total of more than 50 gall midge species. *Quercus robur* L., *Q. petraea* (Matt.) Liebl. and *Q. pubescens* Willd. are Euro-Siberian species with broad distributions. *Q. cerris* L. is a southeastern European species whose natural distribution extends from the Mediterranean through Central Europe to the southern part of the Czech and Slovak Republics. *Quercus coccifera* L., *Q. ilex* L,. and *Q. fruticosa* Brot. (= *Q. lusitanica* Lam.) occur along the coast of the Mediterranean Sea. *Quercus ostryaefolia* Borbas (= *Q. macedonica* DC) occupies a smaller area spread mainly in the Balkan Peninsula. *Quercus acutissima* Corr., *Q. dentata* Thunb., *Q. variabilis* Blume, *Q. mongolica* Fisch. ex Ledeb., and *Q. serrata* Thunb. are typical east-Asian species.

History

Gall midge galls developing on oak leaves had already drawn the attention of researchers by the middle of the 19th century. V. Kollar (1850) reared adults from leaf galls on *Quercus cerris* occurring abundantly in forests surrounding Vienna and described the causal agent as *Lasioptera cerris* (now *Janetia cerris*). In the same area, also on leaves of *Quercus cerris,* Giraud (1861) observed interesting galls, reared adults and described the causal agent as *Cecidomyia circinans* (now: *Dryomyia circinans*). Binnie (1877) reared adults from leaf bud galls on *Quercus robur* which he found in Scotland and described them as *Cecidomyia quercus* (now: *Arnoldiola quercus*). F. Löw (1877), who studied galls around Vienna, observed small galls on leaves of *Quercus cerris* which differed from those induced by *Janetia cerris* and described the insect as *Cecidomyia homocera* (now *Janetia homocera*). In the same paper he described also the agent of leaf margin galls on *Quercus robur* as *Diplosis dryobia* (now: *Macrodiplosis dryobia*). A year later, Löw (1878) described the agent of leaf galls on *Quercus ilex* (sent by J. Lichtenstein from Montpellier in southern France) as *Cecidomyia lichtensteini* (now: *Dryomyia lichtensteini*). Löw (1889) discovered larvae of gall midges living under the galls of cynipid gall wasp, *Neuroterus lenticularis* (now known as the asexual generation of *Neuroterus quercusbaccarum*), and described the adults as *Diplosis galliperda* (now: *Parallelodiplosis galliperda*). He also briefly mentioned the peculiar biology of this gall midge species. Several gall midge species causing or inhabiting oak galls were described by Rübsaamen (1890, 1891, 1899) who worked mainly in southern Germany.

In the last decade of the 19th century the French entomologist, J.J. Kieffer, described several gall midge species developing in galls or associated with various species of oak. He seemed to have wished to name all gall midge species before publishing his greatest work (Kieffer 1913: Cecidomyiidae in Genera Insectorum), and named about 170 gall midge species only on the basis of gall habits (Kieffer 1909). These names remain valid according to the ICZN. This total includes 14 species developing on oaks. Adults of many these species are undescribed and their systematic position remains uncertain.

The majority of gall midge species associated with oaks were described from 1850-1909, with few additions in the 20th century (Del Guercio 1918, Tavares 1919, Kovalev 1972, Kritskaja and Mamaeva 1981). Two newly-discovered species have been described recently (*Contarinia cerriperda*, Skuhravá 1991, and *Schueziella quercicola*, Stelter 1994).

THE BIOLOGY OF GALL MIDGES ASSOCIATED WITH OAKS

Cecidomyiid gall midges may be divided into three large biological groups on the basis of larval feeding habits: phytophagous, zoophagous and mycophagous (Skuhravá et al. 1984). Each of these groups may be divided into subgroups. Gall midges associated with oaks include representatives of all of these groups. Larvae of the majority of species are gall makers, while smaller numbers of species are either inquilines, cambium-feeders, xylophilous or mycophagous (table 1).

Gall Inducing Gall Midges

Larvae of cecidogenic (gall-inducing) gall midges attack various oak organs, including leaf buds, leaves, and twigs (figs. 1 and 2). This group also includes several species which do not induce galls but whose larvae damage and deform oak flowers and male catkins, or cause atrophy of acorns. More than 40 cecidogenic gall midge species have been described from oaks in the Palaearctic Region. Many species causing pustule or spot galls were described by Kieffer (1909) based on the shape of the gall, sometimes together with a note about the color of the larva. Although the shapes of gall midge galls are well known and their occurrence is recorded in many European countries, the adults of many gallers remain unknown because it is very difficult to catch larvae in galls and to rear adults of these univoltine gall midges. Generic classifications of such species thus remain tentative.

1. Species associated with *Q. robur* and *Q. petraea*

Thirteen gall midge species have been described from the leaves of *Quercus robur* and *Q. petraea*, of which eight are gall inducers and five are inquilines.

Larvae of *Macrodiplosis* spp. cause galls on the leaf margin. The gall of *M. dryobia* consists of a thickened, downwards-folded marginal leaf lobe which is depigmented, often spotted with red or yellow. Each larval chamber contains from 1 to several yellowish-white larvae. The gall of *M. volvens* consists of an upwardly-rolled and swollen section of leaf margin located between two leaf lobes. Each chamber contains from 1-5 larvae. At the beginning of their development the larvae are whitish, later becoming orange at both ends of the body.

In Central Europe, the adults of both *Macrodiplosis* species fly in May. Shortly after mating the males perish and females lay their eggs on bursting oak leaves. Larvae which hatch from the eggs after several days and start suck the leaf tissue, which induces gall formation. The larvae grow relatively quickly, and are fully developed by the end of June. Mature larvae leave their galls and drop to the soil, where they remain until the following spring. Both species have a univoltine cycle. Galls of both *Macrodiplosis* species occur abundantly on leaves of *Quercus robur* and *Q. petraea* in central Europe and can be found from Portugal, Spain, and England in the west eastwards almost as far as the Urals. Both *Macrodiplosis dryobia* and *M. volvens* may be found on oaks from an altitude of 180 m up to 800 m in submontane areas (Skuhravá 1987, 1991, 1994).

Larvae of *Polystepha* spp. and *Arnoldiola libera* cause pustule or parenchyma galls on oak leaves. Larvae of *Dasineura panteli* cause folded galls along lateral leaf veins. Little is known of their biology: all three are univoltine species whose adults fly in May.

The yellowish-white larvae of *Contarinia quercina* induce galls in oak buds, each of which typically contains several larvae. Larvae of both the gall-forming

Table 1.—*Gall midge species associated with oak species in the Palaearctic Region*

1. Gall inducing gall midges and their inquilines according to host plant trees

Gall midge species	Gall type	Distribution*
Quercus robur L. and ***Q. petraea*** (Matt.) Liebl.		
Macrodiplosis dryobia (F.Lw.)	leaf gall	Europe
Macrodiplosis volvens Kieffer	leaf gall	Europe
Monodiplosis liebeli (Kffr.)	inquiline	Europe
Contarinia quercina (Rübs.)	swollen bud	Europe
Arnoldiola quercus (Binnie)	inquiline	Europe
Dasineura dryophila Rübs.	inquiline	GB,G,RU
Moreschiella roburella D.Gue.	inquiline	I
Schueziella quercicola Stelter	inquiline	G
Polystepha quercus Kieffer	leaf gall	Europe
Polystepha malpighii Kieffer	leaf gall	Europe
Polystepha rossica H.Mam.	leaf gall	RU
Arnoldiola libera (Kieffer)	leaf gall	Europe
Dasineura panteli (Kieffer)	leaf fold	S,E
Quercus pubescens Willd.		
Contarinia amenti Kieffer	deformed catkin	I
Quercus cerris L.		
Contarinia quercicola (Rübs.)	swollen bud	CS,A,H,RU,YU,UK
Arnoldiola dryophila (Kffr.)	inquiline	F
Clinodiplosis cilicrus (Kffr.)	inquiline	A
Moreschiella ilicicola D.Gue.	inquiline	I
Janetia cerris (Kollar)	leaf gall	Med. up to TR
Janetia homocera (F.Lw.)	leaf gall	Med.
Janetia szepligetii Kieffer	leaf gall	CS,H,R,I,YU,BG,TR,IS
Janetia nervicola (Kieffer)	leaf gall	Med.
Janetia plicans (Kieffer)	leaf fold	A
Janetia pustularis (Kieffer)	leaf gall	CS,A,H,YU
Arnoldiola trotteri Kieffer	leaf gall	I
Contarinia subulifex Kieffer	leaf gall	F,CS,A,H,R,YU,IS
Dasineura tubularis (Kieffer)	leaf gall	CS,A,H,R,YU
Dryomyia circinans (Giraud)	leaf gall	Med. up to TR
Contarinia cerriperda Skuhravá	deformed flower	Slovakia
Quercus ithaburensis Boiss.		
Contarinia subulifex Kieffer		IS
Janetia szepligetii Kieffer		IS
Quercus ostryaefolia Borbs		
Arnoldiola baldratii (Kieffer)	leaf gall	I
Quercus coccifera L.		
Dryomyia cocciferae (Marchal)	leaf gall	F,P,AL,TU
Contarinia cocciferae (Tavares)	bud gall	GB,P,E,I,YU,AL,MO

(table 1 continued on next page)

(table 1 continued)

1. Gall inducing gall midges and their inquilines according to host plant trees

Gall midge species	Gall type	Distribution*
Quercus ilex L.		
Dryomyia lichtensteini (F.Lw.)	leaf gall	F,P,I,YU,MO,AL
Contarinia ilicis Kieffer	leaf gall	P,I,AL
Contarinia luteola Tavares	twig gall	P,I,YU
Dasineura ilicis Tavares	inquiline?	P
Contarinia minima (Kieffer)	leaf gall	I
Arnoldiola tympanifex (Kieffer)	leaf gall	I
Quercus fruticosa Brot. (= *Q. lusitanica* Lam.)		
Dasineura squamosa (Tavares)	atrophied acorns	S,GB
Kiefferiola panteli (Kieffer)	leaf gall	F,G,A,P,E,YU
Quercus mongolica Fisch. ex Ledeb.		
Macrodiplosis flexa Kovalev	leaf gall	Russia (Far East)
Quercus acutissima Carr., *Q. dentata* Thunb. and *Q. variabilis* Blume		
Ametrodiplosis acutissima (Monzen)	leaf and stem gall	Japan
Quercus serrata Thunb.		
Silvestriola quercifoliae Shinji	leaf gall	Japan

* **Distribution in countries is indicated by the following abbreviations**: A Austria; AL Algeria; BG Bulgaria; CS Czech and Slovak Republics; DK Denmark; E Spain; F France; G Germany; GB England; H Hungary; I Italy; IS Israel; Med. Mediterranean area; MO Morocco; N The Netherlands; P Portugal; RU Romania; S Sweden; SF Finland; TR Turkey; TU Tunis; UK Ukraine; YU the former Yugoslavia

2. Inquilines in cynipid galls

Gall wasp species	Gall midge species	Distribution
Neuroterus quercusbaccarum	*Parallelodiplosis galliperda* (F.Lw.)	Europe
Neuroterus albipes	*Xenodiplosis laeviusculi* (Rübs.)	GB, D, CS
Biorhiza pallida	*Clinodiplosis cilicrus* (Kieffer)	F, NL
Andricus fecundator	*Clinodiplosis cilicrus* (Kieffer)	F, G
	Arnoldiola gemmae (Rübsaamen)	S, GB, G, CS
	Lestodiplosis necans (Rübsaamen)	G
Dryocosmus australis Mayr	*Lasioptera nigrocincta* Kieffer	AL

3. Cambium-feeding gall midges

Resseliella quercivora Mamaev   Russia, G, CS

4. Xylophilous gall midge species

Xylodiplosis nigritarsis (Zetterstedt)   SF, DK, GB, NL, F, G, CS

Xylodiplosis praecox (Winnertz)   G, CS

Xylodiplosis aestivalis Kieffer   F

5. Biology of mycophagous gall midge species

Clinodiplosis cilicrus (Kieffer) larvae in mine caused by (=*C. coriscii* Kieffer) undetermined caterpillars
Coquillettomyia umida (Möhn) larvae among oak leaves
Trisopsis abdominalis Mamaev larvae in rotten wood produced by *Polyporus sulphureus*
Winnertzia maxima Mamaev larvae in mycelium of *Auricula mesenterica* under bark of oak trunk

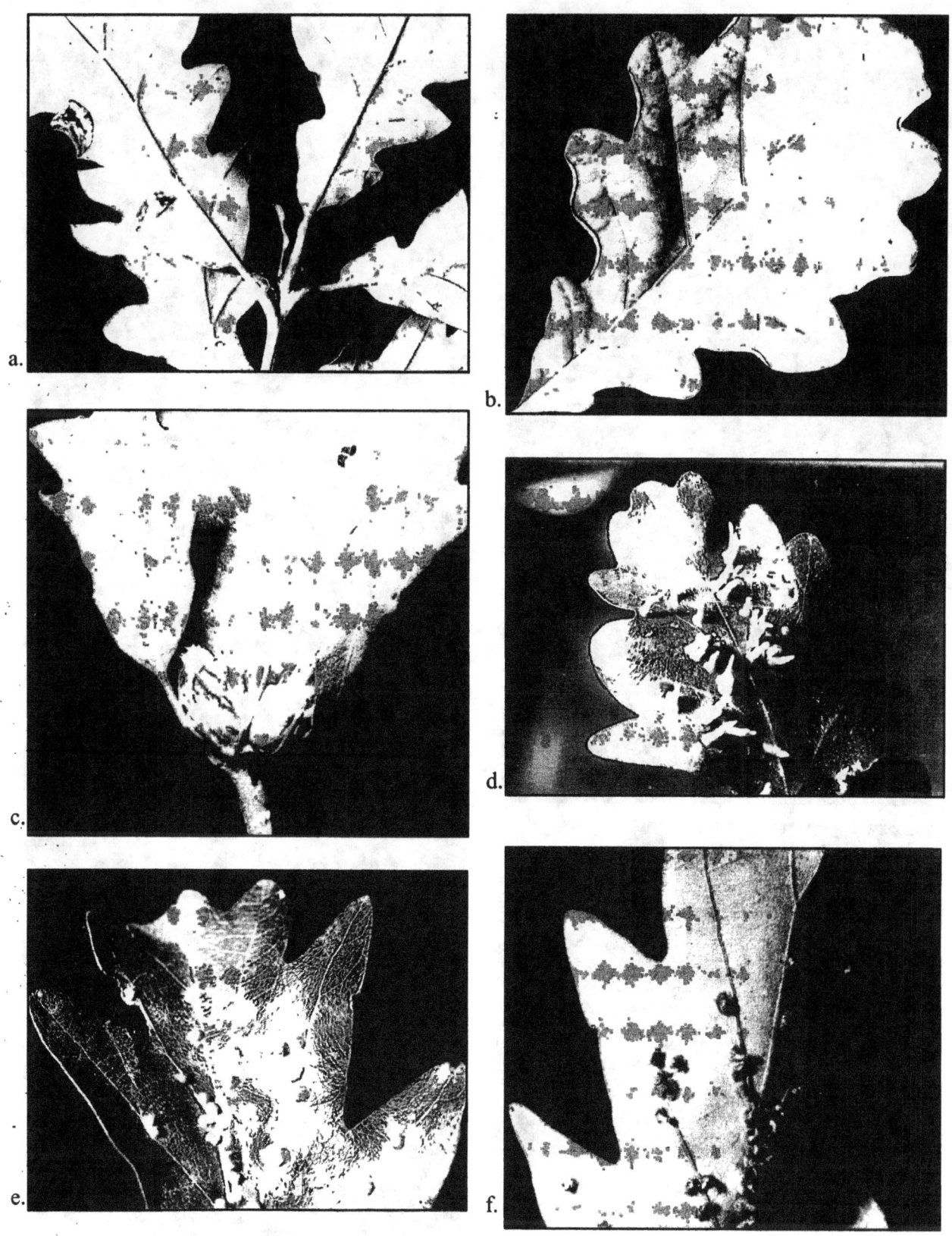

Figure 1.—*Galls of gall midges caused on* Quercus robur: *a.* Macrodiplosis dryobia; *b.* Macrodiplosis volvens; *c.* Contarinia quercicola; *d.* Contarinia subulifex; *e-f.* Dasineura tubularis, *e. upper leaf side, f. lower leaf side.*

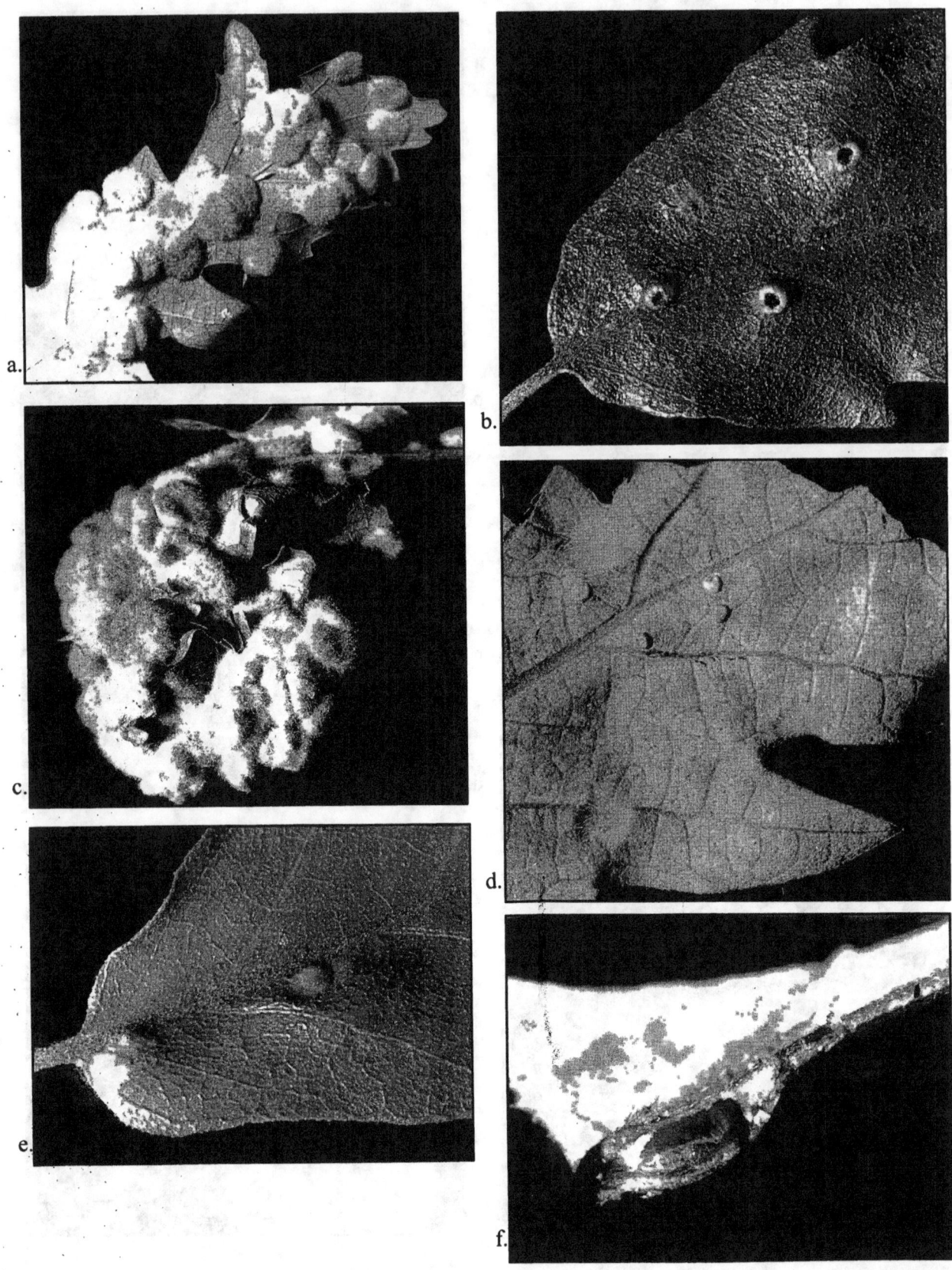

Figure 2.—*Galls of gall midges on leaves of* Quercus cerris: *a-c.* Dryomyia circinans, *a. lower leaf side, b. upper leaf side, c. heavily attacked, deformed leaf; d.* Janetia cerris, *upper leaf side; e.* Janetia homocera, *upper leaf side; f.* Janetia nervicola, *longitudinal section of the main leaf vein.*

species and several inquilines may be found among the small deformed young leaves of the galls. Inquiline larvae differ from those of the gall-causer mainly in the color of the body, and in the shape of the spatula sternalis and last abdominal segment. Larvae leave their galls between August and the beginning of September and drop to the soil where they hibernate until the following spring. *C. quercina* has two generations each year. Adults of the overwintering generation fly in May and adults of the summer generation fly in July.

2. Species associated with *Q. pubescens*

Little is known about *Contarinia amenti*, which induces galls on *Quercus pubescens*. This species was only briefly described by Kieffer (1909): "*Quercus pubescens*. Catkin deformed, shortened and swollen. Larvae red, situated among the flower parts; dorsal papillae without bristles, spatula at the end only little emarginated." No further details are known of the biology of this species.

3. Species associated with *Q. cerris*

Quercus cerris is the host plant with the highest number of gall midge species associated with one oak host species. Of 15 gall midge species, 12 are gall makers and 3 are inquilines. Larvae of *Contarinia quercicola* cause bud galls in which the larvae of three other gall midges, *Arnoldiola dryophila*, *Clinodiplosis cilicrus*, and *Moreschiella ilicicola*, live as inquilines *Janetia cerris*, *J. homocera*, *Contarinia subulifex*. *Dasineura tubularis* and *Dryomyia circinans* cause unilocular galls of various shapes, pubescent or hairless, on the leaves of *Quercus cerris*. These galls usually protrude on both sides of the leaves. *Janetia szepligetii*, *J. pustularis,* and *Arnoldiola trotteri* cause pustule or parenchyma galls on leaves. Larvae of *Janetia nervicola* inhabit galls formed from swollen leaf veins and larvae of *J. plicans* cause folds on the leaves. Larvae of *Contarinia cerriperda* malform single flowers on male catkins.

Gall midges associated with *Quercus cerris* have, along with their host plant, a Mediterranean and sub-Mediterranean distribution, the majority extending north as far as the most southern part of the Czech and Slovak Republics (Skuhravá 1986, 1991, 1994). In these areas, larvae of the gall midge species mentioned above leave their galls to hibernate in the soil, whereas in areas nearer the Mediterranean, the larvae hibernate inside their galls and also pupate in them in the following spring. Some gall midge species occur to the north of the natural boundaries of the *Q. cerris* distribution, attacking trees and shrubs planted in botanical gardens and castle parks in the Czech Republic and Germany (Skuhravá *et al.* 1984).

All gall midge species associated with *Quercus cerris* have only one generation a year. Adults fly from the second half of April to early May. Females lay their eggs on, or in oak organs where larvae hatch from their eggs after several days. Larval development lasts from several weeks up to several months. Larvae of *Janetia pustularis* leave their galls by the end of May, larvae of *Dasineura tubularis* a little later. Larvae of the remaining species leave their galls between the end of October and the beginning of November (Skuhravy and Skuhravá 1971, Skuhravá and Skuhravy 1973, Schremmer 1991).

3. Species associated with *Q. coccifera, Q. ilex,* and *Q. lusitanica*

Little is known about the gall midges associated with *Quercus coccifera*, *Q. ilex,* and *Q. lusitanica* which occur in the Mediterranean area. Larvae of 10 gall midge species cause galls of various types on organs of these evergreen oaks. They have also only one generation in a year and their larvae hibernate in their galls on the leaves rather than dropping to the soil.

4. Species associated with oaks in the Eastern Palaearctic

Little is known about the biology of gall midges associated eastern palaearctic oaks. Larvae of *Macrodiplosis flexa* galls the leaves of *Quercus mongolica* and were found at Sichote-Alin in the Russian far east. Galls of *Ametrodiplosis acutissima* causing leaf and stem galls on *Quercus acutissima*, *Q. dentata,* and *Q. variabilis,* and galls of *Silvestriola guercifoliae*, were found in Japan (Yukawa 1971).

Subglobular galls of an undetermined *Contarinia* species gall the leaves of *Quercus glauca* in Japan. Yukawa and Tsuda (1986) studied the development of this species in detail and came to the conclusion that it causes premature leaf fall.

Relations Between Gall Midges and Their Host Plants

It is usually accepted that gall midges are tightly associated with particular host plant species. Some oak gall midges, however, attack species in addition to the oak from which they were first described, usually related sympatric oak species or their hybrids. For example, galls of *Macrodiplosis dryobia* and *M. volvens* are usually found on *Quercus robur* and *Q. petraea,* but they are also occasionally found on other oaks including *Quercus pubescens* and even *Quercus cerris* (Skuhravy *et al.* 1997). The full range of host plants, including occasional hosts, is given in identification keys for galls (Houard 1908-1909, Buhr 1964-1965).

INQUILINES

Larvae of inquilines gall midges are phytophagous but neither induce galls nor live freely in plant tissue. Females of inquiline gall midges lay their eggs on or in galls produced by other species—usually gall midges or gall wasps. Hatched larvae feed by sucking sap from the tissue of the gall, as do those of gall inducing species. Nine inquiline gall midge species have been described to date in galls produced by gall midges on oaks. Inquiline species are usually not as abundant as their gall-inducing hosts.

Galls induced by oak cynipid gallwasps also sometimes contain gall midge larvae. The orange-yellow larvae of *Parallelodiplosis galliperda* develop under the asexual spangle galls of *Neuroterus quercusbaccarum*, sucking sap from the gall tissues. This alters the shape of the galls, which are less regularly circular than unattacked galls. Some are so damaged that they fall prematurely, leading to the death of the gallwasp larvae (hence the name *galliperda*). Similarly, the larvae of *Xenodiplosis laeviusculi* develop under the asexual spangle galls of *Neuroterus albipes*. Nothing more is known about the biology of this species.

Adults of *Clinodiplosis cilicrus* have been reared from galls of two gall wasp species—the sexual generation galls of *Biorhiza pallida* and the asexual generation galls of *Andricus fecundator*. Originally these gall midges were considered to be separate and named *Clinodiplosis biorrhizae* and *Clinodiplosis gallicola*, respectively. On the basis of extensive experiments and an analysis of morphological characters of larvae and adults, Skuhravá (1973) demonstrated there is only one species, *Clinodiplosis cilicrus*. It is a phytosaprophage whose larvae develop in decaying plant matter.

Two other gall midge species have been reared from the asexual generation galls of *Andricus fecundator*. Larvae of *Arnoldiola gemmae* are probably inquilines between the scales of the gall where they pupate in white cocoons. Larvae of *Lestodiplosis necans*, which in common other member of the genus *Lestodiplosis* are probably zoophagous, attack the larvae of *Arnoldiola gemmae*.

All these inquiline gall midge species were found in Central Europe. Only one species of inquiline gall midge is known from the Mediterranean area: adults of *Lasioptera nigrocincta* were reared from galls of the cynipid *Dryocosmus australis* on leaves of *Quercus ilex* in Algeria.

CAMBIUM-FEEDING GALL MIDGES

Cambium necroses and bark damage to the trunks of young oaks (sometimes called oak-cancer or T-necroses) have been recorded from several locations in Europe. Initially, either abiotic (either frost or drought) or biotic (particularly fungi) causes for such damage have been considered. Gibbs (1982) was the first to provide evidence that such canker formation is the result of infestation by the larvae of an undetermined species of gall midge in the genus *Resseliella*. Females were observed to lay eggs in recently formed wounds made by the woodpecker *Dendrocopus major* (Picidae, Aves) in oak trunks during the months of July and August.

In 1990, Dengler found thousands of young oak trees in forest stands near Rottenburg, southern Germany, to be severely damaged by these gall midges. Analysis of morphological characters of the larvae and later also of reared adults, led us to identify this species as *Resseliella quercivora* (Mamaev and Krivosheina 1965). Original uncertainty with our identification resulted from the fact that *R. quercivora* was described only briefly by Mamaev (in Mamaev and Krivosheina 1965) and in the original description the host plant and biology are not given. The injury to oak trees may be derived only from the species name "*quercivora*". The larvae of *Resseliella quercivora* are responsible for the origin of the necrosis on oak trunks. These necroses only develop secondarily following primary damage by woodpeckers or through other causes. Females of *R. quercivora* are probably attracted to such wounds by olfactory cues present in the cambium exudate.

Twenty-three gall midge species are included in the genus *Resseliella* Seitner, 1906 in the Palaearctic Region (Skuhravá 1986). Gagné (1973) considered the genera *Thomasiniana* Strand, 1927, *Profeltiella* Kieffer, 1912 and *Wichmanniella* Möhn, 1955 to be synonyms of *Resseliella*. Members of this genus usually do not cause galls, but are often associated with the resin of conifers or with damaged trunks of deciduous trees, where larvae have been found under the damaged bark. Species of this genus are considered to be host-plant specific. Only *Resseliella oculiperda* (Rübsaamen), whose larvae have been found between bud grafts and the stock of roses, is also known to occur in similar locations on several fruit tree species in other rosaceous genera.

Using a knife or die, Dengler intentionally wounded the trunks of 37 host plant species belonging to 12 plant families in the arboretum at Rottenburg am Neckar. Gall midge larvae and necrosis of a form similar to that induced by *Resseliella querciperda* on oaks were found on several other tree species. We are currently trying to discern whether a single *Resseliella* species is responsible for the development of necroses on many host tree species, or whether several *Resseliella*-species are responsible for the development of necrosis on specific host tree species.

XYLOPHILOUS GALL MIDGE SPECIES

Adults of midges in this group are attracted to wood and woody materials. No details are known on their biology, including what kind of food their larvae are eating. Typically, single specimens have been collected flying near such material or sitting on a log, sometimes without exact determination of the host plant species. Some of these gall midges are probably mycophagous. More than 20 xylophilous gall midge species are known from the Palaearctic region, belonging to the genera *Xylodiplosis* Kieffer, 1894; *Karschomyia* Felt, 1908; *Ledomyia* Kieffer, 1895; *Lauthia* Kieffer, 1912; *Microperrisia* Kieffer, 1912; and *Holobremia* Kieffer, 1912 (Kieffer 1900, 1904; Skuhravá *et al.* 1984).

The most interesting of these genera is *Xylodiplosis* Kieffer, 1894. Three European species are known: *X. nigritarsis* (Zetterstedt, 1850), *X. praecox* and *X. aestivalis* Kieffer, 1904 (Skuhravá 1986). All of these species were very insufficiently described and their biology is little known. It remains unclear whether they are truly independent species or whether one or more are synonyms of *Xylodiplosis nigritarsis*, the oldest and most important species in this group. Zetterstedt (1850) found two females in Denmark and described them very briefly as *Cecidomyza nigritarsis*. Kieffer (1894) established the genus *Xylodiplosis* for this species and studied its biology (Kieffer 1900). The larvae develop in freshly cut oak wood. *X. nigritarsis* is found—according to scant information—in Finland, England, the Netherlands, Germany, and France.

We observed adults—only females—of *X. nigritarsis* for the first time in the Czech Republic in 1984, at the beginning of March. Females were found sitting on oak-stumps or on the cut-surfaces of piled wood. The population of *X. nigritarsis* was so abundant that it was possible to study the diurnal activity and the flight period.

Diurnal Activity

Xylodiplosis nigritarsis has a morning activity which is influenced by temperature. Females are attracted by the fresh cut oak wood. At the beginning of March, 1984, the first females started to fly to oak stumps and cut surfaces of freshly cut oak trunks at 9 o'clock when the temperature rose from freezing point to 4°C and the sun began to shine. Females flew slowly round the fresh-cut logs. Many of them sat on the line of the circle containing the xylem vessels. Some females extended their ovipositor, as far as several times their body length, and inserted it into xylem vessels. Oviposition may take several hours. Some females perish in the course of egg-laying because they cannot withdraw their ovipositor from the xylem vessels. The number of females rose with rising temperature, peaking around 3:00 pm. Numbers declined thereafter until by 6:00 pm only dead females remained.

Flight Period

Single females of *Xylodiplosis nigritarsis* were observed in flight as early as late February, when temperatures rise above freezing the sun shines. The flight period lasted one month. From the beginning of March the number of females rose, peaking around the 20th March when up to 1,400 females of *X. nigritarsis* were observed sitting or flying round one surface of freshly cut oak. Thereafter the number of females fell abruptly. Single females occurred sporadically in April are were absent in May.

Dengler collected females of *Xylodiplosis* (probably *X. aestivalis,* Kieffer 1904) at freshly cut oak trunks at the beginning of October 1991 near Rottenburg am Neckar. His work suggests that flight activity by *X. aestivalis* is crepuscular. Females flew to exposed cut oak logs from October to December. The flight period lasted three months. We found that in central Europe, females of *Xylodiplosis nigritarsis* flew only to freshly cut *Quercus robur*, *Q. petraea*, and *Q. rubra*, and did not respond to another 15 coniferous and deciduous tree species belonging to various genera and families (*Picea, Pinus, Larix, Pseudotsuga, Betula, Alnus, Carpinus, Fagus, Castanea, Ulmus, Populus, Tilia,* and *Fraxinus*).

In North America, Gagné (1985) and Rock and Jackson (1985, 1986) also studied the gall midges associated with xylem vessels. Because females of the American species, *Xylodiplosis longistylus*, lay eggs into freshly cut wood of various tree species, they came to the conclusion that the most important cue for oviposition was the size of the xylem vessels.

MYCOPHAGOUS GALL MIDGES

The mycophagous gall midges include those whose larvae are associated in various ways with fungi (Mycophyta). In the Palaearctic Region about 240 gall midge species are associated with various fungal groups. Only four gall midge species are known to be associated with fungi or with processes of decay of oak wood. Larvae of *Clinodiplosis cilicrus* (*C. coriscii*) were found in a mine caused by an undetermined caterpillar; larvae of *Coquillettomyia umida* were found among oak leaves. Larvae of *Trisopsis abdominalis* lived in rotten wood infected by the fungus *Polyporus sulphureus*. Larvae of *Winnertzia maxima* were found in the mycelium of *Auricula mesenterica* under the bark of an oak trunk.

LITERATURE CITED

Binnie, F.G. 1877. Further notes on the Cecidomyidae, with description of three new species. Proceedings of the Natural History Society of Glasgow. 3: 178-186.

Buhr, H. 1964-1965. Bestimmungstabellen der Gallen (Zoo- und Phytocecidien) an Pflanzen Mittel- und Nordeuropas. Vol. 1 and 2. VEB Gustav Fischer Verlag Jena. 1,572 p.

Bytinski-Salz, H.; Sternlicht, M. 1967. Insects associated with oaks (*Quercus*) in Israel. Israel Journal of Entomology. 2: 107-143.

Del Guercio, G. 1918. I moscerini della ghiande delle Querce e del Leccio in Italia. Agricoltura Coloniale. 12: 358-369.

Gagné, R.J. 1973. A generic synopsis of the Nearctic Cecidomyiidae (Diptera: Cecidomyiidae: Cecidomyiinae). Annals of the Entomological Society of America. 66(4): 857-889.

Gagné, R.J. 1985. Descriptions of new Nearctic Cecidomyiidae (Diptera) that live in xylem vessels of fresh-cut wood, and a review of *Ledomyia* (s.str.). Proceedings of the Entomological Society of Washington. 87(1): 116-134.

Gibbs, J.B. 1982. An oak canker caused by a gall midge. Forestry. 55: 67-78.

Giraud, J. 1861. Fragments entomologiques. II. Supplement a l'histoire des Dipteres gallicoles. Verhandlungen der Kaiserlichen Königlichen Zoologisch-botanischen Gesellschaft Wien. 11: 470-491.

Houard, C. 1908-1909. Les zoocécidies des plantes d'Europe et du Bassin de la Méditerranée. A. Hermann et fils, Paris. 1,247 p.

Kieffer, J.J. 1894. Sur le groupe des *Diplosis*. Genres nouveaux. Bulletin du Societé Entomologique de France. 63: CCLXXX.

Kieffer, J.J. 1900. Monographie des Cécidomyides d'Europe et d'Algerie. Annals du Societé Entomologique de France. 69: 181-472.

Kieffer, J.J. 1904. Nouvelles Cécidomyies xylophiles. Annales du Societé Scientifique de Bruxelles. 28: 367-410.

Kieffer, J.J. 1909. Contribution a la connaissance des insectes gallicoles. Bulletin du Societé d' Histoire Naturelle du Département de Metz. (3) 2 (26): 1-35.

Kieffer, J.J. 1913. Diptera, Fam. Cecidomyiidae. In: Wytsman, P., ed. Genera Insectorum, Fasc. 152: 1-346.

Kollar, V. 1850. Naturgeschichte der Zerr-Eichen-Saummücke (*Lasioptera cerris*) eines schädlichen Forstinsekts. Denkschrift der Akademie der Wissenschaften Wien. 1: 48-50.

Kovalev, O.V. 1972. New species of gall midges (Diptera, Cecidomyiidae) from the southern Far East of the USSR. Review of Entomology. 51: 412-429. (In Russian with English summary.)

Kritskaja, I.G.; Mamaeva, H.P. 1981. New data on systematics of the gall midges of the tribe Asphondylini (Diptera: Cecidomyiidae) from the USSR. Review of Entomology. 60(2): 401-406. (In Russian)

Löw, F. 1877. Über Gallmücken. Verhandlungen der Kaiserlichen Königlichen Zoologisch-botanischen Gesellschaft Wien. 27: 1-38.

Löw, F. 1878. Mitteilungen über Gallmücken. Verhandlungen der Kaiserlichen Königlichen Zoologisch-botanischen Gesellschaft Wien. 28: 387-406.

Löw, F. 1889. Beschreibung zweier neuer Cecidomyiden-Arten. Verhandlungen der Kaiserlichen Königlichen Zoologisch-botanischen Gesellschaft Wien. 39: 201-204.

Mamaev, B.M.; Krivosheina, N.P. 1965. Lichinki gallic (Diptera, Cecidomyiidae). (Larvae of gall midges). Moskva. 1-278. (In Russian)

Rock, E.A.; Jackson D. 1985. The biology of xylophilic Cecidomyiidae (Diptera). Proceedings of the Entomological Society of Washington. 87: 135-141.

Rock, E.A.; Jackson D. 1986. Host selection in xylophilic Cecidomyiidae (Diptera): vessel size and structure. Proceedings of the Entomological Society of Washington. 88(2): 316-319.

Rübsaamen, E.H. 1890. Die Gallmücken und Gallen des Siegerlandes. Verhandlungen des Naturhistorischen Vereins Preussischen Rheinlandes. 47: 18-58, 231-264.

Rübsaamen, E.H. 1891. Neue Gallmücken und Gallen. Berlinere Entomologische Zeitschrift. 36: 393-406.

Rübsaamen, E.H. 1899. Ueber die Lebensweise der Cecidomyiden. Biologisches Centralblatt. 19: 529-549, 561-570, 593-607.

Schremmer, F. 1991. Zwei Gallmücken-Gallen verschiedener Art an den Blättern der Zerreiche (*Quercus cerris*) - Beobachtungen im westlichen Wienerwald. Entomologische Nachrichten und Berichte Berlin. 35(4): 227-235.

Skuhravá, M. 1973. Monographie der Gattung *Clinodiplosis*, Kieffer, 1894 (Cecidomyiidae, Diptera). Studie CSAV, Academia, Praha, Nr. 17: 1-80.

Skuhravá, M. 1986. Cecidomyiidae. In: Soós, A.; Papp, L., eds. Catalogue of Palaearctic Diptera. Akadémiai Kiadó, Budapest: 4: 72-297.

Skuhravá, M. 1987. Analysis of areas of distribution of some Palaearctic gall midge species (Cecidomyiidae, Diptera). Cecidologia Internationale. 8(1+2): 1-48.

Skuhravá, M. 1991. Gallmücken der Slowakei (Cecidomyiidae, Diptera). VI. Die Zoogeographie der Gallmücken. Zbornik Slovenskeho Národniho Múzea Prírodni Vedy. 37: 85-178.

Skuhravá, M. 1994. The zoogeography of gall midges (Diptera: Cecidomyiidae) of the Czech Republic I. Evaluation of faunistic researches in the 1855-1990 period. Acta Societatis Zoologicae Bohemicae. 57: 211-293.

Skuhravá, M.; Skuhravy, V. 1973. Gallmücken und ihre Gallen auf Wildpflanzen. Ziemsen Verlag, Wittenberg. 1-118.

Skuhravá, M.; Skuhravy, V.; Brewer J.W. 1984. Biology of gall midges. In: Ananthakrishnan, T.N., ed. Biology of gall insects. New Delhi, Bombay, Calcutta: Oxford & IBH Publishing Co.: 169-222.

Skuhravy, V.; Hrubjk, P.; Skuhravá, M.; Pozgaj, J. 1997. Occurrence of insects (Insecta) associated with nine *Quercus* species (Fagaceae) in cultural plantations in southern Slovakia during 1987-1992. Journal of Applied Entomology. (In press).

Skuhravy, V.; Skuhravá, M. 1971. Die Gallmücken (Diptera, Cecidomyiidae) an der Zerreiche (*Quercus cerris* L.). Marcellia. 37: 75-101.

Stelter, H. 1994. Untersuchungen über Gallmücken XC: Die Arten der Gattung Schueziella Möhn, 1961. Beiträge zur Entomologie Berlin. 44(2): 399-402.

Tavares, J.S. 1919. Espécies novas de Cynipides e Cecidomyias da Peninsula Ibérica. II. Série. Brotéria. 17: 85-101.

Tutin, T.G.; Heywood, V.H.; Burges, N.A.; Moore, D.M. 1964. Flora Europaea. Cambridge: Cambridge University Press: 1: 464 p.

Yukawa, J. 1971. A revision of the Japanese gall midges (Diptera: Cecidomyiidae). Memoirs of the Faculty of Agriculture, Kagoshima University. 8(1): 1-203.

Yukawa, J.; Tsuda, K. 1986. Leaf longevity of *Quercus glauca* Thunb., with reference to the influence of gall formation by *Contarinia* sp. (Diptera: Cecidomyiidae) on the early mortality of fresh leaves. Memoirs of the Faculty of Agriculture, Kagoshima University. 22: 73-77.

Zetterstedt, J.W. 1850. Diptera scandinaviae disposita et descripta. Lundae. 9: 3367-3710.

THE ZOOGEOGRAPHIC SIGNIFICANCE OF EUROPEAN AND ASIAN GALL MIDGE FAUNAS (DIPTERA: CECIDOMYIIDAE)

Marcela Skuhravá[1], Václav Skuhravy[2]

[1] Czech Zoological Society, CZ-128 00 Praha 2, Vinicna 7, Czech Republic
[2] Institute of Entomology, Czech Academy of Sciences, CZ-370 05 Ceske Budejovice, Branisovská 31, Czech Republic

Abstract.—Between 1955 and 1997 we carried out extensive studies of gall midge at 1,600 localities in Europe and in 17 localities in Asia (Siberia). Occurrence data collected by earlier authors are summarized and compared with our data. Our analyses show that the gall midge species composition in Europe is strongly influenced by geographical and altitudinal position, climatic factors, vegetation type and human activities. Analysis of our collected data make it possible to produce zoogeographic distribution maps for gall midge species, to recognize long-term changes in population dynamics and to determine which species should be considered threatened with regard to nature conservation.

INTRODUCTION

In zoogeographical studies of gall midges, it is essential to gather extensive data on the horizontal and vertical occurrence of each species throughout a larger study area. The more data are collected, the more reliable and meaningful the conclusions that can be drawn. It is possible to begin in smaller areas but it is necessary to extend such investigations gradually to larger and larger areas. We have therefore broadened our investigations from a starting point in 1955 in the former Czechoslovakia to include other countries in Europe and Asia. The following approaches are very important in addressing zoogeographical questions:

1. Summary of scattered gall midge occurrence data gathered by earlier authors in various parts of Europe in the past.
2. Obtaining new data on the present occurrence of gall midges species in the same regions.
3. Extension of investigations to little-explored areas.

MATERIALS AND METHODS

We use the term fauna to indicate the sum of all gall midges species recorded from a particular area, country or period. Other authors have described the same concept using terms including species richness, species diversity and faunal composition. We generated lists of gall midge species forming partial faunas of several European countries in which we included not only the summarization of all data gathered by earlier authors but also which we enriched by results of our own investigations (table l). The gall midge fauna of each territory was evaluated from two points of view: geographic and zoological. Geographic data present two features of gallwasp distributions: (a) Spatial distribution is represented by the gall midge species found at individual localities within a larger region. (b) Vertical occurrence is shown by the average numbers of gall midge species which occur in the rising altitudinal zones. When possible, past and present distributions are compared to examine long-term changes in population dynamics.

We used the same collecting method at each sample location. We walked slowly through various biotopes in each location for between one and several hours, searching and collecting all galls on plants, or plants inhabited by mites, aphids or other insects, or rusts, in which larvae of gall midges may develop. All findings were recorded, including notes about the local abundance of species. Each locality was investigated only once by this method.

In 1955, at the beginning of our studies we developed a strategy for systematic investigations of the gall midge fauna of the Czech Republic. Localities in which we intended to collect galls were identified throughout the country using the map with Ehrendorfer's network, as far as possible with one locality per 11 x 12 km area (described in detail in Skuhravá 1994). We used the same grid-based method in Slovakia (Skuhravá 1991) but this systematic approach could not be applied to the whole of Europe due to time constraints.

Table 1.—*A summary of studies of gall midge faunas in Europe and Asia*

Investigated country	Year(s) of research	Number of localities	Number of species	Publication (M.S.=M.Skuhravá) (V.S.=V.Skuhravy)
Czech Republic	1955-1979	670	500	15 contributions of M.S.1957-1982 (all cited in M.S.1994)
Slovak Republic	1969-1976	336	350	5 contributions of M.S.1972-1989 (all cited in M.S.1991)
Yugoslavia	1963	15	290	M.S., V.S.1964
Slovenia	review	—	219	Simova-Tosic, M.S.,V.S.1996
Austria	1967	32	98	M.S.,V.S.1967
	review	—	284	M.S., Franz 1989
	1991	44	311	M.S.,V.S. 1991/1992
	1992,1993	198	374	M.S.,V.S.1995
Romania	1969	23	310	M.S.,V.S.,Neacsu 1972
Poland	(fauna) review	—	320	M.S.,V.S., Skrzypczynska 1977
	(pests) review	—	28	M.S., Skrzypczynska 1983
Siberia	1988	17	110	M.S.,V.S.1993
Germany, Harz	1988	7	90	M.S.,V.S.1988
Fichtelgeb.	1991	15	68	M.S.,V.S.1992
Bayer. Wald	1991	4	29	M.S.,V.S.1992
Bulgaria	1978-1987	46	240	M.S., V.S., Doncev, Dimitrova 1991
	pests review	—	38	M.S.,V.S., Doncev, Dimitrova 1992
Liechtenstein	1993	9	65	M.S.,V.S.1993
Italy	review	—	324	M.S.,V.S.1994
Spain (Aragon)	1997	4	20	unpublished
	determination of material and review	17	—	M.S., Blasco-Zumeta,V.S.1993
Spain+Portugal	review	—	220	M.S.,V.S., Blasco-Zumeta, Pujade 1996
Switzerland	1993-1996	56	227	M.S.,V.S.,1997
Hungary	1988,1997	22	318	in preparation
France	review	—	600	in preparation
Norway	1995	5	35	unpublished
Greece	1994-1996	56	140	in preparation
Crete	1996	10	39	in preparation
Mallorca	1997	7	23	in preparation
Sardinia	1997	15	35	in preparation
Cyprus	1997	17	21	in preparation

In the period 1955-1997 we investigated the composition of gall midge faunas at 1,600 European localities ranging from Norway north of the Arctic Circle (Harstad, 68.48 N.L., 16.30 E.L.) south to Crete in the Mediterranean (Iraklion, 35.20 N.L., 25,08 E.L., Greece), from Switzerland in the west (Geneva, 46.13 N.L., 6.09 E.L.) to easternmost Greece (Drama, 25.20 N.L., 25.08 E.L.) and southern Central Siberia (Kyzyl, 51.45 N.L., 94.28 E.L., Tuvin Autonomous Republic, Russia).

RESULTS

Skuhravá (1986) has described 2,200 gall midge species from the Palaearctic region, of which about 1,500 species occur in Europe. Based on our long-term research on the distribution of gall midges in Europe we came to conclusion that gall midge species richness is influenced mainly by the following factors: geographical and altitudinal position (both of which are associated with changing levels of climatic variables including sunshine, temperature, and rainfall), by the type of vegetation cover, and by human activity.

Geographical Position

Gall midge species richness changes with geographical position, mainly along a north-south axis. The greatest richness occurs in natural biotopes of Central Europe situated at latitudes between 40 and 60 degrees, with a rich vegetation cover of mixed forest growing in mild weather conditions. From these latitudes the average species number falls to both north and south (fig.1).

From Central Europe northwards, the average number of gall midge species falls dramatically. In the southern part of Sweden, for example, the gall midge fauna is relatively rich, falling in boreal habitats further north. Only two or three gall midge species are known from the coniferous forests of northern inland Sweden (the continental sector of Noirfalise 1987). In contrast, the gall midge fauna of more coastal regions of northern Norway (the so called oceanic sector) is relatively rich: 35 gall midge species were found in 1995 around Harstad, a town lying on Hinnoya island in northern Norway (unpublished). Such a rich gall midge fauna may be explained by a relatively high plant species diversity. Many trees, shrubs and herbs grow well on coastal rocks, cliffs, slopes and other landscapes due the influence of the warm Gulf-stream.

From Central Europe southwards, the gall midge species composition changes and the average number of gall midge species also falls. In the continental parts of southern Europe and in European mountains where cold and wet conditions predominate, the number of gall midge species remains relatively high—gall midge larvae cause galls on leaves of trees and shrubs—but in mediterranean Europe where hot and dry weather predominates and mostly evergreen shrubs occur—gall midge species richness is low.

Altitudinal Position

Gall midge species composition changes, and species richness falls, from lowlands to mountains in Slovakia. (Skuhravá 1991), the Czech Republic (Skuhravá 1994) and Switzerland (Skuhravá and Skuhravy 1997).

Climatic Factors

The most important climatic factor appears to be temperature. For example, the long-term population dynamics of *Thecodiplosis brachyntera*, larvae of which cause galls at the base of needle pairs of various species of *Pinus*, correlate with long-term fluctuations in temperature (Skuhravy 1991).

Vegetation Type

Vegetation species composition and structure are influenced by climatic factors and change gradually with geographical position, so forming the vegetational zones of the Earth. Vegetation composition also changes with increasing altitude from lowlands to mountains, forming altitudinal belts of vegetation. Each type of vegetation includes characteristic host plant species and a corresponding gall midge fauna. Some gall midges are found throughout the distribution of their host plant, while others occupy only a part of their host plant distribution. Some gall midges are patchily distributed within the continuous distribution of their host plant, while others extend beyond the natural distribution of their host by attacking introduced plants in botanical gardens (Skuhravá 1987, Skuhravá et al. 1984).

Human Activity

Mankind has long had a substantial impact on nature, particularly in recent decades. Many plant and animal species have already disappeared or are in the process of disappearing as a result of human activities such as burning of trees and shrubs to obtain new land for agriculture, grazing, development of industry, and use of chemicals in agriculture which damage the environment.

Our long term analysis of the population dynamics of gall midges through the 20th century in the Czech Republic showed that 64 species (12 percent of the Czech fauna) which were once abundant are now less abundant, and in some cases undetectable (Skuhravá 1994). Such species may be designated as threatened in the terms of the International Union for Conservation of

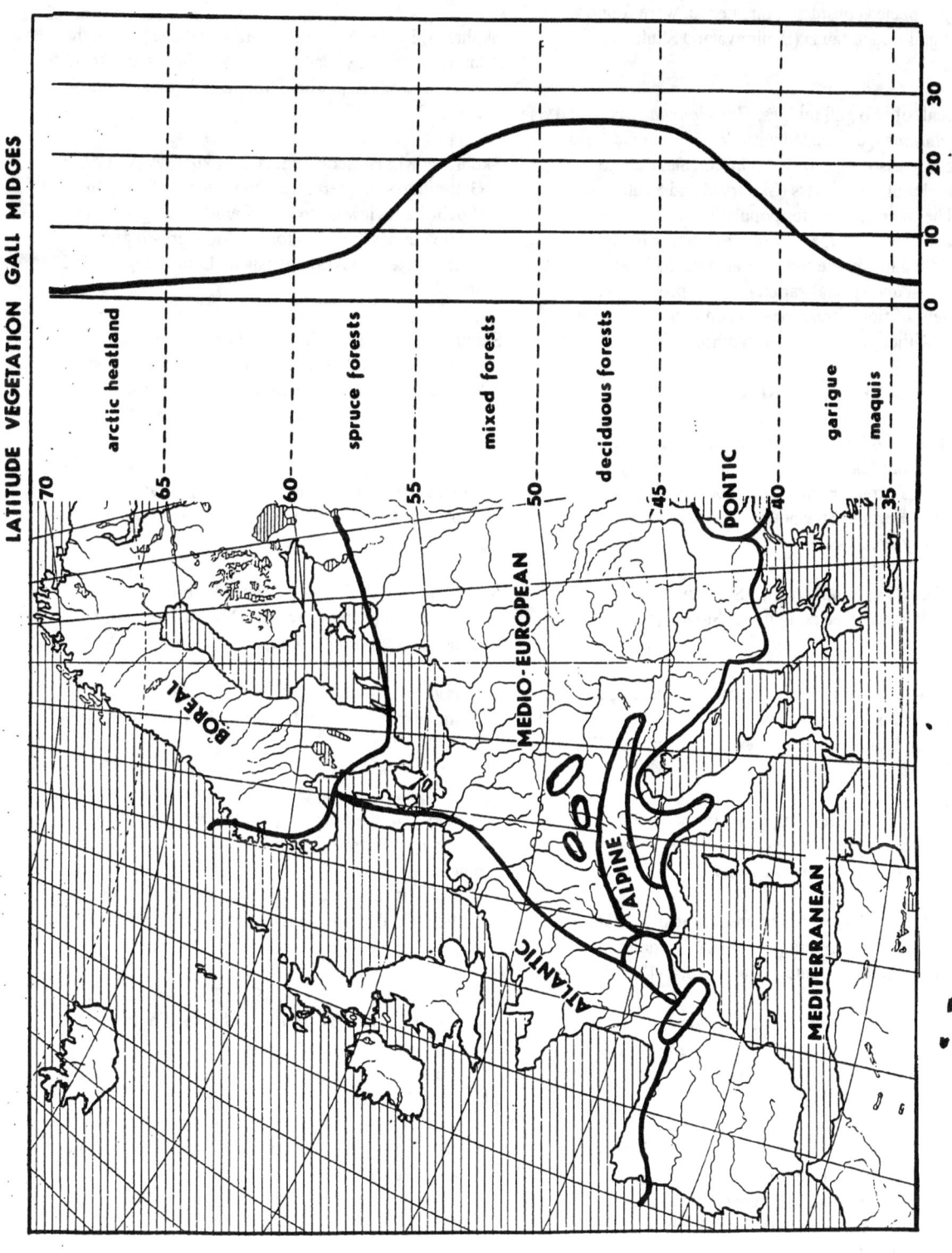

Figure 1.—*The relationship between the gall midge species richness and geographical position in Europe. Geobotanical divisions of Europe according to Noirfalise (1987) (left), the main character of vegetation and the latitudinal gradient expressed in degrees (in the middle) and the average number of gall midge species (right).*

Nature and Natural Resources and may be placed into three categories indicating the degree of threat: extinct, endangered or vulnerable (Skuhravá 1994). Similarly, 18 gall midge species considered threatened, were found in the swiss gall midge fauna (Skuhravá and Skuhravy 1997).

An outbreak of two gall midges, *Harrisomyia vitrina* and *Drisina glutinosa*, associated with *Acer pseudoplatanus*, was probably associated with environmental deterioration through pollution stress (Skuhravá and Skuhravy 1986). The main reasons for population increases and outbreaks of *Haplodiplosis marginata*, larvae of which cause saddle galls on the stems of cereals, include concentration and specialization of plant production, enlargement of fields and repeated consecutive planting of cereals without crop rotation (Skuhravy et al. 1993).

LITERATURE CITED

Noirfalise, A. 1987. Map of the natural vegetation of the member countries of the European Community and the Council of Europe. Explanatory notes by A. Noirfalise. Council of Europe, Strasbourg. Rolled map and explanatory text: 1-80.

Simova-Tosic, D.; Skuhravá, M.; Skuhravy, V. 1996. Gall midges (Diptera: Cecidomyiidae) Slovenije. Scopolia. 36: 1-23.

Skuhravá, M. 1982. The use of faunal data in zoogeographic studies. Folia Facultatis Scienciarum Naturalium Universitatis Purkynianae Brunensis 23, Biol. 74: 127-131.

Skuhravá, M. 1986. Cecidomyiidae. In: Soós, Á.; Papp, L., eds. Catalogue of Palaearctic Diptera. Vol. 4. Akadémiai Kiadó, Budapest: 72-297.

Skuhravá, M. 1987. Analysis of areas of distribution of some Palaearctic gall midge species (Cecidomyiidae, Diptera). Cecidologia Internationale. 8: 1-48.

Skuhravá, M. 1991. Gallmücken der Slowakei (Cecidomyiidae, Diptera) VI. Die Zoogeographie der Gallmücken. Zbornik Slovenskeho Národniho Múzea Prírodni Vedy. 37: 85-178.

Skuhravá, M. 1994. The zoogeography of gall midges (Diptera: Cecidomyiidae) of the Czech Republic I. Evaluation of faunistic researches in the 1855-1990 period. Acta Societatis Zoologicae Bohemicae. 57(1993): 211-293.

Skuhravá, M. 1994. The zoogeography of gall midges (Diptera: Cecidomyiidae) of the Czech Republic II. Review of gall midge species including zoogeographical diagnoses. Acta Societatis Zoologicae Bohemicae. 58: 79-126.

Skuhravá, M. 1995. A zoogeographical analysis of the family Cecidomyiidae (Diptera) in the Czech Republic and in Slovak Republic. Diptera Bohemoslovaca. 7: 159-163.

Skuhravá, M.; Blasco-Zumeta, J.; Skuhravy, V. 1993. Gall midges (Diptera, Cecidomyiidae) of Aragón (Spain): a review of species found in the period 1890-1990 with new records for the Monegros region. Zapateri Revista Aragónensis de Entomologica. 3: 27-36.

Skuhravá, M.; Franz, H. 1989. Familie Cecidomyiidae (Itonididae). In: Franz, H., ed. Die Nordost-Alpen im Spiegel ihrer Landtierwelt. B.VI/1. Innsbruck: Universitätsverlag Wagner, Innsbruck: 67-97.

Skuhravá, M.; Skrzypczynska, M. 1983. A review of gall midges (Cecidomyiidae, Diptera) of Poland. Acta Zoologica Cracov. 26(12): 387-420. (In Polish, with English summary.)

Skuhravá, M.; Skuhravy, V. 1964. Verbreitung der Gallmücken in Jugoslawien (Diptera, Itonididae). Deutsche Entomologische Zeitschrift. 11: 449-458.

Skuhravá, M.; Skuhravy, V. 1967. Beitrag für Gallmückenfauna von Oesterreich. Marcellia. 34(3-4): 213-225.

Skuhravá, M.; Skuhravy, V. 1986. Outbreak of two gall midges, *Harrisomyia* n. gen. *vitrina* (Kieffer) and *Drisina glutinosa* Giard (Cecidomyiidae, Diptera) on maple, *Acer pseudoplatanus* L. in Czechoslovakia, with descriptions of the two genera and species. Zeitschrift für Angewandte Entomologie. 101(3): 256-274.

Skuhravá, M.; Skuhravy, V. 1988. Die Gallmücken (Cecidomyiidae, Diptera) des Harzes. Entomologische Nachrichten Berlin. 32: 29-33.

Skuhravá, M.; Skuhravy, V. 1991-1992. Die Gallmücken (Cecidomyiidae, Diptera) der Kalkalpen und des Waldviertels in Ost-Österreich. Sitzungsberichten der Österreichischen Akademie der Wissenschaften, Mathematische - Naturwissenschaften Klasse, Abteilung I. 199: 27-57.

Skuhravá, M.; Skuhravy, V. 1992. Die Gallmücken (Cecidomyiidae, Diptera) des Naturparkes Fichtelgebirge. Entomologische Nachrichten Berlin. 36: 1-8.

Skuhravá, M.; Skuhravy, V. 1992. Zwei an Ahorn lebende Gallmücken und weitere Gallmücken (Diptera, Cecidomyiidae) im Naturpark Bayerischer Wald. Entomologische Nachrichten Berlin. 36: 97-101.

Skuhravá, M.; Skuhravy, V. 1993. Gall midges (Cecidomyiidae, Diptera) of the southern part of Central Siberia. Diptera Bohemoslovaca. 5: 93-100.

Skuhravá, M.; Skuhravy, V. 1993. The distribution areas of several Euro-Siberian gall midge species (Cecidomyiidae, Diptera). Diptera Bohemoslovaca. 5: 101-108.

Skuhravá, M.; Skuhravy, V. 1993. Die Gallmücken (Diptera: Cecidomyiidae) des Fürstentums Liechtenstein. Praha, Vaduz: 1-16.

Skuhravá, M.; Skuhravy, V. 1994. Gall midges (Diptera: Cecidomyiidae) of Italy. Entomologica, Bari. 28: 45-76.

Skuhravá, M.; Skuhravy, V. 1995. Die Gallmücken (Cecidomyiidae, Diptera) von Österreich II. Sitzungsber. Österr. Sitzungsberichten der Österreichischen Akademie der Wissenschaften, Mathematische - Naturwissenschaften Klasse, Abteilung I. 201: 3-34.

Skuhravá, M.; Skuhravy, V. 1997. Gall midges (Diptera: Cecidomyiidae) of Switzerland. Mitt. Schweizes Entomologisches Geselschaft. 70: 133-176.

Skuhravá, M.; Skuhravy, V.; Blasco-Zumeta, J.; Pujade, J. 1996. Gall midges (Diptera: Cecidomyiidae) of the Iberian Peninsula. Boletin del Asociacion Española Entomologica. 20(1-2): 41-61.

Skuhravá, M.; Skuhravy, V.; Brewer, J.W. 1984. Biology of gall midges. In: Ananthakrishnan, T.N., ed. Biology of gall insects. New Delhi, Bombay, Calcutta: Oxford and IBH Publishing Co.: 169-222.

Skuhravá, M.; Skuhravy, V.; Doncev, K.D.; Dimitrova, B.D. 1991. Gall midges (Cecidomyiidae, Diptera) of Bulgaria I. Faunistic researches in the 1978-1987 period. Acta Zoologica Bulgarica. 42: 3-26.

Skuhravá, M.; Skuhravy, V.; Doncev, K.D.; Dimitrova, B.D. 1992. Gall midges (Cecidomyiidae, Diptera) of Bulgaria II. Host plant relations and economic importance. Acta Zoologica Bulgarica. 43: 23-42.

Skuhravá, M.; Skuhravy, V.; Neacsu, P. 1972. Verbreitung der Gallmücken in Rumänien (Diptera, Cecidomyiidae). Deutsche Entomologische Zeitung N.F. 19: 375-392.

Skuhravá, M.; Skuhravy, V.; Skrzypczynska, M. 1977. Gall midges causing damage of cultivated plants and orchard trees in Poland. Rocznika Nauk Rolniczuch, Ser. A. 7(1): 11-31. (In Polish, with English summary.)

Skuhravy, V. 1991. The needle-shortening gall midge *Thecodiplosis brachyntera* on the genus *Pinus*. Rozpravy CSAV Nr.10, Praha, Academia: 1-104.

Skuhravy, V.; Skuhravá, M.; Brewer, J.W. 1993. The saddle gall midge *Haplodiplosis marginata* (Diptera: Cecidomyiidae) in Czech and Slovak Republic from 1971-1989). Acta Societatis Zoolicae Bohemicae. 57: 117-137.

GALL MIDGES INFESTING CHENOPODIACEAE: BIOLOGY AND TAXONOMY

Netta Dorchin

Department of Zoology, The George S. Wise Faculty of Life Sciences, Tel Aviv University, Tel Aviv 69978, Israel

Abstract.—Gall midges (Cecidomyiidae) are dominant among the relatively few insect taxa that are known to infest plants of the beet family (Chenopodiaceae). More than 300 gall midge species have been reported from chenopod hosts, and about 115 species of Chenopodiaceae are known to be hosts for gall midges. This extensive speciation of gall midges on chenopods is demonstrated by current data on plant-midge relationships, with special focus on the Lasiopterini, which is the largest group involved, and a possible explanation for this evolutionary process is discussed. Taxonomic difficulties concerning the Lasiopterini are also reviewed.

CHENOPODS AS HOSTS FOR GALL MIDGES

Chenopodiaceae is a cosmopolitan family of plants, mainly distributed in the deserts of central Asia, the Middle East, Africa, Australia and in south American prairies. The family comprises about 100 genera and 1,300 species, with the most prominent genera being *Anabasis*, *Atriplex*, *Salsola*, *Suaeda* and *Haloxylon*, together comprising more than 600 species worldwide (Mabberly 1997) (fig. 1a). Chenopods are annual or perennial herbaceous plants, shrubs, or rarely, trees, often dominating a variety of harsh biotopes such as salines and deserts, where almost no other plants grow. It is in these biotopes that chenopods form the characteristic landscape of salt marshes, Irano-turanian prairies, and sand dunes. Among the several anatomical and physiological adaptations that enable chenopods to grow in these extreme conditions, the best studied are their means for plant disposing of excess salts: concentrating them in its tissues or excreting them through epidermal salt-secreting hairs or salt glands (Mozafar and Goodin 1970, Waisel 1972). These mechanisms possibly also play a role in the unpalatability of chenopods to most insects.

While other arthropods seldom infest chenopods, more than 300 cecidomyiid species are exclusive chenopod feeders, and the great majority of them are gall formers. More than 100 species of Chenopodiaceae are known to be hosts of gall midges, and about 60 percent of them belong to one of the five prominent genera mentioned above (*Anabasis*, *Atriplex*, *Salsola*, *Suaeda*, and *Haloxylon*, fig. 1b). However, many additional chenopod-infesting gall midges and chenopod hosts are still awaiting discovery, as is indicated by a recent research on the chenopod-infesting gall midges in Israel (Dorchin

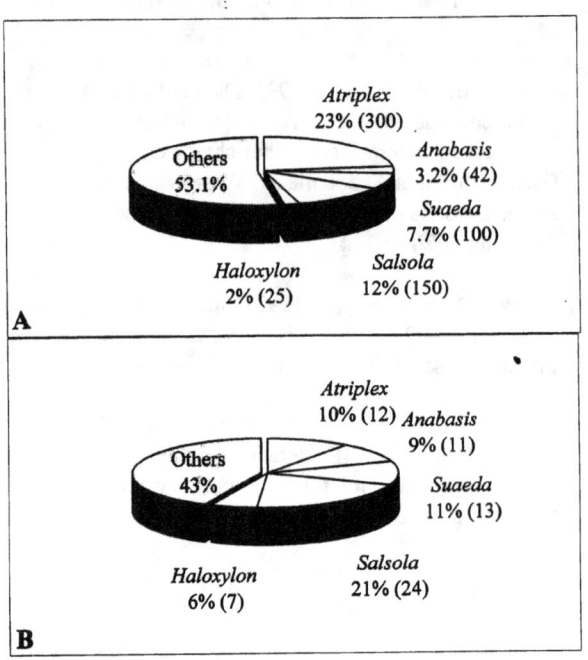

Figure 1.—*Five prominent genera within the Chenopodiaceae. A. number of species (total no. 1300), B. number of host species (total no. 117).*

1997). In that study, I screened 41 out of the 76 chenopod species of Israel, and found 28 of them to be hosts of gall midges. Seven of these plants were not previously known to be hosts. Out of these 28 hosts 55 species of gall midges were reared, most of which are probably undescribed. The number of gall midge species per host averaged 2 in that study. However, the number of gall midge species infesting a given host varies greatly among the hosts, with many hosts yielding only one gall midge species, several hosts yielding three or four species and one host (*Salsola tetrandra* Forssk.) yielding seven species. Although this distribution pattern is in accordance with the overall statistics for the chenopod infesting gall midges as a group (fig. 2), examples from Central Asia are much more impressive than the Israeli ones, especially that of *Haloxylon aphyllum* with 28 gall midge species associated with it (Marikovskij 1955, Mamaev 1972). I speculate that the successful relationship between gall midges and chenopods, as reflected by these data, is due to the almost unique ability of the midges to overcome the physical and physiological obstacles presented by the plants, making it possible for them to occupy an under-utilized niche. The latter evolutionary event was then followed by extensive speciation of the midges, a process which is probably still going on today.

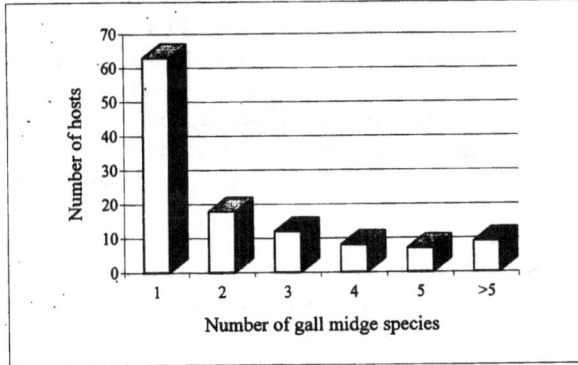

Figure 2.—*Distribution of number of gall midge species per chenopod host.*

THE CHENOPOD GALL MIDGES

Within the guild of chenopod infesting gall midges, about 170 species (ca. 54 percent) belong to the tribe Lasiopterini (fig. 3), and within this tribe mainly to the chenopod-restricted genera *Baldratia* and *Stefaniola*. Over 80 additional species belong to the Cecidomyiini (mainly the genus *Halodiplosis*), and the rest belong to small genera or species groups in the tribes Asphondyliini (e.g., *Asphondylia* and *Kochiomyia*), Oligotrophini (*Dasyneuriola*) and Alycaulini (*Neolasioptera*). My own findings on the Israeli fauna (Dorchin 1997) are in accordance with these data, and out of the 55 reared species of chenopod-infesting gall midges, 27 (64 percent) belong to the Lasiopterini. This, together with the continual discovery of new species in central Asia (Fedotova 1989-1995) strengthens the impression that our knowledge of this group is far from complete.

CHENOPOD INFESTING LASIOPTERINI

Biology and Ecology

The chenopod infesting Lasiopterini constitute the most interesting and diverse group within the guild both with respect to their biology and taxonomy. Like other Lasiopterini, which most commonly gall the vegetative rather than reproductive plant organs (Roskam 1992), most of the chenopod infesting species induce the formation of twig or stem galls, many form leaf galls, but only a few form bud and flower galls. Galls can be rather large or barely visible, and they vary considerably in structure and hairiness. Some galls are rare, while others are found in great numbers. Some species have also been reared that live in the plant tissue without

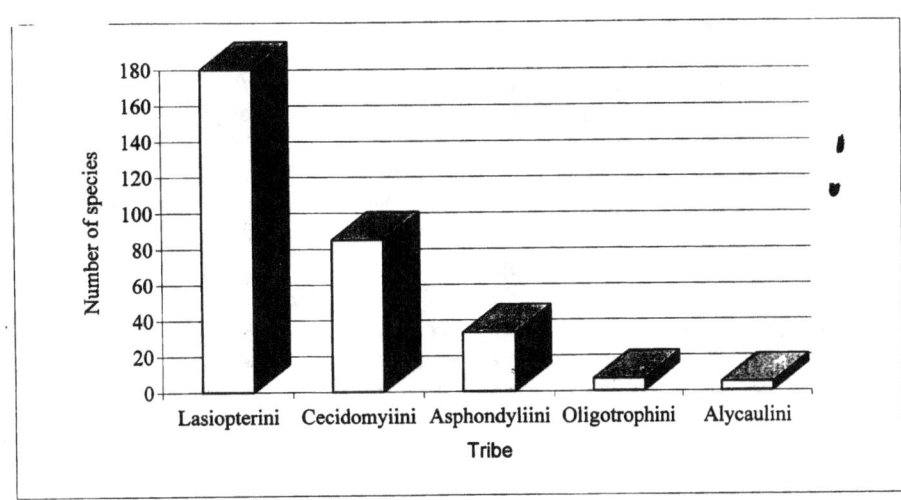

Figure 3.—*Number of chenopod infesting species within cecidomyiid tribes.*

causing visible deformation, a habit apparently much more widespread in the Lasiopterini than previously thought. The presence of the midges in such cases is often discovered only after exuviae are found protruding from normal looking leaves or stems (Dorchin 1997; Fedotova 1992a, b, 1993). Larvae may be solitary or gregarious, and when gregarious, may cause considerable damage to shoots (Dorchin 1997) or to fruits and seeds (Fedotova 1991, 1992b).

The great majority of the chenopod infesting Lasiopterini complete their life cycle inside their galls, but some exceptions are known for which the mature larvae fall to the ground and pupate in it (e.g., *Stefaniola ustjurensis* Fedotova (1991) and *Baldratia jaxartica* Fedotova (1992b)). It was previously thought that most species have only one generation per year (Möhn 1971). However, this impression might be the result of insufficient biological and ecological data, as many species recently studied by Fedotova (1991, 1992b, 1993, 1993a, 1995) proved to have two or even three generations per year, and this is also implied by partial data I have obtained on the Israeli fauna (Dorchin 1997).

An interesting ecological adaptation found in my study reveals that despite their delicacy and fragility, some Israeli desert species are active throughout the hottest months of the year, when temperatures may reach 45°C (~113°F). Thus, Cecidomyiidae are an exception to most Nematocera, which prefer cool and moist biotopes, and are probably the largest and most diverse nematoceran family to be found in desert areas.

Taxonomy and Taxonomic Problems

In view of the potential in the Lasiopterini for many new and interesting taxonomic discoveries, it is important to mention that the tribe exhibits several complicated taxonomic problems and is in urgent need of comprehensive taxonomic revision and phylogenetic analysis. However, these tasks confront some serious, technical difficulties. The first difficulty is related to the revision of the Palaearctic Lasiopterini, published by Dr. Edwin Möhn between 1966 and 1971 in the series Die Fliegen der Palaearktischen Region. This revision, including the keys and description of 72 new species, is entirely based on larvae recovered from galls found in dried plants, most of which were chenopods collected years earlier in north Africa, central Asia, and the Mediterranean basin. The identification of cecidomyiid larvae is extremely difficult, and the usefulness of any revision based on larval characters alone is therefore rather limited.

Another complication concerns Möhn's types. Thirty-five new species of *Stefaniola* and 19 new species of *Baldratiaa*—two of the largest genera of Lasiopterini—were described, among other taxa, in Möhn's revision. During my research, I reared 26 species belonging to these genera, some of them from hosts recorded by Möhn. A study of Möhn's collection was therefore essential. Unfortunately, the present condition of this collection, which is deposited in the Staatliches Museum für Naturkunde in Stuttgart, renders it impossible to locate and study most of the type material, as the microscope slides are still temporary, with glycerol as a mounting medium, and the types are actually not labeled. The problem is not limited to *Baldratia* and *Stefaniola* alone, but also concerns the genera *Ozirhincus* and *Lasioptera* which were revised in the same work, as well as the Palaearctic species of the tribe Trotteriini. Collecting the relevant galls all over again and obtaining from them larvae as well as associated adults seems to be the best solution to this problem. Dr. Fedotova, now at the Agricultural Academy in Ust Kinelskij, Russia, has recently described the adults of few of Möhn's species from Kazakhstan (Fedotova 1989a, b, c, 1992) and I have undertaken this task in Israel. However, further surveys throughout north Africa, the Middle East, and central Asia are necessary, although at this time seem somewhat unrealistic.

Another difficulty is created by the inaccessibility of many relevant types of species described by Russian workers. Conditions presently prevailing in Russia, especially communication problems, form a great hindrance to any study relying on the availability of these types.

LITERATURE CITED

Dorchin, N. 1997. Gall midges infesting Chenopodiaceae (Diptera: Cecidomyiidae) in Israel. Tel Aviv, Israel: Tel Aviv University. 131 p. M.Sc. thesis. (In Hebrew)

Fedotova, Z.A. 1989a. Gall midges of the genus *Stefaniola* in Kazakhstan. communication 1. Zoologicheskiy Zhurnal. 68(5): 59-71. (In Russian)

Fedotova, Z.A. 1989b. Gall midges of the genus *Stefaniola* in Kazakhstan USSR. communication 2. Zoologicheskiy Zhurnal. 68(6): 78-86. (In Russian)

Fedotova, Z.A. 1989c. Gall midges of the genus *Stefaniola* in Kazakhstan USSR. communication 3. Zoologicheskiy Zhurnal. 68(8): 57-66. (In Russian)

Fedotova, Z.A. 1990. New species of gall midges (Diptera, Cecidomyiidae) found on pastures with *Anabasis* spp. and *Nanophyton erinaceum* (Chenopodiaceae) in Kazakhstan. Entomologicheskoye Obozreniye. 69: 223-232. (In Russian)

Fedotova, Z.A. 1991. Gall midges (Diptera, Cecidomyiidae) damaging Orache (*Atriplex* spp.) in Kazakhstan. Entomological Review. 70(3): 1-10.

Fedotova, Z.A. 1992. New species of gall midges (Diptera, Cecidomyiidae) from Chenopodiaceae in Kazakhstan. Izvestia Akademii Nauk Respubliki Kazakhstan, Seriya Biologicheskikh. (1): 44-52. (In Russian)

Fedotova, Z.A. 1992a. Gall midges of the Genus *Baldratia* (Diptera, Cecidomyiidae) of Kazakhstan. Entomological Review. 71(2): 25-39.

Fedotova, Z.A. 1992b. Gall midges of the subtribe Baldratiina (Diptera, Cecidomyiidae, Lasiopterini) in Kazakhstan. Entomological Review. 71(6): 145-154.

Fedotova, Z.A. 1993. New species of gall midges (Diptera, Cecidomyiidae, Lasiopterini) developing on Chenopodiaceae in Kzakhstan and Turkmenia. Entomological Review. 72(1): 66-77.

Fedotova, Z.A. 1993a. Review of species of gall midges (Diptera, Cecidomyiidae) on Saksaul (*Arthrophytum* spp.) from Kazakhstan. Entomological Review. 72(1): 78-88.

Fedotova, Z.A. 1993b. New species of gall midges of the genus *Halodiplosis* (Diptera, Cecidomyiidae) from Kazakhstan. Entomological Review. 72(7): 121-133.

Fedotova, Z.A. 1995. New species of gall midges (Diptera, Cecidomyiidae) on Chenopodiaceae from Kazakhstan and Turkmenia. Entomological Review. 74(1): 79-93.

Mabberly, D.J. 1997. The plant-book: a portable dictionary of the higher plants. Cambridge, UK: Cambridge University Press. 858 p.

Mamaev, B.M. 1972. New gall midges (Diptera, Cecidomyiidae) of the desert zone. Entomological Review. 51: 526-534.

Marikovskij, P.I. 1955. New species of gall midges of the genus *Asiodiplosis* Marikovskij (Diptera, Itonididae) from Kazakhstan deserts. Zoologicheskii Zhurnal. 34: 336-346. (In Russian)

Mozafar, A.; Goodin, J.R. 1970. Vesiculated hairs: a mechanism for salt tolerance in *Atriplex halimus* L. Plant Physiology. 45: 62-65.

Möhn, E. 1966-1971. Cecidomyiidae (=Itonididae). Vol. 2(2), part 6. L: I-V. In: Lindner, E., ed. Die fliegen der palaearktischen region. Schweizer bart'sche, Stuttgart: 1-248 and pls.

Roskam, J.C. 1992. Evolotion of the gall-inducing guild. In: Shorthouse, J.D.; Rohfritsch, O., eds. Biology of insect-induced galls. New York: Oxford University Press: 34-49.

Waisel, Y. 1972. Biology of Halophytes. London, UK: Academic Press. 395 p.

ACTILASIOPTERA (DIPTERA: CECIDOMYIIDAE), A NEW GENUS FOR AUSTRALASIAN AND ASIAN GALL MIDGES OF GREY MANGROVES, *AVICENNIA* SPP. (AVICENNIACEAE)

Raymond J. Gagné[1] and Laraine J. Law[2]

[1] Systematic Entomology Laboratory, PSI, Agricultural Research Service, USDA, c/o U. S. National Museum NHB 168, Washington, DC 20560, USA
[2] Department of Entomology, The University of Queensland, Brisbane, QLD 4072, Australia

Abstract.—A new genus of Lasiopterini, *Actilasioptera*, is described for five new species of gall midges that form galls on *Avicennia marina* along the coast of Queensland, Australia, and for *Stefaniella falcaria* Felt from leaf galls of *Avicennia officinalis* in Java. The five new species, each forming a distinct gall, are named: *Actilasioptera coronata*, *A. pustulata*, *A. subfolium*, *A. tuberculata*, and *A. tumidifolium*. Keys are given for the recognition of galls, adults, pupae, and larvae.

This study is a result of a survey on the diversity and host associations of arthropod galls on mangroves in Moreton Bay, SE Queensland, Australia (Law 1996). Eight foliar galls were found on grey mangrove, *Avicennia marina* (Forssk.) Vierh. Five of the galls were each caused by a new species of Cecidomyiidae described in this paper, another by an eriophyid mite (Acarina), one by a coccid (Homoptera), and one by an unknown insect or mite. No galls were found on the other mangroves studied, which are: river mangrove, *Aegiceras corniculatum* (L.) (Myrsinaceae); blind-your-eye, *Excoecaria agallocha* (L.) (Euphorbiaceae); black mangrove, *Lumnitzera racemosa* (Willd.) (Combretaceae); red mangrove, *Rhizophora stylosa* (Griffith) (Rhizophoraceae); yellow mangrove, *Ceriops tagal* (Perr.) Robinson (Rhizophoraceae); and large-leafed orange mangrove, *Bruguiera gymnorrhiza* (L.) Savigny (Rhizophoraceae).

The Cecidomyiidae from grey mangrove belong to an apparently monophyletic group of gall midges of the tribe Lasiopterini (sensu Gagné 1994) for which a new genus, *Actilasioptera*, is erected. In addition to the five new species from the east coast of Queensland, this genus is otherwise known from one previously described species from Java, *Stefaniella falcaria* Felt, that is redescribed and transferred to the new genus, and a species from Papua New Guinea known only from adults and so left undescribed. The genus probably occurs throughout the range of grey mangroves in Asia and Australasia. Galls made by these species are either simple or complex (figs. 1-8). The simple galls (figs. 1-4) are swellings of either both leaf surfaces or only the under surface. The complex galls (figs. 5-8) are characteristic of the particular gall maker and are of a fundamentally different shape from any structure normally found on the host. In the Western Hemisphere only one species of gall midge, *Meunieriella avicenniae* (Cook), is known from *Avicennia* mangroves (Gagné and Etienne 1996). It forms simple leaf blister galls and belongs to a tribe of gall midges, the Alycaulini, that is restricted to the Americas. No other galls are known on *Avicennia* spp. in the Western Hemisphere. We note here to prevent possible confusion, that *Avicennia* spp. of Australia are called grey mangroves, but Western Hemisphere *Avicennia* spp. are called black mangroves.

The gall midge tribe Lasiopterini is known from about 230 species from the Old World and about 40 from North and Central America (Gagné 1989a, 1994). Most Lasiopterini form simple swellings on stems, petioles, and leaf ribs of various plants, but some live in galls or galleries of other insects as inquilines or successors. The three species treated here that form complex galls are the first of the tribe known to do so. Many Lasiopterini, but not the present species, are associated with a symbiotic fungus. Larvae of most Lasiopterini pupate in the galls, as do these, but a species in Japan (Yukawa and Haitsuka 1994) and one in North America (Gagné, unpubl.) are exceptions and drop to the soil to pupate.

Only ten species of Lasiopterini are known from the Australasian Region, all from Australia and at present assigned to the genus *Lasioptera* (Gagné 1989b). None is from mangrove. Nine of these species are poorly known, not having been noted except in catalogs since their original descriptions by Skuse (1888, 1890). The remaining species, *Lasioptera uncinata* Gagné, was recently described from paperbark, *Melaleuca quinquenervia* (Cav.) S.T. Blake (Gagné et al. 1997).

Actilasioptera species differ radically from other Lasiopterini in having the ovipositor modified for piercing plant tissue, the whole ninth segment being glabrous and rigid and nearly devoid of setae (figs. 13-16). Pupae have an enlarged or otherwise modified vertex and have lost the vertexal papillae, both modifications unique to the tribe. Larvae have a reduced spatula and have lost some papillae, characters that occur elsewhere, presumably independently, in some Lasiopterini. The closest relative of this genus among the Lasiopterini is not apparent, not surprising when so little is known of this tribe and the family in general.

METHODS AND MATERIALS

Most of the material used in this study was discovered, collected, and reared by L.J. Law, the taxonomy was done by R. J. Gagné. The new species were collected by L. J. Law from *Avicennia marina* growing in coastal areas of Moreton Bay in the vicinity of Brisbane, Queensland. Moreton Bay is shallow with a mud and sand substrate. It extends south from the Noosa River (26°48'S, 153°08'E) to Southport on the Gold Coast (27°58'S, 153°25'E). The bay has well-protected shores and extensive stands of mangroves. Additional material of one species was collected by D. Burrows in Townsville, Queensland. Galls from the field were either preserved in alcohol or kept in small plastic containers that were checked daily for emergence of adults. Larvae and pupae were excised from some galls, adults were reared from others, and all specimens were preserved in 70 percent ethanol. Some larvae and adults were mounted for microscopic study in Canada balsam, using the method outlined in Gagné (1989a, 1994), and some were prepared for SEM viewing. In the following descriptions, anatomical terminology of the adult stage follows McAlpine (1981) and that of the larval stage follows Gagné (1989a). The holotypes and most of the specimens of the new species described in this study are deposited in the Australian National Insect Collection, Canberra (ANIC), the remainder in the U.S. National Museum of Natural History, Washington, DC. (USNM).

ACTILASIOPTERA GAGNE, NEW GENUS

Adult

Head (figs. 9-10). Antenna with scape cylindrical, longer than wide; pedicel spheroid; flagellomeres 10-12, each longer than wide, the first and second connate, the apicalmost usually fused. Palpus 1-2 segmented. Labella elongate ovoid, with 0-many scales.

Thorax. R_5 0.6-0.7 length of wing (figs. 31-32). Scutum with two lateral and two dorsocentral rows of setae, otherwise completely covered with scales; scutellum with group of setae on each side, elsewhere covered with scales; anepisternum with scales on dorsal half; mesepimeron with row of setae and no scales; remainder of pleura without vestiture. Tarsal claws (fig. 11) strong, pigmented, with large, more or less sinuous basal tooth; empodia as long as claws; pulvilli about 2/3 as long as empodia.

Male abdomen. First through sixth tergites rectangular, much wider than long, with mostly single, uninterrupted row of setae, a pair of anterior trichoid sensilla, and covered with scales; seventh tergite as for sixth or sclerotized only on anterior half, with only anterior pair of trichoid sensilla and few scales; eighth tergite sclerotized only on anterior half, with only anterior pair of trichoid sensilla and few scales posteriorly. Genitalia (figs. 12, 33-36): robust; gonocoxite cylindrical; gonostylus abruptly tapered beyond bulbous base, mostly setulose, ridged apically; cerci convex; hypoproct convex or bilobed; parameres elongate, narrow beyond anterodorsal lobe; aedeagus elongate, narrow, pointed apically, apex bent ventrally in some species.

Female abdomen (figs. 13-16). First through sixth tergites rectangular, much wider than long, with single, uninterrupted row of setae, a pair of anterior trichoid sensilla, and covered with scales; seventh tergite as for sixth but with two or three rows of posterior setae; eighth tergite weakly sclerotized, about as wide as long, the only vestiture the pair of trichoid sensilla; intersegmental membrane between eighth and ninth segments with dorsolateral group of robust, elongate setae on each side, the remainder of the surface covered with elongate, fine setulae; ninth segment and cerci modified into a sclerotized, glabrous piercing organ 2-3 times as long as seventh tergite and with only a few pairs of apicoventral setae.

Pupa

Vertex (figs. 17-24) more or less prominent, conical to convex, without vertexal papillae. Antennal bases not prominent, rounded to acutely angled. Face smooth or with pair of slight, rounded projections. Facial papillae absent. Prothoracic spiracle short, two to three times as long as wide. Tergal disks covered with uniformly short microsetulae.

Larva

Third instar (figs. 25-30). White. Body shape ovoid. Integument completely or mostly covered with conical, pointed verrucae. Antenna about twice as long as wide. Spatula short, broad, variously shaped. Anus caudoventral, the terminal segment in dorsoventral view appearing cleft. Papillar pattern as for Lasiopteridi and Lasiopterini (Gagné 1989a, 1994) with, when discernable (reduced and usually not visible in *A. tumidifolium* and *A. subfolium*), the lateral papillae reduced to three on each side, two of them setose on prothorax, all setose on meso- and metathorax (fig. 26); thoracic sternal papillae without or with very short setae on prothorax, setose on meso- and metathorax; terminal segment with only two papillae, both setose.

Etymology

Actilasioptera (feminine gender) combines the Greek akte, meaning seashore, with *Lasioptera*, the genus in which most Lasiopterini are now placed.

Remarks

The new genus belongs to the tribe Lasiopterini with which it shares the following apomorphies: the gynecoid male antennae, the greatly shortened and nearly straight R5 wing vein, the reduced number of four lateral setae on each side of the larval thorax, and the two dorsolateral groups of enlarged setae posterior to the female eighth tergite. Most female Lasiopterini have hooked setae on the cerci, but these are not present in *Actilasioptera* from which they are presumably lost due to the great modification of the ovipositor.

Apomorphies of *Actilasioptera* are: the narrow, elongate parameres and aedeagus, the reduction of the palpi to one or two segments, the modification of the ovipositor into a piercing organ, the developed pupal vertex, the reduction in size of the larval spatula, and the reduction of the larval thoracic lateral papillae to three pairs from four on each side and of the larval terminal papillae to one pair instead of four.

KEY TO GALLS OF *Actilasioptera* ON GREY MANGROVE IN EASTERN AUSTRALIA

1. Soft, simple leaf swelling, 10 or more mm in diameter, with texture similar to that of normal leaf (figs. 1-4) 2
   Hard, ridged, warty, or tuberculate leaf growths less than 5 mm diameter (figs. 5-8).................................3
2. Swelling apparent on both leaf surfaces (figs. 1-2) .... *A. tumidifolium*
   Swelling apparent only on lower leaf surface (figs. 3-4) .... *A. subfolium*
3. Unevenly hemispheroid and warty on upper leaf surface, craterlike on lower surface (figs. 5-6) .... *A. coronata*
   Either circular and flat or with one or more, smooth, elongate, conical projections on upper leaf surface, not craterlike on lower surface (figs. 7-8) 4
4. Circular and flat on upper leaf surface, nearly evenly hemispherical and usually covered with tiny pustules below (fig. 7) .... *A. pustulata*
   With one or more smooth, elongate, conical, projections on upper leaf surface, circular and flat below (fig. 8) .... *A. tuberculata*

KEY TO SPECIES OF *Actilasioptera*

The pupa and larva of the Javanese *A. falcaria* are unknown; available specimens of adults of that species are indistinguishable from those of *A. tumidifolium*. *A. pustulata* and *A. tuberculata* are indistinguishable except in their galls.

1. Adult with second palpal segment segment smaller than first or absent; paramere of male genitalia glabrous apically (figs. 33-35); ovipositor bilaterally flattened, broad and curved in lateral view (figs. 13-14); pupal vertex conical or strongly convex (figs. 17-20); larval setae barely longer than wide when apparent (fig. 25) 2
   Adult second palpal segment larger than first (fig. 9); paramere of male genitalia setulose apically (fig. 12); ovipositor narrowing abruptly at base into long, cylindrical tube (fig. 15-16); pupal vertex weakly convex and produced ventrally between antennal bases (figs. 21-24); larval setae all many times longer than wide (figs. 27-28) 3
2. Male hypoproct bilobed (fig. 33); pupal vertex conical (unknown for *A. falcaria*) (figs. 17-18); larval spatula much wider on posterior half than anteriorly, with two or more weak teeth anteriorly (unknown for *A. falcaria*) (fig. 25) .... *A. tumidifolium* and *A. falcaria*
   Male hypoproct simple (fig. 36); pupal vertex convex (figs. 19-20); larval spatula parallel sided, with two large anterior teeth (fig. 29) .... *A. subfolium*
3. Pupal antennal bases rounded, pupal face without projections (figs. 21-22); larval spatula more or less parallel sided, the middle pair of 4 anterior teeth smaller (fig. 27) .... *A. coronata*
   Pupal antennal base pointed apicoventrally, pupal face with slight projections (figs. 23-24); larval spatula indented laterally, all 4 anterior teeth subequal (fig. 30) .... *A. pustulata* and *A. tuberculata*

Actilasioptera tumidifolium GAGNE, NEW SPECIES

Adult

Head. Antenna with 12 flagellomeres, occasionally, especially in females, the two distalmost segments partially fused. Palpus 2 segmented, the second segment much smaller than the first and occasionally not completely separate, both segments with several long setae and covered with scales. Labella with a few apical setae, elsewhere covered with scales.

Thorax. Wing length, 2.3-2.7 mm (n = 5) in male, 2.4-2.8 mm (n = 4) in female; R_5 about 0.7 length of wing (fig. 31).

Male abdomen. Seventh tergite fully sclerotized, its vestiture similar to that of sixth. Genitalia (figs. 33-35): Hypoproct bilobed posteriorly; parameres without setulae beyond anterodorsal lobe; aedeagus narrow, with two sensory pits and curved ventrally at apex.

Female abdomen (figs. 13-14). Intersegmental membrane between eighth and ninth segments with lateral group of robust, elongate setae on each side, with no other setation but surface covered with elongate, fine setulae; combined ninth segment and cerci strongly sclerotized, glabrous, bilaterally flattened, dorsally curved.

Pupa

Vertex conical, greatly exceeding antennal bases (figs. 17-18). Antennal bases rounded anteroventrally. Face smooth, without projections.

Larva

Third instar (figs. 25-26). Integument with conical, pointed verrucae present mesally on venter and laterally on dorsum of mesothorax and metathorax and circling completely or almost completely (absent dorsomesally on first through sixth) middle of first through eighth abdominal segments. Spatula almost blunt anteriorly, with pair of small anterolateral teeth, much widened on posterior half. Papillae usually not visible, the setae when present very diminutive; when visible, agreeing with generic diagnosis.

Holotype and Specimens Examined

Holotype, male, reared from leaf gall on *Avicennia marina*, Australia, Queensland, Sandgate, 23-VIII-1996, L. Law, deposited in Australian National Insect Collection, Canberra (ANIC).

Other specimens examined (all from *Avicennia marina*, Queensland, Australia, and collected by L. Law unless otherwise indicated; most deposited in ANIC, a representative sample in USNM).—Sandgate, 23-VIII-1996, 2 males, 1 female; Donnybrook, 16-IV-1994, 5 pupae, 5 larval exuviae; Toorbul, 20-III-1994, 9 females, 12 pupae, 3 third instars; Brighton, 4-IX-1995; 6 males, 1 female, 2 pupae, 7 third instars; Townsville, Saunders Beach, 8-VII-1995, D.W.Burrows, 4 males, 2 females, 4 pupal exuviae.

Etymology

The name *tumidifolium* (noun in apposition) means "swollen leaf," for the gall typical of this species.

Gall

This species is responsible for a large, green to brown, soft, simple, convex leaf swelling that is apparent on both leaf surfaces (figs. 1-2). Galls are 10-40 mm across and up to 15 mm thick. They occasionally completely engulf a leaf. Galls are generally similar in exterior texture to the normal leaf. Inside each gall are several to many ovoid chambers each harboring a single larva. Pupae break through either leaf surface just before adult emergence.

Remarks

Adults of this species and two others, *A. falcaria* and *A. subfolium*, are similar in several ways, distinguishing them from the remaining three species (see contrasting remarks about the other group under *A. coronata*): the antennae are usually 12 segmented, the second palpal segment is smaller than the first (the second is occasionally absent in *A. subfolium*), the R_5 wing vein is about 0.7 wing length, the male seventh tergite is similar to the preceding in sclerotization and vestiture, the parameres are mostly glabrous, and the ovipositor is bilaterally flattened, broad and curved in lateral view. The pupae and larvae of this species and *A. subfolium* (those of *A. falcaria* are unknown) share the following characters: the pupal vertex is strongly convex, the larval papillae are diminutive, often lost, and the larval integument has extensive areas without verrucae.

Adults of *A. tumidifolium* are indistinguishable from those of *A. falcaria*, but the pupa and larva of *A. falcaria* are unknown. Their galls are apparently similar also, but the hosts of the two species are different and their provenance widely separated, *A. tumidifolium* from eastern Australia and *A. falcaria* from Java.

Actilasioptera tumidifolium differs from *A. subfolium* in the following ways: scales cover the labella of *tumidifolium* but are sparse on those of *subfolium*; the male hypoproct is bilobed in *tumidifolium*, simple in *subfolium*; the pupal vertex is much more swollen in *tumidifolium* than in *subfolium*; and the larval spatula of the two species are shaped differently (compare figs. 25 and 29). The galls of these two species are similar in shape but that of *subfolium* shows only on the abaxial surface of the leaf rather than on both surfaces.

Actilasioptera falcaria FELT, NEW COMBINATION

falcaria Felt 1921: 141, *Stefaniella*

Adult

Head. Antenna with 12 flagellomeres. Palpus 2 segmented, the second segment much smaller than the first.

Thorax. Wing length, 1.8 mm (n = 1) in male, 1.8 mm (n = 2) in female; R_5 about 0.7 length of wing (as in fig. 31).

Male abdomen. Seventh tergite fully sclerotized, its vestiture similar to that of sixth. Genitalia: Hypoproct bilobed posteriorly; parameres without setulae beyond anterodorsal lobe; aedeagus narrow and curved ventrally at apex.

Female abdomen. As for *A. tumidifolium*.

Pupa and Larva

Unknown.

Types

Lectotype, female, here designated, from *Avicennia officinalis*; Indonesia: Java, Semarang; 27-IV-1914; J. and W. Docters van Leeuwen Reijnvaan; Felt Collection a3089; on long-term loan to Systematic Entomology Laboratory, USDA, Washington, DC from the New York State Museum in Albany. Paralectotypes: 1 male and 4 females with data as for lectotype; 1 female, same data except 29-IV-1914 and Felt Collection a3090.

Gall

According to J. and W.D. Docters van Leeuwen-Reijnvaan (1910), the gall is a 1 cm diameter swelling situated on or very near the leaf midvein of *Avicennia officinalis* L., is unevenly round and apparent on both leaf surfaces, the surface green and shiny above, gray and matte below. Affected leaves are often bowed. The gall has several chambers, each with a single larva. Their figures of this gall (their no. 96), a whole gall on a leaf and one in cross section, show a gall similar to that formed by *A. tumidifolium* (figs. 1-2). They show also a cross section of another gall, no 97, that is described as a slight leaf swelling about 3 mm wide by 1 mm thick, yellowish green above, yellow beneath, with two larval chambers. This may be caused by a different species (see discussion under Remarks).

Remarks

Felt (1921) placed *Actilasioptera falcaria* in the genus *Stefaniella* for convenience, writing that, "it seems best for the present ... to consider [this species] as simply an extreme type of specialization rather than erect a new genus." *Stefaniella* species, all associated with leaf, stem, and bud galls on Palearctic Chenopodiaceae, have partially sclerotized ovipositors but all have discrete, partially soft cerci bearing hooked setae and setulae. There is no reason to regard *Stefaniella* and *Actilasioptera* as particularly closely related.

Felt (1921) wrote that he had a considerable series of females, at least 2 males, and several pupae from the large globular leaf galls collected by W. Docters van Leeuwen in 1914. There remain only 1 male and 4 females, all originally placed in Canada balsam without first passing them through a clearing agent. Two females were remounted to good effect for this study so that their ovipositors are visible, but a new series would best serve comparative purposes. Felt (1921) also referred to this taxon another lot from the small leaf blister gall reported as gall no. 97 in J. and W.D. Docters van Leeuwen-Reijnvaan (1910). Felt wrote that the lot from this second kind of gall was with little question referable to the same species as the first, but the single remaining available specimen is mounted with the ovipositor in dorsoventral view, still retracted into the uncleared abdomen, and so not directly comparable to females of the first lot. Whether this female is similar to the females in the other series cannot be determined at present.

As pointed out under the Remarks heading of *A. tumidifolium*, this species is similar in many ways to *A. tumidifolium* and *A. subfolium*, and indistinguishable from the former. Because the pupa and larva of *A. falcaria* are unknown, the two species cannot be compared. Until the immature stages are found, we choose to treat *A. tumidifolium* as distinct because the hosts of the two species are different and *A. tumidifolium* is known from eastern Australia and *A. falcaria* from Java. If the immature stages of the two species are one day found to be similar, the name *tumidifolium* can simply be synonymized under *falcaria*; if they prove to be different, *A. tumidifolium* will already be properly described.

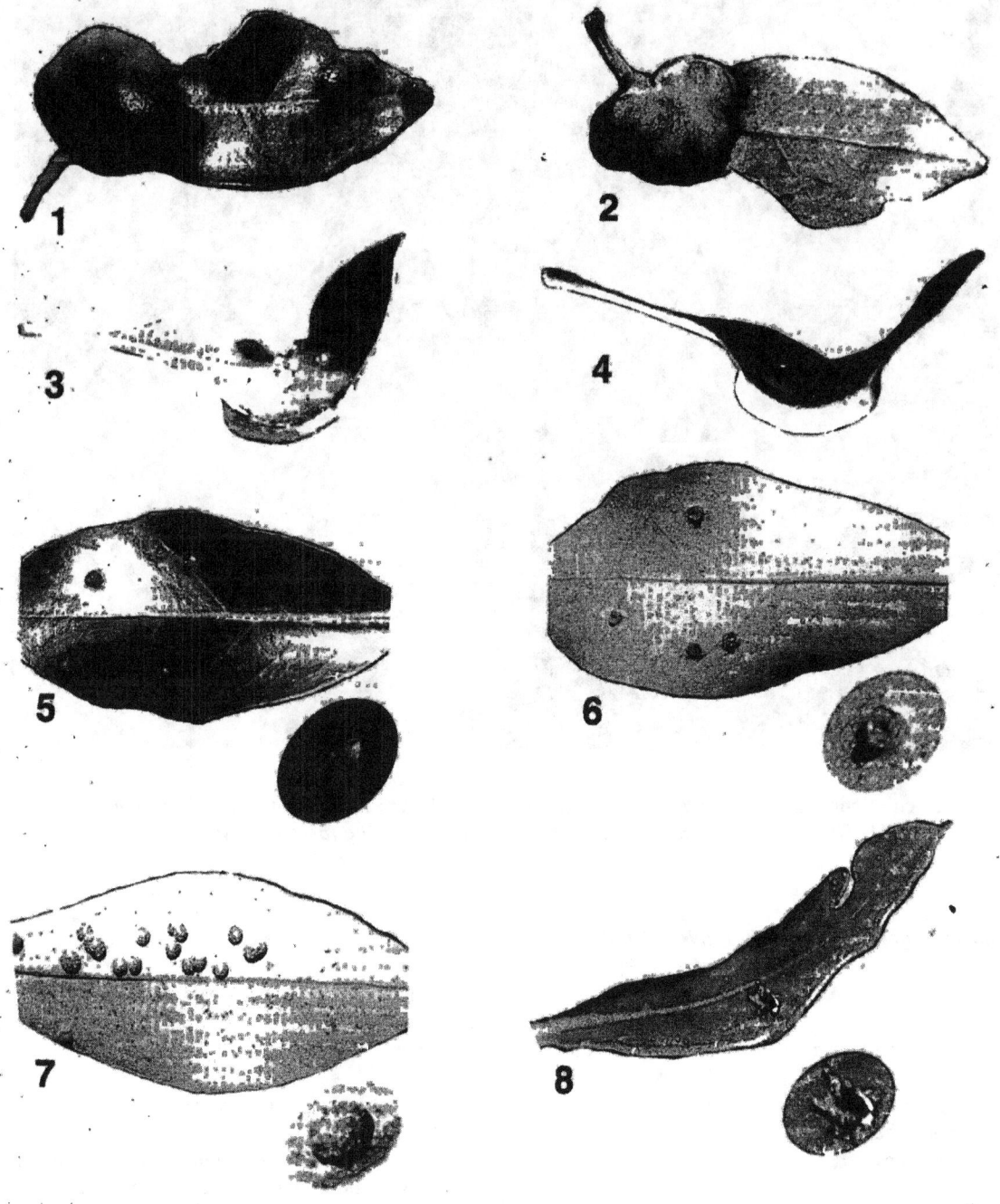

Figures 1-8, *Galls of* Actilasioptera *spp. on* Avicennia marina: Figures 1-2, *Gall of* A. tumidifolium, *apparent on both sides of leaf (1, adaxial leaf surface; 2, abaxial leaf surface)*; Figures 3-4, *Gall of* A. subfolium *(3, adaxial; 4, lateral)*; Figures 5-6, *Galls of* A. coronata *(5, adaxial; 6, abaxial)*; Figure 7, *Galls of* A. pustulata *(adaxial)*; Figure 8, *Gall of* A. tuberculata *(abaxial)*.

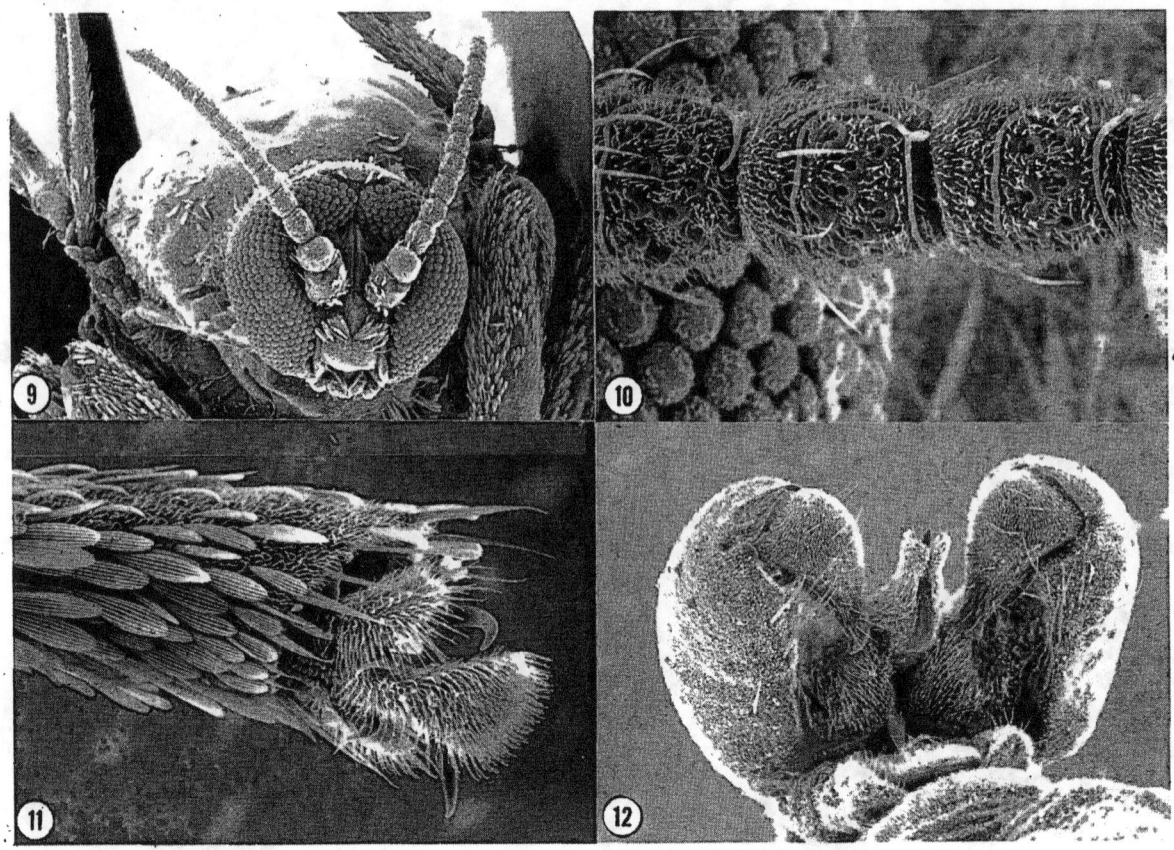

Figures 9-12, Actilasioptera pustulata: Figure 9, *Head (frontal view)*; Figure 10, *Basal antennal flagellomeres (frontal)*; Figure 11, *Fifth tarsomere (lateroventral)*; Figure 12, *Male genitalia (posterodorsal)*.

Actilasioptera subfolium GAGNE, NEW SPECIES

Adult

Head. Antenna with 12 flagellomeres, usually with the 2-3 distalmost segments partially or wholly fused. Palpus 1-2 segmented, the second segment when present, much smaller than the first. Labella covered with apical setae and a few scales.

Thorax. Wing length, 2.0-2.4 mm (n = 5) in male, 1.8-2.3 mm (n = 3) in female; R_5 about 0.7 length of wing (as in fig. 31).

Male abdomen. Seventh tergite fully sclerotized, its vestiture similar to that of sixth. Genitalia: Hypoproct convex posteriorly (fig. 36); parameres without setulae beyond anterodorsal lobe; aedeagus narrow, with two sensory pits and curved ventrally at apex.

Female abdomen. As for *A. tumidifolium* (figs. 13-14).

Pupa

Vertex convex, exceeding antennal bases anteriorly (figs. 19-20). Antennal bases anteroventrally rounded. Face smooth, without projections.

Larva

Third instar. Integument with conical, pointed verrucae present mesally on venter and laterally on dorsum of mesothorax and metathorax and circling completely or almost completely (missing dorsomesally on first through sixth) middle of first through eighth abdominal segments. Spatula (fig. 29) parallel-sided, with two large anterior teeth. Papillae usually not visible, the setae when present very diminutive; when visible, agreeing with generic diagnosis.

Holotype and Specimens Examined

Holotype, male, reared from leaf gall on *Avicennia marina*, Australia, Queensland, Brighton, 4-IX-1995, L. Law, deposited in Australian National Insect Collection, Canberra (ANIC).

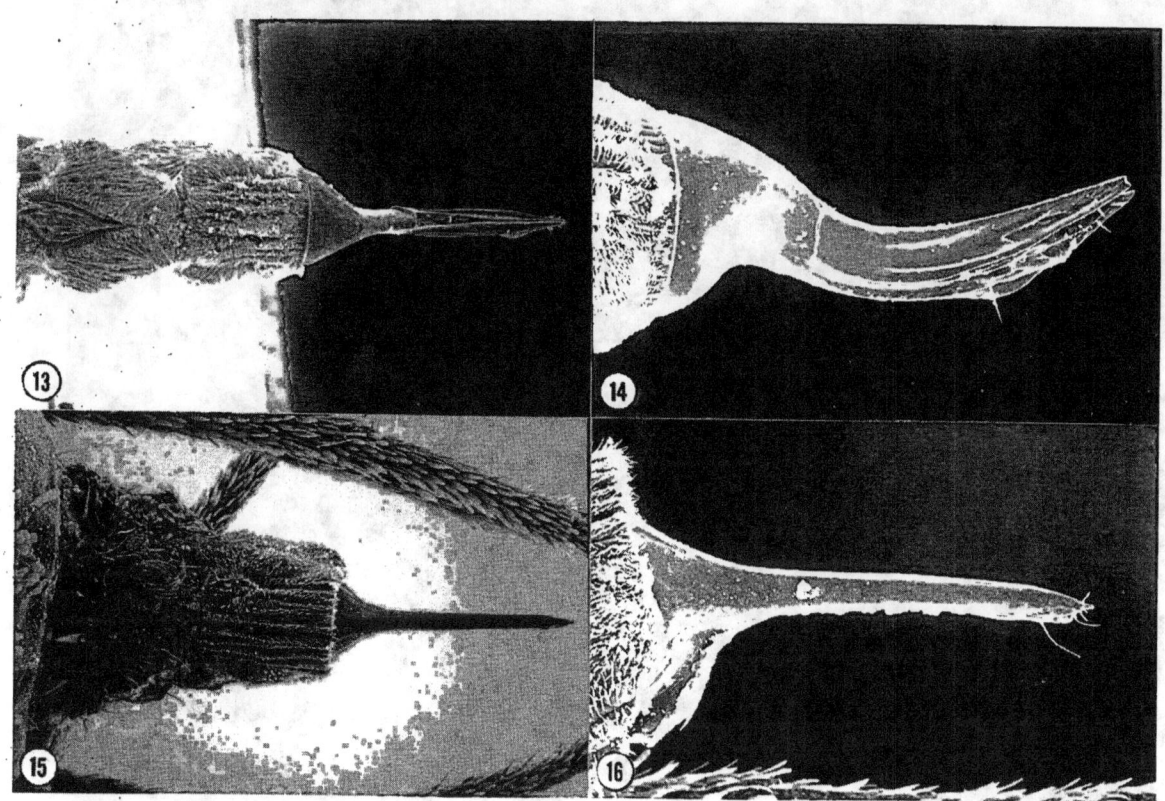

Figures 13-16, *Postabdomens of* Actilasioptera *spp.* Figures 13-14, A. tumidifolium: Figure 13, *Eighth segment to end (dorsal);* Figure 14, *Detail of ovipositor made up of ninth and tenth segments (lateral).* Figures 15-16, A. pustulata: Figure 15, *Eighth segment to end (dorsal);* Figure 16, *Detail of ovipositor made up of ninth and tenth segments (lateral).*

Other specimens examined (all from *Avicennia marina*, Queensland, Australia, and collected by L. Law).—Brighton, 4-IX-1995, 8 males, 8 pupae, 2 third instars; Whyte Island, 10-IX-1993, 1 female; Wynnum, 10-IX-1993, 1 adult without abdomen; Doboy, 18-III-1994, 3 last instars; Doboy, 24-III-1994, 3 pupae, 2 last instars.

Etymology

The name *subfolium* (noun in apposition) means "under the leaf," where the swelling typical of this species is found.

Gall

This species is responsible for a large, green to brown, mostly soft, simple leaf swelling that is apparent on only the lower leaf surface (figs. 3-4). Galls are similar in exterior texture to the normal leaf. They vary from 10-70 mm in length and may encompass the entire leaf. Inside each gall are several ovoid chambers each harboring a single larva. The margins of infested leaves are characteristically hardened and crimped. The main difference between this gall and that of *A. tumidifolium* is that the gall of *A. subfolium* is apparent only on the lower leaf surface. Pupae break out through the abaxial leaf surface.

Remarks

Adults of this species generally resemble *A. tumidifolium* and *A. falcaria*. Their similarities are listed under *A. tumidifolium*. Adults of *A. subfolium* differ from those of the other species in that scales are sparse on the labella instead of completely covering the surface, and the male hypoproct is simple instead of bilobed. The pupal vertex is much less swollen in *subfolium* than in *tumidifolium* and the larval spatulae of the two species are shaped differently. The galls of these two species are similar in shape but that of *subfolium* is present only on the abaxial surface of the leaf rather than showing on both surfaces.

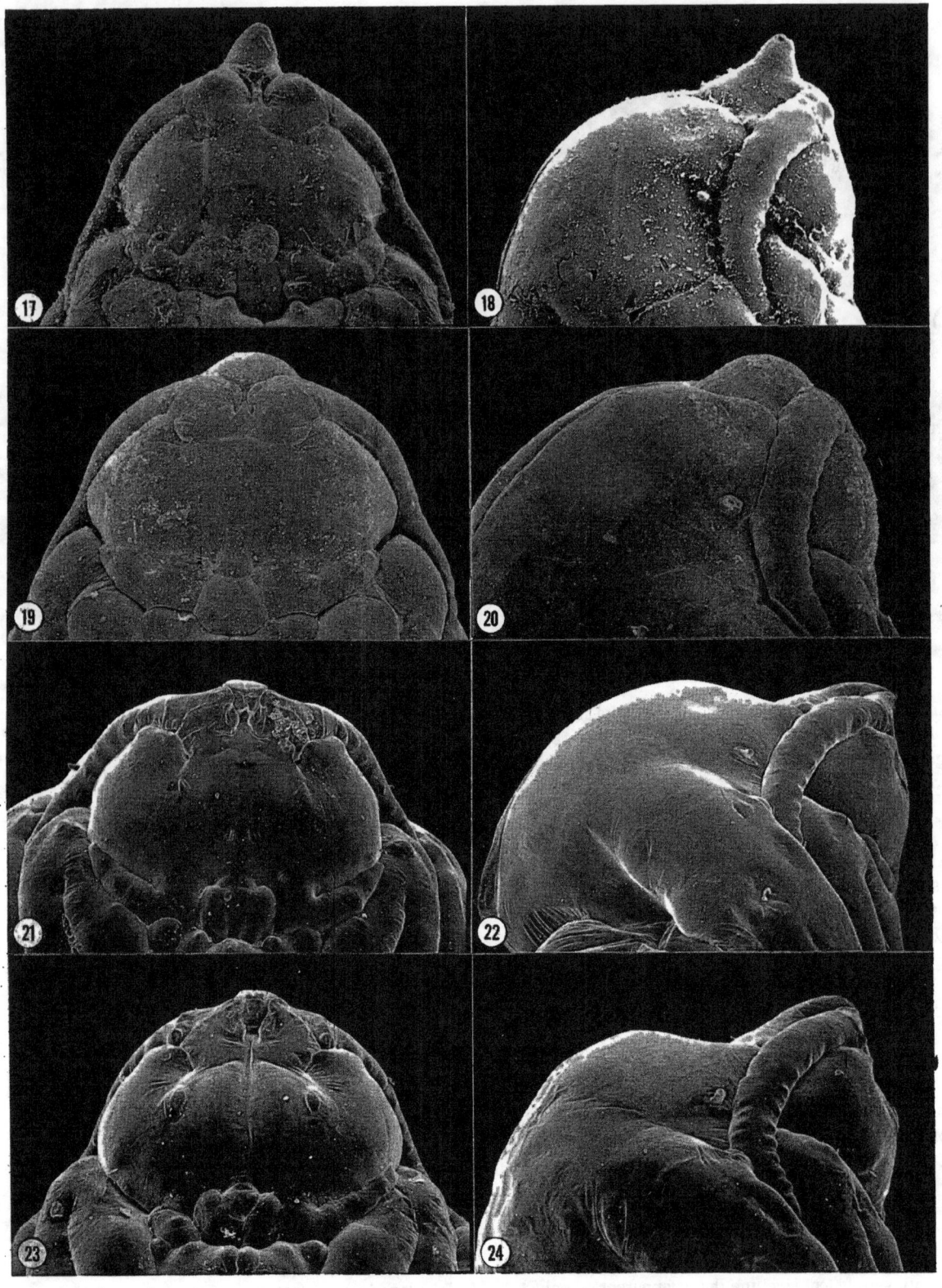

Figures 17-24, *Pupal heads of* Actilasioptera *spp. (odd numbers, ventral view, even numbers, lateral):* Figures 17-18, A. tumidifolium; Figures 19-20, A. subfolium; Figures 21-22, A. coronata; Figures 23-24, A. tuberculata.

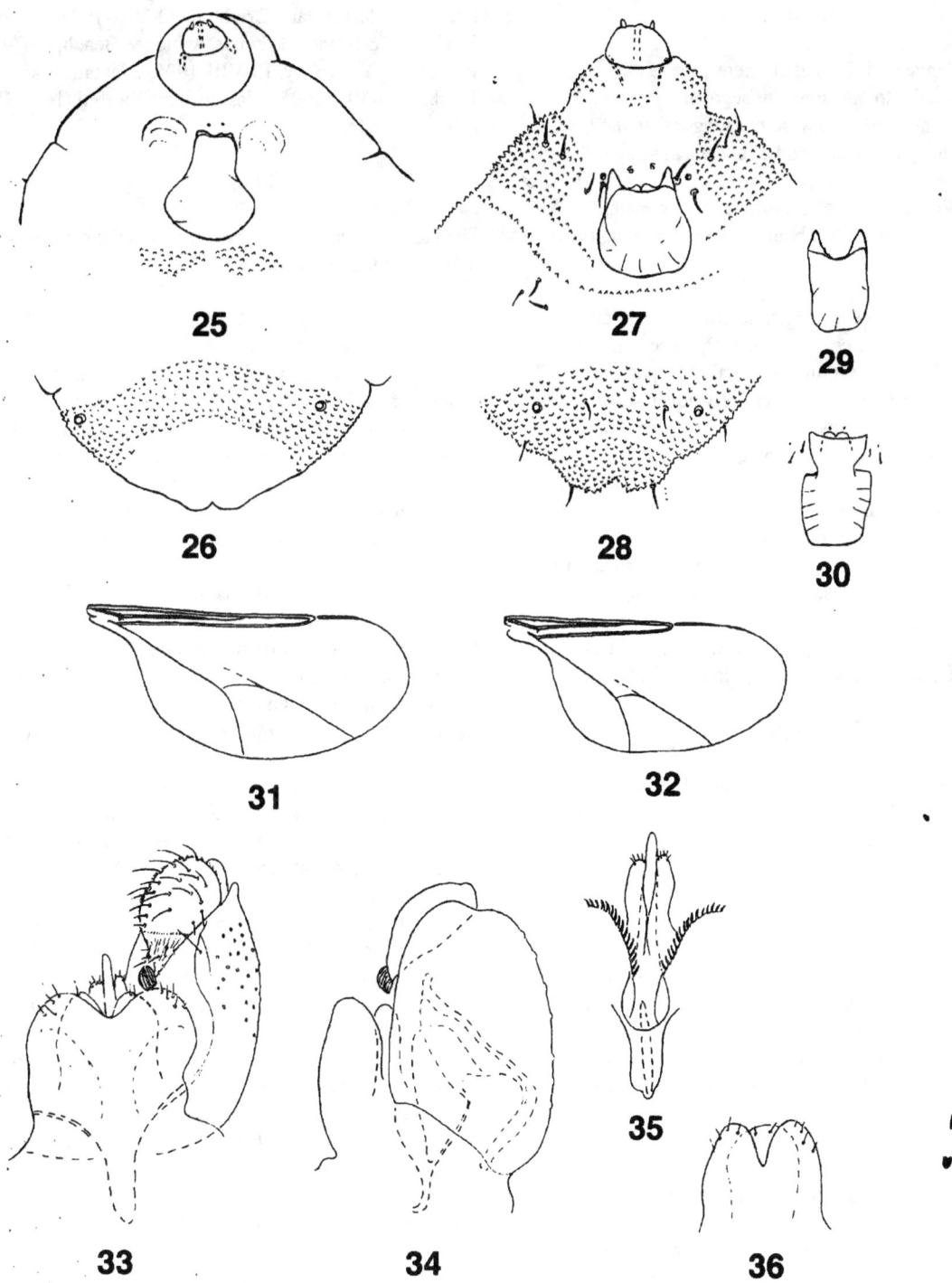

Figures 25-34, *Parts of larvae, the wings, and male genitalia of* Actilasioptera *spp.* Figures 25-26, A. tumidifolium: Figure 25, *larval head and prothorax (ventral);* Figure 26, *larval eighth and terminal segments (dorsal).* Figures 27-28, A. coronata. Figure 27, *larval head, prothorax, and lateral setae of one side of mesothorax (ventral);* Figure 28, *larval eighth and terminal segments (dorsal).* Figure 29, *larval spatula* of A. subfolium. Figure 30, *Larval spatula and associated papillae* of A. tuberculata. Figure 31, *Wing of* A. tumidifolium. Figure 32, *Wing of* A. pustulata. Figures 33-35, *Male genitalia of* A. tumidifolium; Figure 33, *dorsal, only right gonopod drawn;* Figure 34, *lateral, showing silhouettes of parts between gonopods;* Figure 35, *parameres, aedegus, and aedeagal apodem.* Figure 36, *cerci and hypoproct* of A. subfolium.

Actilasioptera coronata GAGNE, NEW SPECIES

Adult

Head. Antenna with 11 flagellomeres, the tenth and eleventh fused into one apparent segment. Palpus 2 segmented, the second as wide but longer than the first. Labella with sparse scales and a few apical setae.

Thorax. Wing length, ca. 2.2 mm (n = 1) in male, 1.6-1.9 mm (n = 4) in female; R_5 about 0.6 length of wing (as in fig. 32).

Male abdomen. Seventh tergite sclerotized only on anterior half, with anterior pair of trichoid sensilla and few scales the only vestiture. Genitalia: Hypoproct convex posteriorly; parameres entirely covered with setulae; aedeagus extremely narrow, the sensory pits not apparent, not curved ventrally at apex.

Female abdomen (as in figs. 15-16). Intersegmental membrane between eighth and ninth segments with lateral group of robust, elongate setae on each side, with no other setation but surface covered with elongate, fine setulae; combined ninth segment and cerci strongly sclerotized, glabrous, abruptly narrowed beyond base, cylindrical, straight, needle-like, pointed apically.

Pupa

Vertex weakly convex, slightly produced ventrally between antennal bases (figs. 21-22). Antennal bases rounded anteroventrally. Face smooth, without projections.

Larva

Third instar (figs. 27-28). Integument with conical, pointed verrucae covering entire surface of body except venter of collar segment and area immediately around spatula. Spatula slightly logner than broad, with pair of pointed anterolateral teeth and smaller pair of rounded mesal teeth. Papillae well developed, as for generic diagnosis.

Holotype and Specimens Examined

Holotype, male, reared from leaf gall on *Avicennia marina*, Australia, Queensland, Sandgate, 23-VIII-1996, L. Law, deposited in Australian National Insect Collection, Canberra (ANIC).

Other specimens examined (all from *Avicennia marina*, Queensland, Australia and collected by L. Law).— Brighton, 2-III-1995, L. Law and R.J. Gagné, last instar; Brighton, 3-VII-1995; 1 last instar; Brighton, 4-IX-1995, 1 female, 5 last instars; Brighton, 22-VIII-1996, 2 pupae; Brighton, 16-I-1997, 3 females; Nudgee Beach, 13-VIII-1995, 9 pupae; Doboy, 13-VIII-1995, 2 last instars; Toorbul, 13-VIII-1995, 2 last instars; Toorbul, 16-VIII-1996, 1 female.

Etymology

The name *coronata* (adjective) refers to the crown-like ridge surrounding the gall.

Gall

This is small, unevenly hemispheroid and warty on the upper leaf surface, and craterlike on the lower leaf surface, the edge raised about 5 mm in height and uneven, suggesting a crown (figs. 5-6). Galls can be close together and become partially congruent, but each larva is in a separate cell. Pupae break through the tissue at the base of the crater on the lower leaf surface.

Remarks

Adults of this species and two others, *A. pustulata* and *A. tuberculata*, are similar in several ways, setting them off as different from the remaining three species (see remarks under *A. tumidifolium*): the antennae are 10-11 segmented, the second palpal segment is larger than the first, the R_5 wing vein is about 0.6 wing length, the male seventh tergite is mostly unsclerotized with little vestiture remaining, the parameres are setulose, and the ovipositor narrows abruptly at its base into a long, cylindrical tube. The pupae and larvae of these three species share the following characters: the pupal vertex is slightly produced ventrally between the antennal bases, the larval papillae are apparent, and the larval integument is almost entirely covered with conical, pointed verrucae.

A. coronata differs only slightly from *A. pustulata* and *A. tuberculata*, in the slightly wider apex of the aedeagus, the evenly rounded pupal antennal bases, the lack of projections on the pupal face, and in the particular shape of the larval spatula (compare figs. 27 and 30).

Actilasioptera pustulata GAGNE, NEW SPECIES

Adult

Head (figs. 9-10). Antenna with 11 flagellomeres, the tenth and eleventh fused into one apparent segment. Palpus 2 segmented, the second as wide but longer than the first. Labella with sparse scales and a few apical setae.

Thorax. Wing length, 2.1-2.3 mm (n = 5) in male, 2.1-2.3 mm (n = 4) in female; R_5 about 0.6 length of wing (fig. 32).

Male abdomen. Seventh tergite sclerotized only on anterior half, with only anterior pair of trichoid sensilla and few scales. Genitalia: Hypoproct convex posteriorly; parameres entirely covered with setulae; aedeagus narrow, the two sensory pits fused, and not curved ventrally at apex.

Female abdomen (figs. 15-16). Intersegmental membrane between eighth and ninth segments with lateral group of robust, elongate setae on each side, with no other setation but surface covered with elongate, fine setulae; combined ninth segment and cerci strongly sclerotized, glabrous, abruptly narrowed beyond base, cylindrical, straight, needle-like, pointed apically.

Pupa

Vertex weakly convex, slightly produced ventrally between antennal bases (as in figs. 23-24). Antennal bases angular anteroventrally. Face anteromesally with pair of slightly wrinkled projections.

Larva

Third instar. Integument with conical, pointed verrucae covering entire surface of body. Spatula (as in fig. 30) narrowing posterior to anterior edge to about one-third length, then widening abruptly and tapering slightly to posterior margin, with pair of short, splayed anterolateral teeth and mesal pair of rounded, usually serrate teeth apically, no longer than lateral teeth. Papillae well developed, as for generic diagnosis.

Holotype and Specimens Examined

Holotype, male, reared from leaf gall on *Avicennia marina*, Australia, Queensland, Sandgate, 23-VIII-1996, L. Law, deposited in Australian National Insect Collection, Canberra (ANIC).

Other specimens examined (all from *Avicennia marina*, Queensland, Australia, and collected by L. Law).—Sandgate, 23-VIII-1996, 2 males, 5 females, pupa; Doboy, 10-IX-1993, 3 males; Doboy, 13-VIII-1995, 5 last instars; Toorbul, 21-III-1994, 1 female; Toorbul, 16-VIII-1996, 4 males, 5 females, 8 pupae, and 3 last instars; Donnybrook, 26-VI-1994, 2 pupae, 3 last instars; Brighton, 4-IX-1995, 1 male, 1 pupa, and 1 pupal exuviae; Brighton, 22-VIII-1996, 1 male, 2 females, 1 pupa.

Etymology

The name *pustulata* (adjective) refers to the uniformly bumpy surface of the gall.

Gall

The gall of this species is circular and flat on the upper leaf surface and nearly evenly hemispherical on the lower surface, 3-6 mm in diameter, about 5 mm in height (fig. 7). The epidermis of the gall on the lower surface is covered with uniformly tiny bumps or pustules, but with age the epidermis may peel off and the bumps with it. Galls may sometimes be closely adjacent, but the larvae are always in separate cells. Pupae break out through the flat, upper surface of the gall.

Remarks

See remarks under *A. coronata* on the differences among that species, *A. pustulata*, and *A. tuberculata*, and between them and the remaining species in the genus. Although galls of *A. pustulata* and *A. tuberculata* are quite distinct, we have found no other characters to distinguish the two species. We consider the difference between their galls to be the expression of at least physiological differences between them and so consider them distinct species.

Actilasioptera tuberculata GAGNE, NEW SPECIES

Adult

Head. Antenna with 11 flagellomeres, the tenth and eleventh fused into one apparent segment. Palpus 2 segmented, the second nearly as wide but longer than the first. Labella with sparse scales and a few apical setae.

Thorax. Wing length, 1.8-1.9 mm (n = 4) in male, 1.7 mm (n = 1) in female; R_5 about 0.6 length of wing (as in fig. 32).

Male abdomen. Seventh tergite sclerotized only on anterior half, with only anterior pair of trichoid sensilla and few scales. Genitalia: Hypoproct convex posteriorly; parameres entirely covered with setulae; aedeagus narrow, not curved ventrally at apex, the two apical sensory pits fused.

Female abdomen (as in figs. 15-16). Intersegmental membrane between eighth and ninth segments with lateral group of robust, elongate setae on each side, with no other setation but surface covered with elongate, fine setulae; combined ninth segment and cerci strongly sclerotized, glabrous, abruptly narrowed beyond base, cylindrical, straight, needle-like, pointed apically.

Pupa

Vertex weakly convex, slightly produced ventrally between antennal bases (figs. 23-24). Antennal bases angular anteroventrally. Face anteromesally with pair of slightly wrinkled projections.

Larva

Third instar: Integument with conical, pointed verrucae covering entire surface of body. Spatula (fig. 30) narrowing posterior to anterior edge to about one third length, then widening abruptly and tapering slightly to posterior margin, with pair of short, splayed anterolateral teeth and mesal pair of rounded, usually serrate teeth apically, no longer than lateral teeth. Papillae well developed, as for generic diagnosis.

Holotype and Specimens Examined

Holotype, male, reared from leaf gall on *Avicennia marina*, Australia, Queensland, Wellington Pt., 10-IX-1993, deposited in Australian National Insect Collection, Canberra (ANIC).

Other specimens examined (all from *Avicennia marina*, Queensland, Australia, and collected by L. Law).— Wellington Pt., 10-IX-1993, 1 male; Toorbul, 10-IX-1993, 1 female (ovipositor missing); Toorbul, 23-III-1994, 3 pupae; Toorbul, 27-III-1994, 3 pupae, 2 last instars; Toorbul, 13-VIII-1995; 3 last instars; Brighton, 10-IX-1993, 2 males; Brighton, 4-IX-1995, 2 last instars; Donnybrook, 27-VIII-1994, 2 males.

Etymology

The name *tuberculata* (adjective) refers to the one to several elongate tubercles on each gall.

Gall

On the upper leaf surface the gall of this species is circular basally, with one or more elongate, conical projections arising from it, 3-10 mm in height, 3-8 mm in diameter (fig. 8). On the lower surface the gall is simply circular and flat. There is one conical projection for each individual larval cell. The pupa breaks out of the lower surface of the leaf.

Remarks

See remarks under *A. coronata* on the differences among that species, *A. pustulata*, and *A. tuberculata*, and between them and the remaining species in the genus. Although galls of *A. pustulata* and *A. tuberculata* are quite distinct, we have found no other characters to distinguish between the two species. We consider the difference between their galls to be the expression of at least physiological differences between them and so consider them distinct species.

ACKNOWLEDGMENTS

We are grateful to Nit Malikul for preparing the microscopic slides and to Damien Burrows of the Australian Centre for Tropical Freshwater Research, Townsville, Peter Kolesik of The University of Adelaide, and Allen Norrbom, of the Systematic Entomology Laboratory, Washington, DC, for their comments on drafts of the manuscript.

LITERATURE CITED

Docters van Leeuwen-Reijnvaan J.; Docters van Leeuwen-Reijnvaan, W.D. 1910. Einige Gallen aus Java. Dritter Beitrag. Marcellia. 9: 37-61.

Felt, E.P. 1921. Javanese gall midges. Treubia. 1: 139-151.

Gagné, R.J. 1989a. The plant-feeding gall midges of North America. Ithaca, NY: Cornell University Press. xi and 356 p. and 4 pls.

Gagné, R.J. 1989b. Family Cecidomyiidae. In: Evenhuis, N.L., ed. A catalog of the Diptera of the Australasian and Oceanian Regions. Bishop Museum Press and E.J. Brill, Leiden: 152-163.

Gagné, R.J. 1994. The gall midges of the Neotropical Region. Ithaca, NY: Cornell University Press. xv and 352 p.

Gagné, R.J.; Balciunas, J.K.; Burrows, D.W. 1997. Six new species of gall midges (Diptera: Cecidomyiidae) from *Melaleuca* (Myrtaceae) in Australia. Proceedings of the Entomological Society of Washington. 99: 312-334.

Gagné, R.J.; Etienne, J. 1996. *Meunieriella avicenniae* (Cook) (Diptera: Cecidomyiidae) the leaf gall maker of black mangrove in the American tropics. Proceedings of the Entomological Society of Washington. 98: 527-616.

Law, L.J. 1996. The diversity, host relationships and abundance of gall insects on mangroves in Moreton Bay and the potential of galls as environmental indicators. Brisbane: University of Queensland. Ph.D. thesis.

McAlpine, J.F. 1981. 2. Morphology and terminology—adults. In: McAlpine *et al.*, eds. Manual of Nearctic Diptera, Vol. 1. Research Branch, Agriculture Canada, Monograph 27: 9-63.

Skuse, F.A.A. 1888. Diptera of Australia. Proceedings of the Linnean Society of New South Wales. (2)3: 17-145, pls. 2-3.

Skuse, F.A.A. 1890. Diptera of Australia. Nematocera.- Supplement I. Proceedings of the Linnean Society of New South Wales. (2)5: 373-413.

Yukawa, J.; Haitsuka, S. 1994. A new cecidomyiid successor (Diptera) inhabiting empty midge galls. Japanese Journal of Entomology. 62: 709-718.

TEPHRITID GALLS AND GALL TEPHRITIDAE REVISITED, WITH SPECIAL EMPHASIS ON MYOPITINE GALLS

Amnon Freidberg

Department of Zoology, The George S. Wise Faculty of Life Sciences, Tel Aviv University, Tel Aviv 69978, Israel

Abstract.—Knowledge of tephritid cecidology is summarized and compared to cecidomyiid cecidology. New data about Myopitini cecidology are presented, in particular for the genus *Stamnophora*, and the evolutionary implications of this information is discussed.

INTRODUCTION

The title of this contribution refers primarily to my review of gall Tephritidae, published in 1984 as a chapter in Ananthakrishnan's book, *Biology of Gall Insects* (Freidberg 1984). Because the book has not been widely disseminated, I am taking this opportunity first to summarize briefly the contents of this chapter. I then review the progress made on this topic since the chapter was published; and finally, I provide some unpublished data, and conclude with some evolutionary remarks.

Tephritid galls and gall Tephritidae are of interest primarily because they represent dipterous gall communities second in number and diversity only to cecidomyiid gall communities. Furthermore, some gall Tephritidae also have practical or potential economic importance, of benefit to human welfare.

REVIEW OF THE LITERATURE

Review Papers

When preparing my 1984 review (Freidberg 1984), I discovered that no previous such review had ever been made. The reason for this was probably two-fold: Firstly, the study of gall tephritids is overshadowed by that of the extremely economically important frugivorous species, none of which are cecidogenous; secondly, the number of gall tephritids is relatively small, with only about 50 known species in each of the major biogeographical regions.

The chapter includes a list of the 48 cecidogenous genera, with data relating to number of species, distribution, host genera and most important literature. It also includes sections on host plants and biogeography, location and structure of tephritid galls, gall histology and morphogenesis, life histories, ecology, economic importance and evolution. Data for most biogeographical regions were previously amassed by many scientists, while those for the Afrotropical region were primarily the result of work, H.K. Munro, to whom I dedicated the chapter. Munro's death, shortly after the chapter was published, terminated a highly productive career which lasted over 60 years. The most general and interesting conclusions emerging from the review are summarized in table 1, which compares Tephritidae with Cecidomyiidae.

Recent Publications

Little work has been done specifically on tephritid cecidogeny in the last two decades, and the credit for most of the published data should go to Prof. Richard D. Goeden, Riverside, California, to whom I dedicate this article. Goeden has meticulously studied the non-frugiverous tephritids of southern California, and in the last two decades he has published around 30 papers on these flies; about half of them deal with cecidogenous species (Goeden 1988, 1990a,b, 1992, 1993; Goeden and Headrick 1990, 1991; Goeden and Teernick 1996a,b,c, 1997a,b; Goeden et al. 1993, 1994; Green et al. 1993; Headrick and Goeden 1993; Silverman and Goeden 1980). While these papers have not dramatically altered the overall picture of gall tephritids and their galls, they have added many interesting details, in particular the descriptions of immature stages of most of these species, based on SEM studies. Although there are probably not enough cecidogenous tephritids in southern California to rival the hegemony of Munro, it is interesting to devote a few lines to the comparison of the two Tephritidologists. Munro was primarily a taxonomist, who described about

Table 1.—*Comparison between Tephritidae and Cecidomyiidae*

Trait	Tephritidae	Cecidomyiidae
No. of described spp.	~4200	~5000
Percent cecidogenous	~5%	~60%
Plant parts galled (in decreasing frequency)	Stem>Flowerhead>Leaf>Root	Leaf>Stem>Bud>Flower
Host families	Asteraceae (90%) + 4-5 other families	Many families; several dominant
Distribution	25-60 N; 15-40 S	Similar?
Independent evolution of cecidogeny	Many times (especially in Tephritinae)	A few or several times

half of the many Afrotropical taxa known to date. He enjoyed field work and believed in the importance of studying the biology of the species he discovered and investigated. His descriptions of immature stages were at the best brief. Goeden's approach tended towards the biology of tephritids from the point of view of biological control of weeds, for which life histories and immature stages are important. Recently, however, he has discovered the joys of taxonomy and extended a few of his papers to include descriptions of new species.

Goeden and his collaborators are not the only scientists to have contributed to tephritid cecidology in recent years. There are also other papers (e.g., Dodson 1985, 1986, 1987a, b; Dodson and George 1986; Shorthouse 1980, 1989; Merz 1992; Silva et al. 1996), most of them dealing with the holarctic fauna. A recent paper by Hardy and Drew (1996) on the Australian Tephritini, contains many new host records including galls, but the paper is a taxonomic revision of the tribe, and the biological data reported in it are very meagre. Hence, the overall contribution of these papers to a better understanding of tephritid galls is rather small.

My own work on cecidogenous tephritids was mostly conducted in the Afrotropical region and Israel in the last two decades (Freidberg 1979, 1984; Freidberg and Kugler 1989; Freidberg and Kaplan 1992, 1993; Freidberg and Mansell 1995). While it does not drastically change most of the features presented above, it clearly shows that the above-suggested pattern of distribution was a collection error, and that many additional galls are to be found nearer the equator than was previously thought. A recent paper (Freidberg and Kaplan 1992) practically doubled the Afrotropical fauna of Ditrycini by describing nine new species, almost all from equatorial Africa. All Old-World members of this group with known biology form stem galls on Asteraceae. I believe that all 19 species of Afrotropical Ditrycini are gall formers, although this faculty has hitherto been documented for only seven of them.

MYOPITINI AND THEIR GALLS

The most interesting discoveries are probably related to the Afrotropical Myopitini, in particular to the genus *Stamnophora* Munro. Therefore, some background about Myopitini and their galls seems appropriate. This relatively small group of less than 200 species is currently under generic and cladistic revision by Dr. Allen L. Norrbom, USDA, Washington, DC and myself. We believe that this group is monophyletic, and we hope that a better understanding of the group and its biology, including cecidogeny, will be possible after this revision is published.

The composition and some attributes of the Palaearctic Myopitini are presented in table 2. *Urophora* Robineau-Desvoidy is the largest myopitine genus, with many European (White and Korneyev 1989) and Asian (Korneyev and White 1993a, b) species, and the biology of some European species is fairly well known. Several species have been introduced to North America, South Africa and Australia to combat noxious weeds there. *Myopites* Blot is a smaller and less well known genus (Freidberg 1979), and most of its species are concentrated around the Mediterranean. Most *Urophora* species with known biology gall the achenes (=flowers) (fig. 1),

Table 2.—*The Palaearctic Myopitini*

Genus	No. of species	Hosts	Galls
Urophora[1]	About 60	Cardueae	flowerhead; 1 sp. in stem
Myopites	About 12	*Inula, Pulicaria*	flowerhead
Asimoneura	1	*Helichrysum*?	?
Nearomyia	1	?	?

[1] Based on Korneyev, personal communication

sometimes together with the receptacle, of their Cardueae (Asteraceae) hosts. An interesting exception is *U. cardui* Linnaeus, the type species, which galls the stems of its hosts. *Myopites* species with known biology gall the flowerheads of their hosts (fig. 2), all of which belong to *Inula* L., *Pulicaria* Gaertn. and possibly a few other, related taxa. Common to both genera, and apparently only to them, are the inclusion of the achenes in the gall; the generally parallel orientation of the individual tunnels and cells; and the emergence of each adult fly (or its parasitoids) singly through an opening in its individual achene. Other members of the Myopitini, especially the rather numerous species of the New World fauna, develop in flowerheads, but cecidogeny is known with certainty for only a small number of them. Indiginous Myopitini are almost unknown from the Oriental and Australasian regions.

The Afrotropical fauna of Myopitini, in particular, has been poorly studied. Although only seven species have been described to date, the material I have collected or borrowed comprises at least an additional 40 species, all undescribed. The composition of the fauna, including the as-yet-undescribed taxa, and some of its attributes are presented in table 3.

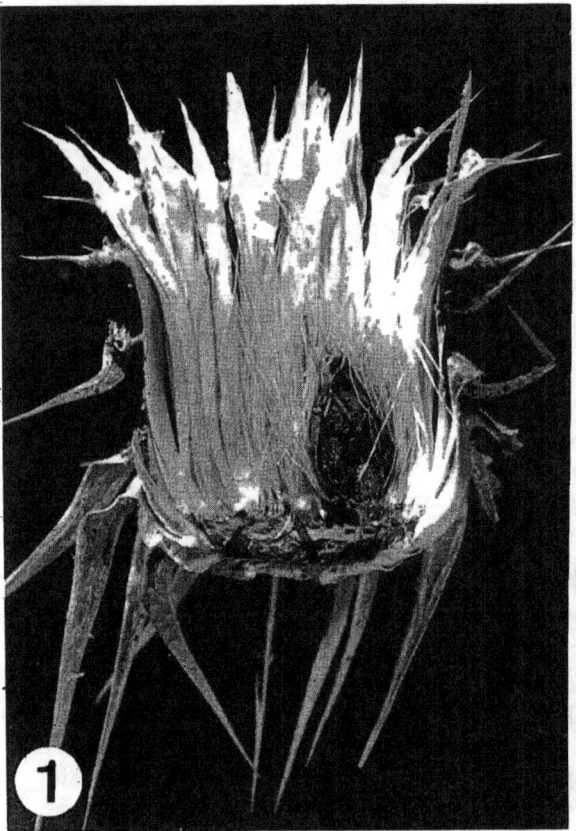

Figures 1-2.—*Myopitine galls (actual diameter in parenthesis)*. Figure 1. *Gall of* Urophora stylata *on* Cirsium phyllocephalum *(15mm)*. Figure 2. *Gall of* Myopites stylata *on* Inula viscosa *(8mm)*.

Table 3.—*The Afrotropical Myopitini*

Genus	No. of species	Hosts	Galls
Asimoneura	About 23	*Helichrysum*	no?
Stamnophora	About 12	*Vernonia*	flowerhead and stem
Myopites	3	*Inula*	flowerhead
Urophora	1	*Centaurea*?	?
New Genus A	About 7	*Vernonia, Berkheya*	no?
New Genus B	1	*Echinops*	?

To date, only one species of *Stamnophora* Munro, *S. vernoniicola* (Bezzi) is known. This is a rather large and robust, blackish species, with wing pattern reduced almost entirely. It is known to gall several species of *Vernonia* Schreb., inducing flowerhead galls in some, stem galls in others. This host-fly relationship is extremely unusual in the Tephritidae, but my own findings corroborate it (see below). In the course of an on-going revision of *Stamnophora*, I have accummulated a large amount of material of this genus that comprises an additional 12 or more species, all undescribed. Fortunately, I was able to discover the host plants and galls of several of the species and to rear the flies out of them. All hosts belong to the genus *Vernonia*. All the species reared so far from *Vernonia* appear to form a monophyletic group.

All reared species of *Stamnophora* have the general habitus of *Myopites*, with similar coloration and wing pattern, although they differ in their short proboscis (which is extremely long in *Myopites*), slightly different wing venation and terminalia. In our forthcoming revision, Norrbom and I will probably retain these genera as separate because no intermediate forms are known. All these new, reared species are extremely similar to each other, and although the relationships between them are not yet fully clarified, it appears that at least three species form flowerhead galls, and two other species form stem galls.

Before describing the galls induced by the undescribed species of *Stamnophora*, I offer a brief revision of our knowledge of the galls induced by *S. vernoniicola*. Galls formed on the flowerheads of certain *Vernonia* spp., such as *V. hymenolepis* A. Rich. and *V. adoensis* Schulz Bip. ex Walp. are large (diameter up to about 3 cm), and contain many individual cells (up to 50-100) each leading to a common opening at the tip (figs. 3-4). The immature stages are found one in each cell, and all the flies (or their parasitoids) escape through the same common opening (Munro 1955). Unlike the parallel orientation of the tunnels in *Urophora* and *Myopites* galls, the orientation here is radial. Galls formed on the stems of *V. lasiopus* O. Hoffm. are very conspicuous, resemble potatoes, and are even larger than the flowerhead galls, although they may contain fewer individuals. The orientation of the tunnels, at least in some galls, is similar to that found in the flowerhead galls, with one common opening (figs. 5-6). However, it is unclear whether this is consistently so.

The flowerhead galls formed by the three undescribed species are basically **similar** to each other and to those formed by *S. vernoniicola*, although they are much smaller, and contain fewer individuals (usually 1-15, depending on the species) (fig. 7). There are nevertheless differences between the galls formed by the three species, with those formed on *Vernonia auriculifera* Hiern. being small and rounded, those formed on *V. ampla* O. Hoffm. being larger and conical, and those formed on *V. myriantha* Hook. f. being the largest, elongate, often sausage-like (fig. 7). The three species of hosts are similar to each other, and might be difficult to tell apart in the field.

The stem galls formed by the undescribed species are **unlike** those of *S. vernoniicola*, and differ also from any other tephritid galls known to me (figs. 8-9). The generally cylindrical stem is only somewhat swollen along the infested part, up to 20 cm long, and has slits from 0.5-1 cm deep. The immature stages are located individually in cells oriented sagitally which open at either of the slit walls. At the end of the season, when the stem dries, it tends to break at the galled part.

Finally, I would like to touch on the subject of the number of galls per plant. This ratio is not only very variable in tephritids, but often averages very small. In 1984 the largest figure I recorded was 135 (Freidberg 1984). Subsequently, Dodson (1987a) recorded over 2,000 galls of *Aciurina bigeloviae* (Cockerell) from a single plant of *Chrysothamnus nauseosus* (Pallen) Britton with a diameter of 1 m. I now have a record of about 10,000 galls of *Stamnophora* n. sp. on one plant of *Vernonia auriculifera* in Kenya. These plants may reach the size of a small tree, and during the blooming season may carry hundreds of thousands, if not millions, of small flowerheads, many of which are galled.

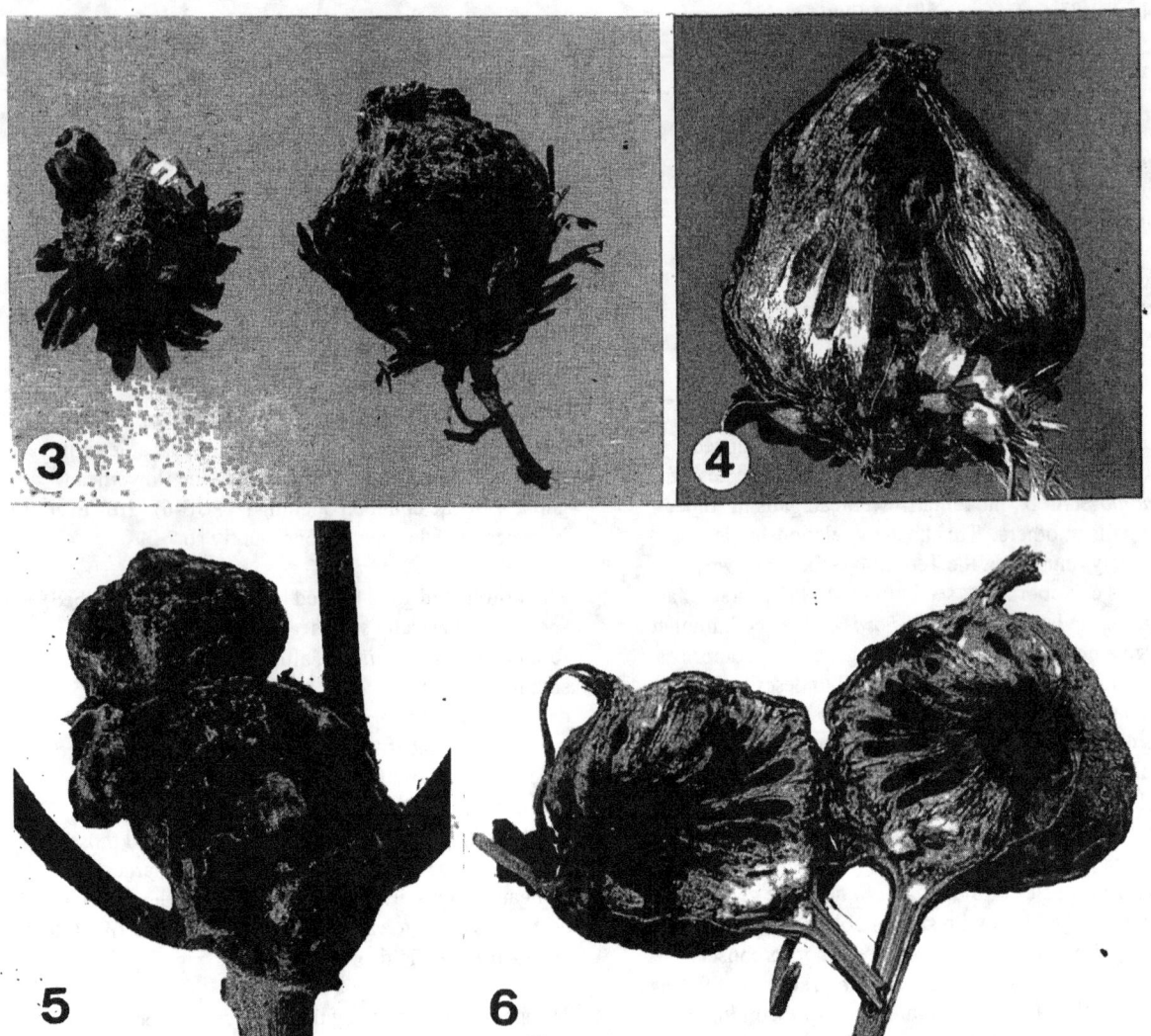

Figures 3-6.—*Galls of* Stamnophora vernoniicola *on various* Vernonia *spp. (actual diameter in parenthesis).* Figure 3. *On flowerhead of* V. hymenolepis; *the left gall (15mm) showing the common opening at the tip.* Figure 4. *Same, dissected sagitally (22mm).* Figure 5. *On stem of* V. lasiopus *(40mm).* Figure 6. *Same, dissected sagitally (45mm).*

EVOLUTIONARY CONCLUSIONS

The new findings presented here shed some light on the possible evolution of tephritid galls, especially myopitine Tephritinae (fig. 10). The hypothesis of the evolutionary transition from development in flowerheads to development in stems (Freidberg 1984) is now supported by the following three facts:
- ***Urophora cardui, which is truely congeneric with many additional species, galls stems, whereas all other congeners gall flowerheads.
- ****Stamnophora vernoniicola* galls both stems and flowerheads.
- ***Other species of *Stamnophora* gall either stems or flowerheads.

Other Tephritinae might have undergone a somewhat different route, which did not involve flowerhead galls, but a direct transition instead from ungalled flowerheads to galled stems. This is so hypothesized because flowerhead galls are practically unknown or nearly so in the other tribes of Tephritinae, most notably in the Tephritini, Ditrycini and Eutretini, which contain the great majority of gall inducing species in the family.

ACKNOWLEDGMNETS

I am grateful to Amikam Shoob for the photographs, and to Netta Dorchin and Tuvit Simon for discussions and technical assistance.

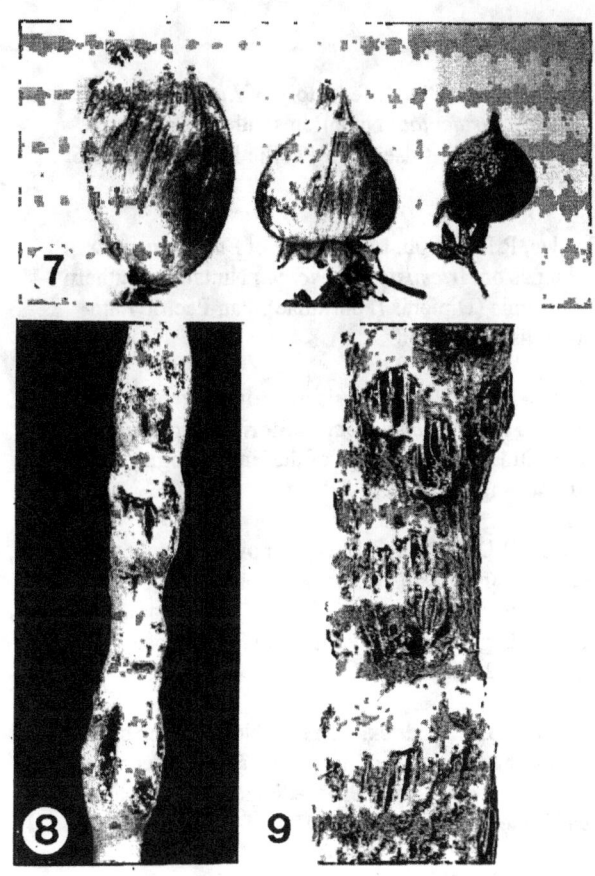

Figures 7-9.—*Galls of undescribed* Stamnophora spp. *on various* Vernonia spp. *(actual diameter in parenthesis).*

Figure 7. *Flowerhead galls; from left: on* V. myriantha *(8mm),* V. ampla, V. auriculifera.

Figure 8. *Stem galls on* V. auriculifera *(25-35mm).*

Figure 9. *Stem galls on* V. sp. sp. *(20mm).*

Hypothetical evolution of cecidogeny in Tephritidae

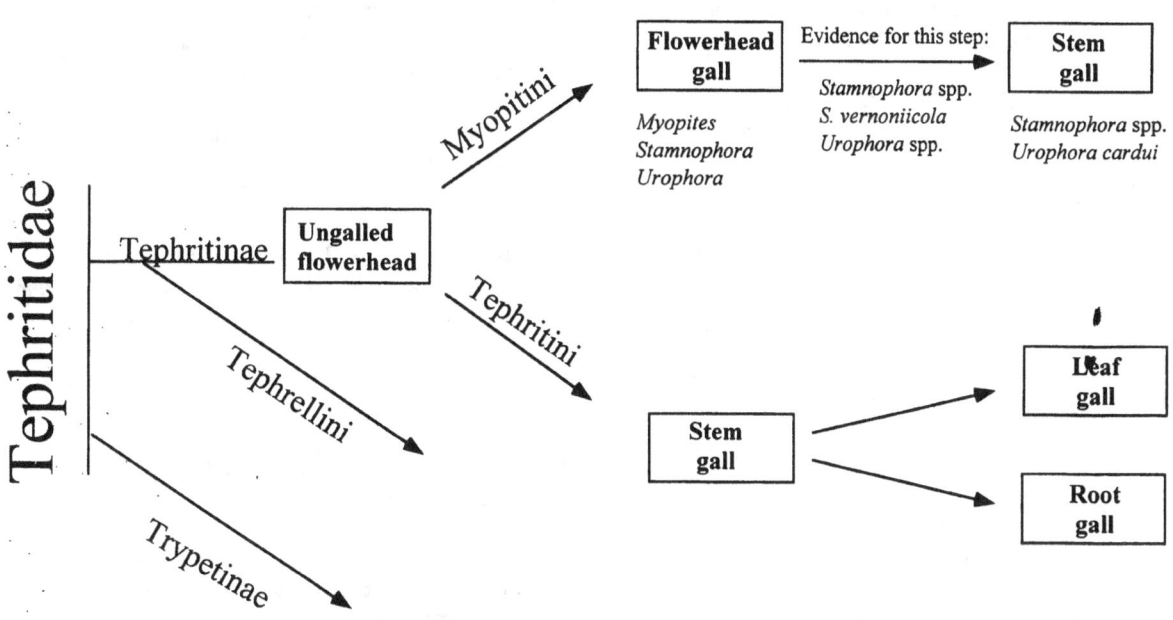

Figure 10.—*Hypothetical evolutionary pathways in tephritid cecidogeny. Only three tribes of Tephritinae (Myopitini, Tephritini, and Tephrellini) are given as examples.*

LITERATURE CITED

Dodson, G. 1985. The significance of sexual dimorphism in the mating system of two species of tephritid flies (*Aciurina trixa* and *Valentibula dodsoni*) (Diptera: Tephritidae). Canadian Journal of Zoology. 65: 194-198.

Dodson, G. 1986. Lek mating system and large male aggressive advantage in a gall-forming tephritid fly (Diptera: Tephritidae). Ethology. 72: 99-108.

Dodson, G. 1987a. Host-plant records and life history notes on New Mexico Tephritidae (Diptera). Proceedings of the Entomological Society of Washington. 89: 607-615.

Dodson, G. 1987b. Biological observation on *Aciurina trixa* and *Valentibula dodsoni* (Diptera: Tephritidae) in New Mexico. Annals of the Entomological Society of America. 80: 494-500.

Dodson, G.; George, S.B. 1986. Examination of two morphs of gall-forming *Aciurina* (Diptera: Tephritidae): ecological and genetic evidence for species. Biological Journal of the Linnean Society. 29: 63-79.

Freidberg, A. 1979. On the taxonomy and biology of the genus *Myopites* (Diptera: Tephritidae). Israel Journal of Entomology. 13: 13-26.

Freidberg, A. 1984. Gall Tephritidae. In: Ananthakrishnan, T.N., ed. Biology of gall insects. New Delhi: Oxford & IBH: 129-167.

Freidberg, A.; Kaplan, F. 1992. Revision of the Oedaspidini of the Afrotropical region (Diptera: Tephritidae: Tephritinae). Annals of the Natal Museum. 33: 51-94.

Freidberg, A.; Kaplan, F. 1993. A study of *Afreutreta* Bezzi and related genera (Diptera: Tephritidae: Tephritinae). African Entomology. 1: 207-228.

Freidberg, A.; Kugler, J. 1989. Fauna Palaestina, Insecta IV - Diptera: Tephritidae. The Israel Academy of Sciences and Humanities, Jerusalem. 212 p.

Freidberg, A.; Mansell, M.W. 1995. *Procecidochares utilis* Stone (Diptera: Tephritidae: Oedaspidini) in South Africa. African Entomology. 3: 89-91.

Goeden, R.D. 1988. Gall formation by the capitulum-infesting fruit fly, *Tephritis stigmatica* (Diptera: Tephritidae). Proceedings of the Entomological Society of Washington. 90: 37-43.

Goeden, R.D. 1990a. Life history of *Eutreta diana* on *Artemisia tridentata* Nuttall in southern California (Diptera: Tephritidae). Pan-Pacific Entomologist. 66: 24-32.

Goeden, R.D. 1990b. Life history of *Eutreta simplex* Thomas on *Artemisia ludoviciana* Nuttall in southern California (Diptera: Tephritidae). Pan-Pacific Entomologist. 66: 33-38.

Goeden, R.D. 1992. Analysis of known and new host records for *Trupanea* from California (Diptera: Tephritidae). Proceedings of the Entomological Society of Washington. 94: 107-118.

Goeden, R.D. 1993. Analysis of known and new host records for *Tephritis* from California, and description of a new species, *T. joanae* (Diptera: Tephritidae). Proceedings of the Entomological Society of Washington. 95: 425-434.

Goeden, R.D.; Headrick, D. 1990. Notes on the biology and immature stages of *Stenopa affinis* Quisenberry (Diptera: Tephritidae). Proceedings of the Entomological Society of Washington. 92: 641-648.

Goeden, R.D.; Headrick, D. 1991. Life history and descriptions of immature stages of *Tephritis baccharis* (Coquillett) on *Baccharis salicifolia* (Ruiz & Pavon) Persoon in southern California (Diptera: Tephritidae). Pan-Pacific Entomologist. 67: 86-98.

Goeden, R.D; Teerink, J.A. 1996a. Life histories and descriptions of adults and immature stages of two cryptic species, *Aciurina ferruginea* (Doane) and *A. Michaeli*, new species (Diptera: Tephritidae), on *Chrysothamnus vicidiflorus* (Hooker) Nuttall in southern California. Proceedings of the Entomological Society of Washington. 98: 415-438.

Goeden, R.D.; Teerink, J.A. 1996b. Life history and descriptions of adults and immature stages of *Aciurina semilucida* (Bates) (Diptera: Tephritidae), on *Chrysothamnus vicidiflorus* (Hooker) Nuttall in southern California. Proceedings of the Entomological Society of Washington. 98: 752-766.

Goeden, R.D.; Teerink, J.A. 1996c. Life history and description of immature stages of *Valentibulla californica* (Coquillett) (Diptera: Tephritidae), on *Chrysothamnus nauseosus* (Pallas) Britton in southern California. Proceedings of the Entomological Society of Washington. 98: 681-694.

Goeden, R.D.; Teerink, J.A. 1997a. Life history and descriptions of immature stages of *Procecidochares anthracina* (Doane) (Diptera: Tephritidae), on *Solidago californica* Nuttall in southern California. Proceedings of the Entomological Society of Washington. 99: 180-193.

Goeden, R.D.; Teerink, J.A. 1997b. Life histories and descriptions of adults and immature stages of *Procecidochares kristineae* and *P. lisae* new species (Diptera: Tephritidae), on *Ambrosia* spp. in southern California. Proceedings of the Entomological Society of Washington. 99: 67-88.

Goeden, R.D.; Headrick, D.H.; Teerink, J.A. 1993. Life history and descriptions of immature stages of *Tephritis arizonensis* Quisenberry (Diptera: Tephritidae) on *Baccharis sarothroides* Gray in southern California. Proceedings of the Entomological Society of Washington. 95: 210-222.

Goeden, R.D.; Headrick, D.H.; Teerink, J.A. 1994. Life history and description of immature stages of *Procecidochares flavipes* Aldrich (Diptera: Tephritidae) on *Brockellia* spp. in southern California. Proceedings of the Entomological Society of Washington. 96: 288-300.

Green, J.F.; Headrick, D.H.; Goeden, R.D. 1993. Life history and description of immature stages of *Procecidochares stonei* Blanc & Foote on *Viguiera* spp. in southern California (Diptera: Tephritidae). Pan-Pacific Entomologist. 69: 18-32.

Hardy, D.E.; Drew, R.A.I. 1996. Revision of the Australian Tephritini (Diptera: Tephritinae). Invertebrate Taxonomy. 10: 213-405.

Headrick D.H.; Goeden, R.D. 1993. Life history and description of immature stages of *Aciurina thoracica* (Diptera: Tephritidae) on *Baccharis sarothroides* in southern California. Annals of the Entomological Society of America. 86: 68-79.

Korneyev, V.A.; White, I.M. 1993a. Fruitflies of the genus *Urophora* R.-D. (Diptera, Tephritidae) of East Palearctic. II. Review of species of the subgenus *Urophora* s. str. (communication 1). Entomological Review. 72: 35-47.

Korneyev, V.A.; White, I.M. 1993b. Fruitflies of the genus *Urophora* R.-D. (Diptera, Tephritidae) of East Palearctic. II. Review of species of the subgenus *Urophora* s. str. (communication 1). Entomological Review. 72: 82-98.

Merz, B.F. 1992. Revision der westpalaearktischen Gattungen und Arten der *Paroxyna* - Gruppe und Revision der Fruchtfliegen der Schweiz (Diptera: Tephritidae). ETH, Zurich. 342 p. Ph.D. dissertation.

Munro, H.K. 1955. The influence of two Italian entomologists on the study of African Diptera and comments on the geographical distribution of some African Trypetidae. Bolletino del Laboratorio di Zoologia Generale e Agraria "Filippo Silvestri" Portici. 33: 412-426.

Shorthouse, J.D. 1980. Modifications of the flower heads of *Sonchus arvensis* (family Compositae) by the gall former *Tephritis dilacerata* (Order Diptera, family Tephritidae). Canadian Journal of Botany. 58: 1534-1540.

Shorthouse, J.D. 1989. Modifications of flowerheads of diffuse knapweed by the gall inducers *Urophora affinis* and *Urophora quadrifasciata* (Diptera: Tephritidae). Proceedings of the 7th International Symposium on Biological Control of Weeds, Rome: 221-228.

Silva, I.M.; Andrade, G.I.; Fernandes, G.W.; Filho, J.P.L. 1996. Parasitic relationships between a gall-forming insect *Tomoplagia rudolphi* (Diptera: Tephritidae) and its host plant (*Vernonia polyanthes*, Asteraceae). Annals of Botany. 78: 45-48.

Silverman, J.; Goeden, R.D. 1980. Life history of a fruit fly *Procecidochares* sp., on the ragweed, *Ambrosia dumosa* (Gray) Payne, in southern California (Diptera: Tephritidae). Pan-Pacific Entomologist. 56: 283-288.

White, I.M.; Korneyev, V.A. 1989. A revision of the western Palaearctic species of *Urophora* Robineau-Desvoidy (Diptera: Tephritidae). Systematic Entomology. 14: 327-374.

THE DIVERSITY OF INSECT-INDUCED GALLS ON VASCULAR PLANTS IN TAIWAN: A PRELIMINARY REPORT

Man-Miao Yang[1] and Gene Sheng Tung[2]

[1] Associate Curator of Entomology, Division of Zoology, National Museum of Natural Science, Taichung 404, Taiwan, Republic of China
[2] Graduate Student, Department of Forestry, National Taiwan University, Taipei 107, Taiwan, Republic of China.

Abstract.—Preliminary results of a long-term gall fauna study are presented here based on a collection between July 1995 and June 1996, covering the whole Taiwan Island, Green Island, and Orchid Island with 527 records of insect galls. We examined the data in four aspects: (1) HOST PLANT.—Gall-forming insects attack a wide range of plants in Taiwan. Within each plant family, the number of host species ranged from 1 to 23. Galls occur more in Lauraceae and Fagaceae than others. (2) TYPES OF GALLS.—There were many more enclosed types of gall (63 percent) than open types (37 percent). The shape of the open type of galls was less modified and mostly non-specific whereas the shape of the enclosed type varied extensively but was species-specific. A third of host plant species had two or more, up to 11, types of gall. (3) GALLING POSITION.—Galls were found on the leaf blade more frequently than any other parts of the plant. Among leaf galls, more were found between veins than on other areas of a leaf blade for both open and enclosed types. It was also common for the open type of galls to occur on the leaf margin. The galls produced on the mid-rib are exclusively the enclosed type. The enclosed type of galls was found on the veins, either primary or secondary veins, more often than the open type. (4) GALL INDUCER.—Diptera (39 percent) and Homoptera (35 percent) were two major galling groups found in this study. Thirteen percent of gallers belong to Hymenoptera, 4 percent Lepidoptera, with one record of a coleopteran gall and some unknown gallers.

INTRODUCTION

Insect galls, the abnormal growth of plants caused by insects, have been widely known and used by people in China, India, and Europe for more than a thousand years (Mani 1992). The adelgid galls of *Schlechtendalia chinensis* Bell on *Rhus semialata* Murray, commonly known as "wu pei tzu" in Chinese, was applied in traditional Chinese medicine (Li 1596, Read 1982). The tannin content of the *Rhus* gall, known as gallotannic acid, is a powerful astringent. Galls, such as nutgalls on *Rhus semialata*, are also used for dyeing silk black. However, Chinese paid little attention to galls other than for their medical and industrial uses. Takahashi (1933, 1934, 1935, 1936, 1939) was virtually the only person who documented the taxonomy of some galling insects and their biology in Taiwan in early days. Recently, Yang (1996) studied the insect-induced galls in the Guandaushi forest ecosystem as part of a long-term ecological research project. The aim of this study is to establish basic information on the gall fauna of Taiwan and henceforth for ecological and evolutionary studies.

MATERIALS AND METHODS

Galls were collected from 55 sites all over Taiwan (fig. 1) between July 1995 and June 1996. Branches bearing galls were cut, placed in sealed plastic bags, and kept in the refrigerator before further handling. In addition to general field notes, host characteristics, habitat, position of galls on the plant, gall shape, and abundance are also recorded (Appendix 1). Each type of gall on each host plant at each site was given a distinct record number. Photos of galls were taken before and after dissection. Both the insects dissected from galls and the opened gall tissue were preserved in 70 percent alcohol. Slide-mounted insect specimens were further prepared for identification when necessary. Twigs bearing galls were trimmed and buried in fine sand for 1 to 3 weeks to make dry gall specimens. Host plants were pressed and preserved as voucher specimens of host records. If the insects inside the galls were found reaching the stage of emergence, some galls were kept in transparent plastic boxes with a paper towel at the bottom and sealed with paraffin film. Emergent insects were collected and preserved in 70 percent alcohol for further examination.

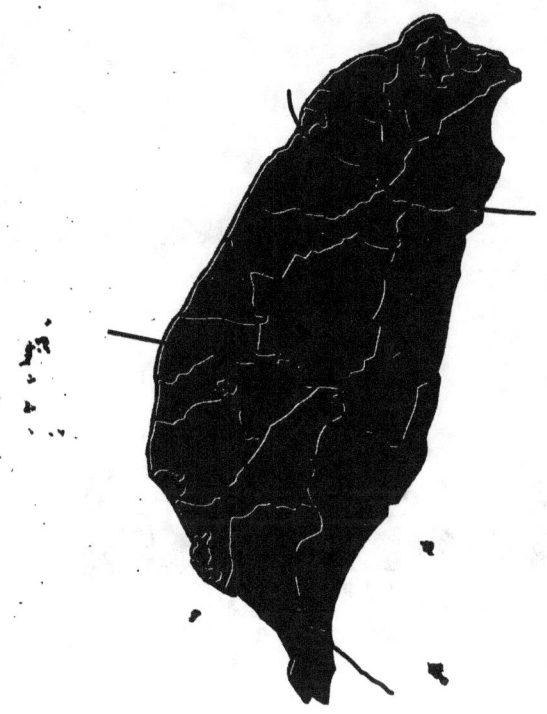

Figure 1.—*The distribution of number of collecting sites in different areas of Taiwan.*

The focus of this survey was the vascular plant galls, although a few galls were also found on non-vascular plants, such as ferns. We generally group the galls into two major categories, the open and the enclosed types. The open type of galls can be easily distinguished from the enclosed type by seeing insects partially exposed outside the gall, usually the dorsal part, or an obvious opening of the gall where the insects can move in and out freely. Subsequent finer categories recognized each type of gall based on both gall shape and the gall inducer. Galls from different areas at different times were counted as separate records. The study did not follow a specific sampling method and the gall type designation was somewhat arbitrary.

RESULTS

Seven hundred and seventy-three records of galls were obtained between July 1995 and June 1996. Among these, 137 were mite galls, 39 fungus galls, 3 bacteria galls, and 67 insect galls with unknown gallers. Therefore, a total of 527 records of insect galls were analyzed in the following four aspects.

Host Plants

Gall-forming insects attack a wide range of plants in Taiwan. The taxonomic distribution of the host plants is given in figure 2. The number of host species within individual plant families ranges from 1 to 23. Among these, the Lauraceae have the highest number of plant species (23) bearing galls (table 1). The second highest is Fagaceae, with 13 species, and Moraceae the third, with 8 species. Other major groups with four or more gall-forming species are Myrtaceae, Euphorbiaceae, Rosaceae, Theaceae, and Verbenaceae.

Types of Galls

There were many more enclosed types of gall than open types collected with a ratio of 71 to 29 (fig. 3), respectively. The shape of galls was less modified in the open type and mostly non-specific while in the enclosed type of galls, the shape varied extensively but was species-specific (fig. 4). The open type was only found on leaves so far, including pit, curling, and folding. The enclosed type usually had a distinct shape when induced by the same insect species and varied a lot among those formed by different galling species. Many are spherical, round shaped with a glabrous or pubescent surface; some are oval with long or short tails; some are needle-like, throne-like, disk-like, conical, bell-shaped, pouch-like, and mushroom-like, etc. A third of host plant species had two or more types of gall (fig. 5). Up to 11 types of galls occurred on *Machilus zuihoensis* var. *zuihoensis* (Lauraceae) and *Cyclobalanopsis glauca* (Fagaceae) (table 2).

Galling Position

Regarding the plant parts attacked, galls were found on leaf blades more frequently than any other part (fig. 6). Among the galls collected, 71 percent of galls were found on leaf blades, 2 percent on petioles, 6 percent on buds, 19 percent on stems, 1 percent on fruit, and 1 percent were non-specific, i.e., found on more than one part of the plant. Among the 156 records of leaf galls (fig. 7), more were found in between veins than other areas on a leaf blade for both open (15.9 percewnt) and enclosed (21.7 percent) types. It is also common for the open type of galls to occur on the leaf margin (15.3 percent). The galls produced on the mid-rib (12.7 percent) are exclusively the enclosed type. The enclosed type of galls was found more often on the veins, either primary or secondary veins, than the open type.

Gall Inducer

Diptera (39 percent) and Homoptera (35 percent) are two major galling groups found in this study (fig. 8). Thirteen percent of gallers belong to Hymenoptera, 4 percent Lepidoptera, with one record (0.4 percent) of a coleopteran gall and some unknown gallers.

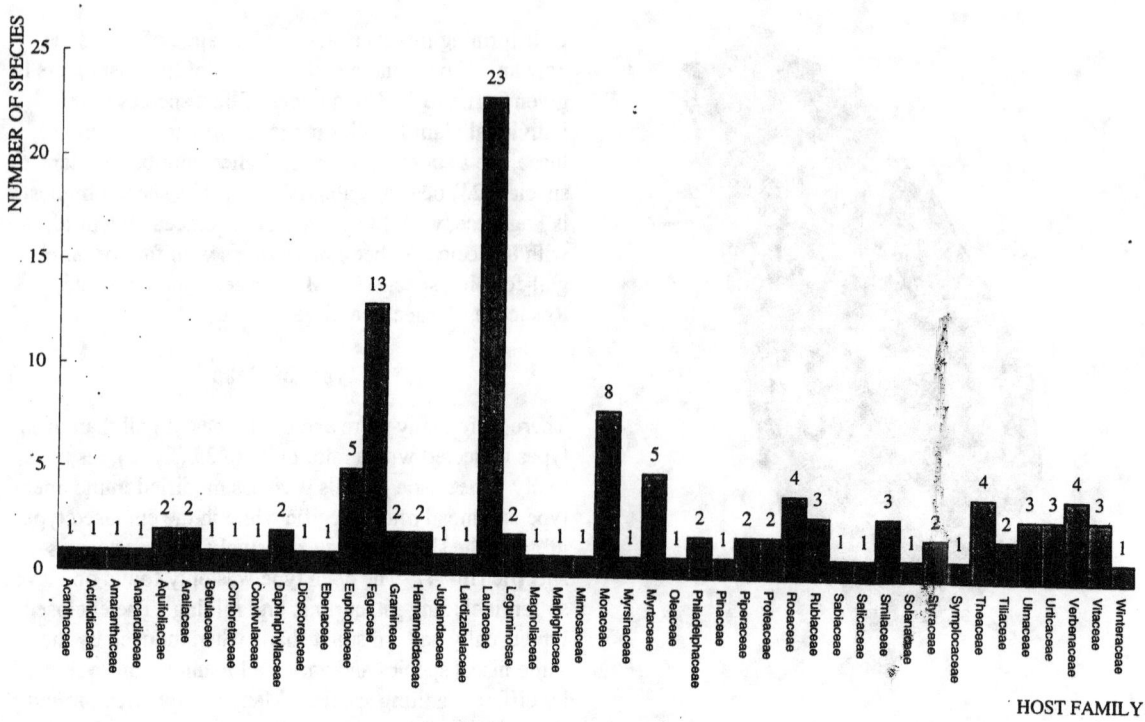

Figure 2.—*The number of species found as galling host in each plant family in Taiwan, based on a survey from July 1995 to June 1996.*

Table 1. The species richness of gall-bearing plant families in Taiwan.

Host Species Number Within Each Family	Number of Families	Host Family (# of species; # of gall types)
>21	1	Lauraceae (23; 77)
11–21	1	Fagaceae (13; 31)
5–10	3	Moraceae (8; 8) Myrtaceae (5; 7) Euphorbiaceae (5; 7)
<5	41	Rosaceae (4; 4) Theaceae (4; 4) Verbenaceae (4; 4)

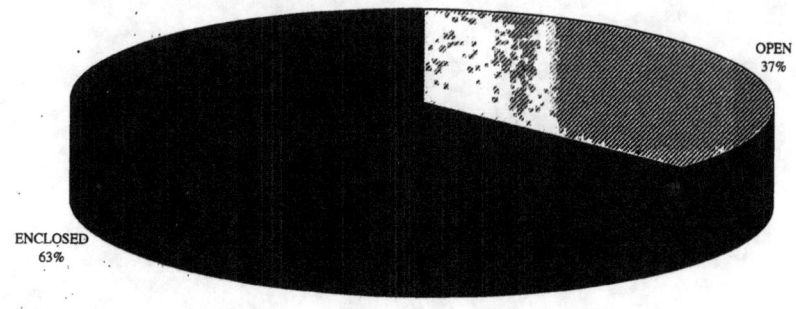

Figure 3.—*A pie chart of open and enclosed type of galls occurred in Taiwan.*

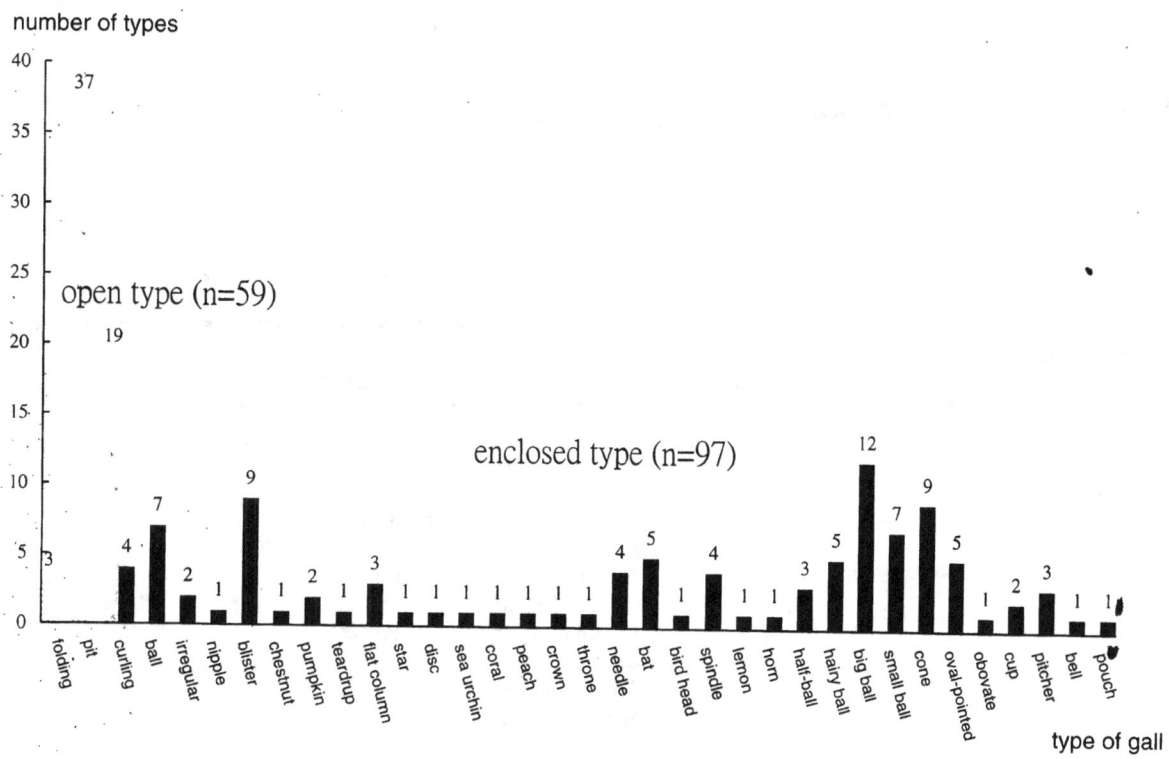

Figure 4.—*Frequency of leaf gall types found in Taiwan, based on a survey from June 1995 to July 1996.*

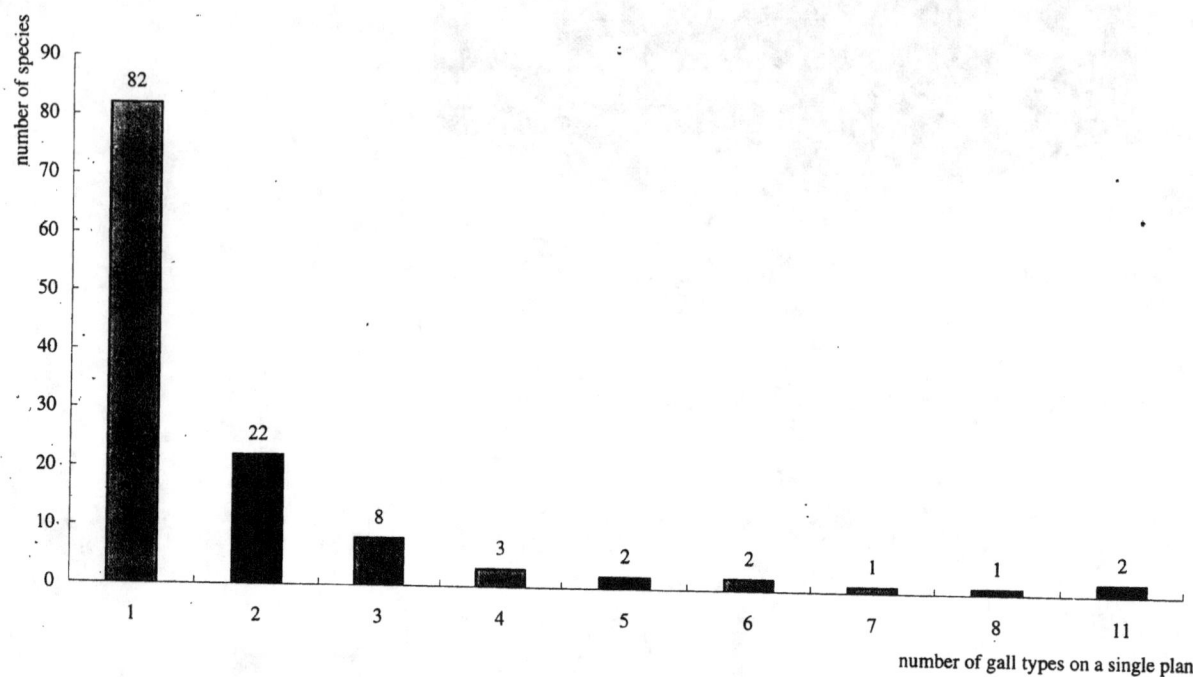

Figure 5.—*The frequency of gall type number occurred on each plant.*

Table 2. The distribution of gall type number per host species and the main gall-forming plants in Taiwan.

Number of Gall Types per Host Species	Number of Host Species	Host Plant Species (Family)
11	2	*Machilus zuihoensis* var. *zuihoensis* (Lauraceae)
		Cyclobalanopsis glauca (Fagaceae)
8	1	*Litsea acuminata* (Lauraceae)
7	1	*Machilus thunbergii* (Lauraceae)
6	2	*Machilus zuihoensis* var. *mushaensis* (Lauraceae)
		Machilus japonica (Lauraceae)
5	2	*Cinnamomum osmophloeum* (Lauraceae)
		Fraxinuns formosana (Oleaceae)
4	3	*Cinnamomum insularimotamum* (Lauraceae)
		Cinnamomum subavenium (Lauraceae)
		Machilus kusanoii (Lauraceae)
3	8	*Pueraria lobata* (Leguminosae) etc.
2	22	*Helicid formosana* (Proteaceae) etc.
1	82	*Miscanthus floridulus* (Poaceae) etc.

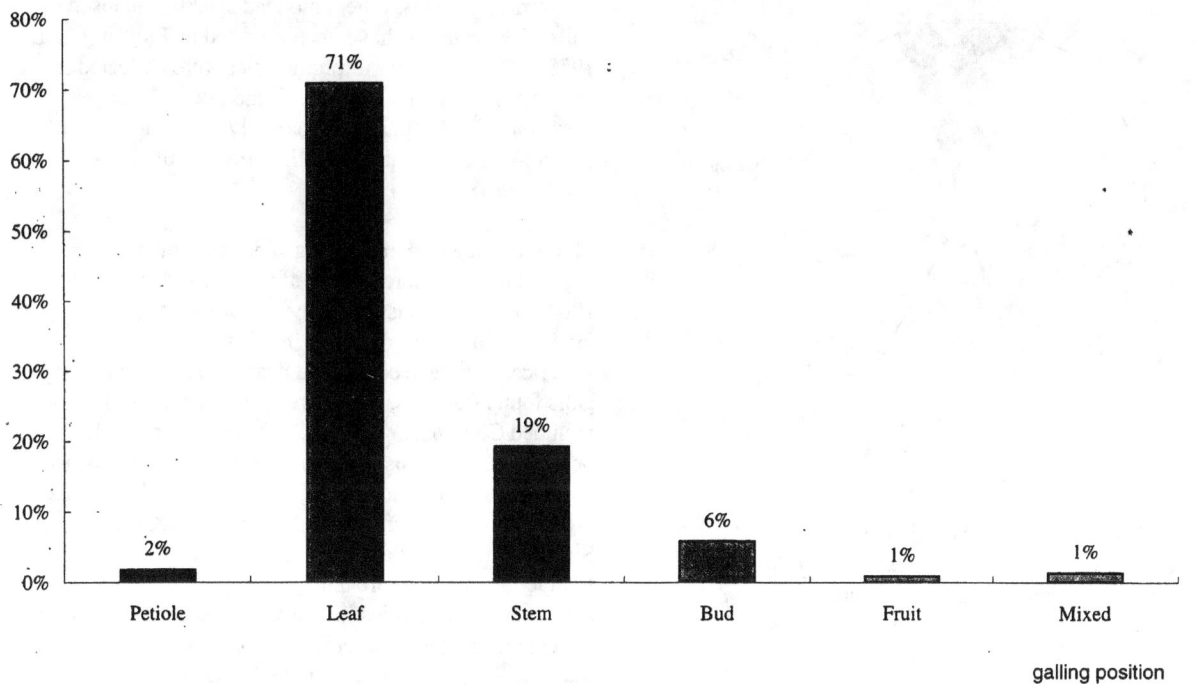

Figure 6.—*The occurrence of various galling position on the host plants.*

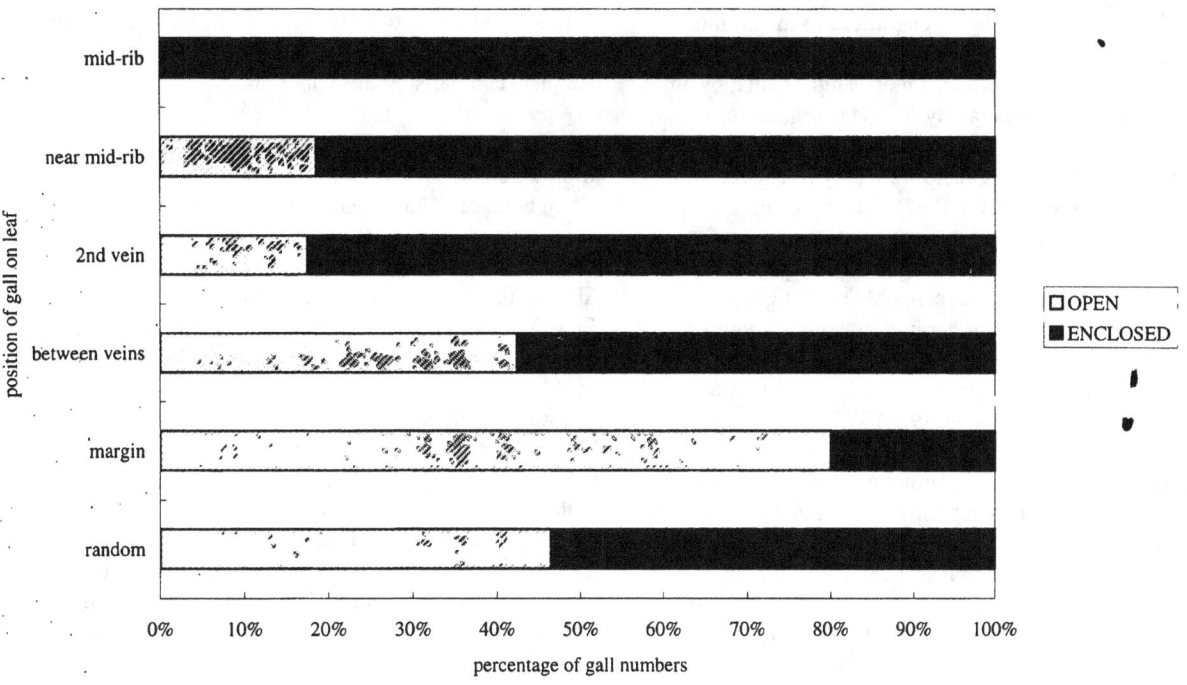

Figure 7.—*The position of galls formed on host plant leaves.*

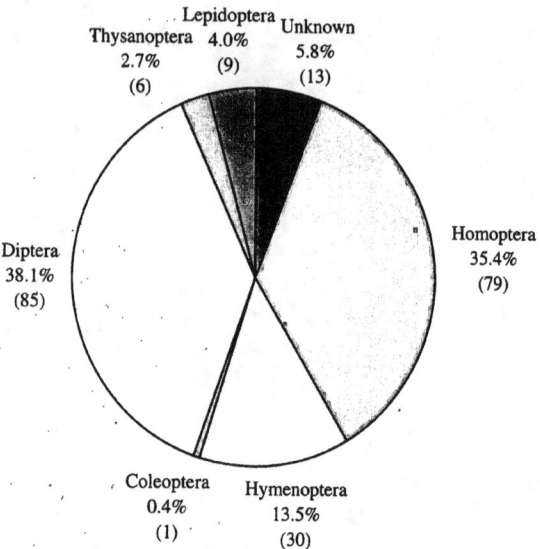

Figure 8.—*A pie chart showing the percentage of galling insects in each order and the number of galling morpho-species (in paranthesis) found in Taiwan, based on a survey from July 1995 to June 1996.*

DISCUSSION

Seven orders of insects were reported forming galls (Dreger-Jauffret and Shorthouse 1992). We found six of them except Hemiptera, true bugs. Regardless of the problem of ambiguous higher classification in Himiptera and Homoptera, the Hemiptera s.s. or Heteroptera is a less significant galling group, which is represented by an exclusive cecidogenous family, Tingidae or lace bugs. Although several genera of Tingidae have been found forming galls on flowers, there were only five species in Europe (Mani 1964, Meyer 1987). In Japan, eight species of galling Tingidae were reported (Yukawa and Masuda 1996). The flower as a galled plant part can easily be overlooked. Since some of the host genera are also present in Taiwan, further investigation is required.

Different galling insects predominate in different areas on a worldwide basis (Gagne 1984). While scale insects are mainly tropical or Australian, galling thrips are basically distributed in the warmer parts of the Eastern Hemisphere. Cynipid gall wasps are common attackers of the oaks in North America and Europe. We found that Diptera and Homoptera represent two major galling groups in Taiwan. Among dipterans, Cecidomyiidae is the most important gall-forming group in terms of the abundant galling species, wide host-plant range, and high gall-shape specificity. Homopterans are the second most important galling group. While many of the homopterans formed pit galls, others induced specific enclosed galls. Few cynipids have been reported in Taiwan (Chiu 1958, 1966). Eleven morpho-species were collected in this very study and more were found after the analyses were done for this paper. Apparently, there are more to be explored and detailed identification should be conducted.

The frequency of gall-forming species on plant groups varies through different parts of the world. Mani (1955, 1964) provided statistics on the major galling plant families in different regions. In Europe and North America, Fagaceae comprised nearly 50 percent of the galls found (table 3). Other major groups in this region included Compositae (20 percent) and Rosaceae (15 percent). Leguminosae are the dominant host plants for galls in South America, India, and part of Africa. Galls are dramatically abundant on Myrtaceae in Australia. Similar to the Himalayas region, our study shows that Lauraceae contain more gall-forming plant species than any other plant families in Taiwan. However, the known host plants are nearly evenly distributed among several major galling host families in the Himalayas (table 3). In contrast, gall-inducers are relatively widespread in Taiwan. Most of the plant families include less than 10 percent of species among the total host species, except Lauraceae (table 3, fig. 2). A similar widespread pattern could also be seen in Japan. However, the result may be related to the extent of effort of studies. Since data from the above literature were not collected on the same basis, complete comparisons are impossible.

Not many research works have covered comprehensively information on both flora and gall fauna in Asia, as in the recent book of Yukawa and Masuda (1996), *Insect and Mite Galls of Japan in Colors*. A comparison of the gall fauna between Japan and Taiwan is given in table 4. Although Japan is more than nine times bigger than Taiwan, they share similar floral diversities. There are more plants bearing galls in Japan than in Taiwan.

According to the Japanese Flora, there are 23 species of Lauraceae (Ohwi 1961) of which 16 were reported supporting galls (Yukawa and Masuda 1996). Nine out of these 23 laural species are also found in Taiwan and seven of them are also galled, showing a high overlap of the host plant taxa. Because the records of Japan are based on 15 years of thorough studies whereas our data of Taiwan came from just a year's survey, there is a lot of potential for further discoveries in Taiwan.

Table 3. Major galling plant families in different geographic regions reported by previous authors and from the current study.

	Europe and N. America (Mani, 1964[1])	Africa and S. America (Mani, 1964[1])	Australia (Mani, 1964[1])	Himalayas (Mani, 1955[1])	Japan (Yukawa & Masuda, 1996)	Taiwan (Yang & Tung, 1997[2])
Fagaceae	50%			20%		7%
Compositeae	20%			22%	9%	4%
Rosaceae	15%			13%	10%	
Myrtaceae			50%			
Lauraceae				24%		13%
Leguminosae		dominant		21%	5%	6%
Euphorbiaceae						5%
Moraceae					4%	
Salicaceae					4%	
Poaceae					4%	
Lamiaceae						

1 No data for the blanks.
2 Blanks indicate data less than 1%.

Table 4. A comparison of the gall faunas of Japan and Taiwan.

	Japan	Taiwan	Ratio
Area	337,801 sq km	36,179 sq km	9.34
Flora	4022 spp	4061 spp	0.99
Galled Plants	569	123	4.62
Galled Plant Ratio	0.14	0.03	4.67
Insect Fauna			
Named	29,000	17,600	1.65
Estimated	100,000	140,000–200,000	0.5–0.7
Galling Insects	1249	223	5.6
Lepidoptera	58	9	6.4
Diptera	647	85	7.6
Hymenoptera	222	30	7.4
Coleoptera	20	1	20
Homoptera	284	79	3.6
Hemiptera	8	0	–
Thysanoptera	10	6	1.7
Unknown	–	13	–

ACKNOWLEDGMENTS

We thank S. F. Yen, C. M. Wang, T. Y. A. Yang, and Y. J. Wang for the identification of host plants and fungi, and Y. H. Hwang for a review of the manuscript. Many people, such as K. Y. Wang, C. Y. Lin, J. L. Huang, M. C. Lin, S. F. Yen, M. L. Chan, W. T. Yang, and S. T. Chiu provided precious gall samples, and museum volunteers helped with pretreatment of samples. The study would not have been completed without their kind help.

Funding support was provided by the National Museum of Natural Science and National Science Council (87-2914-I-178-002-A1), Taiwan, R.O.C.

REFERENCES

Chiu, Shui-Chen. 1958. Bibliography of entomology in Taiwan (1684–1957). Taiwan Agriricultural Research Institute, Spec. Pub. No.1. p. 219

Chiu, Shui-Chen. 1966. Bibliography of entomology in Taiwan supplement I (1957–1966). Taiwan Agricultural Research Institute, Spec. Pub. No. 8. p. 51.

Dreger-Jauffret, F.; Shorthouse, J.D. 1992. Diversity of gall-inducing insect and their galls. In: Shorthouse, J.D.; Rohfritsch, O., eds. Biology of insect-induced galls. Oxford: Oxford University Press: 8-33.

Gagne, R.J. 1984. The geography of gall insects. In: Ananthakrishnan, T.N., ed. Biology of gall insects. Oxford: Oxford and IBH Publishing Co.: 305-322.

Li, Shih-Chen. 1596. Pen-ts'ao kang mu (the Great Systematic Materia Medica) Vol. 39. Reissued by Chinese Medicine Research Institute. (1994): 1270–1274.

Mani. M.S. 1955. Entomological survey of Himalayas. Part 8. Plant galls. Agra University Journal of Research Science. 4(1): 187–208.

Mani, M.S. 1964. The ecology of plant galls. Monographiae Biologicae. The Hague: W. Junk.

Mani, M.S. 1992. Introduction to Cecidology. In: Shorthouse, J.D.; Rohfritsch, O., eds. Biology of insect-induced galls. Oxford: Oxford University Press: 3-7.

Meyer, J. 1987. Plant galls and gall inducers. Gebruder Borntraeger, Berlin. 291 p.

Ohwi, J. 1961. Flora of Japan. Lauraceae. Shibundo, Tokyo: 553-559. (In Japanese)

Read, B.E. 1982. Chinese materia medica: insect drugs, dragon and snake drugs, fish drugs. Chinese medicine series 2. Southern Materials Center. Inc.: 1–213.

Takahashi, R. 1933. *Pemphigella asdificator* Brokton produce galls in Formosa (Aphididae, Hemiptera). Trans. n. Hist. Soc. Formosa. 23: 352–353.

Takahashi, R. 1934. Species co-habitat in Thysanoptera galls. Plants and Animals. 2: 1827–1835. (In Japanese)

Takahashi, R. 1935. Galls of Taiwan. Science of Taiwan. 3(5): 3–8. (In Japanese)

Takahashi, R. 1936. A new scale insect causing galls in Formosa (Homoptera). Trans. n. Hist. Soc. Formosa. 26: 426–428.

Takahashi, R. 1939. A new Aphid producing galls in Formosa. Journal of Zoology. 51: 425–427. (In Japanese)

Yang, S.Y. 1996. The relatiohship between understory and insect fauna at Guandaushi forest ecosystem. Nat. Chung-Hsing University. M.S. thesis. 105 p.

Yukawa J.; Masuda, H. 1996. Insect and mite galls of Japan in colors. Japan: Association of Japanese Agricultural Education: 632–635.

Appendix 1. *Data Sheet for Galls Collected in Taiwan*

Date:	Rec. #: GA	
Locality:	-longitude: -latitude:	-galls from same tree: GA
Collector:	-ALT: -Temp., RH:	

Host plant and habitat:
1. common name: family:
2. species name:
3. tree/shrub/vine/herb/others, hight: (   )m
4 single/aggregation
5. isolated/mixed
6. open area/semi-shaded /undershade
7. wet/dry/median toward (wet/dry)
8. Distribution of galls on plant:
    sunny side/shade/higher part/lower part/not obviouse/others
	Type of habitat:
	coniferous forest
	mixed coniferous forest
	broad-leaved forest
	alpine grassland/grassland
	cultivar(orchard、bamboo forest..)
	wetland/pond
	humman environment
	  (park/urban forest/homeland/
	   waste land/graveyard....)
	others( )	

The position of galls on plant: leaf/petiole/bud(axillary/terminal)/stem/twig/flower/fruit **The position of leaf gall:** 1. mid-rib/near mid-rib/2nd vein/between veins/margin/random 2. main structure on: upper side/lower side/even	**Number of galls:** ( ) n≤5, 5<n≤10, 10<n≤100, n>100 per leaf/part n≤5, 5<n≤10, 10<n≤100, n>100 per plant	
Gall morphology: open type: curling/folding/pit/others ( ) enclosed type: shape ( )	**Color of gall:**	

MMY gallrcd-eng.doc

Associate organisms	Notes (Treatments)
Organisms associate with gall: 1. single cell/ multiple cell 2. one ind. per cell/ multiple ind. per cell 3. inside the gall: 4. outside the gall: 5. emergent hole (present/ absent), shape ( )	**Date of treatment:** / / 19 **Photo no.:** R# M#
Parasitic rate: no. of healthy galls: no. of parasitized galls: **Position of galler inside the gall:** body position: basal/middle/terminal head direction: basal/terminal/lateral	-gall: 70% EtOH/dry/other -plant: 70% EtOH/press -insect: 70% EtOH/pin/slide/ freeze/Carnoy/rear -other material holder:
Drawings and measurements: 	

GALL-FORMING AND SEED-EATING INSECTS, AND THEIR ASSOCIATES, ON JOINT PINES (*EPHEDRA* SPP.) IN SPAIN, A RICH AND NEWLY DISCOVERED FAUNA

R. R. Askew and J. Blasco-Zumeta

RRA: 5 Beeston Hall Mews, Beeston, Tarporley, Cheshire CW6 9TZ, England
JB-Z: C/ Hispanidad 8, 50750 Pina de Ebro, Zaragoza, Spain

Abstract.—We present the findings to date of a continuing study of the rich, and largely hitherto unknown, insect fauna inhabiting galls and seeds on Joint Pines (Gnetales: Ephedraceae, *Ephedra* spp.) in Spain. Galls are formed by *Eurytoma* (Hymenoptera: Chalcidoidea) and Cecidomyiidae (Diptera), and seeds are fed upon by larvae of *Blascoa* (Hymenoptera: Chalcidoidea). The phytophages are attacked by communities of parasitoid Hymenoptera.

INTRODUCTION

Gymnospermae, in contrast to Angiospermae, are hosts to relatively few internally-feeding phytophagous insects, but Joint Pines (Gnetales: Ephedraceae, *Ephedra* spp.) are exceptional. In the arid steppe country of Monegros in Aragon, eastern Spain, species of *Ephedra* have recently been found to support rich faunas principally centered upon gall-forming and seed-eating species of Chalcidoidea. *Ephedra* are placed in Gnetales, a group apparently of ancient lineage suggested by some as the sister group to Angiospermae. Knowledge of the insects that depend upon them is therefore likely to be of wide interest.

Previous Information

There are few literature citations to gall-forming or seed-eating insects attacking *Ephedra*. Mani (1964) mentions that 'Globose or fusiform galls on the stem of *Ephedra* spp., caused by Coleoptera, Lepidoptera and Diptera, are known from different parts of the world', but he gives no details except the figure of a plurilocular stem gall attributed to a gall midge. Zerova et al. (1988) describe *Eurytoma flaveola* (Zerova 1976) (Chalcidoidea: Eurytomidae) as forming elongated clusters of galls on roots of several species of *Ephedra* in Asia, and Zerova and Seryogina (1994) refer only to *Eurytoma ermolenkoi* Zerova as phytophagous in seeds of *Ephedra procera* in Armenia. With the exception of a record of *Nikanoria ephedrae* Steffan (Eurytomidae) reared from seeds of *E. distachya* L. in France as a possible parasitoid of *Nemapogon* (Lepidoptera: Tineidae) (Steffan 1961), almost nothing was known about the European fauna until 1994. Then one of us (J B-Z), looking at *Ephedra nebrodensis* Tineo in the arid Monegros (Zaragoza) region of eastern Spain, noticed insect emergence holes in the seeds and found globular galls on the shoots.

Seed-eaters

An extensive study of the seeds of *E. nebrodensis* (Askew and Blasco-Zumeta 1997) showed that the primary phytophage was a miscogasterine pteromalid, *Blascoa ephedrae* Askew, and it attacked over 60 percent of seeds in the study area. Phytophagy is rare in Pteromalidae and has not been previously reported in Miscogasterinae. Five species of Chalcidoidea, *Mesopolobus semenis* Askew and *M. arcanus* Askew in Pteromalidae, *Aprostocetus lutescens* Askew and *Baryscapus aenescens* Askew in Eulophidae, a species in the *Eupelmus urozonus* Dalman aggregate (Eupelmidae) and a species of Braconidae, are parasitoids of *Blascoa*. The trophic relations of these species are depicted as a food web (fig. 1).

Blascoa has also been found in seeds of *E. distachya* L. and *E. fragilis* Desf. On the latter plant, it appears from limited data to have a similar fauna of parasitoids to that found on *E. nebrodensis*, but on *E. distachya* it is heavily attacked by a species of *Adontomerus* (Torymidae) that is possibly absent from *E. nebrodensis*.

Gall-formers

An investigation of the globular galls (Askew and Blasco-Zumeta, in press), involving dissection and rearing, revealed that the gall-former is a species

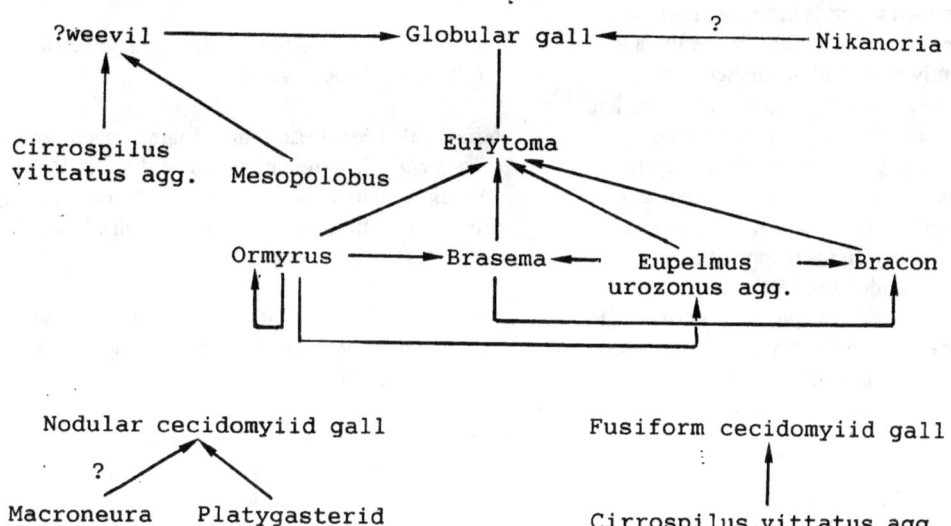
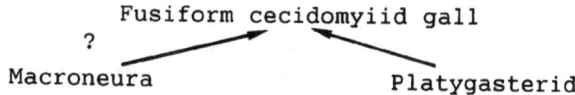
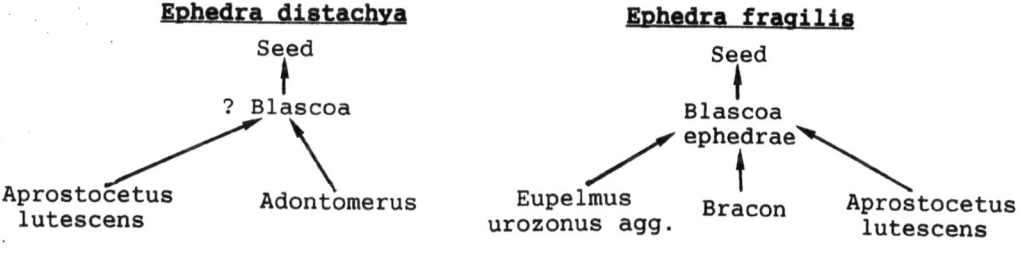

Figure 1.—*Trophic relations of the insects inhabiting seeds and galls on species of* Ephedra *in Spain.*

assigned to *Eurytoma* although showing some characters of *Bruchophagus* (endophagous in seeds of angiosperms). Its unilocular galls are quite different to those of *E. flaveola* on *Ephedra* in Asia (Zerova 1995), and very conspicuous. It is very surprising, therefore, that they do not appear to be mentioned in the literature. They have not recently colonized Spain; herbarium material in Madrid was collected in 1907. Three chalcid species (an undescribed *Brasema*, a species in the *Eupelmus urozonus* aggregate (Eupelmidae), and an undescribed *Ormyrus* (Ormyridae)) and one species of Braconidae are parasitoids of the gall-former, and a species of *Nikanoria* (Eurytomidae) appears to be an inquiline in the galls. In addition, beetle (Curculionidae?) larvae feed beneath the epidermis of the galls and these are parasitized by two chalcid species. The community of insects associated with galls of *Eurytoma* on *E. nebrodensis* is depicted as a food web (fig. 1).

Unidentified species of Cecidomyiidae also form galls on *Ephedra* in Spain. Fusiform, plurilocular galls have been found on *E. nebrodensis* and *E. distachya* (perhaps formed by different gall-midge species), and small unilocular galls at the shoot nodes of *E. nebrodensis* appear usually to be young stages of the *Eurytoma* galls, but sometimes to be formed by another species of Cecidomyiidae. The few species of Hymenoptera so far detected as parasitoids of these Cecidomyiidae are indicated in figure 1.

LITERATURE CITED

Askew, R.R.; Blasco-Zumeta, J. 1997. Parasitic Hymenoptera inhabiting seeds of *Ephedra nebrodensis* in Spain, with descriptions of a phytophagous pteromalid and four other new species of Chalcidoidea. Journal of Natural History. 31: 965-982.

Askew, R.R.; Blasco-Zumeta, J. In press. Insects associated with galls of a new species of Eurytomidae (Hymenoptera: Chalcidoidea) on *Ephedra nebrodensis* in Spain. Journal of Natural History.

Mani, M.S. 1964. Ecology of plant galls. The Hague: Dr. W. Junk, Publishers. 434 p.

Steffan, J.R. 1961. Description d'une nouvelle espèce de *Nikanoria* Nik. et remarques sur l'identité du "Bruchophagus sativae Ashm." (Hym. Eurytomidae). Bulletin du Muséum National d'Histoire Naturelle. 33: 197-201.

Zerova, M.D. 1976. Khaltsidy sem. Eurytomidae podsemieistva Rileyinae i Harmolitinae. Fauna SSSR, n.s. VII. 6. 230 p.

Zerova, M.D. 1995. Parasitic Hymenoptera - Eurytominae and Eudecatominae of Palaearctics. Kiev: National Academy of Sciences of Ukraine. 460 p.

Zerova, M.D.; Diakontshuk, L.A.; Ermolenko, V.M. 1988. Nasekomyie-galloobra zovateli kulturnykh i dikorastushchikh rastenij evropejskoj chasti SSSR. Kiev: Naukova dumka. 160 p.

Zerova, M.D.; Seryogina, L. Ya. 1994. The seed-eating Chalcidoidea of Palaearctics. Kiev: National Academy of Sciences of Ukraine. 238 p.

IS THERE A COOL-WET TO HOT-DRY GRADIENT OF INCREASING GALL-FORMING INSECT DIVERSITY?

Ros Blanche and John Ludwig

CSIRO Wildlife and Ecology, Tropical Ecosystems Research Centre, PMB 44 Winnellie Darwin NT Australia 0821

Abstract.—Some studies have suggested that gall species richness increases as environments become hotter and drier because there are fewer gall insect enemies in hot, dry environments. Other work has shown that low soil fertility and certain host plant groups, not hot, dry conditions, are related to high numbers of gall species. High levels of secondary chemicals and long-lived plant parts, adaptations of some plants growing on infertile soils, may favor galls by providing long-term sites for gall formation and protection from enemies. The main function of galls may be to gain a better food supply from low nutrient, well-defended plant tissues.

INTRODUCTION

This paper evaluates some of the evidence regarding the existence of a cool-wet to hot-dry gradient of increasing gall-forming insect species diversity. The presence or absence of such a gradient is interesting because it sheds light on the possible function of insect-induced galls.

Evidence for Increasing Gall Species With Increasing Temperature and Dryness

The data which first suggested that hot-dry environments favor gall insects were collected by Fernandes and Price (1988) along two altitudinal gradients—one from the top of the San Francisco Peaks (3843 m) to the lower Sonoran Desert (305 m) near Phoenix, Arizona, USA, and the other from the top (1350 m) to the base (650 m) of the Serra do Cipo in Minas Gerais, Brazil. At both locations sites became hotter and drier as altitude decreased. Samples were taken, where possible, both away from and near watercourses at each elevational site.

In both USA and Brazil gall species richness was found to increase as sites became hotter and drier and altitude decreased. The increase in gall species richness was attributed to the increase in temperature and dryness rather than the decrease in altitude because gall species richness was higher in samples taken away from watercourses than in those taken near watercourses at the same elevation, and there was no gradient in gall species richness when mesic sites were considered alone.

Later work by Fernandes and Price (1992) suggested that hot-dry environments have fewer enemies (e.g., fungal diseases and predators) which attack gall insects and this may be why there are more gall species in such environments. Galls would then function primarily to protect insects from suffering 'hygrothermal stress' (Fernandes and Price 1988, 1991; Price et al. 1986, 1987) in the relatively enemy free but hotter, drier environments.

Indications of Alternative Explanations

During the course of our studies in Australia several findings caused us to question whether the gradient in temperature and moisture was the only explanation for the gradients in gall species richness. The first author had estimated regional gall species richness on certain Australian eucalypts and acacias, some with geographic ranges in semi arid western NSW, and others with geographic ranges in the more mesic coastal areas of the State (Blanche and Westoby 1996). If galling is more prevalent in hot-dry environments than cooler-wetter environments, one might expect that gall species richness would be higher on plant species with ranges in the semi-arid western regions than in the coastal regions. There was no indication that this was so. Comparison of gall species richness at three widely spaced NSW locations with different temperature/rainfall conditions also failed to produce the expected pattern.

The first author had also compared gall species diversity on vegetation growing at infertile soil sites with that on vegetation at fertile soil sites (Blanche and Westoby

1995). There were more gall species at infertile soil sites than fertile soil sites but the effect was concentrated in myrtaceous trees, especially eucalypts.

The link between gall species richness and certain host plant species on infertile soils in Australia made us wonder if the altitudinal gradients in the USA and Brazil could also have been soil fertility gradients and perhaps the sites away from watercourses less fertile than those near watercourses. Also, the plant species composition had been very different along the altitudinal gradients. For example it varied from alpine-tundra, through spruce-fir forest, ponderosa pine forest, pinyon-juniper woodland and chaparral, to desert vegetation in the USA. So we felt that the effect of soil fertility and plant species composition were worth exploring further.

METHODS AND RESULTS

Evidence That Soil Fertility and Host Plant Groups, Not Temperature and Moisture, Influence Gall Species Richness

To test the hypothesis more rigorously, gall-forming insect diversity was determined at five Northern Territory locations along a strong temperature/rainfall gradient (mean January maximum temperature 33°C to 39°C and median annual rainfall 1500 mm to 550 mm) from the north coast, near Kakadu National Park, inland to the Tanami Desert, a distance of 500 km, with no significant change in elevation. All sites were in eucalypt woodland/open forest on infertile soils (low phosphorus). No cool-wet to hot-dry gradient in gall-forming insect diversity was detected (fig. 1).

In an additional test of the hypothesis, gall species richness, plant species richness, and soil fertility were assessed at five elevational locations from the top of the Chisos Mountains (2400 m) to near the Rio Grande River (600 m) in Texas, USA. This elevational gradient is also a temperature/rainfall gradient (cool at higher altitudes to hot at lower altitudes, mean annual rainfall 530 mm to 257 mm). Gall species richness was greatest at mid-elevations along the altitudinal gradient and lowest at both low and high elevations, that is, in dry desert and mesic mountain locations (fig. 2). Mesic plots (in washes) and dry plots (adjacent uplands) at the same altitude were not significantly different in gall species richness.

The pattern of gall distribution was related to the distribution of woody plant species along the gradient (fig. 3). Richness of woody plant species was negatively related to soil phosphorus levels (fig. 4).

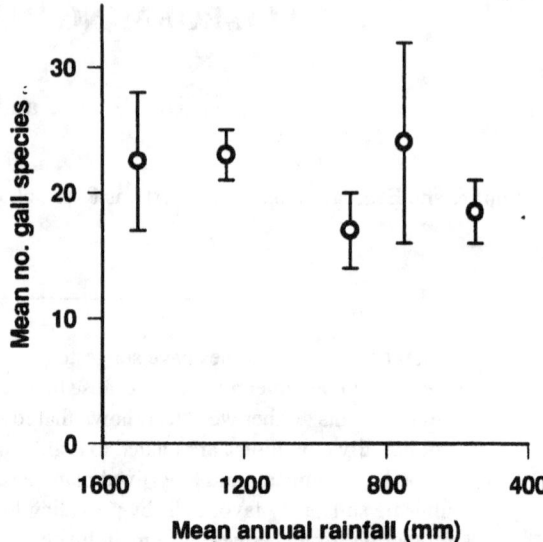

Figure 1.—*Number of gall species at sites with different mean annual rainfall in the Northern Territory, Australia. Error bars are ± SE. Temperature increases as rainfall decreases. Shows no significant increase in the number of gall species as sites become hotter and drier.*

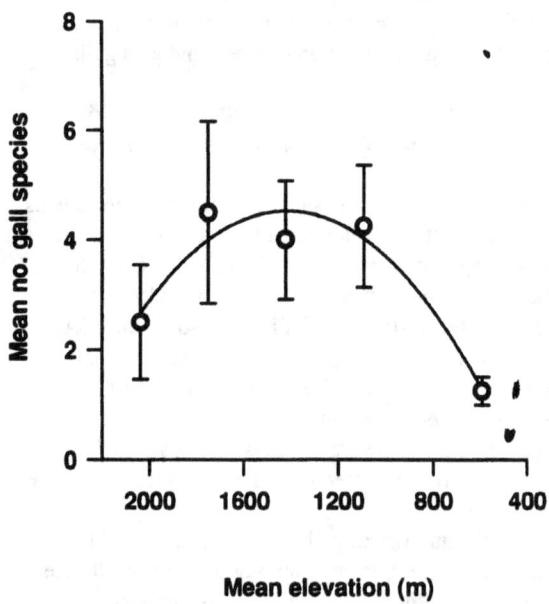

Figure 2.—*Number of gall species at different elevations at Big Bend NP, Texas. Error bars are ± SE. Showing no increase in the number of gall species as conditions become hotter and drier with decreasing elevation.*

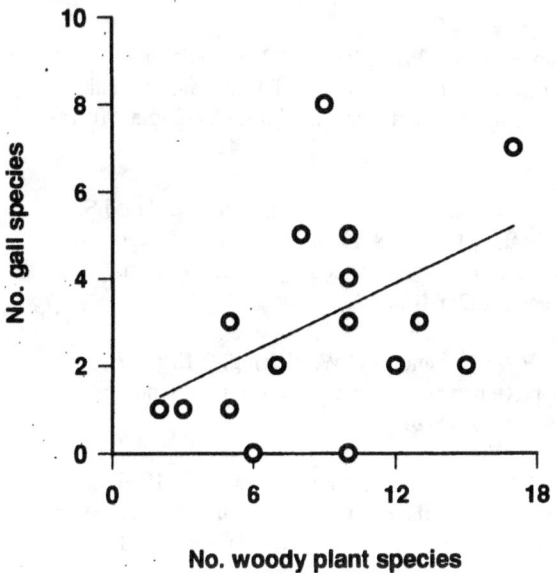

Figure 3.—*Showing the positive relationship between the number of gall species and the number of woody plant species at Big Bend NP, Texas ($r^2 = 0.23$, $P = 0.03$).*

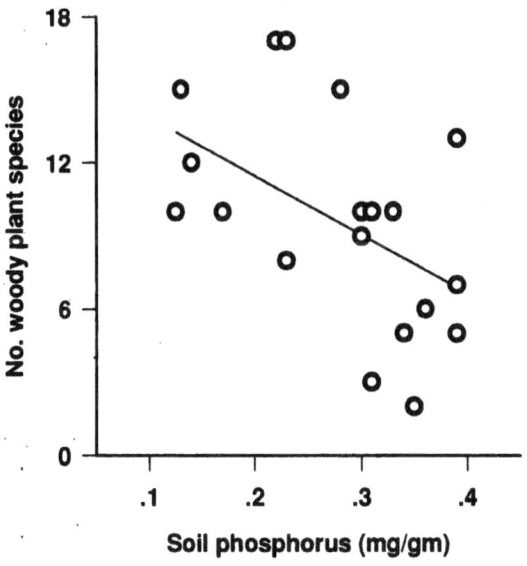

Figure 4.—*Showing the negative relationship between the number of woody plant species and soil phosphorus levels at Big Bend NP, Texas ($r^2 = 0.25$, $P = 0.02$).*

These findings suggest that hotter-drier conditions do not favor gall-forming insects. Gall species richness may be related more to changes in soil phosphorus and plant community composition than changes in temperature and moisture.

DISCUSSION

Relevance to Gall Function

Many plant groups growing on infertile soils have long-lived plant parts (Coley et al. 1985) and high concentrations of secondary compounds (McKey et al. 1978). Long-lived plant parts provide reliable, long term sites for gall formation and it has been suggested that plant secondary compounds, which are considered to help defend plant tissues against herbivores, also protect galls from chewing herbivores and fungi (Cornell 1983).

Galls also act as sinks for nutrients and other plant compounds. Concentration of secondary compounds in the outer gall wall, away from the insect inhabitants, means gall insects can avoid ingestion of the undesirable secondary compounds while benefiting from the flow of nutrients into the nutritive cells of the gall. So galling may function primarily as a means to obtain a quality food supply from otherwise low nutrient, well defended plant tissues, rather than providing a means of avoiding hygrothermal stress.

ACKNOWLEDGMENTS

We would like to thank Carl Fleming, Big Bend National Park Scientist, for arranging permits and accommodation and providing background material required by us to complete part of this study. We also thank Carl White, from the University of New Mexico, for carrying out the analysis of soils from Big Bend National Park.

LITERATURE CITED

Blanche, K.R.; Westoby, M. 1995. Gall-forming insect diversity is linked to soil fertility via host plant taxon. Ecology. 76(7): 2334-2337.

Blanche, K.R.; Westoby, M. 1996. The effect of the taxon and geographic range size of host eucalypt species on the species richness of gall-forming insects. Australian Journal of Ecology. 2: 332-335.

Coley, P.D.; Bryant, J.P.; Chapin, F.S., III. 1985. Resource availability and plant antiherbivore defense. Science. 230: 895-899.

Cornell, H.V. 1983. The secondary chemistry and complex morphology of galls formed by the Cynipinae (Hymenoptera): why and how? American Midland Naturalist. 110: 225-234.

Fernandes, G.W.; Price, P.W. 1988. Biogeographical gradients in galling species richness. Oecologia. 76: 161-167.

Fernandes, G.W.; Price, P.W. 1991. Comparison of tropical and temperate galling species richness: the roles of environmental harshness and plant nutrient status. In: Price, P.W.; Lewinsohn, T.M.; Fernandes, G.W.; Benson W.W., eds. Plant-animal interactions: evolutionary ecology in tropical and temperate regions. New York, NY: John Wiley and Sons: 91-115.

Fernandes, G.W.; Price, P.W. 1992. The adaptive significance of insect gall distribution: survivorship of species in xeric and mesic habitats. Oecologia. 90: 14-20.

McKey, D.; Waterman, P.G.; Mbi, C.N.; Gartlan, J.S.; Struhsaker, T.T. 1978. Phenolic content of vegetation in two African rainforests: ecological implications. Science. 202: 61-64.

Price, P.W.; Fernandes, G.W.; Waring, G.L. 1987. Adaptive nature of insect galls. Environmental Entomology. 16: 15-24.

Price, P.W.; Waring, G.L.; Fernandes, G.W. 1986. Hypotheses on the adaptive nature of galls. Proceedings of the Entomological Society of Washington. 88(2): 361-363.

ROLE OF *PERICLISTUS* (HYMENOPTERA: CYNIPIDAE) INQUILINES IN LEAF GALLS OF *DIPLOLEPIS* (HYMENOPTERA: CYNIPIDAE) ON WILD ROSES IN CANADA

Joseph D. Shorthouse

Department of Biology, Laurentian University, Sudbury, Ontario, P3E 2C6, Canada

Abstract.—Cynipid wasps of the genus *Diplolepis* are restricted to inducing galls on roses (*Rosa* species). Six *Diplolepis* galls found in Canada and induced from leaf tissues are attacked and structurally modified by inquiline cynipids of the genus *Periclistus*. Immature larvae of *Diplolepis* are killed by ovipositing *Periclistus* and *Diplolepis*-induced gall cells quickly lose their characteristics. *Periclistus* larvae then induce their own layers of nutritive and sclerenchyma cells and are encased within individual chambers. *Periclistus* are important mortality factors for *Diplolepis* and the presence of their larvae increase the diversity of associated component communities. *Periclistus* are likely driving forces in the evolution of the *Diplolepis* complex.

INTRODUCTION

Approximately 30 species of cynipid wasps of the genus *Diplolepis* induce galls on the wild roses (*Rosa* spp.) of North America (Beutenmuller 1907, Burks 1979, Shorthouse 1993) and about half of these are found in Canada. Galls of each species of *Diplolepis* are structurally distinct and organ specific to either leaves, stems or roots. Like the galls of all inducers (Rohfritsch 1992), those of *Diplolepis* undergo a series of developmental events referred to as initiation, growth, and maturation as the inducers gain control of the development and physiology of attacked organs. Immatures of gall inducers become surrounded with concentric layers of modified plant cells which provide food and shelter. The arrangement of these layers varies between inducers. In all cynipid galls, there are cytoplasmically dense cells lining the larval chambers, referred to as nutritive cells (Bronner 1992), which contain high concentrations of fats and proteins and are the inducer's sole source of food. Much has been written about the anatomy of cynipid galls (see references in Meyer and Maresquelle 1983, Meyer 1987, Shorthouse and Rohfritsch 1992); however, little is known about how gall inhabitants other than the inducers influence developmental events.

Because cynipid galls are so apparent, are composed of cells with more nutrients than unmodified host organs, and the inducers are stationary within their chambers, they are attractive to numerous species of parasitoids and inquilines. Inquilines are wasps that live in a close spatial relationship with the larvae of gall inducers, never feeding on tissues of the inducers, but nevertheless frequently destroying it (Askew 1971). Most galls of cynipids are host to such distinct assemblages of inducers, parasitoids and inquilines that they are considered component communities (Shorthouse 1993). All cynipid inquilines are phytophagous and feed on specialized nutritive cells that they themselves induce from tissues of the host gall. Host galls are usually enlarged and anatomically modified (Shorthouse 1973, 1980); however, some inquilines retard the growth of their host galls (Washburn and Cornell 1981, Wiebes-Rijks 1982). All inquilines associated with *Diplolepis* galls are cynipids of the genus *Periclistus*, but their role in component communities and influence on gall anatomy is poorly known.

All *Periclistus* have their life cycles obligatorily associated with galls induced by *Diplolepis*. The inducer dies in the association and the inquiline larvae redirect the growth and development of attacked galls such that each inquiline is enclosed in its own chamber. Askew (1971, 1984) considered inquilinism a form of commensalism intermediate between parasitism and symbiosis and provided an overview of inquiline biology. Because inquilines feed in a manner so similar to the inducers, Hawkins and Goeden (1982) called the structures made by inquilines 'endogalls'. Ronquist (1994) proposed the term 'agastoparasitism' and emphasized that cynipid inducers and inquilines are closely related.

If we are to understand the factors influencing speciation of inducers and the structural complexity of their galls, we need to learn more about the roles of all community members. One approach is to compare the activites of one guild of gall inhabitants associated with a single genus of inducers. Galls induced by *Diplolepis* on wild roses are ideal for such a study because they are diverse, their component communities are complex and several species occur sympatrically in some areas. In this paper, the roles of *Periclistus* in the galls of six species of *Diplolepis* found on the leaves of wild roses in Alberta and Saskatchewan, Canada are compared. The six gallers are *D. polita* (Ashmead), *D. bicolor* (Harris), *D. rosefolii* (Cockerell), *D. nebulosa* (Bassett), *D. ignota* (Osten Sacken) and *D. gracilis* (Ashmead). The gross anatomical features of inducer and inquiline-modifed galls will be illustrated for each species along with the role of *Periclistus* as a mortality factor of the inducers.

BIOLOGY OF *DIPLOLEPIS* AND THEIR GALLS

Cynipid gall inducers are divided into five tribes, based on morphological similarities and partly on host plant preferences (Ronquist 1994). Most species of cynipids are in the tribe Cynipini, all of which gall Fagaceae, whereas all species of *Diplolepis* are found in the Tribe Rhoditini and gall roses. *Diplolepis* are univoltine and do not exhibit alternation of generations. Females of most species exit their galls in the early spring and search for suitable oviposition sites without mating (Shorthouse 1993). Populations of most species are predominantely female. Eggs of all leaf-gall inducers are deposited on the surface of immature leaflets still within leaf buds (fig. 1A). Presence of eggs cause nearby tissues to lyse during initiation and the freshly hatched larva enters a small chamber. The larva is quickly surrounded by rapidly proliferating cells with those nearest the larva becoming cytoplasmically dense nutritive cells (fig. 1B). Vacuolate parenchymatous cells adjoining the nutritrive cells also appear early in gall growth (fig. 1B). During the growth phase, the layers of gall parenchyma cells increase in thickness and are responsible for the gall's rapid increase in size (fig. 1C). The cytoplasmically dense nutritive cells are the larva's sole source of food and new nutritive cells are differentiated from gall parenchyma as they are consumed. Vascular bundles appear among the gall parenchyma and join those of the host organ (fig. 1C). Most larval feeding occurs once galls enter the maturation phase when gall parenchyma is differentiated into nutritive cells. The larval chamber is largest at the maturation phase (fig. 1D) providing the larva ample room to maneuver and feed on nutritive cells. A band of lignified sclerenchyma (fig. 1E) is differentiated within the layer of gall parenchyma during the maturation phase. By the end of larval feeding, all gall parenchyma inside the sclerenchyma layer has been converted to nutritive cells and consumed such that larval chambers of mature galls are lined with sclerenchyma. The inducers overwinter in the prepupal stage within their galls. Most mature galls abscise in late summer or fall along with non-galled leaves and are buried by snow. For references on the development of cynipid galls, see Rohfritsch (1992).

BIOLOGY OF *PERICLISTUS*

Five genera of cynipids are known to be inquilines within the galls of other cynipids: *Ceroptres*, *Synergus*, *Saphonecrus*, *Periclistus*, and *Synophromorpha*, whereas *Synophrus* and *Rhoophilus* are suspected of being inquilines (Roskam 1992, Ronquist 1994). Roskam (1992) placed all inquiline genera in the Tribe Aylacini and Ronquist (1994) showed that they form a monophyletic group. Ronquist (1994) also suggested that inquilines evolved from gall inducers in or related to the genus *Diastrophus*, as was also hypothesized by Ritchie (1984), and that *Diastrophus* is the genus of gall inducers most closely related to the inquilines. A recent, more comprehensive analysis of high-level cynipid relationships, comprising representatives from all but the Aylacinia genera, indicates that inquilines are closely related to three Aylacini genera that are all associated with rosaceous hosts (Liljeblad and Ronquist 1998): *Xestophanes* and *Gonaspis* on *Potentilla*, and *Diastrophus* on *Potentilla*, *Rubus*, and a few other plants (one *Diastrophus* species occurs on *Smilax* (Smilacaceae), but this is obviously due to secondary colonization of a distant host plant). Of these genera, *Xestophanes* is most closely related to the inquilines, not *Diastrophus*, and this conclusion is strongly supported by the analysis of Liljeblad and Ronquist (1998). Thus, it appears that inquilines evolved from gall inducers associated with *Potentilla*. However, the results of Liljeblad and Ronquist (1998) do support the notion that the inquilines were originally attacking *Rubus*-galling species in the genus *Diastrophus*.

Ten species of *Periclistus* occur in North America (Ritchie 1984). In his dissertation where he revised the genus, Ritchie (1984) found that only four of the previously described species were valid; three existing species were either synonyms of other *Periclistus* species or belonged to other genera and described six new species. Unfortunately, the new names cannot be used here as they are still not validly published in the sense of the ICZN; therefore, no specific names of *Periclistus* are used in this publication. Even so, it is useful to note that Ritchie found a new species of *Periclistus* associated with both galls of *Dipolepis polita* and *D. bicolor*, another from galls of *D. rosaefolii*, another from galls of

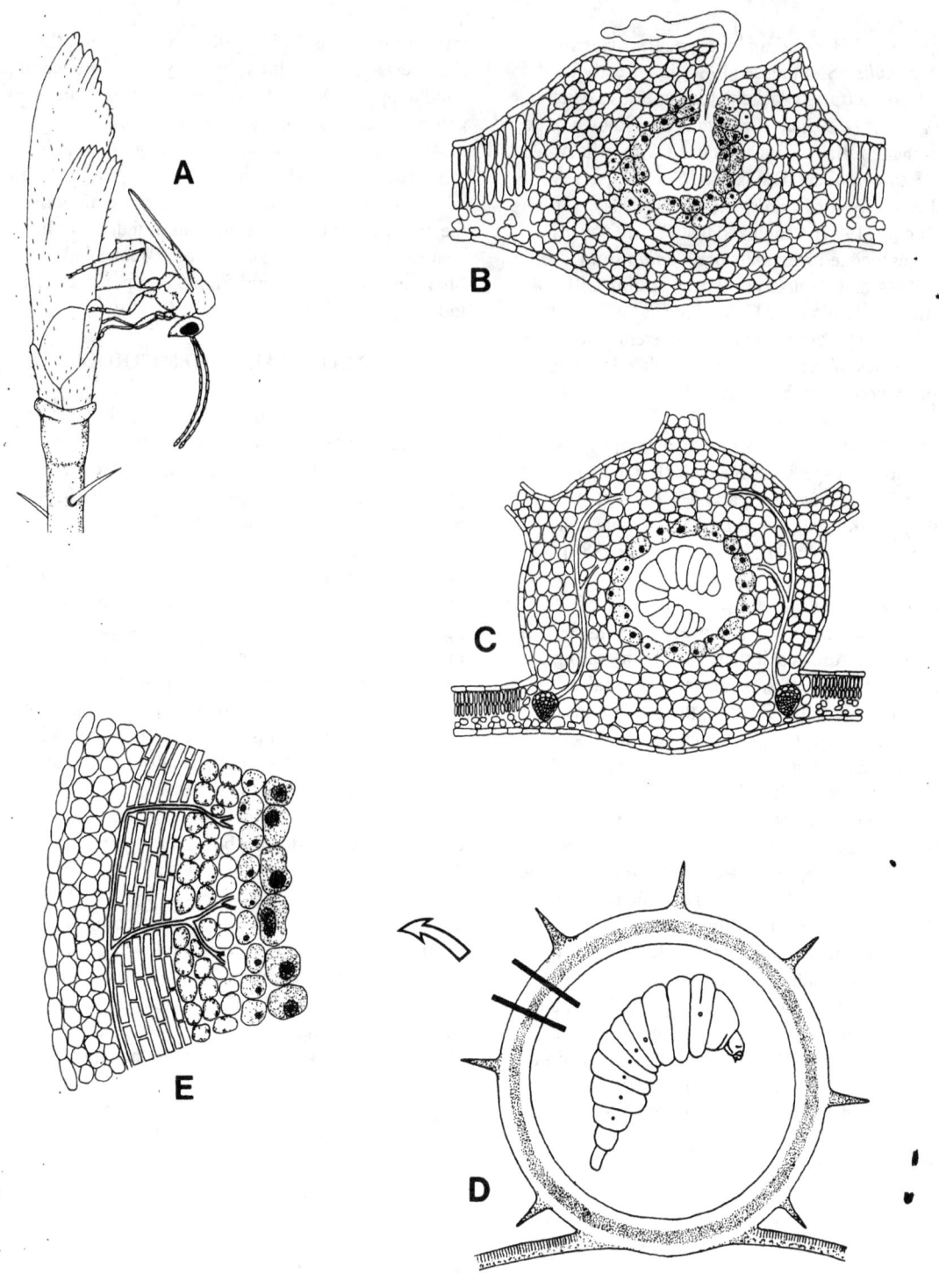

Figure 1.—*Schematic representation of developmental events of a typical* Diplolepis *gall. **A**. Female* Diplolepis *ovipositing in an immature unforced leaf bud of wild rose. **B**. Initiation phase showing first-instar larva surrounded by gall parenchyma and freshly formed nutritive cells. Note remains of egg shell protruding from larval entrance channel. **C**. Gall in growth phase. Note the larva is completely surrounded by nutritive cells lining the chamber and gall parenchyma. Strands of vascular bundles extend into the gall parenchyma. **D**. Mature gall showing amount of space inside the larval chamber. **E**. Section of mature gall in D showing nutritive cells lining larval chamber and their enlarged nucleoli, a layer of gall parenchyma with starch granules, a thick layer of brick-shaped sclerenchyma cells and vascular bundles extending towards the nutritive cells. A layer of gall cortex is found between the sclerenchyma and epidermis.*

both *D. ignota* and *D. nebulosa*, and a new species from galls of *D. gracilis*. Shorthouse (1980) incorrectly used the name *P. pirata* (Osten Sacken) for a species associated with galls of *D. polita*. *P. pirata* is now known to be the correct name for a species associated with galls of *D. nodulosa* (Beutenmuller) (Brooks and Shorthouse 1997a) and 11 other species of *Diplolepis* (Ritchie 1984). Of interest, Ritchie (1984) recognized that *P. californicus* Ashmead is associated with galls of *D. polita* from California, USA and *P. piceus* Fullaway is from galls of *D. polita* from Wyoming and Colorado, USA. The other species of *Periclistus* are associated with stem galls. Thus, four species of *Periclistus* are associated with the six galls of *Diplolepis* highlighted in this publication.

All species of *Periclistus* attack monothalamous galls, except the European species *P. brandtii* (Ratzburg) which attacks the polythalamous leaf gall of *D. rosae* (L.) (Stille 1984). In this gall, *P. brandtii* has no apparent negative effects on survival of inducers (Stille 1984).

Life cycles of all *Periclistus* are similar to those of *Diplolepis*; they are univoltine and overwinter inside host galls as prepupae. Adults exit their galls in the spring a few weeks after the inducers, mate, and the females oviposit in immature galls (fig. 2A); the proportion of females to males is equal. Eggs are laid on the inner surface of the chambers and the inducer is always killed by the ovipositing female (fig. 2B). Multiple ovipositions are common and galls such as *D. bicolor* may receive up to 25 eggs; each stem gall of *D. nodulosa* (Beutenmuller) receives an average of 78 eggs with some galls having as many as 174 (Brooks and Shorthouse 1997a). Eggs hatch 7-10 days after oviposition. *Diplolepis*-induced nutritive cells revert to gall parenchyma soon after the inducer is killed, but the gall continues to grow. Galls where the inducer is killed but no inquiline eggs are laid stop growing and die. Freshly hatched *Periclistus* larvae feed on gall parenchyma lining the chamber and induce the formation of additional cell proliferation such that galls at this stage are larger than inducer-inhabited galls of the same age. *Periclistus* larvae disperse about the chamber surface and additional proliferation of gall parenchyma results in each larva becoming encapsulated in its own chamber (figs. 2C and 2D), the *Periclistus*-equivalent of growth phase. As the larvae become encapsulated, *Periclistus*-induced nutritive cells near feeding sites are differentiated into nutritive cells. *Periclistus*-induced nutritive cells are smaller than *Diplolepis*-induced nutritive cells, but they are more numerous and densely packed (fig. 2E). *Periclistus* larvae do most of their feeding as galls enter the inquiline-equivalent of maturation phase. Sclerenchyma cells are differentiated in a broad band circumscribing the zone of *Periclistus* influence and around each chamber (fig. 2E). Similar to inducer-inhabited galls, *Periclistus* continue converting gall parenchyma into nutritive cells during gall maturation until the chambers are lined with sclerenchyma. More vascular bundles appear within tissues of *Periclistus*-modified galls than in inducer-inhabited galls. Some galls inhabited by *Periclistus* are much larger than inducer-inhabited galls, whereas galls of other species are smaller when inquilines are present. *Periclistus* commonly cause the mortality of a large proportion of inducers. In a 2 year study of the *D. nodulosa* gall, Brooks and Shorthouse (1997a) found from 55-65 percent of the inducer population was killed by *P. pirata*.

MATERIALS AND METHODS

Information on the biology of the six galls came from numerous field seasons and collections in Alberta and Saskatchewan, along with additional collections of all galls, except those of *D. ignota* and *D. gracilis*, from central Ontario in Canada. Galls of some species are from roses growing in the forested regions of central and northern Alberta and Saskatchewan, whereas galls of other species are from roses growing on the open prairies. To obtain adult inhabitants, large collections of mature galls were made in the fall and stored for 3 months at 3°C. Galls were either placed in jars at room temperature and adults removed, counted and curated daily as they exited, or prepupae were placed individually in gelatin capsules and the adults allowed to emerge, after the same cold treatment as the galls. Time in days for the emergence of adults from pupae stored in gelatin capsules was recorded for both *Diplolepis* and *Periclistus*.

To obtain data on seasonal changes in composition of component communities associated with galls of each species, large number of galls were collected monthly during the summer in a single season by walking haphazardly through large patches of roses and harvesting every gall observed. Galls were then dissected in the laboratory and contents identified. For the purposes of this study, gall contents were identified as being either inducers, inquilines or parasitoids; however, parasitoids associated with *Diplolepis* leaf galls are in the genera *Eurytoma*, *Torymus*, *Aprostocetus* and *Pteromalus*, along with the torymid *Glyphomerus stigma* (F.) and *Ormyrus rosae* (Ashmead), the only species of ormyrid associated with *Diplolepis* galls (Hanson 1992). Representatives of all gall inhabitants are deposited in the Canadian National Collection of Insects in Ottawa, Ontario.

Antomical studies of inducer- and inquiline-modified galls were undertaken by collecting galls in the field, dissecting them in the laboratory to identify inhabitants, then fixing and sectioning tissues using standard histological techniques (O'Brien and McCully 1981). Schematic drawings of gall anatomy were made after examining sections from at least 25 mature galls of each species.

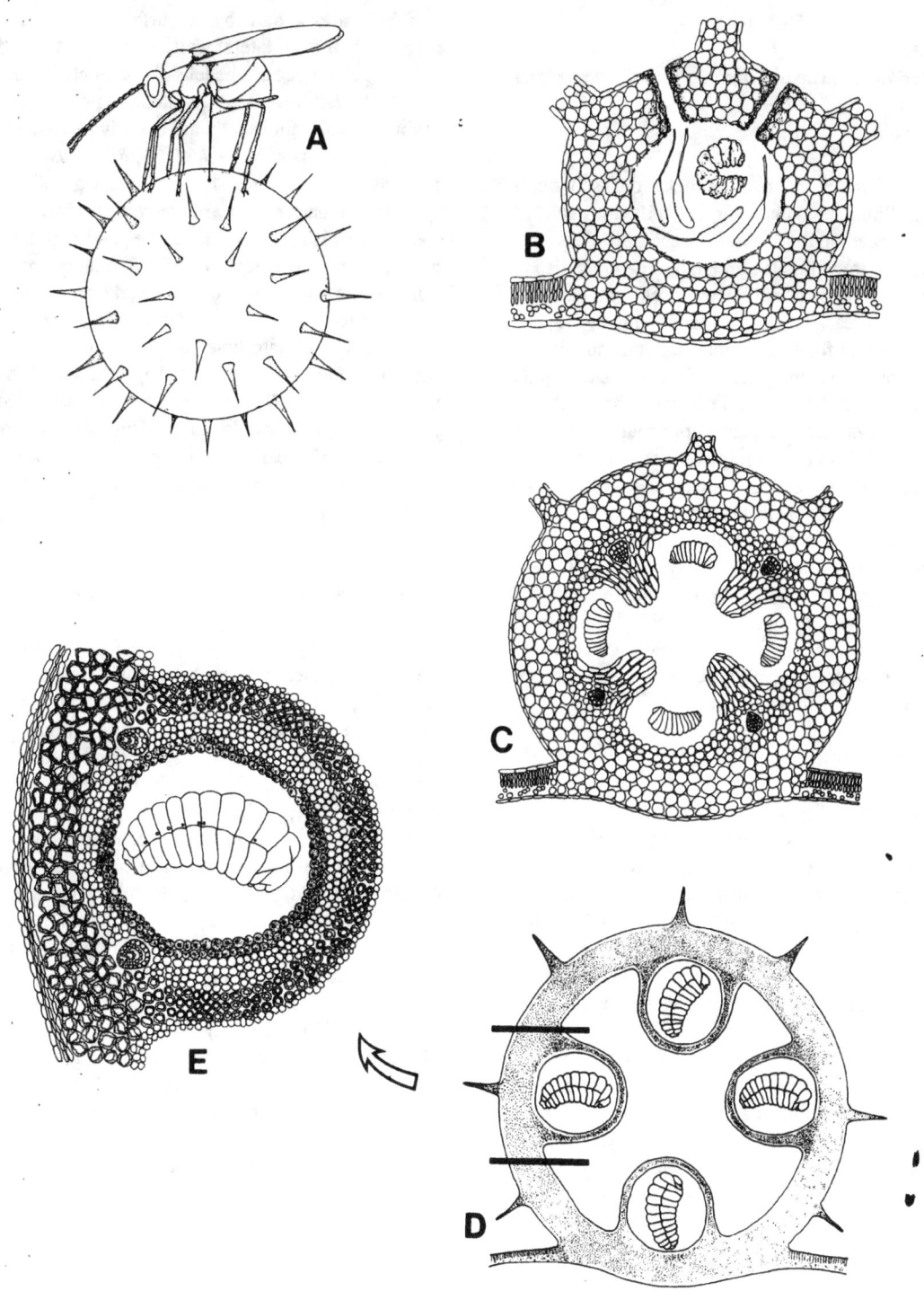

Figure 2.—*Schematic representation of developmental events of a typical* Diplolepis *gall attacked and modified by* Periclistus. ***A***. *Female* Periclistus *ovipositing in an immature* Diplolepis *gall*. ***B***. *Gall in early growth phase with eggs of* Periclistus *and dead* Diplolepis *larva. Note that the* Diplolepis-*induced nutritive cells have disappeared and the two oviposition channels*. ***C***. *Gall with* Periclistus *larvae in early stage of chamber formation. Note that each larva is partially surrounded with* Periclistus-*induced gall parenchyma and there has been an increase in vascular bundles*. ***D***. *Mature gall with each* Periclistus *larva enclosed in its own chamber*. ***E***. *Section of mature gall in D showing showing* Periclistus *chamber lined with inquiline-induced nutritive cells, gall parenchyma and sclerenchyma cells*.

RESULTS

Characteristics of the Six *Diplolepis* and Their Galls

Diplolepis polita

Galls of *D. polita* (fig. 3) are monothalamous, spherical, 3.5 to 5.0 millimeters in diameter, clothed with weak spines and formed on the adaxial surface of leaflets. Immature galls are yellowish-green, but mature galls are often bright red. They usually occur in clusters with several galls per leaflet. They are found only on *Rosa acicularis* Lindl. in forested regions of Alta. and Sask. growing in partly shaded habitats. Walls of mature galls are thin (about 1 millimeter) and easily crushed with fingers. The larvae feed over the entire chamber surface. Adults removed from cold storage take an average of 26.8 days to emerge (fig. 9) and are present in the field in early May with the first galls present by about May 20 (fig. 10). The sex ratio of 51 adults was 0.411. *Periclistus* take an average of 23.4 days to emerge after cold storage (fig. 11). *Periclistus* are found in about 71 percent of the galls by mid-season (table 1) and are the major source of inducer mortality. An average of 5.4 *Periclistus* are found per gall (table 2). Galls are mature and on the ground by mid-August.

Diplolepis bicolor

Galls of *D. bicolor* (fig. 4) are spherical, 7.0 to 11.0 millimeters in diameter, clothed with sharp, stiff spines, and are induced on the adaxial surface of leaflets. They often appear in tight clusters such that the host leaf is obscured. Most galls remain on the host plant throughout the winter. Immature galls are yellow-green, mature galls are green to bright red. The most common host is *R. woodsii* Lindl.; however, they are also found on *R. arkansana* Porter, both growing in open, grassland habitats. Walls of mature galls are 1.5 - 2.0 millimeters in thickness and cannot be crushed with fingers. Larvae feed over the entire chamber surface. Adults removed from cold storage take an average of 22.4 days to emerge (fig. 9) and are present in early May with the first galls present by about May 15 (fig. 10). The sex ratio of 98 adults was 0.857. *Periclistus* take an average of 25.6 days to emerge after cold storage (fig. 11). *Periclistus* are found in about 59 percent of the galls by mid-season (table 1) and are the major source of inducer mortality. An average of 9.3 *Periclistus* are found per gall (table 2). This is the fastest of the six galls to mature with most mature by mid-July.

Diplolepis rosaefolii

Mature galls (fig. 5) are lenticular, 2.0 to 2.5 millimeters thick and 3.0 to 5.5 millimeters in diameter and protrude from both adaxial and abaxial surfaces. Both surfaces are smooth and the lateral walls are thicker than adaxial and abaxial surfaces. The lens-shape of chambers restricts larval movements to lateral regions. Galls can be found singly but usually are densely packed and coalesced. Hosts are *R. acicularis*, *R. woodsii*, and *R. arkansana* growing from forested to dry grassland habitats. Immature galls are green, mature galls are green to reddish purple. Larvae lie parallel to the leaflet surface with most feeding restricted to lateral regions. Adults removed from cold storage take an average of 21.9 days to emerge (fig. 9) and are present in the field from late April to late June. Immature galls are found from late June to late July (fig. 10). The sex ratio of 609 adults was 0.582. *Periclistus* take an average of 36.5 days to emerge after cold storage (fig. 11). *Periclistus* are found in 63.2 percent of the galls by mid-season (table 1) and are a major source of inducer mortality. An average of 1.4 *Periclistus* are found per gall (table 2).

Diplolepis nebulosa

Mature galls (fig. 6) are spherical, 5.0-7.0 millimeters in diameter, spineless and on the abaxial surface. They occur singly or in rows. They are most common on short *R. woodsii* on dry exposed prairie. Immature galls are green to light brown with scaly, thick walls and a prominent depression distal to the point of attachment. Mature galls are pale brown and the depression is less prominent. Walls are thin and easily crushed between fingers. Larvae feed over the entire chamber surface. Adults removed from cold storage take an average of 26.3 days to emerge (fig. 9) and are present in the field from mid-May to mid-June. Immature galls are found from early July to mid-August. Both immature and maturing galls appear together suggesting a lengthy period of gall initiation (fig. 10). The sex ratio of 315 adults was 0.647. *Periclistus* take an average of 58.0 days to emerge after cold storage (fig. 11). *Periclistus* are found in 66.6 percent of the galls by mid-season (table 1) and are a major source of inducer mortality. An average of 3.3 *Periclistus* are found per gall (table 2).

Diplolepis ignota

Mature galls (fig. 7) are either spherical, irregularly globose, or kidney-shaped, spineless and on the abaxial surface. Sperical galls average 6.0 to 9.0 millimeters in diameter whereas those kidney-shaped may be up to 20.0 millimeters in length, making it the largest gall in the prairie region. They are soft, succulent and light green when immature, but become light brown and hard when mature and usually do not abscise in the fall. They are only found on *R. arkansana* in open habitats or

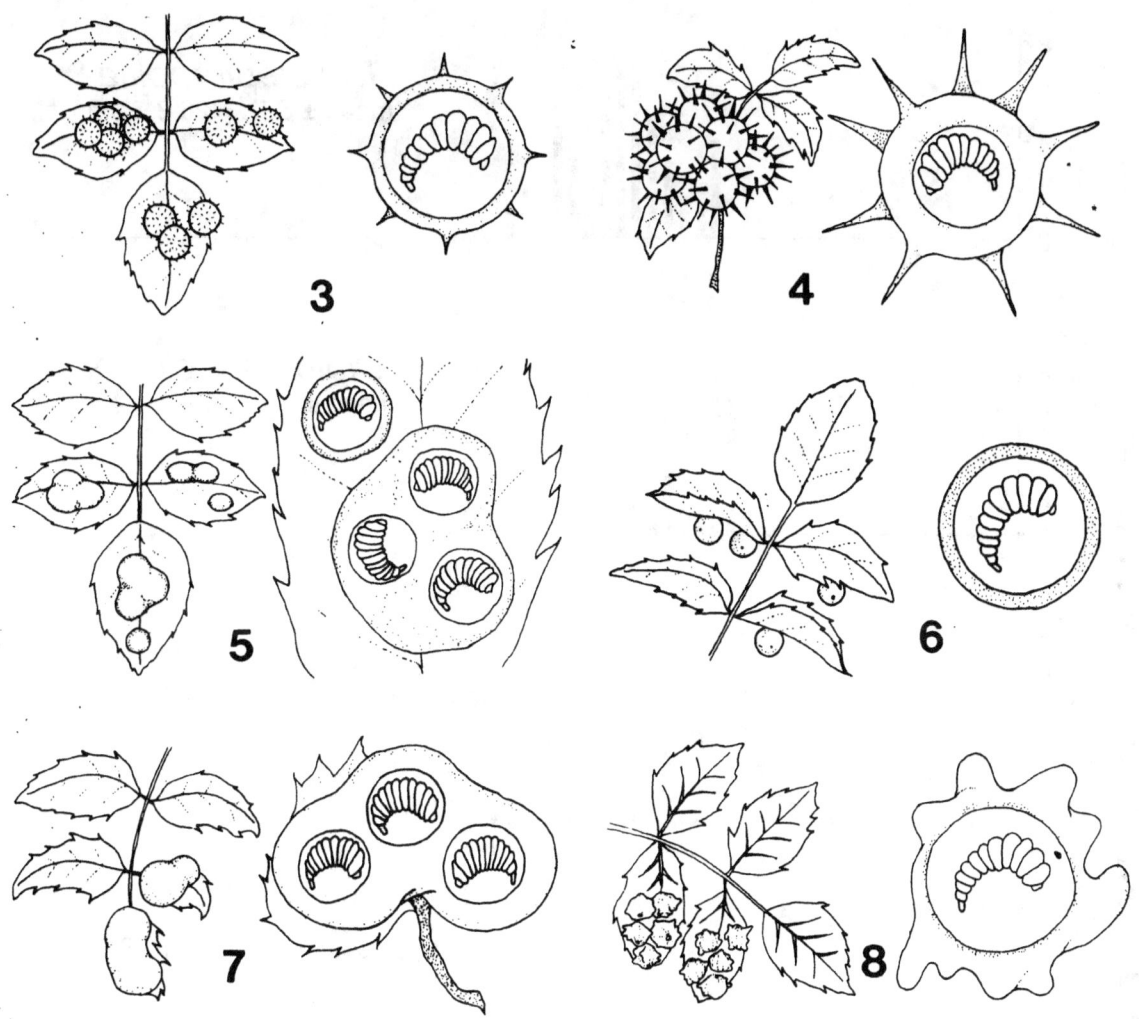

Figures 3-8.—*Schematic drawings of six* Diplolepis *galls with mature galls to the left and dissected galls to the right.* 3. *Galls of* D. polita. 4. *Galls of* D. bicolor. 5. *Galls of* D. rosaefolii. 6. *Galls of* D. nebulosa. 7. *Multichambered gall of* D. ignota. 8. *Galls of* D. gracilis.

Table 1.—*Percentage of galls from central Saskatchewan modified by* Periclistus *larvae*

Gall	Mid-season maximum	Mature galls
Diplolepis polita	71.0	12.0
Diplolepis bicolor	58.8	28.1
Diplolepis rosaefolii	63.2	14.7
Diplolepis nebulosa	66.6	13.1
Diplolepis ignota	11.4	5.8
Diplolepis gracilis	20.0	19.1

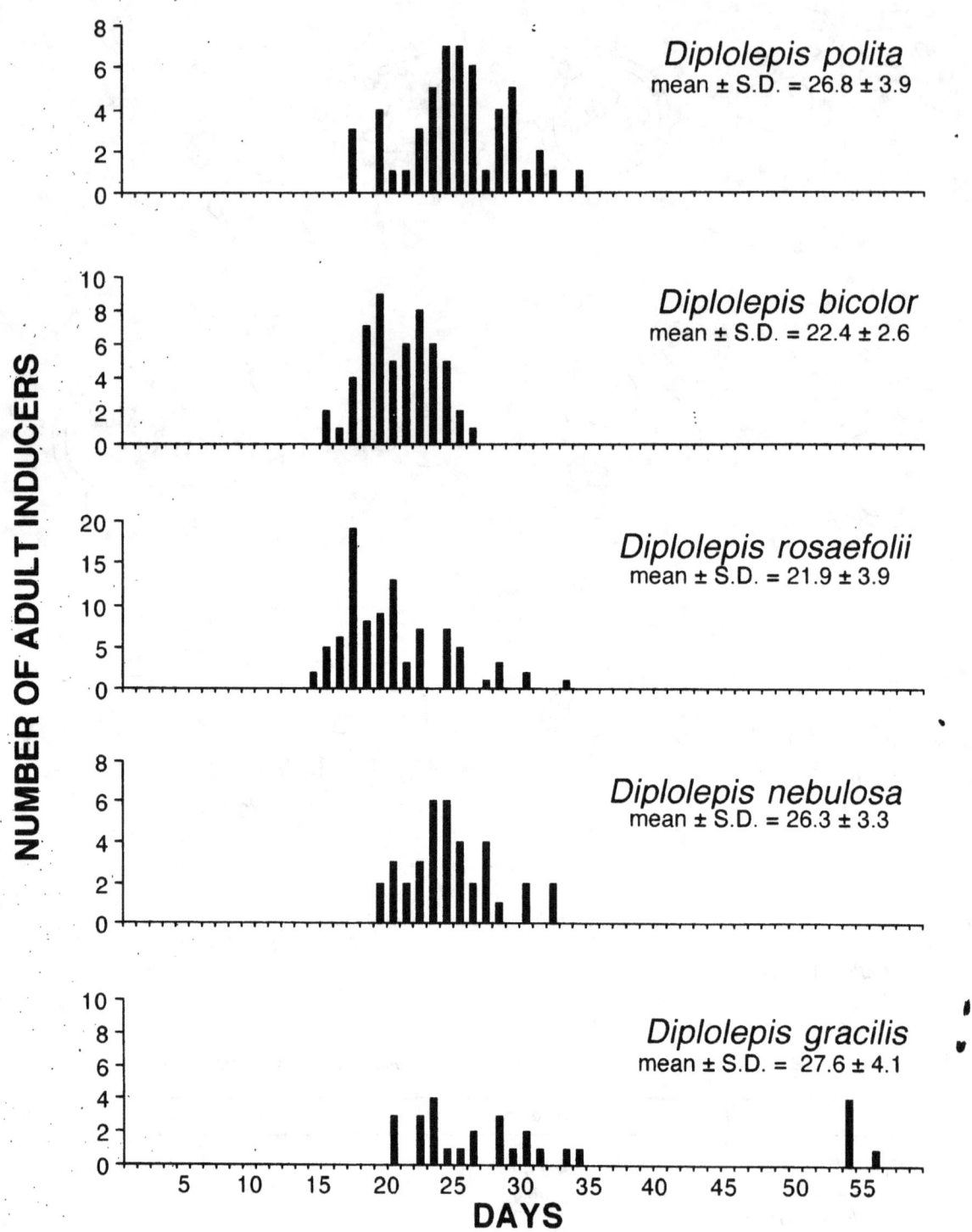

Figure 9.—*Time taken for laboratory emergence of* Diplolepis *females from larvae stored in gelatin capsules at 3°C for 3 months then transferred to 22°C. Day 1 is first day at 22°C.*

Table 2.—*Number of* Periclistus *larvae per modifed* Diplolepis *gall without parasitoid inhabitants. All galls from central Saskatchewan*

Species of inducer	Number of galls	Number of Periclistus larvae (± S.D.)
Diplolepis polita	350	5.4 (±4.4)
Diplolepis bicolor	53	9.3 (±4.3)
Diplolepis rosaefolii	47	1.4 (±0.6)
Diplolepis nebulosa	31	3.3 (±3.1)
Diplolepis ignota	31	2.9 (±1.4)
Diplolepis gracilis	78	1.8 (±1.0)

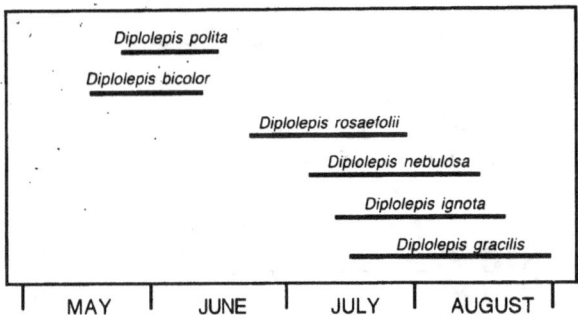

Figure 10.—*Seasonal occurrence of immature* Diplolepis *galls (less than 2.0 millimeters in diameter) in central Alberta and Saskatchewan.*

on the sides of coulees. Most spherical galls are single-chambered whereas all others are composed of several coalesced chambers. Larvae feed throughout the chamber surface. No data on time for emergence after cold storage were obtained for this species. Adults are found in the field from early to late May. Both immature and maturing galls appear together suggesting a lengthy period of gall initiation. Immature galls are found from mid-July to late August (fig. 10). The sex ratio of 577 adults was 0.662. *Periclistus* take an average of 51.3 days to emerge after cold storage (fig. 11). *Periclistus* are found in 11.4 percent of the galls by mid-season (table 1) and are not a major source of inducer mortality. An average of 2.9 *Periclistus* are found per gall (table 2).

Diplolepis gracilis

Mature galls (fig. 8) are ellipsoid, averaging 4.0 to 6.0 millimeters in diameter, have blunt protuberances extending from their lateral circumference, and form on the abaxial leaf surface. The protuberances are of various lengths and thickness, giving the gall the appearance of a large beet seed. They are found on *R. acicularis* growing in shaded, thick woods. Maturing galls are white, single-chambered and usually found in non-coalesced clusters. Walls of the gall are thin and soft to maturity and easily crushed between the fingers. Galls abscise with their host leaves in late fall. Adults removed from cold storage take an average of 27.6 days to emerge (fig. 9) and are present in the field from mid-June to mid-July. Immature galls are found from mid-July to early September (fig. 10) making this the latest maturing gall of the six. The sex ratio of 44 adults was 0.568. *Periclistus* take an average of 55.2 days to emerge after cold storage (fig. 11). *Periclistus* are found in 20.0 percent of the galls by mid-season (table 1) and are not a major source of inducer mortality. An average of 1.8 *Periclistus* are found per gall (table 2).

Anatomy of Inducer and Inquiline-Inhabited Galls In Mid-Maturation Phase

Diplolepis polita

Mature, inducer-inhabited galls consist of a single layer of nutritrive cells (NC) lining the larva chamber (fig. 12A), followed by a layer of gall parenchyma (GP), a thick layer of sclerenchyma cells (SC) (fig. 13A) with cytoplasm occluded, a thin layer of cortex parenchyma cells (CP) and a distinct epidermis (EP). Sparse vascular bundles (VB) are found within the GP with strands crossing the SC and extending to the NC.

Inquiline-modified galls are 2-3 times the diameter of inducer-inhabited galls. Inquiline chambers protrude into the former inducer chamber (fig. 14A). Walls are much thicker than those of inducer-inhabited galls with a thick layer of GP and CP beyond the chambers. A layer of 1-4 inquiline-induced NC line the interiors of each chamber (fig. 14A). The layer of SC is thicker than in inducer galls and circumscribes the gall; as well, inquiline-induced sclerenchyma circumscribe each *Periclistus* chamber (fig. 15A). VB are much denser in the GP than in inducer galls. There is a distinct EP.

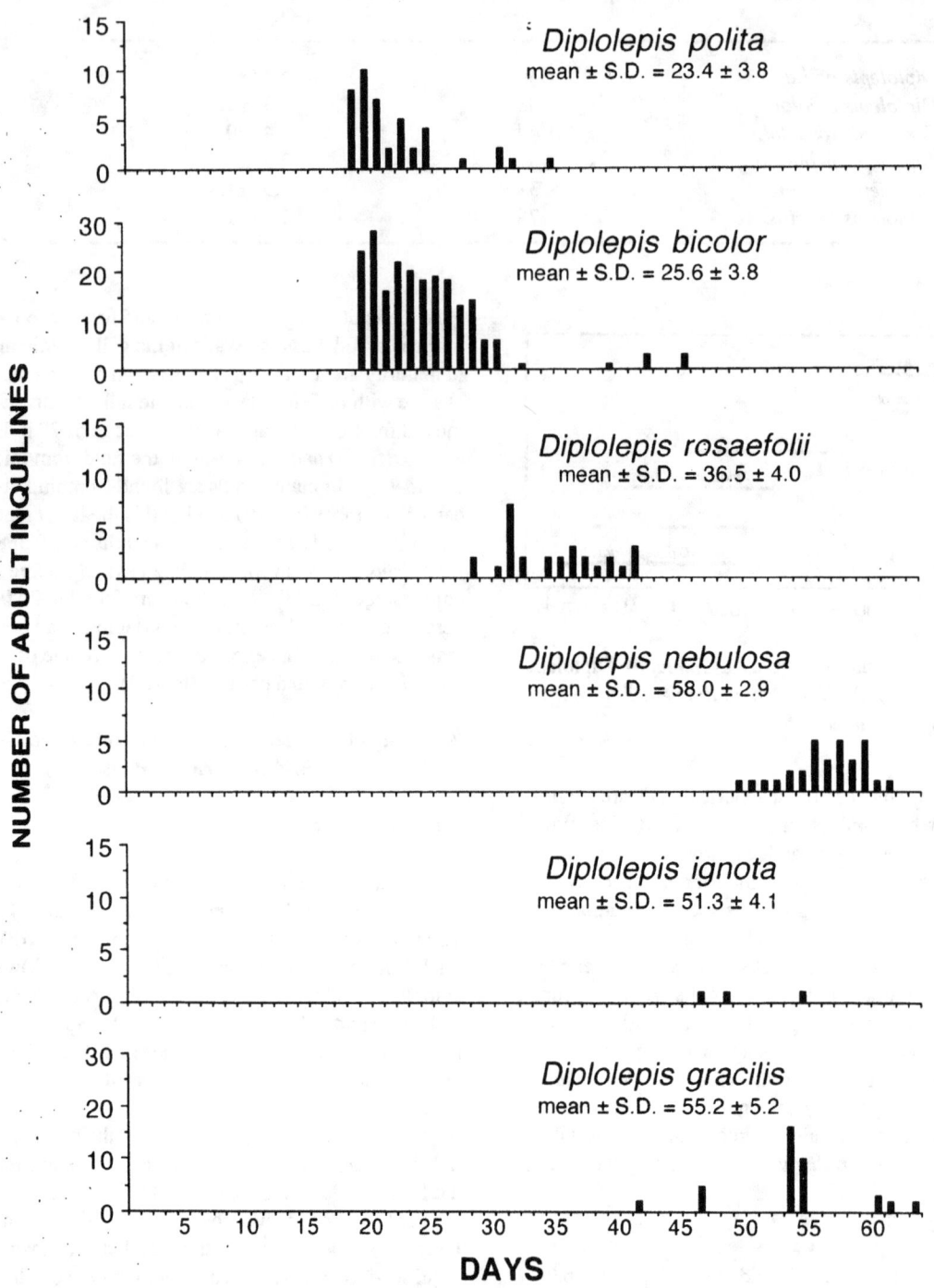

Figure 11.—*Time taken for laboratory emergence of* Periclistus *females removed from* Diplolepis *galls and stored in gelatin capsules. All larvae were stored at 3°C for 3 months then transferred to 22°C. Day 1 is first day at 22°C.*

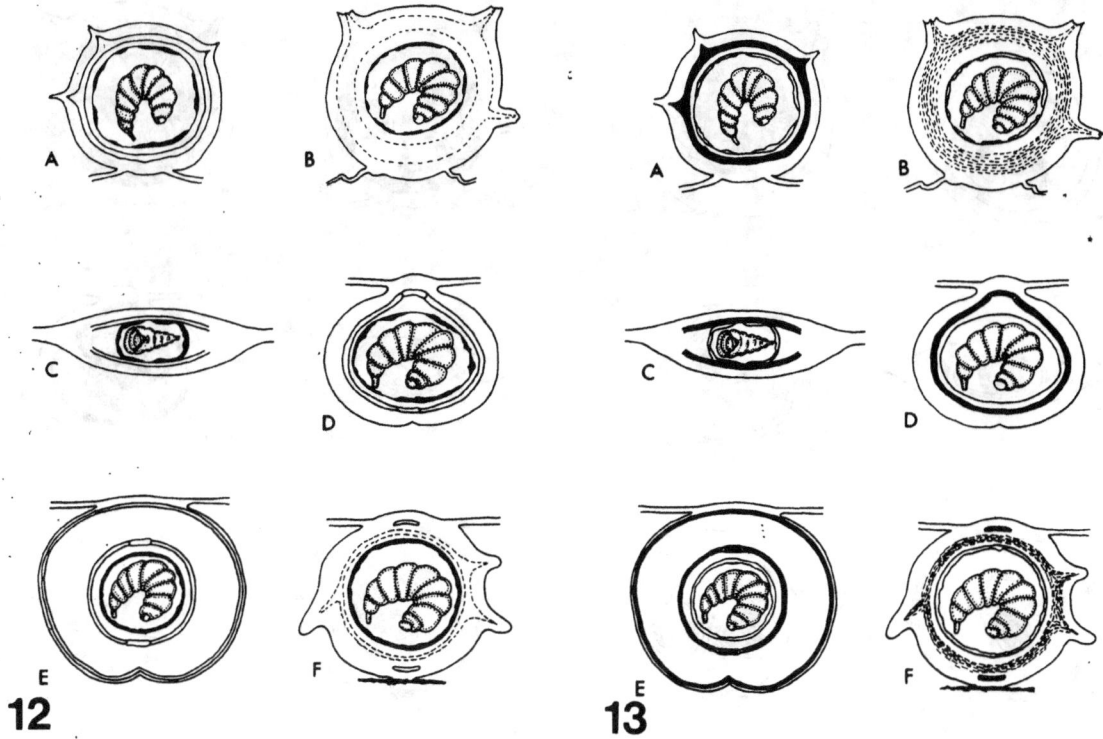

Figures 12.—Schematic drawings of sections of mature Diplolepis-inhabited galls. 12. Arrangement of nutritive cells (in black). **A.** Gall of D. polita. **B.** Gall of D. bicolor. **C.** Gall of D. rosaefolii. **D.** Gall of D. nebulosa. **E.** Gall of D. ignota. **F.** Gall of D. gracilis.

Figure 13.—Arrangement of sclerenchyma layers (in black). **A.** Gall of D. polita. **B.** Gall of D. bicolor. **C.** Gall of D. rosaefolii. **D.** Gall of D. nebulosa. **E.** Gall of D. ignota. **F.** Gall of D. gracilis.

Diplolepis bicolor

Walls of inducer-inhabited galls are twice the thickness (fig. 12B) of those of *D. polita* galls consisting of patches of single NC (fig. 12B), a thick layer of GP, a thick layer of barely perceptable SC (fig. 13B), the walls of which have a thin layer of lignin (the lumen of these cells are not occluded), a thick layer of CP and a distinct EP. VB are much denser than in *D. polita* galls.

Inquiline-modified galls are 2-3 times the diameter of inducer galls and the chambers fill the former inducer chamber (fig. 14B). A layer of 1-4 inquiline-induced NC line the chambers (fig. 14B) followed by a thick layer of GP. The layer of CP beyond the inquiline chambers is thicker than in inducer galls. Each of the inducer chambers are surrounded by a layer of SC (fig. 15B) with the lumen of all cells occluded. The circumscribing layer of SC in tissues beyond the inquiline chambers is thicker than in inducer galls and the cell walls have more lignin. VB are denser than in inducer galls and there is a distinct EP.

Diplolepis rosaefolii

The lenticular shape of this gall results in walls on adaxial and abaxial surfaces being thinner than regions lateral to the inducer's chamber. This is the only gall of the six with a lenticular chamber (figs. 12C and 13C) and as such the inducer is restricted to circular movements parallel to the leaflet surface and most NC and larval feeding occur in lateral regions (circular in paradermal section) (fig. 12C). NC in lateral regions (circumscribing the chamber parallel to leaflet surface) consist of thick patches adjoined by thick patches of GP. Two disc-shaped patches of SC, each about 10-15 cells thick, occur in each gall, one below and the other above the chamber with no SC in lateral, circumscribing regions (fig. 13C). Lumen of SC are partially occluded. Periclinally, the SC is 10-15 cells thick. A thin adaxial and abaxial CP lies outside the SC and there is a distinct EP. In coalesced galls a thin SC may form between adjoining chambers. VB are only present in lateral regions suggesting that nutriment comes directly from nearby VB's of the host leaflet.

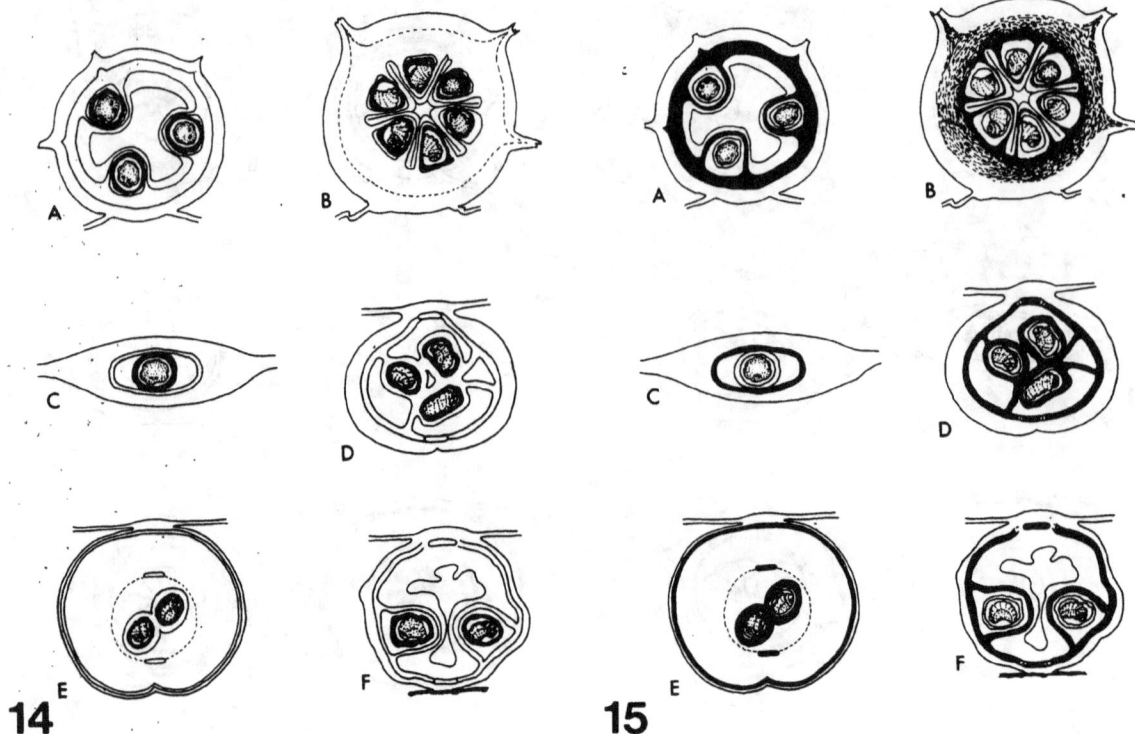

Figures 14.—*Schematic drawings of sections of mature* Periclistus-*inhabited galls. 14. Arrangement of inquiline-induced nutritive cells (in black). **A**. Gall of* D. polita. ***B***. *Gall of* D. bicolor. ***C***. *Gall of* D. rosaefolii. ***D***. *Gall of* D. nebulosa. ***E***. *Gall of* D. ignota. ***F***. *Gall of* D. gracilis.

Figure 15.—*Arrangement of inquiline-induced sclerenchyma layers (in black)*. ***A***. *Gall of* D. polita. ***B***. *Gall of* D. bicolor. ***C***. *Gall of* D. rosaefolii. ***D***. *Gall of* D. nebulosa. ***E***. *Gall of* D. ignota. ***F***. *Gall of* D. gracilis.

Inquiline-attacked galls with one *Periclistus* larva are about half the size of inducer galls; however, when 3-4 larvae are present, gall diameter is greater than that of inducer galls and they are thicker. Inquiline chambers are spherical and a uniformly thick NC lines the chamber surface (fig. 14C). GP occurs beyond the NC and is thickest in lateral regions. Walls between inquiline larvae are thin, but never completely consumed. SC surrounds the inquiline chambers (fig. 15C), regardless of the number of chambers present. The lumen of SC cells are more occluded than those in inducer galls. VB appear in abaxial GP and the EP is normal.

Diplolepis nebulosa

This gall is initiated within spongy mesophyll and early growth is between the epidermal layers. The abaxial EP is ruptured by abaxial growth of GP such that the surface of the spherical gall is no longer covered by an EP. While still immature, discs of sclerenchyma, about 20 cells thick, form in the GP above and below the larval chamber (fig. 12D). Lignification of these cells occurs soon after they are initiated. CP external to the distal disc grows slower than in other regions resulting in a depression in the gall surface. The proximal and distal discs do not expand as the gall matures but remain separated by GP.

Walls of mature galls are thin and the chamber surface is lined with a layer of 1-2 NC (fig.12D) and a layer of GP. A circumscribing layer of SC (fig. 13D) with about one-quarter of their lumen occluded and filled with unidentified phenolic-like substances is found between the two discs. VB are found within the GP.

Periclistus-inhabited galls are smaller than *D. nebulosa*-inhabited galls, regardless of the number of inquilines present. In most cases, the former *Diplolepis* chamber is filled with *Periclistus* chambers (fig. 14D). Tightly packed NC, varying in size and shape, encircle walls of the *Periclistus* chambers (fig. 14D). Thickness of the adjacent GP is variable. *Periclistus*-induced SC appears adjacent to the GP and surrounds each chamber (fig. 15D). The *Diplolepis*-induced sclerenchyma discs remain as does the circumscribing layer and the SC surrounding each chamber (fig. 15D) joins the circumscribing SC.

Diplolepis ignota

Eggs are laid on the abaxial leaflet surface near the midrib with egg clusters resulting in large, coalesced galls. Initial gall growth is within the mesophyll but the abaxial EP is ruptured as the gall becomes spherical and expands abaxially leaving CP on the gall surface. While in early growth phase, the NC is thickest in lateral regions and larval movement is restricted; however, as the gall enlarges, chamber volume increases and the inducer feeds over its entire surface and the NC becomes evenly thickened (fig. 12E). Two disc-shaped patches of SC (fig. 12E) appear above and below the larval chamber early in gall growth, but do not circumscribe the larval chamber. VB are abundant throughout the GP.

Once the gall is about 4.0 millimeters in diameter, a band of GP between the two discs become lignified, forming the first circumscribing SC (fig. 13E). CP near the gall exterior becomes stretched and ruptures as the gall matures. Cells flake off at high humidity and are responsible for the 'powdery' appearance. As the gall matures, a second SC (fig. 13E) develops next to the layer of ruptured cells. Cells of this layer are filled with unidentified phenolic compounds. Intercellular spaces are prominent in the GP giving the gall a corky appearance.

Periclistus oviposit in galls 5.0 millimeters or less in diameter and only those occurring singularly. GP proliferates once larvae begin feeding and the SC discs induced by *D. ignota* are unaffected. The inducer's chamber becomes filled with inquiline-induced GP which are the same size as those induced by *D. ignota*. Each *Periclistus* becomes surrounded by GP and VB are common. A layer of 1 or 2 NC appear along the surface of all chambers once the inquiline larvae are enclosed (fig. 14E). SC forms around each chamber, but a circumscribing layer does not form between the *D. ignota*-induced discs (fig. 15E). A second SC appears in the same external region as normal galls (fig. 15E).

Diplolepis gracilis

Eggs of *D. gracilis* are laid in rows on the abaxial leaflet surface and the resulting galls are equally spaced. Immature galls are easily located because a patch of red cells appears adaxial to each larva. Galls do not develop at uniform rates for freshly initiated galls are found alongside older galls.

Immature lens-shaped galls originate in spongy mesophyll and the first NC appear lateral to the chamber with the larva positioned parallel to the leaflet surface. The gall increases in thickness as the GP proliferates and the EP is stretched. Two sclerenchyma discs (fig. 12F) appear adaxial and abaxial to the larval chamber and are forced apart as the gall grows. The growing gall becomes ellipsoid and the abaxial EP separates from the leaflet with the outside layer of CP becoming the exterior. The ruptured circular patch of EP remains attached to the gall (fig. 12F). NC (fig. 12F) lines the chamber to a thickness averaging 4 cells.

As the gall matures, a layer of 6-8 GP cells near center of the gall wall lignify forming a circumscribing SC (fig. 13F); however, the walls of these cells are much thinner than those of all galls other than *D. bicolor*. Also, the circumscribing layer forms inside the sclerenchyma discs (fig. 13F). VB are common throughout the SC and GP. Intercellular spaces are prominent in GP making walls of the mature galls soft.

Periclistus lay in galls 4.0 millimeter or less in diameter. *Periclistus*-induced GP occupy most of the inducer's chamber (fig. 14F). In contrast to the previous *Periclistus*-inhabited galls, the border between *D. gracilis* and *Periclistus*-induced cells are not easily distinguished. *Periclistus*-induced NC line the chambers (fig. 14F). SC appear around each chamber (fig. 15F). A circumscribing layer of SC is differentiated near the outside of the gall and occurs between the *Diplolepis*-induced sclerenchyma discs (fig. 15F).

Role of Inquilines in Community Composition

Diplolepis polita

Nearly all of the first galls found in late May contained an inducer larva; however, within 2 weeks, the percentage of galls containing a live inducer was quickly reduced (fig. 16A). *Periclistus* emerge throughout the period immature galls were present and by the first week in June, about 70 percent contained inquiline eggs (table 1). More *D. polita* galls are attacked by inquilines than any of the other *Diplolepis* leaf galls (table 1) and the average number of inquiline larvae per gall is also high (table 2). Parasitoids begin ovipositing in both inducer-inhabited and inquiline-inhabited galls in June and by early July, about 70 percent of all galls contained at least one parasitoid (fig. 16A). Parasitoids feed on larvae of both inducers and inquilines. The most common parasitoids of the inducers are *Eurytoma* sp., *Glyphomerus stigma*, *Torymus* sp., *Eupelmus vesicularis* (Retzius), *Ormyrus rosae* and *Aprostocetus* sp. About 12 percent of mature galls contain a *G. stigma* and 5 percent a *Torymus* sp. The most common parasitoid of

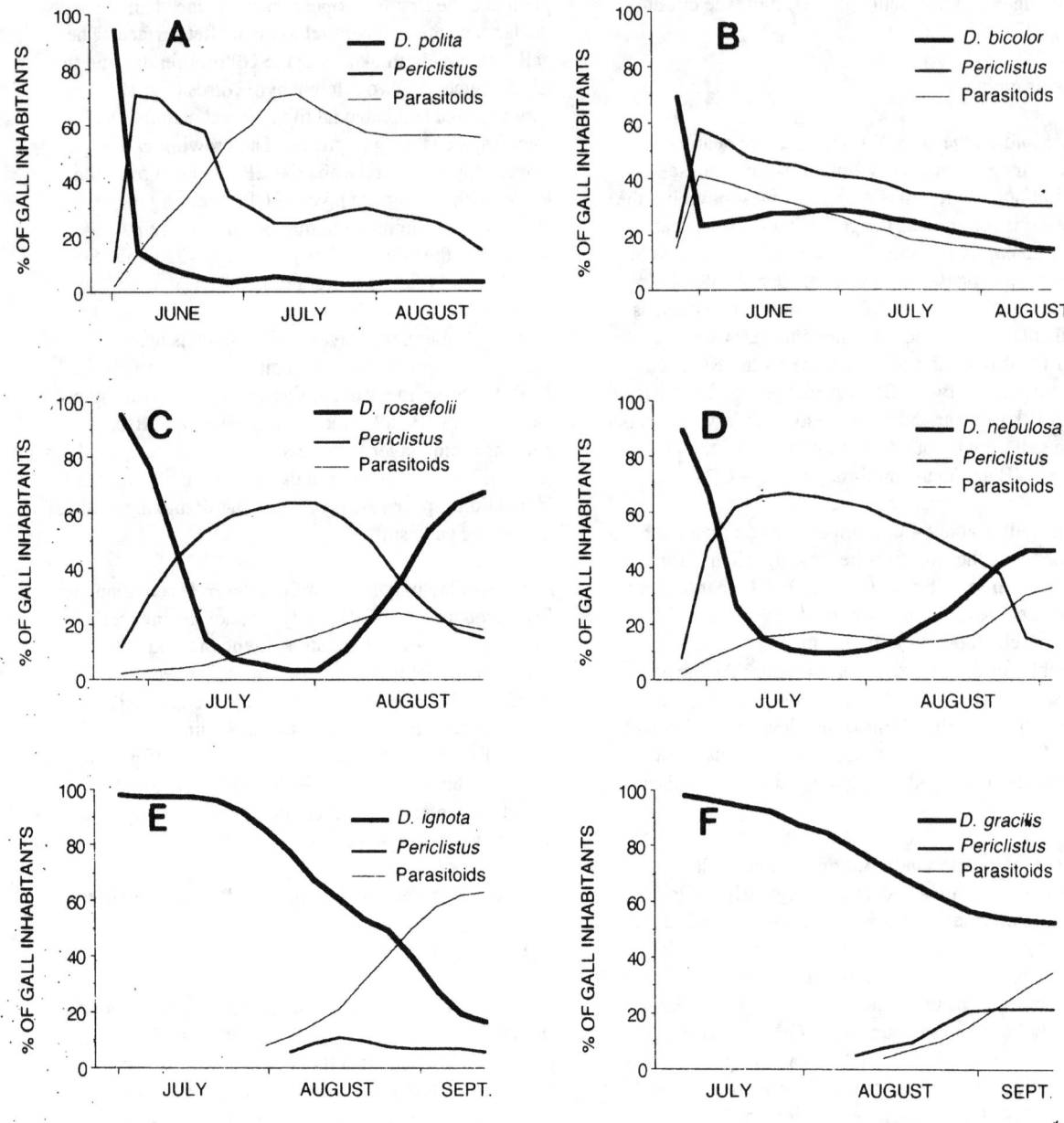

Figure 16.—*Seasonal change in the percentage of gall inhabitants.* **A.** Galls of D. polita *collected at George Lake, Alberta.* **B.** *Galls of* D. bicolor *collected 40 km east of Saskatoon, Saskatchewan.* **C.** *Galls of* D. rosaefolii *collected 7 km west of Langham, Saskatchewan.* **D.** *Galls of* D. nebulosa *collected 7 km west of Langham, Saskatchewan.* **E.** *Galls of* D. ignota *collected 40 km east of Saskatoon, Saskatchewan.* **F.** *Galls of* D. gracilis *collected 24 km southwest of Saskatoon, Saskatchewan.*

Table 3.—Incidence of inhabitants, expressed in percent, in typical component communities associated with large collections of mature galls. Fall emergents are not included. Inhabitants reported here are the assemblages that will overwinter and reestablish the community the following year.

Gall	Inducers	Periclistus	Parasitoids
Diplolepis polita	4.1	26.4	69.5
Diplolepis bicolor	8.1	66.4	25.5
Diplolepis rosaefolii	65.9	18.2	15.9
Diplolepis nebulosa	22.1	18.4	59.5
Diplolepis ignota	4.9	5.6	89.5
Diplolepis gracilis	27.2	9.4	63.4

Periclistus is Pteromalus sp. with each inquiline being host to one Pteromalus. The percentage of parasitoids within the final assemblage of inhabitants in mature galls is high (table 3) because of the number of inquiline larvae attacked by Pteromalus. Some parasitoids, such as Eurytoma sp., Glyphomerus stigma and Torymus sp. are partially phytophagous and eat gall tissues after consuming an inducer or chew into inquiline chambers to reach the inhabitants. Thus the number of galls containing inquilines decreases as the season advances (fig. 16A). Some parasitoids also consume other parasitoids. The slight increase in galls containing inducers or inquilines in mid-July is because of the appearance of later forming galls on sucker shoots. Some parasitoids exit galls in the year they were initiated; however, only inhabitants of mature galls that overwinter and reestablish the community the following year are considered here. D. polita comprise less than 5 percent of the community by the end of the season (table 3). About five times as many Periclistus are present in the final assemblage as D. polita, although their population is much reduced by parasitoids earlier in the season (fig. 16A). Pteromalus comprised the largest population in the community (about 41 percent of all inhabitants) mainly due to mid-season abundance of Periclistus larvae.

Diplolepis bicolor

About 20 percent of the first galls observed in the spring contain eggs of Periclistus; however, Periclistus quickly find the galls and the percentage of galls with an inducer larva drops (fig. 16B). By 3 weeks after the first galls were observed, nearly 60 percent contained Periclistus eggs (table 1). More Periclistus are found per gall than in galls of D. polita (table 2). Few Periclistus eggs were found in galls past mid-June as the thick walls of maturing galls likely deter oviposition. As with the gall of D. polita, the most common parasitoids of the inducers are Eurytoma sp., G. stigma, Torymus sp., Eupelmus vesicularis and Ormyrus rosae. By mid-June, 34 percent of the galls contain Eurytoma with most being found in galls with Periclistus. Eurytoma contribute to the decline of both inducers and inquilines (fig. 16B); however, in contrast to galls of D. polita, it is common for several Periclistus per gall to escape Eurytoma attack. Thus, some galls mature with both Eurytoma and Periclistus resulting in fewer galls of D. bicolor being depleted of all inquilines (fig. 16B) than occurs in galls of D. polita. Sometimes Eurytoma starve after consuming all inquiline larvae before they mature resulting in an empty gall; this is one reason for the decline in galls with parasitoids (fig. 16B). Periclistus are the main hosts of Pteromalus; however, rarely are all larvae per gall attacked and this parasitoid is responsible for only a slight reduction in the number of Periclistus-inhabited galls. G. stigma consume either the inducer or all inhabitants in inquiline-inhabited galls and contributes to the decline in galls with either inhabitant (fig. 16B). Some parasitoids exit galls in late season resulting in empty galls. Data on empty galls are included in figure 16B and are partly responsible for the decrease in galls containing parasitoids. D. bicolor comprised about 8 percent of the community by the seasons end, while the Periclistus population is largest at about 66 percent. Periclistus comprise a higher proportion of the final D. bicolor gall community than they do in the other galls (table 3).

Diplolepis rosaefolii

Galls of D. rosaefolii are initiated over a longer period than those of D. polita and D. bicolor and immature galls are present for well over a month. In eastern Canada, galls of this species are present until late September indicating an even lengthier period of initiation. However, the laboratory stored adults of Saskatchewan galls emerge at the same time as D. bicolor (fig. 9). Galls are attacked by both Periclistus and parasitoids soon after initiation such that by mid-July only 20 percent of the galls contain an inducer (fig. 16C). By the end of July only 3 percent contained an inducer, while 63 percent contained Periclistus. However, in samples collected in mid-August, 32 percent of the galls had an inducer and by early September, the number had risen to 68 percent, with only 14.7 percent with an inquiline. The steady appearance of new galls after most of the *Periclistus* and parasitoids had attacked is likely responsible for the increase in galls containing an inducer later in the season (fig. 16C). Sixty-six percent of inhabitants overwintering in this gall are inducers, the largest percentage of inducers in any of the six galls (table 3). Parasitoids also contributed to the decline in galls containing *Periclistus*. The most common parasitoids of the inducers are *Eurytoma* sp., *Glyphomerus stigma*, *Torymus* sp. and *E. vesicularis*. *Eurytoma* was the most abundant parasitoid comprising 3 percent of the final community. Perhaps the low number of *Periclistus* per gall (table 2) and the relatively small size of inducers makes galls of *D. rosaefolii* less attractive to parasitoids than the other galls (table 3). *Pteromalus* also attacked *Periclistus* larvae; however, they were far less numerous in this community compared to those of *D. polita* and *D. bicolor*.

Diplolepis nebulosa

Immature galls of *D. nebulosa* are found from late June to early August, about 1 month later than the above species (fig. 10). Of interest, adult inducers were found in the spring inside mature galls on the ground from the previous year, suggesting either that adults exit their galls at the same time as the previous species (fig. 10), oviposit, and the initiation of their galls is delayed, or that adults remain in their galls for some time. The later emergence of *Periclistus* (fig. 11) confirms that ovipositing inquilines are synchronized with later appearing host galls. As with galls of *D. rosaefolii*, most of the first galls to appear are attacked by *Periclistus* and parasitoids, but the continuous appearance of new galls after the inquilines and parasitoids have oviposited, results in an increase in galls with inducers by early September (fig. 16D). By mid-season, 66.6 percent contained inquilines (table 1); however, by the last collection, only 13.1 percent contained an inquiline and 48.1 percent contained an inducer (fig. 16D). The most common parasitoids of the inducers are *Eurytoma* sp., *Glyphomerus stigma*, *Torymus* sp. and *Aprostocetus* sp., with *Aprostocetus* comprising 48 percent of all inhabitants by seasons end. Thirteen and seven-tenths percent of all mature galls contained an *Aprostocetus*. Unlike other parasitoids, *Aprostocetus* deposits several eggs in the tissues of inducer larvae and feed communally. It is common to find 15-20 larvae per gall. Although almost half the mature galls contained an inducer, they comprised only 22.1 percent of the inhabitants (table 3), due to the numbers of *Aprostocetus*. Although the number of inquilines per gall was average for the six galls (table 2), few were attacked by *Pteromalus*; most inquilines were consumed by *Eurytoma*.

Diplolepis ignota

As with *D. nebulosa*, galls of *D. ignota* appear late in the season (fig.10) and although all inducers stored in capsules died after their cold storage, they likely would have emerged at the same time as other species. Kinsey (1920) found that adults of *D. ignota* in northeastern U.S.A. exited galls from May to August, but most appeared near the end of May. The later appearance of *Periclistus* adults (fig. 11) illustrates synchronization with galls appearing later in the season than those of *D. polita* and *D. bicolor*. Fewer galls of *D. ignota* are inhabited by *Periclistus* than those of the previous species (table 1). All galls in the first two collections contained inducers; eggs of *Periclistus* were not found until early August. Only 11.4 percent of the galls in mid-August contained *Periclistus* followed by 5.8 percent (table 1) of mature galls by early September revealing that *Periclistus* is not a major cause of inducer mortality in this gall. As with the gall of *D. nebulosa*, the most important parasitoids of the inducer are *Eurytoma* sp., *Glyphomerus stigma*, *Torymus* sp., and *Aprostocetus* sp., with *Aprostocetus* comprising 79 percent of all inhabitants and found in 33.6 percent of all galls by seasons end. *Eurytoma* comprised 7 percent of all inhabitants and were found in 19.7 percent of the mature galls. *Torymus* were found in 6.3 percent of the mature galls. The low percentage of inducers in the final community is caused by the large numbers of *Aprostocetus* and the low percentage of *Periclistus* is due to fewer galls being attacked (table 3).

Diplolepis gracilis

Galls of *D. gracilis* were the last of the six species to be initiated, with immature galls common from mid-July to early September (fig. 10). Twelve adults were collected by sweeping in central Saskatchewan in early July confirming that it is last of the species to oviposit. However, adults emerging after cold storage were the only species to show a bimodal emergence pattern, with most emerging at the same time as the other species (fig. 9). As with the other inducers above, there is likely a delay in gall initiation which is also indicated by the later emergence of *Periclistus* (fig. 11). Few galls are attacked by *Periclistus* either by mid-season or gall maturity (table 1) and the mean number of inquilines per gall is low (table 2). Only 6.8 percent of galls in mid-August contained *Periclistus* and by late-September, only 19.1 percent contained *Periclistus* (fig. 16F). Most parasitoids associated with this gall attack maturing larvae of inducers. The most common species were *Eurytoma*, *Aprostocetus* sp., and *Orthopelma* sp. *Eurytoma* were found in 10.9 percent of the mature galls, but comprised only 0.8 percent of final community. *Orthopelma* sp. comprised 1.2 percent of the final community. As with the previous two galls, *Aprostocetus* were found in 13.9 percent of the mature galls but comprised 61.4 percent of the final community. Far fewer inducers of this gall are attacked by *Perielistus* and parasitoids (fig. 16F); however, inducers comprised only 27.2 percent of the final community (table 3) because of the large number of *Aprostocetus*. Only 9.4 percent of the community was comprised of *Periclistus* (table 3).

DISCUSSION

The findings reported in this paper well illustrate that the relationships gall-inducing cynipids have with their host plants are highly specialized and complex. Gall inducers do not move from place to place seeking food as do other phytophagous insects (Strong *et al.* 1984), but remain stationary and create highly nutritious food at their feeding sites. Although it was relatively simple to compare the elaborate structures of the six galls examined here, we are still no closer to answering two perplexing questions posed by all students of galls. First, what is the mechanism by which these gallers gain control and then manipulate the growth patterns of attacked plant tissues, and second, by what mechanism can a group of distinct species of gall wasps cause such species-specific, anatomically distinct plant structures when the eggs of each are laid on leaflets at the same stage of development? As if questions of this type are not sufficiently challenging, the occurrence of inquilines increases the complications we associate with galls to new heights.

Inquilines such as *Periclistus* are reliant on *Diplolepis* for their livelihood and their life histories and feeding strategies are equally as complex. Indeed, the life histories of inquilines and inducers are intricate and interwoven. Adult *Periclistus* must time their appearance to the presence of immature galls suitable for oviposition and then locate particular galls prior to killing the inducer larva with their ovipositors. Once the gall inducer is killed, original nutritive cells lose their characteristics, but somehow the presence of *Periclistus* eggs causes gall parenchyma to continue to grow and proliferate. *Periclistus* then become galler-like when their larvae begin to feed causing further proliferation where each larva becomes enclosed in its own chamber and induces the formation of its own nutritive cells.

While previous studies have shown that *Periclistus* are important mortality factors for inducers and that inquiline larvae alter the anatomy of attacked galls (Shorthouse 1980, 1993; Brooks and Shorthouse 1997a), it is by the examination of a series of galls, as reported here, that patterns useful in unravelling how these two genera became associated, start to become evident. For example, those *Diplolepis* initiating galls early in the season suffer higher mortality than do those initiating galls later in the season. Furthermore, early season *Diplolepis* galls have many more *Periclistus* larvae per gall than do later season galls (table 2), and their *Periclistus*-inhabited galls become much larger than inducer-inhabited galls. In contrast, late season galls attacked by *Periclistus* remain small and less anatomically complex. Also, the assemblages of parasitoids of early season galls are more diverse than those of later season galls. Together, these observations provide evidence of evolutionary events leading to the *Diplolepis-Periclistus* associations seen today.

The six species of leaf-galling *Diplolepis* examined here are probably closely related, and share a common ancestor. Although the adult wasps are very similar and difficult to distinguish taxonomically, the galls they induce are strikingly different. For example, galls of *D. polita* have a prominant layer of thick-walled sclerencyma cells whereas those of *D. ignota* have two sclerenchyma discs and two layers of sclerenchymya cells filled with phenolic substances (figs 13A and 13E). Furthermore, their galls are separated temporally and by host plant and habitat differences. Some driving force has caused the galls of these six species to become anatomically different and separated in space and time, and *Periclistus* are likely candidates. There can be little doubt that both *Periclistus* and parasitoids have been successful in tracking *Diplolepis* and must have played an important role in their speciation. Others have suggested that parasitoids and the need for enemy-free-space have been driving forces in the evolution of gall types (Cornell 1983, Askew 1984, Waring and Price 1989, Zwolfer and Arnold-Rinehart 1994), but it appears that inquilines may be just as important.

The first *Periclistus* to become associated with *Diplolepis* probably adopted a species with galls that were abundant, widely distributed and appeared early in the spring. Galls of *D. polita* and *D. bicolor* fit these requirements, and it is interesting that the same species of *Periclistus* (as yet undescribed) is associated with both galls across much of Canada. Furthermore, large numbers of inquilines exist in each gall (table 2), and attacked galls are grossly enlarged and anatomically modified. Less than 5 percent of *D. polita* galls and 20 percent of *D. bicolor* galls contain inducers by the end of the season. Yet populations of each are sustained, suggesting that these two *Diplolepis-Periclistus* associations have existed for a long time.

Galls of *D. rosaefolii* and *D. nebulosa* are initiated later and over a longer period of time than galls of *D. polita* and *D. bicolor* (fig. 10), but heavy inquiline attack is restricted to the first galls induced. Galls of *D. rosaefolii* and *D. nebulosa* appearing later escape *Periclistus* attack and as a result, more inducers of these species are present in their communities by the end of the season than occurred in galls of *D. polita* and *D. bicolor*. *Periclistus* appear to cause less structural modification in galls of *D. rosaefolii* and *D. nebulosa* compared to galls of *D. polita* and *D. bicolor*, which implies that their *Periclistus*-inducer associations have occurred over a shorter period of time. An undescribed species of *Periclistus* is restricted to galls of *D. rosaefolii*. Galls of *D. ignota* and *D. gracilis* are induced even later in the season and are even more successful at escaping inquiline attack. In those cases, inquilines cause 20 percent mortality or less and more than half of the galls contain inducers by gall maturation. The same undescribed species of *Periclistus* attacks galls of *D. nebulosa* and *D. ignota* and a separate undescribed species is associated with galls of *D. gracilis* (Ritchie 1984). *Periclistus* are therefore gall specific, providing further evidence that the relationships between inquilines and inducers are well refined. For example, *Periclistus* associated with galls of *D. polita* and *D. bicolor* are never associated with the galls of other species. Furthermore, *P. pirata* found in the stem gall of *D. nodulosa* (Brooks and Shorthouse 1997a) are never associated with galls of *D. polita* and *D. bicolor* even though all three galls are induced concurrently in the same habitat and the various species of *Periclistus* oviposit during the same period.

If inquilines are derived from *Diastrophus* inducers, as suggested by Ronquist (1994), then at some point in time, members of this population began to attack galls induced by other members of their own species, due possibly to the lack of suitable oviposition sites. This led to a change in behavior from inducing their own galls in immature plant tissues to inducing gall-like structures within conspecific galls. Thus, two distinct populations formed; one that continued to induce galls and another that relied on other immature galls for their existence. Once inquilinism of this type evolved, then it is possible a population moved to galls of other genera on different host plants, such as galls of *Diplolepis* on *Rosa*. Perhaps the ancestral species of *Periclistus* first attacked common and prominent clusters of galls such as those induced by *D. polita* and *D. bicolor*. It is also possible that the thick layer of sclerenchyma found in *D. polita* galls and the firm, thick walls of *D. bicolor* galls were in response to *Periclistus* attack, but then to improve chances of escaping the inquilines, subpopulations began delaying the initiation of their galls and forming different structures leading to the appearance of new species such as *D. rosaefolii* and *D. nebulosa*. Similarly, inquiline pressure could have been responsible for the later appearing and structurally distinct galls of *D. ignota* and *D. gracilis*.

It is also likely that one species of *Periclistus* originally became associated with the *Diplolepis* complex and then radiated as they moved to different galls. Once a species of *Periclistus* became associated with galls such as those of *D. polita* or *D. bicolor*, subpopulations could have become associated with galls of other species, leading to the radiation of the *Periclistus* complex. It is unfortunate that the taxonomic study of Ritchie (1984) has yet to be published and the new species he proposed established. Once this occurs, the process of suggesting phylogenetic relationships between species of *Periclistus* can be enhanced and hopefully the information reported here will prove useful in this exercise.

Perhaps the most intriguing aspect of *Periclistus-Diplolepis* associations is the means by which *Periclistus* gain control of attacked galls. For all cases of North American *Diplolepis* galls studied to date, the inducer is killed by *Periclistus* as they deposit eggs on the inner surface of the chamber. Soon after the inducer is killed, nutritive cells lose their characteristics and become vacuoate gall parenchyma. However, these cells receive some type of stimulus from either the ovipositing female, or from the presence of eggs, because the gall parenchyma continues to proliferate prior to the inquiline's hatching. Galls where inducers are killed without eggs being deposited stop growing and die (Rohfritsch 1992). Whereas, galls of most species of *Diplolepis* with *Periclistus* eggs and larvae grow faster and larger than normal galls with just inducers. The rapidly proliferating *Periclistus*-influenced galls cause a substantial increase in vascular-bundle-rich gall tissues which supply the increased number of inhabitants with food. Gall parenchyma proliferates near the freshly hatched *Periclistus* such that each larva is enveloped within its own chamber. *Periclistus*-induced nutritive cells appear in each chamber near the feeding larvae and soon line the surface of the chamber (figs. 14A-F) and become the inquiline's sole source of food. *Periclistus*-induced nutritive cells are smaller and more cuboidal than those induced by *Diplolepis*. However, they similarly have large nuclei and are cytoplasmically dense.

The arrangement and structure of gall sclerenchyma is also influenced by *Periclistus*. In galls such as those of *D. bicolor*, *D. rosaefolii* and *D. gracilis*, there is more sclerenchyma than in inducer-inhabited galls and in galls of *D. ignota*, thick-walled sclerenchyma cells in *Periclistus*-modified galls replace the phenol-filled sclerenchyma cells of inducer-inhabited galls. Sclerenchyma in inquiline-inhabited galls forming in the same region as inducer-inhabited galls is not laid down in uniform columns but rather the cells become arched, remaining attached at either end. Sclerenchyma cells do not form in maturing galls when the inducer is killed by parasitoids (Rohfritsch 1992), confirming that *Periclistus*, just as the *Diplolepis*, influence its differentiation. Furthermore, *Periclistus*-induced sclerenchyma cells, structurally similar to those of normal galls, form within the walls of each *Periclistus* chamber.

Periclistus apparently have many of the plant-stimulating capabilities of *Diplolepis*. Chambers formed by maturing *Periclistus* are anatomically similar to maturing *Diplolepis* galls and the larvae of each are nourished by specialized nutritive cells. Thus *Periclistus* are close to being gall inducers, but they still must rely on galls of other species. It appears that only the *Diplolepis* are capable of stimulating immature tissues of the host plant to initiate the gall formation process. *Periclistus* often lay their eggs in small galls about 1 millimeter in diameter, but they do not lay eggs on normal host plant tissues. Leaf-galling *Diplolepis* and the stem galler *D. spinosa* (Bronner 1985, Shorthouse 1993) attach their eggs to single epidermal cells whereas other stem gallers deposit their eggs at the end of a channel made with their ovipositors, but the eggs are in close contact with plant cells (Brooks and Shorthouse 1997b). Eggs of *Periclistus* are not attached to single cells but instead are laid on the surface of gall cells lining the chamber while some eggs are even deposited on the dead inducer.

It is assumed that *Periclistus* cannot initiate galls from immature tissues of the host plant if they accidently deposited eggs at the same spot as a *Diplolepis*. Thus, *Periclistus* are close to being gall inducers and it is likely that their tissue-stimulating abilities have become more sophisticated and galler-like as they moved away from the *Diastrophus* complex and became associated with *Diplolepis* galls. Perhaps in the future some species of *Periclistus* will surmount this one last hurdle and become true gall inducers.

Galls of each species of cynipid house a species-specific assemblage of inhabitants which have proven useful for ecologists interested in examining attributes of component communities (see references in Brooks and Shorthouse 1997a). Galls of each species of *Diplolepis* similarly have distinct component comunities that are relatively constant from year to year where populations of roses are stable. It is evident that *Periclistus* associated with the six species of leaf gallers highlighted here play a dominant role in their dynamics. *Periclistus* are responsible for the deaths of about 70 percent of *D. polita* and 60 percent of *D. bicolor* populations and then become sources of food for parasitoids such as *Pteromalus*. Parasitoids such as *Eurytoma*, *G.stigma* and *Torymus* species also consume *Periclistus* larvae and then consume inquiline and *Diplolepis*-induced gall tissues. Presumably the additional plant matter in inquiline-modified galls improves the success of these parasitoid/phytophagous inhabitants. Thus, *Periclistus* tend to increase the complexity of food webs in gall communites where they are predominant.

Although the communities associated with the six leaf gallers and the one stem galler Brooks and Shorthouse (1997a) are relatively distinct, there are basic similarities. In all galls, *Periclistus* oviposit soon after galls are formed and the first parasitoids to oviposit are *Eurytoma*. Most of the remaining parasitoids that require maturing larvae of either the inducers or inquilines oviposit after the *Eurytoma*. Galls of *D. ignota* and *D. gracilis* appearing later in the season have a high percentage of their inducers attacked by *Aprostocetus*.

Several factors have been suggested as influencing the species complement of gall communities: period in the season for gall induction, competition, position of galls on host plants, gall size and gall structure (Askew 1961, 1984; Brooks and Shorthouse 1997a). Time of the year for gall initiation appears to be a key factor for *Diplolepis* leaf galls because assemblages associated with the early spring initiated galls of *D. polita* and *D. bicolor* are more complex than those associated with late season galls such as those of *D. ignota* and *D. gracilis*. Size and structure of *Diplolepis* leaf galls appears of little consequence because most inquilines and parasitoids oviposit when galls are immature and soft. Minute parasitoids such as *Aprostocetus* have little difficulty ovipositing in either the thick-walled galls of *D. ignota* or the thin-walled galls of *D. gracilis*.

Wild roses associated with forested regions are early to mid-successional plants and when clearings are formed by natural or human caused events. Roses often become dominant shrubs. Once the forest canopy reforms, populations of roses decrease and may disappear, but during their period of abundance, the inducers and their assemblage of inhabitants quickly colonize the sites. Populations of galls such as *D. polita* and *D. gracilis* which are most abundant on roses in recently cleared areas are probably more cyclical than species such as *D. bicolor*, *D. nebulosa,* and *D. ignota* which are found in more stable open, prairie habitats. Other studies on *D. polita* (Shorthouse 1994) and the stem galler *D. spinosa* (Shorthouse 1988) have illustrated that inducers and their parasitoids are highly vagile and capable of quickly establishing themselves in domestic roses in atypical habitats such as new urban areas. Even *Periclistus* were able to find isolated galls of *D. polita* in urban habitats (Shorthouse 1994). Furthermore, the presence of all community members in widely distributed galls of most species of *Diplolepis* (i.e., galls of *D. polita* occur from central Yukon to central Ontario), illustrates the stability of these intricate insect-plant complexes.

Much remains to be learned about *Periclistus-Diplolepis* associations, their galls and associated component communities. Descriptions of new species and revision of the *Periclistus* complex must come first. Then one can begin more indepth studies of the galls themselves. If gall-inducing morphogens can be identified, then perhaps similar messages can be attributed to either ovipositional fluids or the eggs of *Periclistus*, or to the inquiline larvae themselves. Cynipid galls are known to be physiological sinks for carbon assimilates (Bagatto *et al.* 1996) and mineral nutrients (Paquette *et al.* 1993), and it would be interesting to compare these attributes between inducer-inhabited galls of *Diplolepis* and those modified by inquilines. As well, much remains to be done with the assemblages of parasitoids associated with *Diplolepis* galls. The many species of *Eurytoma*, *Torymus*, *Pteromalus*, and *Aprostocetus* need to be identified, and after this, further fascinating patterns of gall specificity will likely be revealed.

ACKNOWLEDGEMENTS

Studies of the six *Diplolepis* galls examined here began at the University of Saskatchewan, Saskatoon, Saskatchewan in the early 1970's and have continued at Laurentian University to the present. Support and encouragement by D. M. Lehmkuhl and T. A. Steeves while in Saskatchewan is gratefully acknowledged. Subsequent studies were supported by research grants from the Natural Sciences and Engineering Research Council of Canada and the Laurentian University Research Fund. I thank S. E. Brooks for his help with the drawings and for many discussions on this and other gall projects.

REFERENCES

Askew, R.R. 1961. On the biology of the inhabitants of oak galls of Cynipidae (Hymenoptera) in Britain. Transactions of the Society for British Entomology. 14: 237-268.

Askew, R.R. 1971. Parasitic insects. New York: Elsevier Publishing Company. 316 p.

Askew, R.R. 1984. The biology of gall wasps. In: Ananthrakrishnan, T.N., ed. Biology of galls insects. New Dehli: Oxford and IBN: 223-271.

Bagatto, G.; Paquette. L.C.; Shorthouse, J.D. 1996. Influence of galls of *Phanacis taraxaci* on carbon partitioning within common dandelion, *Taraxacum officinale*. Entomologia Experimentalis et Applicata. 79: 111-117.

Beutenmuller, W. 1907. The North American species of *Rhodites* and their galls. Bulletin of the American Museum of Natural History. 23: 629-651.

Bronner, R. 1985. Anatomy of the ovipositor and oviposition behavior of the gall wasp *Diplolepis rosae* (Hymenoptera: Cynipidae). Canadian Entomologist. 117: 840-858.

Bronner, R. 1992. The role of nutritive cells in the nutrition of cynipids and cecidomyiids. In: Shorthouse, J.D.; Rohfritsch, O., eds. Biology of insect-induced galls. New York: Oxford University Press: 118-140.

Brooks, S.E.; Shorthouse, J.D. 1997a. Biology of the rose stem galler *Diplolepis nodulosa* (Hymenoptera: Cynipidae) and its associated component community in Central Ontario. The Canadian Entomologist. In press.

Brooks, S.E.; Shorthouse, J.D. 1997b. Developmental morphology of galls of *Diplolepis nodulosa* (Hymenoptera: Cynipidae) and those modified by the inquiline *Periclistus pirata* (Hymenoptera: Cynipidae) on the stems of *Rosa blanda* (Rosaceae). Canadian Journal of Botany. Submitted.

Burks, B.D. 1979. Cynipoidea. In: Krombien, K.V.; Hurd, P.D., Jr.; Smith, D.R.; Burks, B.D., eds. Catalog of Hymenoptera of North America North of Mexico. Vol. 1. Washington, DC: Smithsonian Institution Press: 1045-1107.

Cornell, H.V. 1983. The secondary chemistry and complex morphology of galls formed by the Cynipinae (Hymenoptera): Why and how? American Midland Naturalist. 110: 225-234.

Hanson, P. 1992. The Nearctic species of *Ormyrus* Westwood (Hymenoptera: Chalcidoidea: Ormyridae). Journal of Natural History. 26: 1333-1365.

Hawkins, B.A.; Goeden, R.D. 1982. Biology of a gall-forming *Tetrastichus* (Hymenoptera: Eulophidae) associated with gall midges on saltbush in southern California. Annals of the Entomological Society of America. 75: 444-447.

Kinsey, A.C. 1920. Life histories of American Cynipidae. Bulletin of the American Museum of Natural History. 42: 319-357.

Liljeblad, J.; Ronquist, F. 1998. A phylogenetic analysis of higher-level gall wasp relationships (Hymenoptera, Cynipidae). Systematic Entomology. In press.

Meyer, J. 1987. Plant galls and gall inducers. Gebruder Borntraeger. Berlin. 291 p.

Meyer, J.; Maresquelle, H.J. 1983. Anatomie des Galles. Gebruder Borntraeger. Berlin. 662 p.

O'Brien, T.P.; McCully, M.E. 1981. The study of plant structure: principles and selected methods. Melbourne: Termacarphi PTY. Ltd. 346 p.

Paquette, L.C.; Bagatto, G.; Shorthouse, J.D. 1993. Distribution of mineral nutrients within the leaves of common dandelion (*Taraxacum officinale*) galled by *Phanacis taraxaci* (Hymenoptera: Cynipidae). Canadian Journal of Botany. 71: 1026-1031.

Ritchie, A.J. 1984. A review of the higher classification of the inquiline gall wasps (Hymenoptera: Cynipidae) and a revision of the Nearctic species of *Periclistus* Forster. Ottawa, Ontario: Carleton University, Ottawa, Ontario. 368 p. Ph.D. dissertation.

Rohfritsch, O. 1992. Pattterns in gall development. In: Shorthouse, J.D.; Rohfritsch, O., eds. Biology of insect-induced galls. New York: Oxford University Press: 60-80.

Ronquist, F. 1994. Evolution of parasitism among closely related species: phylogenetic relationships and the origin of inquilinism in gall wasps (Hymenoptera, Cynipidae). Evolution. 48: 241-266.

Roskam, J.C. 1992. Evolution of the gall-inducing guild. In: Shorthouse, J.D.; Rohfritsch, O., eds. Biology of insect-induced galls. New York: Oxford University Press: 34-49.

Shorthouse, J.D. 1973. The insect community associated with rose galls of *Diplolepis polita* (Cynipidae, Hymenoptera). Quaestiones Entomologicae. 9: 55-98.

Shorthouse, J.D. 1980. Modifications of galls of *Diplolepis polita* by the inquiline *Periclistus pirata*. Bulletin de la Societe Botanique de France (Actualites Botaniques). 127: 79-84.

Shorthouse, J.D. 1988. Occurrence of two gall wasps of the genus *Diplolepis* (Hymenoptera: Cynipidae) on the domestic shrub rose *Rosa rugosa* Thunb. (Rosaceae). The Canadian Entomologist. 120: 727-737.

Shorthouse, J.D. 1993. Adaptations of gall wasps of the genus *Diplolepis* (Hymenoptera: Cynipidae) and the role of gall anatomy in cynipid systematics. Memoirs of the Entomological Society of Canada. 165: 139-163.

Shorthouse, J.D. 1994. Host shift of the leaf galler *Diplolepis polita* (Hymenoptera: Cynipidae) to the domestic shrub rose *Rosa rugosa*. The Canadian Entomologist. 126: 1499-1503.

Shorthouse, J.D.; Rohfritsch, O., eds. 1992. Biology of insect-induced galls. New York: Oxford University Press. 285 p.

Stille, B. 1984. The effect of host plant and parasitoids on the reproductive success of the parthenogenetic gall wasp *Diplolepis rosae* (Hymenoptera: Cynipidae). Oecologia. 63: 364-369.

Strong, D.R.; Lawton, J.H.; Southwood, T.R.E. 1984. Insects on plants. Oxford: Blackwell. 313 p.

Waring, G.L.; Price, P.W. 1989. Parasitoid pressure and the radiation of a gall forming group (Cecidomyiidae: *Asphondylia* spp.) on creosote bush (*Larrea tridentata*). Oecologia. 79: 293-299.

Washburn, J.O.; Cornell, H.V. 1981. Parasitoids, patches, and phenology: their possible role in the local extinction of a cynipid gall wasp population. Ecology. 62: 1597-1607.

Wiebes-Rijks, A.A. 1982. Early parasitation of oak-apple galls (*Cynips quercusfolii* L., Hymenoptera). Netherlands Journal of Zoology. 32: 112-116.

Zwolfer, H.; Arnold-Rinehart, J. 1994. Parasitoids as a driving force in the evolution of the gall size of *Urophora* on Cardueae hosts. In: Williams, M.A.J., ed. Plant galls; organisms, interactions, populations. Oxford: Clarendon Press: 245-257.

SEXUAL SELECTION VIA FEMALE CHOICE IN THE GALL-MAKING FLY *EUROSTA SOLIDAGINIS* FITCH (DIPTERA: TEPHRITIDAE)

Lisa J. Faust and Jonathan M. Brown*

Biology Department, Grinnell College, P.O. Box 805, Grinnell, IA 50112-0806 USA
*to whom correspondence should be addressed

Abstract.—Recent studies have suggested that female choice may focus on characters that are good indicators of overall male fitness, e.g., fluctuating asymmetry. We tested two populations from central Iowa, USA, of the tephritid gallmaker, *Eurosta solidaginis*, for evidence of female choice on male morphological characteristics and their symmetries. Females from two populations ("Krumm" and "Ladora") were allowed to choose between potential mates in cages. "Krumm" females chose males with significantly lower levels of tibial asymmetry, but no other differences in character asymmetries between successful and unsuccessful males in either population were significant. However, "Ladora" females chose males with significantly darker wings. Since these populations lie in a zone of contact (or perhaps hybridization) between described subspecies of *E. solidaginis* differentiated solely by the extent of wing pigmentation, this result suggests that sexual selection via female choice may be acting against hybrids and may thus play a role in the maintenance of reproductive barriers between incipient species.

INTRODUCTION

Sexual selection arises as a result of "the advantage which certain individuals have over other individuals of the same sex and species, in exclusive relation to reproduction" (Darwin 1871). This mating advantage arises through competition between individuals of the same sex, usually males, for access to a mate and through individuals of one sex, usually females, choosing individuals of the opposite sex to mate with. Theoretical models have shown that divergent sexual selection could result in the evolution of reproductive barriers (Lande 1981, 1982) if hybrids between types are at a mating disadvantage (Fisher 1930). This could occur as in classic reinforcement, wherein lower viability of hybrid offspring creates selection favoring increasing prezygotic isolation between hybridizing forms, or could occur as a result of divergence in female preference that has evolved during allopatry. Thus the study of variation in female choice in wild populations is an important component of a better understanding of the processes of species diversification.

Females can discriminate between males in several possible ways. The choice can be based upon direct fitness benefits the male offers the female, such as the provision of a nuptial gift or another indicator of the ability of the male to provide parental care to offspring (Thornhill 1980). However, in species where males do not offer such direct fitness benefits, but instead offer only sperm, females may still show strong preference for male traits. Current theories recognize at least two mechanisms by which this could occur. In the first, the male trait and an arbitrary female preference for that trait become linked and "runaway," with the initial preference not necessarily being adaptive (Fisher 1930). The second theory hypothesizes that females choose males based upon traits that truthfully advertise male fitness and viability. In this case, females are selecting males with traits that indicate heritable qualities, which will thus be passed on to the resulting offspring.

Fluctuating asymmetry (hereafter FA) of morphological characters may be one such truthful indicator of male fitness. FA is a population-level measure of the slight deviations from perfect symmetry that can arise on either side of the body (Liggett *et al.* 1993). It measures developmental stability, reflecting the ability of an organism to develop evenly in response to environmental and genomic stress, an ability that has a genetic basis (Palmer and Strobeck 1986, Parsons 1992, Watson and Thornhill 1994). Thus, FA will theoretically be a good indicator trait for male fitness, as it reflects the genetic quality of the male, and could therefore be a basis for female choice. Female fitness would increase by choosing more symmetrical males because their offspring would inherit the resilience to environmental and genetic stresses of the father, conferring a selective advantage. Previous studies examining the relationship between FA and sexual selection have found that a mating advantage does exist for individuals with low FA and high developmental stability (Allen and Simmons 1996, Liggett et al. 1993, McLachlan and Cant 1995, Møller 1996, Swaddle et al. 1994, Thornhill 1992). Thus if fluctuating asymmetry is an indicator trait for male fitness, it must be demonstrated that females that are given a choice of males with which to mate choose more symmetrical males.

METHODS

Study Organism

We examined the association between levels of fluctuating asymmetry in morphological traits and mating success in *Eurosta solidaginis* Fitch (Diptera: Tephritidae), a native North-American gallmaker. This fly exists in Iowa as two distinct host races, each specialized on one of two species of goldenrod (*Solidago altissima* and *S. gigantea*) upon which it feeds, mates, and oviposits (Craig et al. 1993). This study focused solely on the *S. altissima* host race. After females oviposit into buds of their goldenrod host, the resulting larvae form a gall in which they over-winter. In the spring adults emerge, mate, and oviposit again, living approximately 10 days. By collecting these galls from the field and incubating them it is possible to produce adults in the laboratory (see Abrahamson and Weis 1997 and Uhler 1951 for more details on gallmaker natural history).

Mating behavior and courtship rituals involve both male and female display, which includes walking in tight circles, wing flicking and side-to-side displays (Walton et al. 1990). Males perch on the apices of their host plants, displaying to each other and to females by vigorously rocking their bodies from side to side. Courting males may touch females from the front with their front legs or attempt to mount them from the rear. In our trials, females reacted to the majority of male displays by walking or flying away, or struggling to escape attempted mounts from the rear. If a female is receptive, the male clasps the female from the rear with the two front pairs of legs, bends the end of the abdomen toward the female's ovipositor, and attempts to insert his aedeagus into the ovipositor after gaining contact with it using his claspers (Uhler 1951). Females may break off copulations during any stage of this process, although most attempts are broken off during male mounting attempts. These behaviors suggest that females have the opportunity to exert choice in which male they mate with. Male-male competition for perch sites is also possible, although we observed more male attempts to court and mount each other than overtly aggressive encounters, such as attempts to displace other males from perches.

Female-Choice Tests

Two sets of flies hatched from *S. altissima* galls were used in this experiment. We collected the first set, hereafter "Krumm flies," in November 1996 at Krumm Nature Preserve (41.709° N, 92.776° W) 3 km WSW of Grinnell, IA, USA and the second, hereafter "Ladora flies," were collected from a roadside prairie (41.757° N, 92.183° W) along Highway 6 0.5 km west of Ladora, IA in March 1997. The galls were incubated at approximately 26°C and kept moist by spraying lightly with distilled water, and flies hatched after approximately 17 days. We then utilized two experimental designs to test the hypothesis that FA was a basis for mate choice in these flies. For the Krumm flies, we used a paired-design, placing two males and a female, all randomly chosen virgin flies, in a mating arena which consisted of a plastic cup with wooden toothpicks as perches for the flies. Once mating had occurred, we immediately separated the mated pair from the unmated male and stored all three flies at -20°C. This design yielded 8 matings. For the Ladora flies, we modified our design in an attempt to produce more matings, placing multiple males and females, all randomly chosen virgins, into several mating arenas. The overall density of flies per arena was around 15. When a mating occurred, the pair was removed from the cup and stored at -20°C. Unmated males for this design were randomly chosen from the remaining males that had not mated after the experiment had stopped. This design yielded 17 mated males, and 28 unmated males were chosen.

Trait Measurement and Analysis

We dissected each fly's wings and rear legs, and mounted all four body parts on a piece of filter paper using spray adhesive. Wings and legs were then viewed through a dissecting microscope. Digital grayscale images were then captured via a CCD video camera to a Powermac 8500 using NIH Image v. 1.61 (Rasband 1996). All images were then randomly coded by JMB to avoid measurement bias. LJF then measured and recorded the right and left tibia length, tarsus length, wing width, wing length, medial spot length, apical spot width, and wing pigmentation (defined as the proportion of white on the wing) for the coded images using NIH Image (see figure 1 for measurement definitions). Any individuals that were damaged in the process of mounting or before were eliminated from analyses for whichever traits were not measurable. Subsequent calculations and analyses were

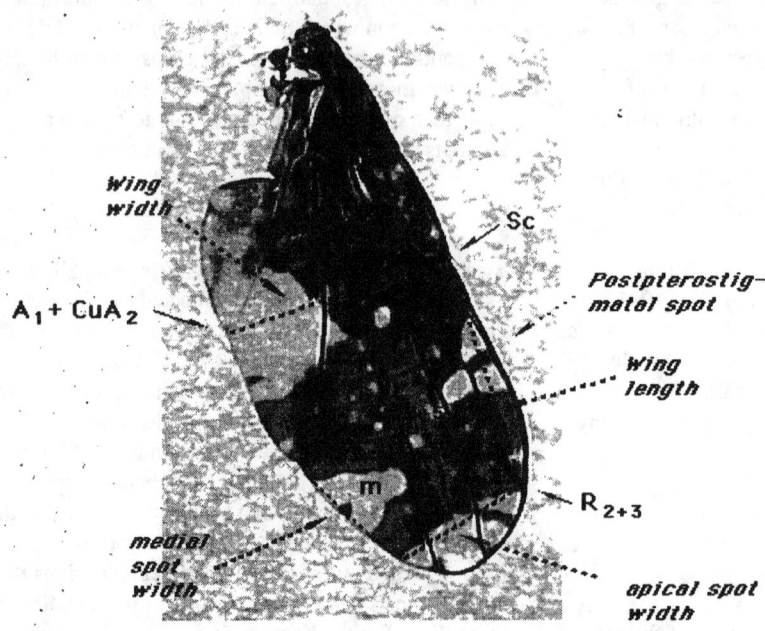

Figure 1.—*Wing characters measured on male flies. $A_1 + CuA_2$, Sc, and R_{2+3} refer to the ends of the branches of anal, cubitus and radius veins. m and r_{4+5} refer to medial and radial cells.*

carried out using Minitab v. 10.5 for Power Macintosh. We calculated the absolute asymmetry (unsigned left-minus-right character) and mean trait sizes (left+right/2) for each trait measured.

Measurement Error and FA

To assess measurement error, we measured the seven traits twice each, on separate days, and calculated the signed asymmetry, calculated as the left-minus-right trait value, for each individual. We then used a one-way ANOVA to determine if the variance in asymmetry between individual flies was significantly greater than the variance due to measurement error (Allen et al. 1996, Swaddle et al. 1994). Significantly greater difference between individuals than between repeated measurements was found for tibia asymmetry ($F = 6.10$, $p < 0.001$), tarsus asymmetry ($F = 36.46$, $p < 0.001$), wing width asymmetry ($F = 9.70$, $p < 0.001$), wing length asymmetry ($F = 4.68$, $p < 0.001$), medial spot asymmetry ($F = 14.09$, $p < 0.001$), and apical spot asymmetry ($F = 14.72$, $P < 0.001$), but not for wing pigmentation asymmetry ($F = 1.63$, $p = 0.076$) (table 1). Therefore wing pigmentation was eliminated from subsequent analyses, since measurement error could highly influence conclusions about asymmetry.

Directional vs. Fluctuating Asymmetry

For asymmetry to be fluctuating, instead of directional, there should be an approximately equal number of individuals with larger right-sided characters to individuals with larger left-sided characters (Liggett et al. 1993).

To determine if asymmetries were truly fluctuating, rather than directional, they had to have normal distribution and not be significantly different from zero (Swaddle et al. 1994). We first tested each asymmetry for normality of distribution using the Anderson-Darling Normality Test. All traits were normally distributed except for tarsus length for the Krumm flies (A-sq. = 3.7736, $p < 0.001$) and medial and apical spot lengths for the Ladora flies (medial spot: A-sq. = 1.8564, $p = 0.001$; apical spot: A-sq. = 0.7535, $p = 0.0454$). Because these traits were non-normal, they were eliminated from subsequent asymmetry analyses. Further analyses using single-sample T-tests revealed that mean signed asymmetries most remaining traits (Krumm: tibia, wing length, medial spot length, apical spot width; Ladora: tarsus, wing length) did not differ significantly from zero, indicating that asymmetry was fluctuating. Wing width for both samples and tibia length for Ladora flies were significantly different from zero (Krumm width: $t = 2.81$, $p = 0.016$; Ladora: width $t = 2.63$, $p = 0.012$, tibia $t = 2.83$, $p = 0.0073$), indicating that these traits exhibited directional asymmetry.

For each of the remaining asymmetries (Krumm: tibia length, wing length, medial spot length, apical spot width; Ladora: tarsus length, wing length) and all average trait sizes, we compared mated with unmated males either with paired T-tests (Krumm flies), or unpaired T-tests (Ladora flies). One-tailed tests are justified for the asymmetry comparison by the *a priori* prediction that mated males should be more symmetrical than unmated males.

Table 1.—*Initial analyses to establish whether asymmetry was truly fluctuating in both sets of flies, showing the final set of traits able to be analyzed for asymmetry.*

Area	Trait	Asymmetry fluctuating?			
		Measurement error?	Distribution normal?	Signed asymm. $\neq 0$?	Final traits analyze
Krumm	Tibia length	No	Yes	No	Yes
	Tarsus length	No	No	—	—
	Wing width	No	Yes	Yes	—
	Wing length	No	Yes	No	Yes
	Medial spot length	No	Yes	No	Yes
	Apical spot width	No	Yes	No	Yes
	Wing percent white	Yes	—	—	—
Ladora	Tibia Length	No	Yes	Yes	—
	Tarsus Length	No	Yes	No	Yes
	Wing width	No	Yes	Yes	—
	Wing length	No	Yes	No	Yes
	Medial spot length	No	No	—	—
	Apical spot length	No	No	—	—
	Wing percent white	Yes	—	—	—

RESULTS

Differences in Asymmetry

There was a significant difference in tibia length asymmetry between mated and unmated males for the Krumm flies, with unmated having significantly higher asymmetry ($t = 3.48$, $p = 0.0052$). However, for other asymmetry traits of Krumm flies there was no significant difference between unmated and mated males. Although not statistically significant, two of the other three characters show a trend towards unmated having greater asymmetry than mated (fig. 2). For the Ladora flies, no significant differences were found in levels of fluctuating asymmetry between mated and unmated for tarsus length and wing length.

Differences in Absolute Traits

None the seven traits differed significantly between mated and unmated Krumm flies (paired *t*-test, $p > 0.05$). However, mated Ladora males had a significantly lower proportion of white (or non-pigmented area) on their wings than did unmated males (Mean$_{Mated}$ = 27.26 ± 0.86 percent, Mean$_{Unmated}$ = 30.1 ± 0.66 percent; $t = -2.60$, $p = 0.015$). None of the other traits differed significantly between mated and unmated Ladora males.

DISCUSSION

Our analyses to determine the correlation between fluctuating asymmetry and mating success in *E. solidaginis* did not strongly indicate that females were choosing based upon levels of FA. Krumm females chose males with significantly lower levels of tibia asymmetry, and also showed a general non-significant trend towards choosing males with lower levels of asymmetry. The Ladora males, however, had no significant differences in asymmetry levels between mated and unmated. The Krumm trend might indicate an association between mating status and asymmetry, but because of small sample sizes (Krumm N = 8, Ladora N = 17) it is difficult to conclude anything from the current study about FA as a basis of mate choice in *E. solidaginis*.

Future studies are needed to further elucidate the relationship between asymmetry and mate choice (if any) in *E. solidaginis*. Experimental manipulation of hypothetically important traits is an important step needed to establish that females are truly looking at the measured traits and not some other character. If FA levels are determined to be traits females pay attention to, and can be correlated with mating success, the next step would be to examine the relationship between levels of FA and aspects of male fitness. Teasing out FA's exact role would involve looking at different aspects of male performance and fitness in relation to levels of FA. Past studies have found that levels of FA are correlated to greater competitive ability in dung flies, implying that it plays a greater role for intra-sexual (male-male competition) than inter-sexual selection (female choice) (Allen and Simmons 1996, Liggett *et al*. 1993). Low levels of FA have also been correlated to quality of nuptial gifts in scorpionflies, supporting the theory of female choice based on direct benefits the male can offer to the female (Thornhill 1992). If female choice based upon FA levels

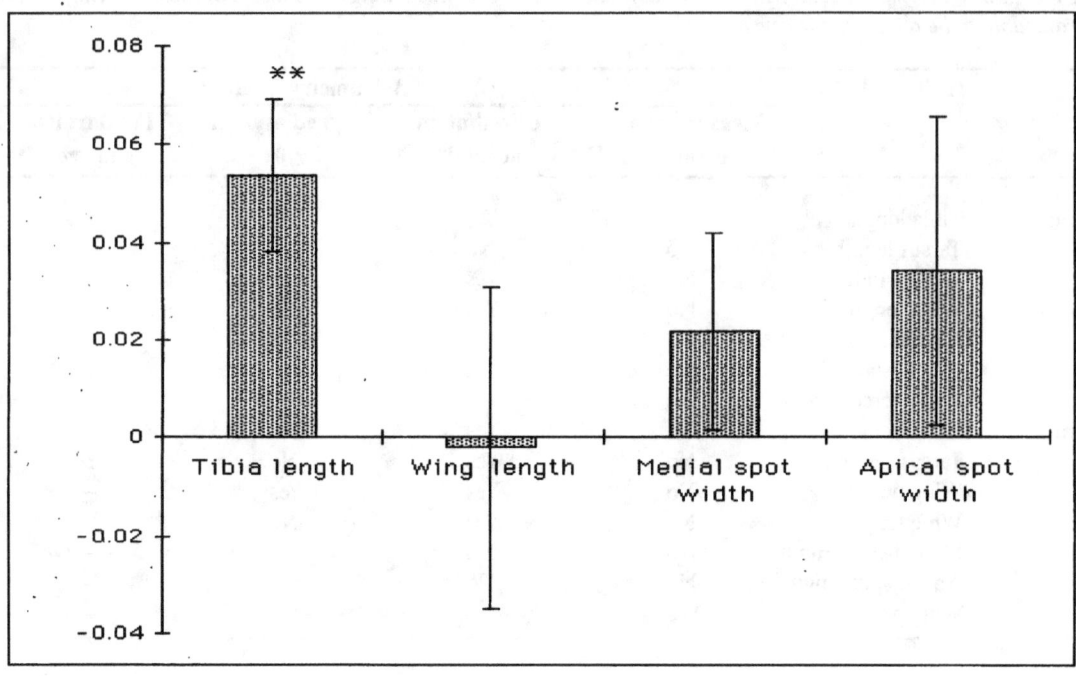

Figure 2.—*Differences in asymmetry between unmated and mated males for 4 leg and wing characters.*

is to be adaptively beneficial to the female, however, FA must serve as an indicator trait of male fitness. Møller (1996), for example, established that in *Musca domestica* levels of FA were correlated with parasite resistance and predator avoidance, therefore serving as appropriate indicators for male fitness. It would also be necessary to estimate heritability for FA, to demonstrate that offspring of symmetrical males are more developmentally stable and inherit the same correlated fitness traits (e.g., parasite resistance, predator avoidance).

Female Choice on Absolute Characters

Our analysis of average trait sizes to determine did not indicate that females were choosing males of larger size. Studies of female choice in some insects have found significant preferences for larger males or larger male traits (Fairbairn and Preziosi 1996, Gilburn *et al.* 1993), although such preferences are not universal (Markow *et al.* 1996).

The discovery of significant differences in wing pigmentation between mated and unmated Ladora males is an intriguing result. Ming (1989) used variation in areas of pigmentation in the wings to distinguish two subspecies of *Eurosta solidaginis*, an "eastern" form (*E. s. solidaginis*) and a "western" form (*E. s. fascipennis* Curran). In the eastern form, the medial and postpterostigmatal spots (non-pigmented sections of the wing) are separated by a pigmented area in cell r_{4+5},

while in the western form "the two hyaline areas are either fully connected forming a complete diagonal subapical band across the wing from anterior to posterior margins, or they are partially joined by a series of pale spots, forming a nearly uninterrupted band" (see figs. 1 and 3). Ming (1989) found no other consistent morphological distinctions in male or female genitalia, larval or pupal features, although she did note slight differences in the chorionic reticulations of the eggs. Mixed populations of these forms are found only from populations in Iowa and western Minnesota, with 'pure' populations found east and west of this zone of overlap. Ming erected these forms to subspecies status based on her inability to recognize no "intermediate" forms; we suggest, however, that her definition of the western form includes potential intermediates, i.e., flies in which the two spots are "partially joined."

In an attempt to quantify the differences between these described forms, we used NIH Image to measure the proportion of white (or non-pigmented) for an area of the wing consisting only of cells *m* and r_{4+5} (fig. 1). We used this smaller area because the extent of hyaline spots in r_{4+5} is the basis of Ming's diagnosis of the two forms, and because our own measurements identified significant variation in the size of the hyaline spot in cell *m*. We included the Ladora and Krumm populations, plus a third local population "Freeway" (located at 41.690° N, 92.794° W., 1.5 km SW of the Krumm site). Figure 3 shows representative wings from individuals that

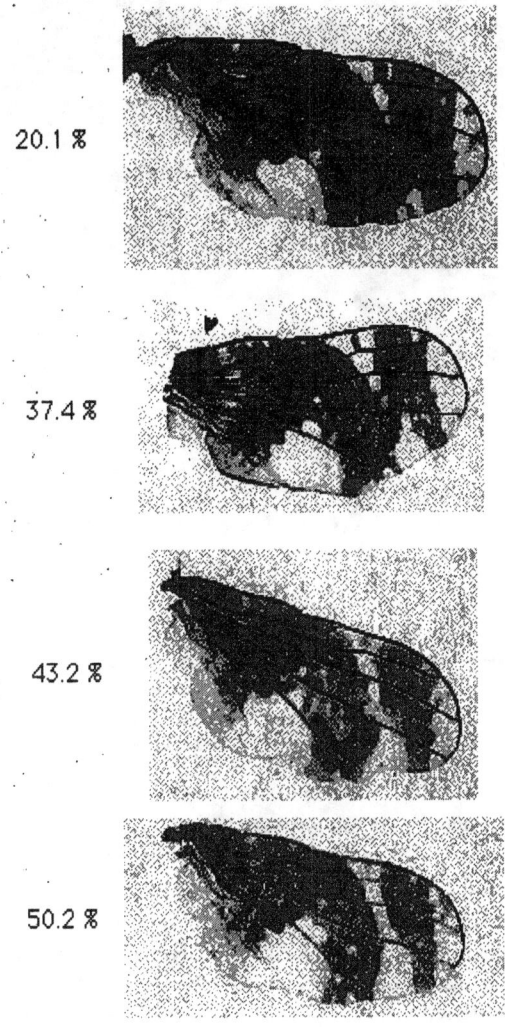

Figure 3.—*Variation in wing pigmentation among individuals in the sampled area. Percent values refer to the percent of the total area of m and r_{4+5} cells that is white (non-pigmented).*

spanned the range of variation found in these populations, from 20-50 percent white. Figure 4 illustrates individual and among-population variation for this measure, showing clear evidence for continuous variation in the trait; eastern and western forms from outside central Iowa are included for comparison. Figure 5 shows distributions of these traits for each population, illustrating that the Ladora population is more "eastern," the Krumm population more "western," and the Freeway population apparently bimodal, with most individuals falling into the extremes of the distribution. We compared mated and unmated males from Ladora and found that mated males had significantly less percent white in these cells, suggesting that females prefer a more "eastern" wing form (Mean$_{Mated}$ = 24.84 ± 0.68 percent, Mean$_{Unmated}$ = 29.08 ± 1.1 percent; t = 3.39, p = 0.002). The difference between mated and unmated males in the Krumm population (in which 2 males were competed against each other) was non-significant (paired T-test p > 0.05). However, in six of seven of these trials the male with larger percent white (more "western" wing) was mated, while in the seventh trial a male with a much higher percent white than his competitor was not chosen [if this mating is excluded, mated males have significantly greater percent white (p = 0.012)]. Unfortunately, females were not retained, so we cannot determine whether there is assortative mating occurring in these populations. [Note: Females were measured in the Freeway population, and no significant differences in wing pigmentation were found between males and females (t = 1.92, p > 0.05)]. Although not significant in the Krumm population due to low sample size, our results suggest that divergent sexual selection via female choice may be acting in a zone of hybridization between these forms, disfavoring individuals of intermediate phenotype. We are continuing studies to test this hypothesis.

CONCLUSIONS

Studying FA as a basis of female choice has merit in some species and for certain traits. For male traits that are not necessarily visible or obvious to females, such as the hind-leg size measurements made in this study, it seems less likely that FA would provide a basis for female assessment of male quality. Rather, FA may simply be correlated with other traits females look at. Levels of asymmetry have also been found to influence predator avoidance and maneuverability, especially in insects, because of the effect of wing asymmetry on locomotive and flight ability (McLachlan and Cant 1995, Møller 1996); thus, it may be possible that females are focusing on some trait other than symmetry to chose successful males. Wing traits, in contrast, are conspicuous during mating displays of tephritids, and thus are potential visible cues for females. Our observations of mating behavior, variation in wing pigmentation patterns, and the significant differences between mated and unmated males in these traits in one population suggest that female choice may play a role in the process of species diversification.

ACKNOWLEDGMENTS

We thank Grinnell College and the Robert N. Noyce Foundation for supporting the development of the image analysis system, Barb Barnett for her assistance with the statistical analyses, and Devin Drown for writing the measurement macro for NIH Image.

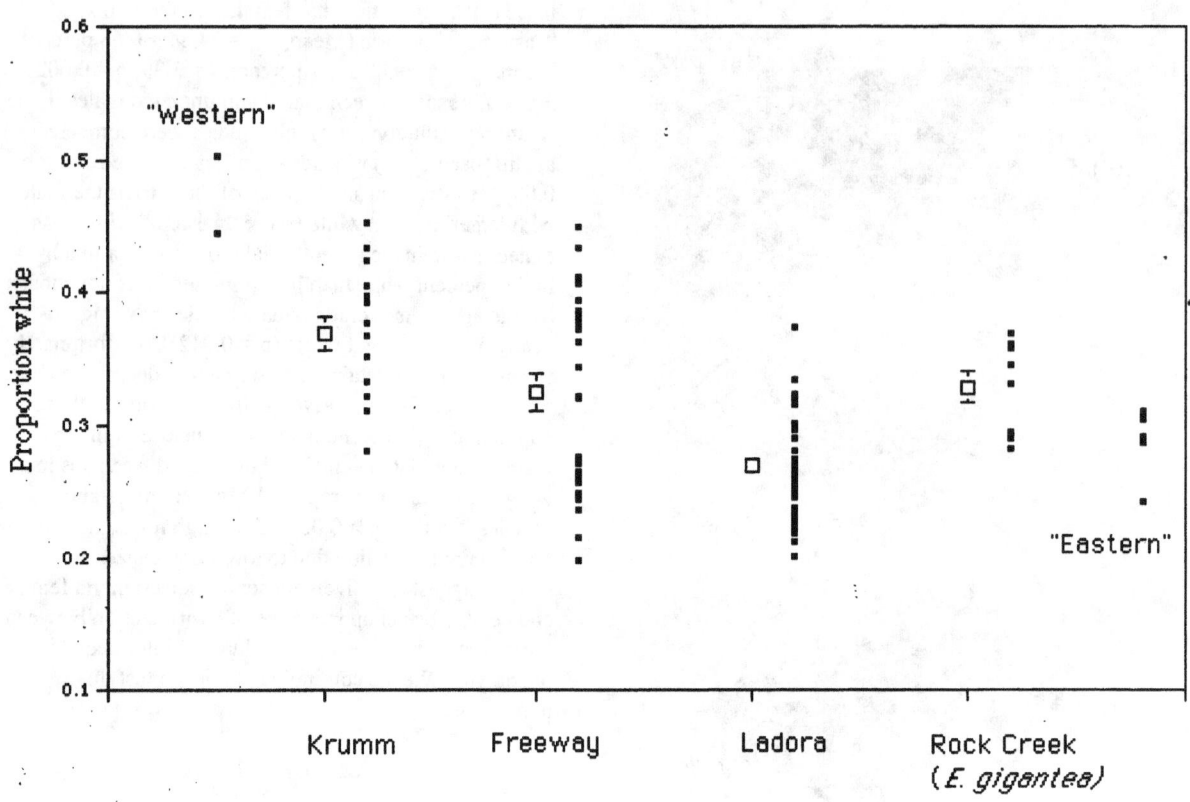

Figure 4.—*Variation within and among local populations in the proportion of the total area of m and r_{4+5} cells that is white (non-pigmented). "Western" flies are from south-central Minnesota; Eastern flies are from New York, Vermont, and Michigan.*

LITERATURE CITED

Abrahamson, W.G.; Weis, A.E. 1997. Evolutionary ecology across three trophic levels — Goldenrods, gallmakers and natural enemies. Princeton, NJ: Princeton University Press. 456 p.

Allen, G.R.; Simmons, L.W. 1996. Coercive mating, fluctuating asymmetry and male mating success in the dung fly *Sepsis cynipsea*. Animal Behavior. 52: 737-741.

Craig, T.P.; Itami, J.K.; Abrahamson, W.G.; Horner, J.D. 1993. Behavioral evidence for host-race formation in *Eurosta solidaginis*. Evolution. 47: 1696-1710.

Darwin, C. 1871. The descent of man, and selection in relation to sex. London: J. Murray. 475 p.

Fairbairn, D.J.; Preziosi, R.F. 1996. Sexual selection and the evolution of sexual size dimorphism in the water strider, *Aquarius remigis*. Evolution. 50: 1549-1559.

Fisher, R.A. 1930. The genetical theory of natural selection. Oxford: Clarendon Press. 291 p.

Gilburn, A.S.; Foster, S.P.; Day, T.H. 1993. Genetic correlation between a female mating preference and the preferred male character in seaweed flies (*Coelopa frigida*). Evolution. 47: 1788-1795.

Lande, R. 1981. Models of speciation by sexual selection on polygenic traits. Proceedings of the National Academy of Sciences USA. 78: 3721-3725.

Lande, R. 1982. Rapid origin of sexual isolation and character divergence in a cline. Evolution. 36: 213-223.

Liggett, A.C.; Harvey, I.F.; Manning, J.T. 1993. Fluctuating asymmetry in *Scatophaga stercoraria* L.: successful males are more symmetrical. Animal Behavior. 45: 1041-1043.

Markow, T.A.; Bustoz, D.; Pitnick, S. 1996. Sexual selection and a secondary sexual characteristic in two *Drosophila* species. Animal Behavior. 52: 759-766.

McLachlan, A.; Cant, M. 1995. Small males are more symmetrical: mating success in the midge *Chironomus plumosus* L. (Diptera: Chironomidae). Animal Behavior. 50: 841-846.

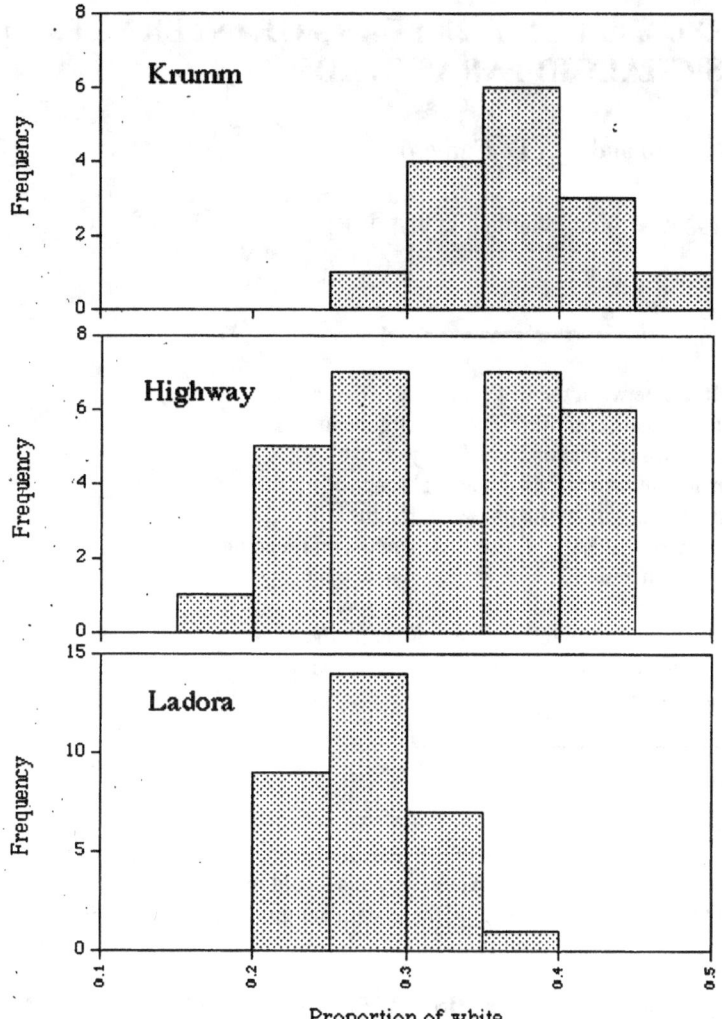

Figure 5.—*Frequency distribution of individuals in wing pigmentation within 3 local populations.*

Ming, Y. 1989. A revision of the genus *Eurosta* Loew, with a scanning electron microscopic study of taxonomic characters (Diptera: Tephritidae). Pullman, WA: Washington State University. 190 p. M.S. thesis.

Møller, A.P. 1996. Sexual selection, viability selection, and developmental stability in the domestic fly *Musca domestica*. Evolution. 50: 746-752.

Palmer, A.R.; Strobeck, C. 1986. Fluctuating asymmetry: measurement, analysis, patterns. Annual Review of Ecology and Systematics. 17: 391-421.

Parsons, P.A. 1992. Fluctuating asymmetry: a biological monitor of environmental and genomic stress. Heredity. 68: 361-364.

Rasband, W. 1996. NIH Image v. 1.61 for the Power Macintosh. National Institutes of Health.

Swaddle, J.P.; Witter, M.S.; Cuthill, I.C. 1994. The analysis of fluctuating asymmetry. Animal Behavior. 48: 986-989.

Thornhill, R. 1980. Mate choice in *Hylobittacus apicalis* (Insecta: Mecoptera) and its relation to models of female choice. Evolution. 34: 519-538.

Thornhill, R. 1992. Fluctuating asymmetry and the mating system of the Japanese scorpionfly, *Panorpa japonica*. Animal Behavior. 44: 867-879.

Uhler, L.D. 1951. Biology and ecology of the goldenrod gall fly, *Eurosta solidaginis* (Fitch). Cornell University Agricultural Station Memior. 300: 1-51.

Walton, R.; Weis, A.E.; Lichter, J.P. 1990. Oviposition behavior and response to plant height by *Eurosta solidaginis* Fitch (Diptera: Tephritidae). Annals of the Entomological Society of America. 83: 509-514.

Watson, P.J.; Thornhill, R. 1994. Fluctuating asymmetry and sexual selection. Trends in Ecology and Evolution. 9: 21-25.

THE YEW GALL MIDGE *TAXOMYIA TAXI*: 28 YEARS OF INTERACTION WITH ITS CHALCID PARASITOIDS

M. Redfern and R.A.D. Cameron

Division of Adult Continuing Education, 196-198 West Street, Sheffield S1 4ET, U.K.

Abstract.—Interactions between the yew gall midge, *Taxomyia taxi*, and its chalcid parasitoids are complex. The 2-year generations of the midge show long-term fluctuations which may be cyclical; periods of high density are associated with near absence of the parasitoid, *Torymus nigritarsus*, alternating with troughs when *T. nigritarsus* is common. Recovery of *T. taxi* is dependent on breeding adults from the 1-year generations which are not attacked by *T. nigritarsus*. One-year *T. taxi* generations are regulated by a second parasitoid, *Mesopolobus diffinis*, which has less effect on the 2-year hosts. Regulation exists in this system but more than 20 years' data are required to demonstrate it.

INTRODUCTION

Taxomyia taxi (Inchbald) is a cecidomyiid midge that induces galls in the leaf and flower buds of yew, *Taxus baccata* L. Its life histories and its population dynamics, and those of its chalcid parasitoids *Torymus nigritarsus* (Walker) and *Mesopolobus diffinis* (Walker), on three yew trees in Kingley Vale National Nature Reserve, West Sussex, England, have been subject to study and regular annual monitoring since 1966 (Cameron and Redfern 1978; Redfern and Cameron 1978, 1993).

The system revealed by this study is highly complex, due to the differing life cycles of the host and parasitoids, and to the effects of gall size on the likelihood of successful attack by the two parasitoids (Redfern and Cameron 1993, 1994). Populations of *T. taxi* fluctuate in size over periods of many years. Density-related effects are hard to detect; many which were not apparent after 10 years' study (Redfern and Cameron 1978) eventually emerged after study over a longer period (Redfern and Cameron 1993).

The addition of data from a further 4 years has generally increased the confidence with which we can identify relationships influencing the density of *T. taxi*. In this paper, we update analyses of population dynamics and, in particular, analyses of factors contributing to long-term fluctuations which may be cyclical in nature.

Details of the study and sampling methods are given in Redfern (1975). An account of life histories and general features of the system are given below to provide a context.

LIFE HISTORIES, MORTALITY FACTORS AND INTERACTIONS

T. taxi is monophagous on yew. Almost the whole life-cycle, except for the brief adult, egg and initial free-living stage of the first larval instar (which together last about 3 weeks), occurs inside a gall on the tree. Individual adults live only about a day and are weak fliers. Most individuals develop from egg to adult in 2 years, inside artichoke galls 1- to 3-cm long. Some, however, complete their life cycle in 1 year, in swollen buds less than 6 mm long. Table 1 summarizes both life cycles and identifies mortality affecting each as a series of k-values.

Two chalcid parasitoids attack *T. taxi*: (i) *M. diffinis* has three generations each year in winter and spring on final instar larvae and pupae of *T. taxi*, and attacks hosts in both the 2-year and 1-year galls. The overwintering generation can be recognized but the two spring generations cannot be separated reliably and so are considered together. It is non-specific, attacking other cecids in other host plants during the summer. (ii) *T. nigritarsus* has one generation per year, on fully fed larvae and pupae of *T. taxi*, almost exclusively on the 2-year life

Table 1.—*Mortality of* T. taxi *expressed as k-values in the 2-year and 1-year life cycles. Note that* k_2-k_5 *occur 1 year later in the 2-year cycle. I, II, III = first, second and third instars, respectively.* Md = Mesopolobus diffinis; Tn = Torymus nigritarsus

k-value		Duration in each life cycle	
		2-year	1-year
k_0	Deaths between adult emergence and formation of new galls (2-year and 1-year cycles); includes deaths of adults, eggs and free-living larvae I.	3-5 weeks, late May-June	3-5 weeks, late May-June
k_1	Deaths in larvae I in gall	14-15 months, June - second August	2-3 months, June - first August
k_2	Deaths in larvae II	1 month, September	1 month, September
k_3	Deaths in larvae III, not fully fed; includes *Md* parasitism, overwintering generation.	6-7 months, October - March	6-7 months, October - March
k_4	Deaths in fully fed larvae III and early pupae; includes *Md* parasitism, spring generations.	1 month, April	1 month, April
k_5	Deaths in late pupae; almost entirely due to *Tn* parasitism; virtually confined to 2-year galls.	1 month, May	1 month, May
K	Total mortality, k_0 to k_5	24 months	12 months

cycle, alternately attacking hosts from odd and even years. It is a specific parasitoid, well synchronized with its host, with only the month-long adult stage spent away from *T. taxi*.

Interactions in this system are influenced by the life cycles of the species concerned. Although most *T. taxi* are biennial, adults emerge each year. There are, therefore, two "leapfrogging" populations on any one tree, independent of one another except for the small (usually less than 10 percent) proportion completing the life cycle in 1 year. Thus, mortality inflicted on one population will have no great effect on the density of the other 1 year later.

M. diffinis, with several generations a year, is capable of both numerical and functional responses to the density of its host within a single host generation. It has the potential to cause direct density-dependent mortality. It attacks 1-year galls of *T. taxi* far more frequently than 2-year galls; the size of the latter appears to offer some protection, albeit in a complex way (Redfern and Cameron 1994).

T. nigritarsus is annual, and confined to *T. taxi*. On an annual host, it might give rise to delayed density-dependent effects, causing either limit cycles or dampening or increasing oscillations. Its numerical response, however, falls on a population of *T. taxi* unaffected by mortality caused by it in the previous year. This pattern is unstable and would normally result in the extinction of one of the two overlapping host generations, followed by loss of the parasitoid (Bulmer 1977), unless some hosts occupied "enemy-free space", in which local extinction of the parasitoid might occur.

RESULTS

General Features of the System

Further details of these features are given in Redfern and Cameron (1993). Figure 1 shows variation in the density of *T. taxi* in 2-year galls over the study period, and of the parasitoids attacking 2-year galls. The pattern in 1-year galls (not shown) is similar but with much lower mean densities and smaller fluctuations.

Density fluctuations in *T. taxi* are correlated between trees, and between 1- and 2-year generations on each tree (table 2). Tree 5 and 63, in particular, show similar patterns, with an initial period in which there are oscillations in density between even and odd years, a middle period of near-equality, stability and high-density in both, and a final period in which oscillations return and density falls. Tree 78 lacks the initial oscillation phase, but follow the others thereafter; this convergence is illustrated in figure 2. There are signs of a recovery over the last 3 years.

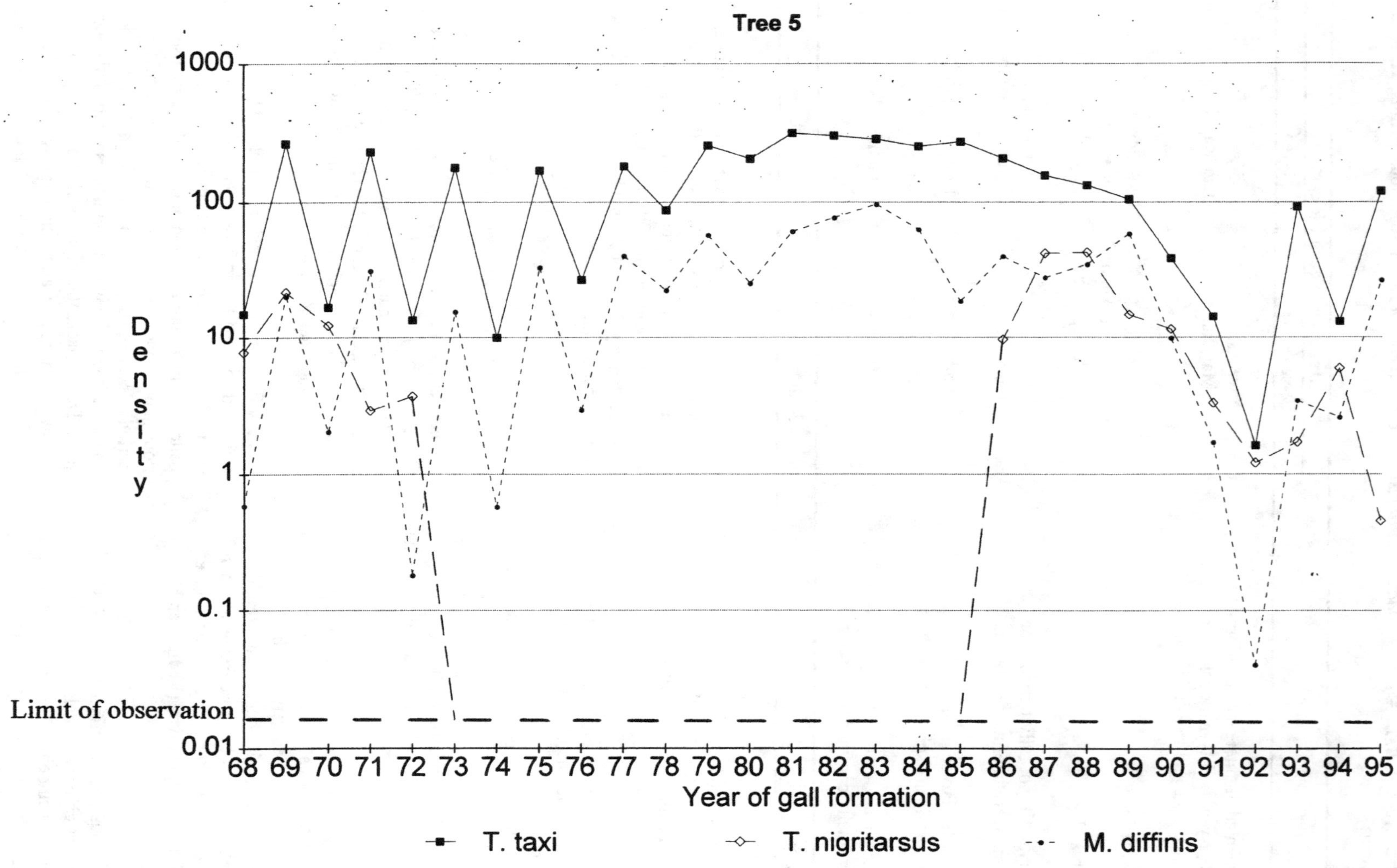

Figure 1.—*Density (number per 1000 shoots) of 2-year* T. taxi *and its parasitoids* T. nigritarsus *and,* M. diffinis *in trees 5, 63 and 78. Note that duration of life cycle of the host is 2 years and of the parasitoids is 1 year or less. Horizontal dashed lines represent limit of observation. ?, galls lost in June 1980.*

(Figure 1 continued on next page)

(*Figure 1 continued*)

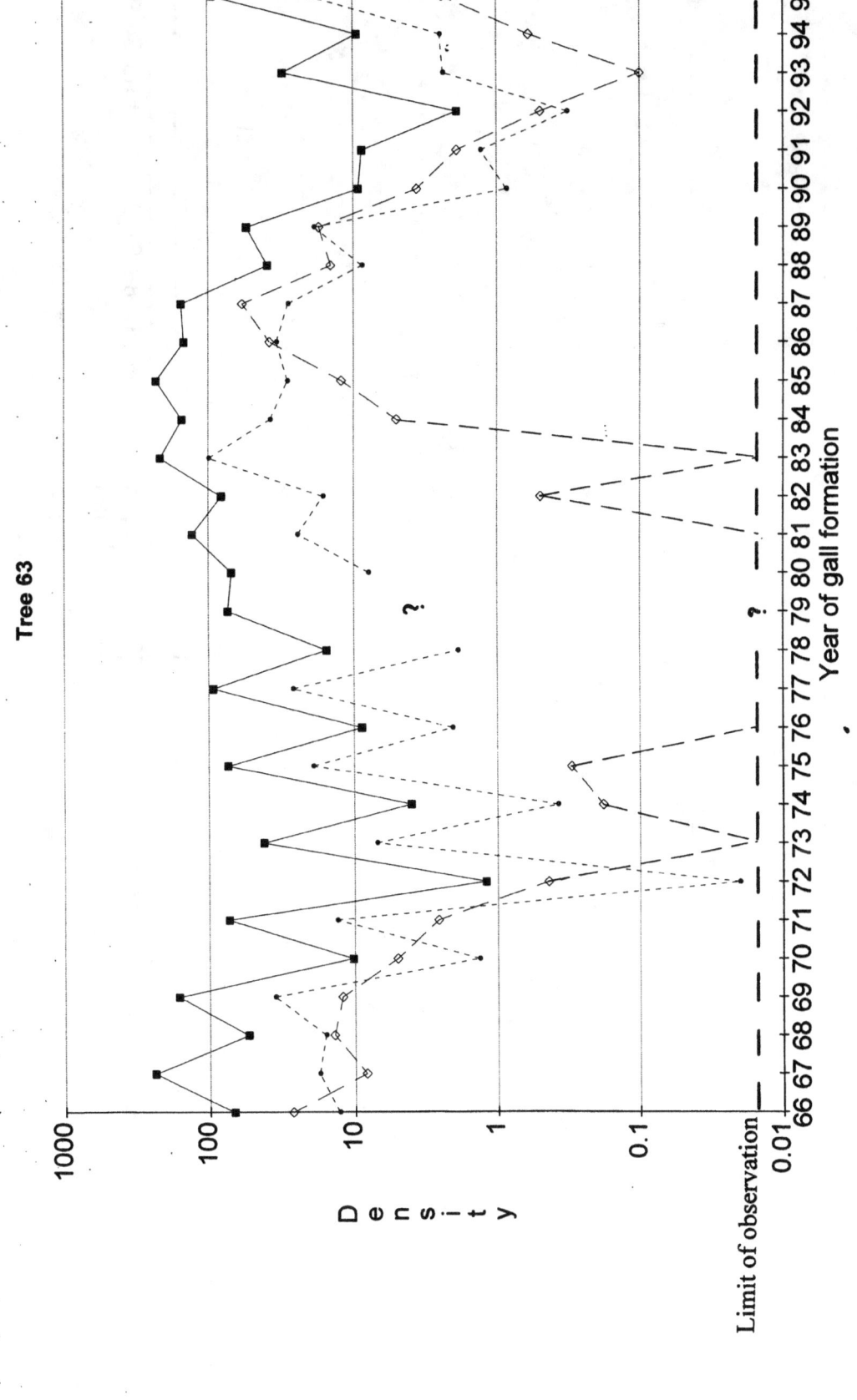

(*Figure 1 continued on next page*)

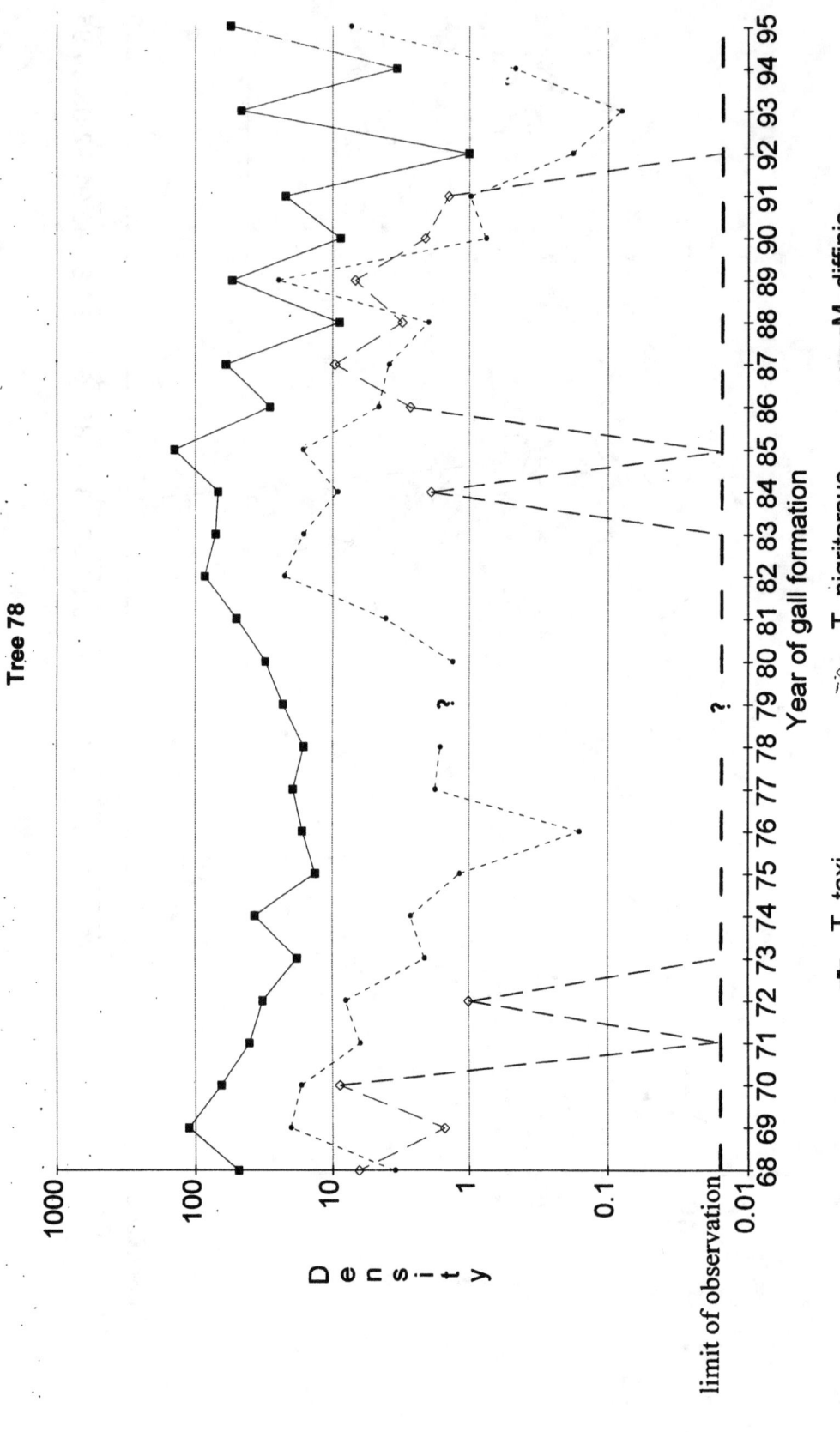

(Figure 1 continued)

Table 2.—*Correlations for number of galls (log.) at the start of the life cycle for (a) 2-year and 1-year generations separately between trees, and (b) between 1- and 2-year generations starting at the same time in each tree. Sample sizes for 1-year galls from tree 78 are often very small; the value entered is based on all years, with notional densities in some cases.* ***p<0.01;* ****p<0.001.*

	(a) Trees	Correlation coefficient	(b) 1-year/ 2-year	Correlation coefficient
2-year	5 and 63	0.892***	Tree 5	0.682***
	5 and 78	0.581**	Tree 63	0.632***
	63 and 78	0.595**	Tree 78	[0.600]
1-year	5 and 63	0.390		
	5 and 78	0.180		
	63 and 78	0.613**		

The density of *M. diffinis* tracks that of its host closely (considered further below). That of *T. nigritarsus* does not: there are periods of absence or extreme rarity in the middle of the study period for all trees (fig. 1).

Table 3 shows the *k*-value analysis of mortality for *T. taxi* on each tree. In trees 5 and 63, k_0 and k_5 have roughly equal influence on total *K* for the 2-year generations but, in tree 78, k_0 is clearly more influential than k_5, which is small and varies less. Factor k_5 is caused almost entirely by *T. nigritarsus*. In 1-year generations, k_4 is clearly the key factor, accounting for most of the variation in total *K*.

Total *K* is higher, on average, in 1-year than in 2-year generations, but this is not true of all years (fig. 3). In trees 5 and 63, total *K* in 2-year generations tends to be greater than that in 1-year generations at times when k_5 is high in the former. As with density, total *K* is correlated between trees over the study period, especially in 1-year galls (Redfern and Cameron 1993).

Density-Related Effects

Table 4 shows the regressions of *k*-values on the densities of *T. taxi* upon which they act. In 2-year generations, overall mortality *K* shows no significant relationship to density in trees 5 and 63, but a significant but under-compensating direct effect in tree 78. Of the individual mortality factors on 2-year generations, k_0 and k_{0-4} combined are significantly but undercompensatingly density-dependent on all trees, while k_5, caused by *T. nigritarsus*, is inversely related to density, significantly so in tree 5.

In 1-year generations, both total *K* and k_4 (caused mainly by spring generations of *M. diffinis*) are significantly density-dependent in all trees and approach full compensation, especially in tree 78.

Thus, *M. diffinis* appears to regulate *T. taxi* in 1-year generations. As shown in table 5, it is itself regulated by the density of available hosts, not only in 1-year galls, but also in 2-year galls, where the incidence of parasitism is much lower. A significant proportion of *T. taxi* in 2-year galls must be protected from parasitism by *M. diffinis*.

The situation is far less clear for *T. nigritarsus* (fig. 4). While there are occasions, generally in even-year generations of *T. taxi*, in which *T. nigritarsus* appears to be limited by the density of its host, there are numerous cases where this is not so, and densities of available hosts in years when *T. nigritarsus* is apparently absent span the whole range.

T. nigritarsus and Host Dynamics

As shown above, even-year populations of *T. taxi* show much more variation in density than those of odd years, and periods of growth or decline extend over several generations. The only factor which appears to promote such variation is k_5, caused by *T. nigritarsus*.

Figure 5 show the variation of N_5 and k_5 over time for each tree, separating even- and odd-year generations. Especially in even years, the figures show an apparently cyclical sequence in which increasing k_5 is followed by decreasing N_5, and low values of N_5 are followed by declining k_5, often with local extinction of *T. nigritarsus*. This effect is confirmed, for trees 5 and 63, by the relationship between k_5 and the starting density (N_1) of galls in the generation descended from the one attacked (table 6). The inverse relationships hold in both odd- and even-year generations separately, and overall. There is only a weak and insignificant trend in tree 78; k_5 is much lower overall in this tree.

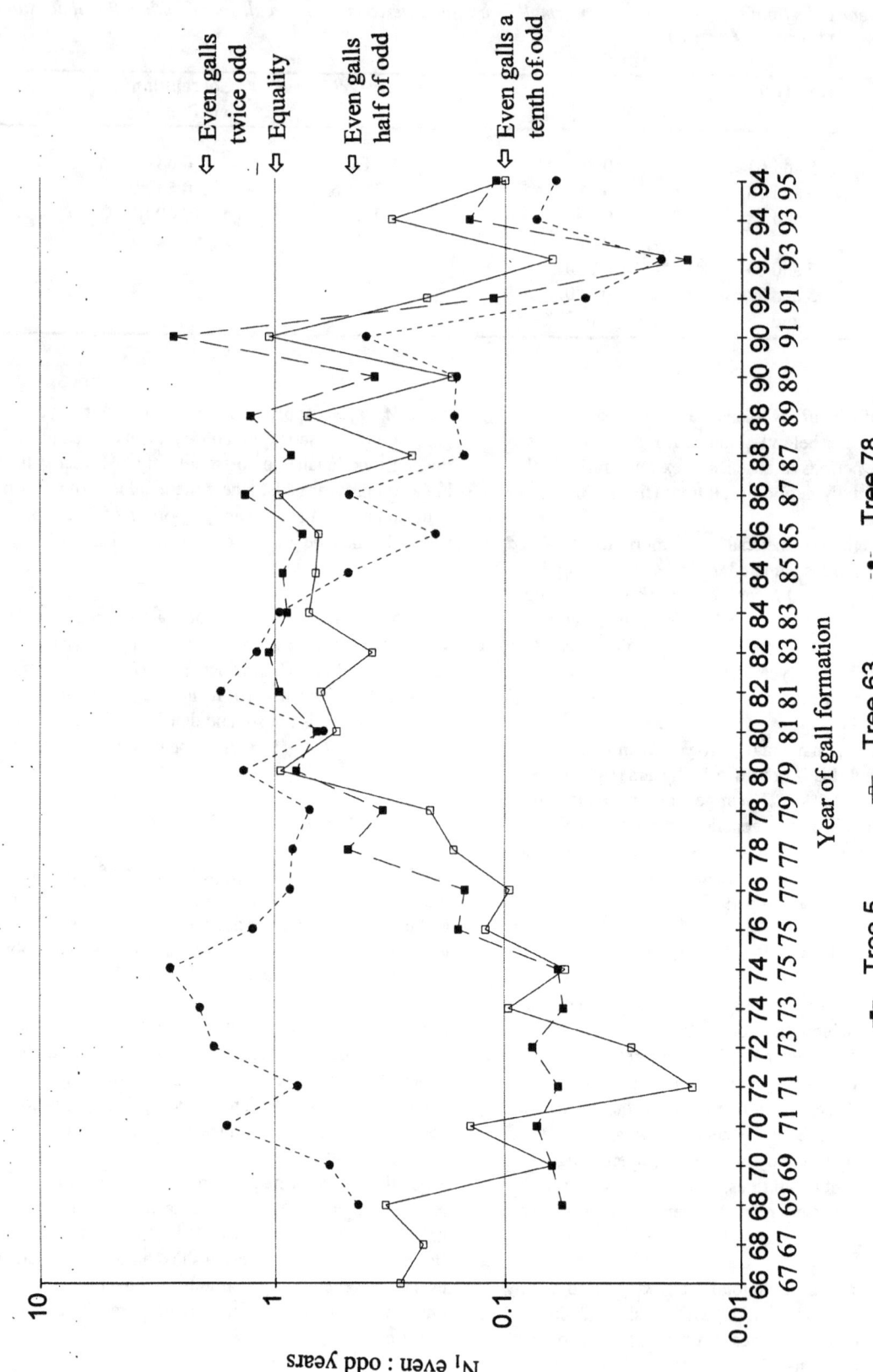

Figure 2.—*The ratio of T. taxi galls formed* ($log.N_1$) *in even and odd years on trees 5, 63, and 78.*

Table 3.—*Mean values for total mortality K and individual k-values, and the slopes of the regressions of k-values on total K. k_1 and k_2 omitted as values usually are very small. k_0 is common to both 1- and 2-year generations; differences within trees are due to a small number of missing estimates, see Redfern and Cameron (1993).* * $p<0.05$; ** $p<0.01$; *** $p<0.001$.

	Tree 5		Tree 63		Tree 78	
	Mean	Slope, b	Mean	Slope, b	Mean	Slope, b
(a) 2-year galls						
Total K	2.062	—	2.102	—	2.063	—
k_0	1.578	0.497***	1.598	0.531***	1.807	0.730***
k_3	0.147	0.055	0.114	0.048	0.080	0.020
k_4	0.140	0.055	0.154	0.014	0.098	0.061
k_5	0.184	0.398***	0.210	0.413***	0.062	0.179**
(b) 1-year galls						
Total K	2.741	—	2.486	—	2.522	—
k_0	1.574	0.297***	1.605	0.190*	1.768	0.062
k_3	0.207	0.183	0.142	0.122*	0.067	0.064*
k_4	0.947	0.505***	0.700	0.622***	0.656	0.793***

Table 4.—*Slope coefficients (b) for regressions of various k-values on the densities (log.) on which they act. Asterisks as in previous tables.*

	Regression slope, b		
	Tree 5	Tree 63	Tree 78
2-year galls			
Total K	0.148	0.215	0.386*
$k_0 - k_5$ incl.	0.344**	0.220*	0.409**
k_0	0.249**	0.202*	0.354**
k_4	0.057*	0.034	0.036
k_5	-0.303**	-0.152	-0.067
1-year galls			
Total K	0.556**	0.773***	0.961
k_3	-0.263	-0.015	0.156
k_4	0.405**	0.421*	1.051*

Table 5.—*Slope coefficients (b) for regressions of log. densities of spring generations of* M. diffinis *on log. densities of available* T. taxi (N_4). *n, number of samples; t_0 value of t for difference of slope from 0; t_1, value of t for difference of slope from 1 (perfect proportionality of densities). Asterisks as in previous tables.*

	n	b	t_0	t_1	Percent parasitism at 10 galls per 1,000 shoots
2-year galls					
Tree 5	28	1.407	12.571***	3.634**	7.41
Tree 6	29	1.199	17.624***	2.925**	14.25
Tree 78	27	1.129	5.255***	0.599	9.09
1-year galls					
Tree 5	26	0.904	8.927***	0.951	55.79
Tree 63	24	1.262	8.050***	1.672	45.16
Tree 78	22	1.181	2.614*	0.401	32.42

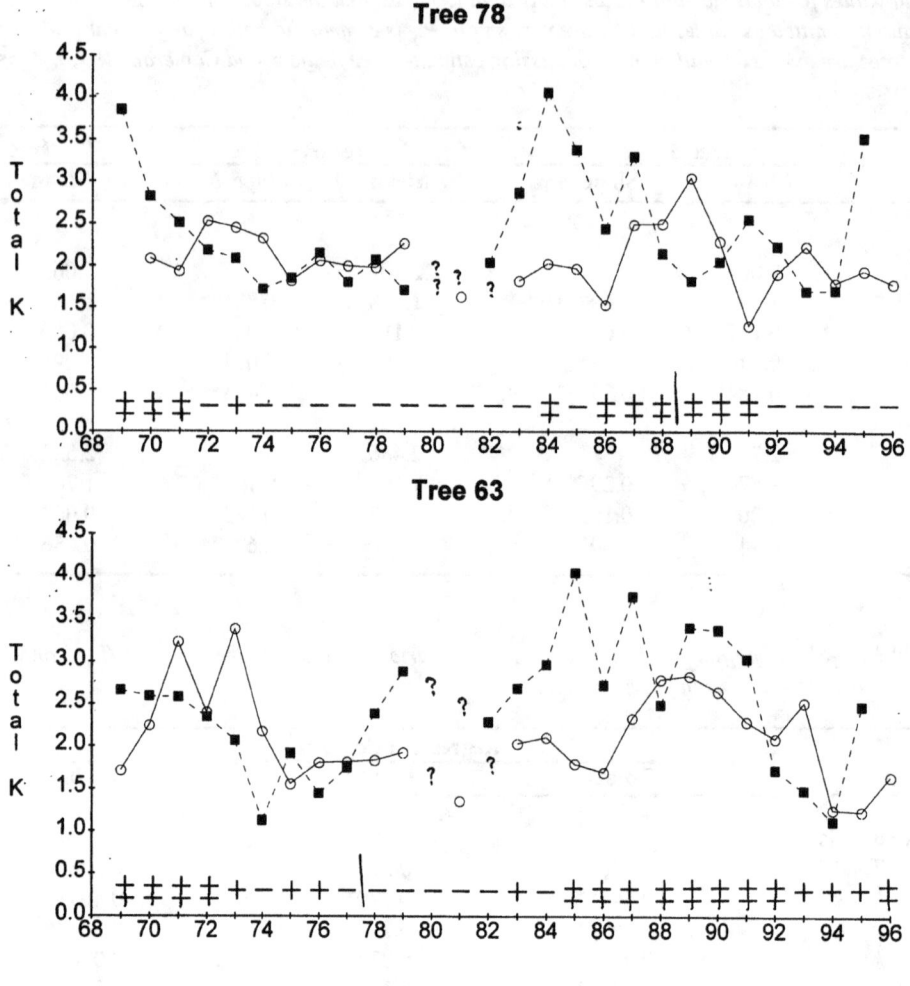
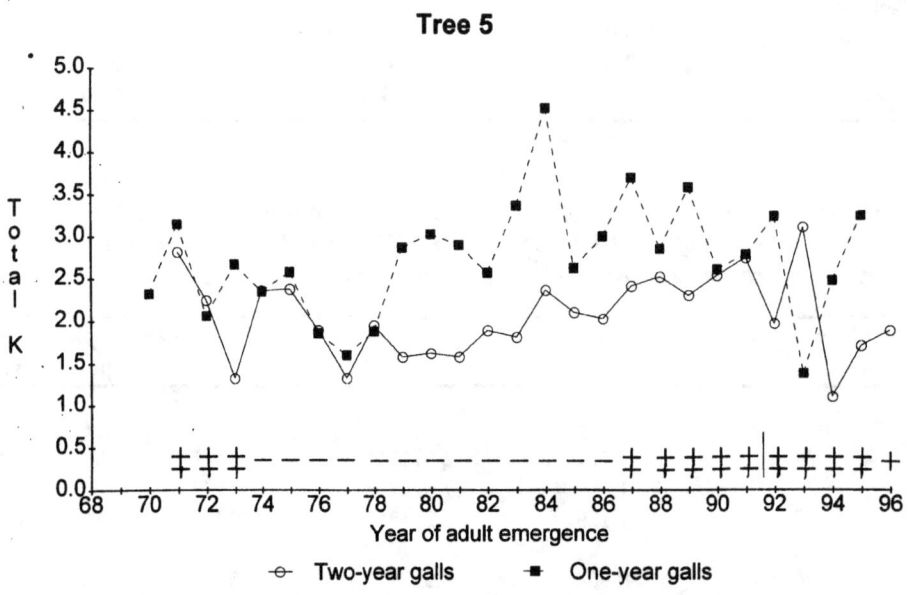

Figure 3.—*Concurrent mortality (total K) each year in 2-year and 1-year galls in trees 5, 63 and 78. Presence (+, 2) and absence (-) of* T. nigritarsus *indicated (2, >1 per 1000 shoots; +, <1 per 1000 shoots).*

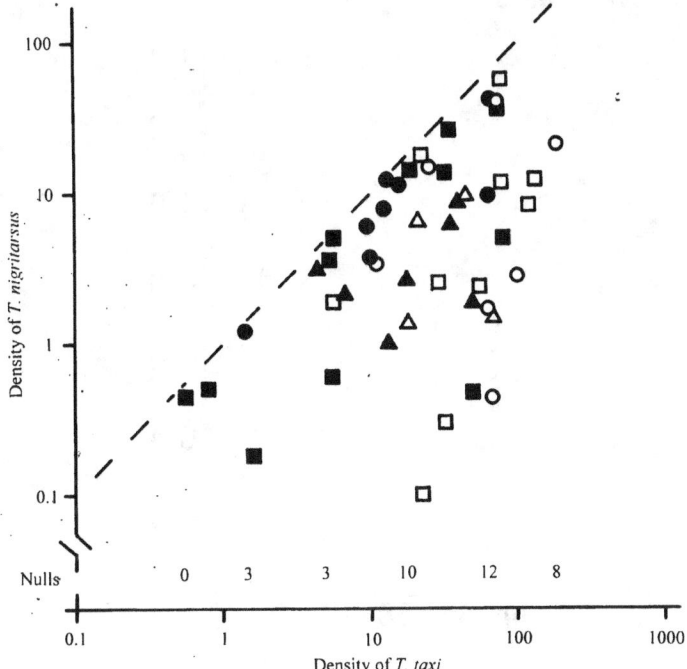

Figure 4.—*Scatter of log. density of* T. nigritarsus *on log. density of available* T. taxi *(N_5). Circles, tree 5; squares, tree 63; triangles, tree 78. Solid symbols, even-year generations; open symbols, odd-year generations. The figures in the nulls row indicate the number of samples of* T. taxi *in which no* T. nigritarsus *were recorded in each 0.5 log. density interval. The dashed line indicates 100 percent parasitism.*

T. nigritarsus does not attack 1-year galls. These usually contribute only a small proportion of the adult *T. taxi* emerging in any one year. Densities in 1-year generations, however, fluctuate less than those in 2-year generations, and it follows that their contribution to the pool of adults should be proportionately greater when the 2-year densities are low. Table 7 shows that this is the case. In particular years, this contribution is clearly necessary for the dramatic recoveries seen in 2-year generation densities. Table 8 presents the data for the two greatest increases each for trees 5 and 63, and the greatest one for tree 78, the latter being much less dramatic. k_0 values required for these increases to be achieved by adults from 2-year galls alone are much smaller than any actually recorded in the study as a whole.

DISCUSSION

Detection of density-dependent relationships from a series of censuses is problematic but becomes less so as more generations are included (Hassell et al. 1989, Manly 1990). Thus we were able to demonstrate some density-dependent relationships after 24 years' study of the *T. taxi* system which were not apparent after 10 years (Redfern and Cameron 1993), and the addition of further data presented here strengthens these relationships.

Mortality of *T. taxi* at the earliest stages of the life cycle (k_0), where competition between hatchlings for buds is known to occur, shows density-dependence in all three trees. This competition is aggravated by the fact that females lay several eggs on a shoot, but larvae cannot move from one shoot to another. It also may be enhanced by limits to the proportion of buds available for infestation. A maximum of 30 percent infestation has been recorded over the whole period, suggesting that not all buds are available.

Once the galls have formed, the patterns of mortality vary with the life cycle of *T. taxi*. In 2-year generations, no other source of mortality adds significantly to density-dependence, and k_5 (parasitism by *T. nigritarsus*) shows a tendency to inverse density-dependence. This might conceal a delayed density-dependent effect: the inverse link between k_5 and the numbers of *T. taxi* in the next generation is even stronger. Parasitism by *T. nigritarsus* is a major cause of instability in *T. taxi* populations (see further discussion in Redfern and Cameron 1993).

In the 1-year cycles, by contrast, parasitism by *M. diffinis* (k_4) is strongly density-dependent and, together with k_0, produces a degree of regulation which constrains population densities effectively; fluctuations are less marked than in the 2-year cycle.

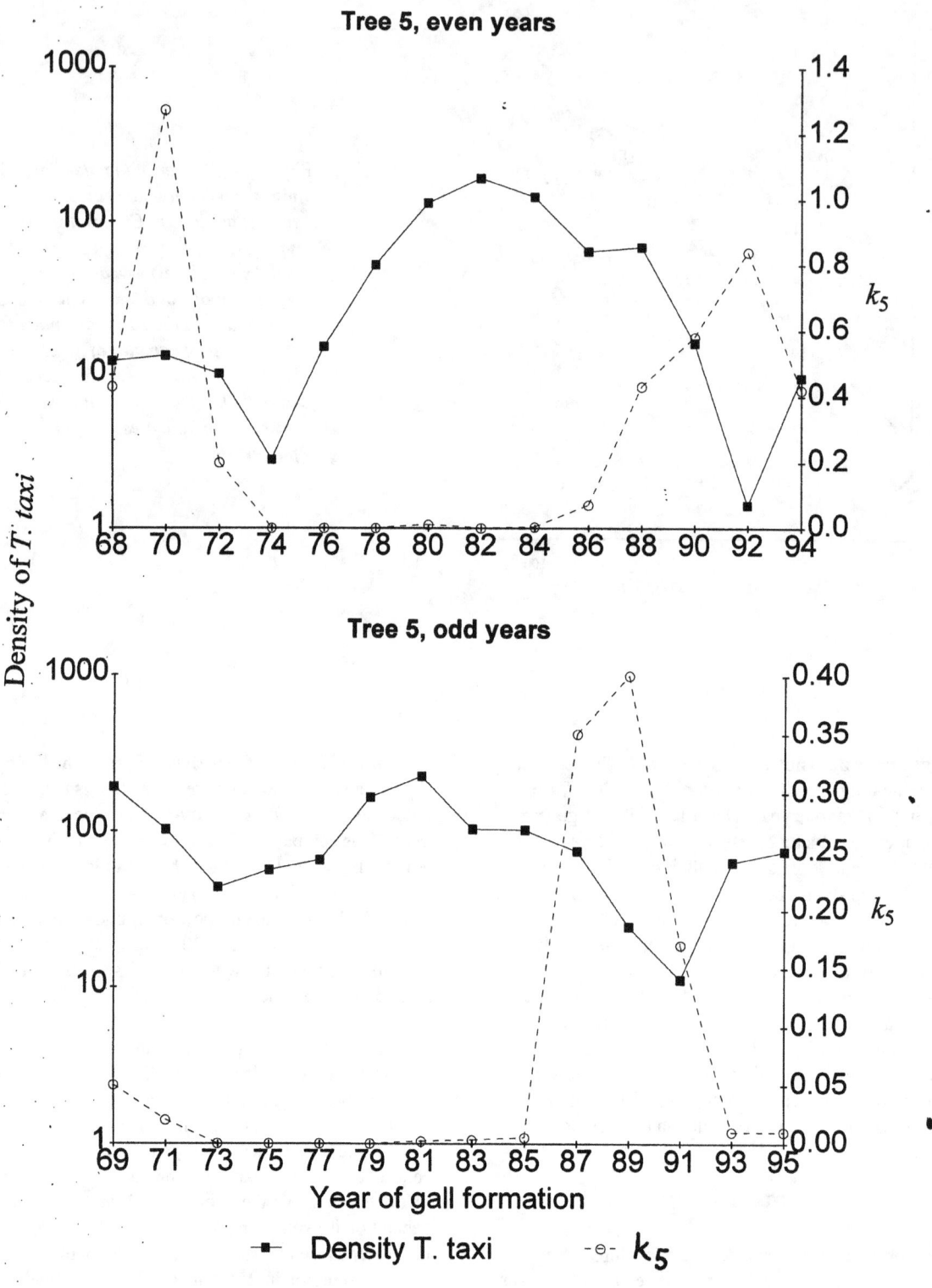

Figure 5.—*Density of available* T. taxi *(log. N_5) and k_5 in even and odd years in trees 5, 63 and 78. ?, galls lost.*

(*Figure 5 continued on next page*)

(*Figure 5 continued*)

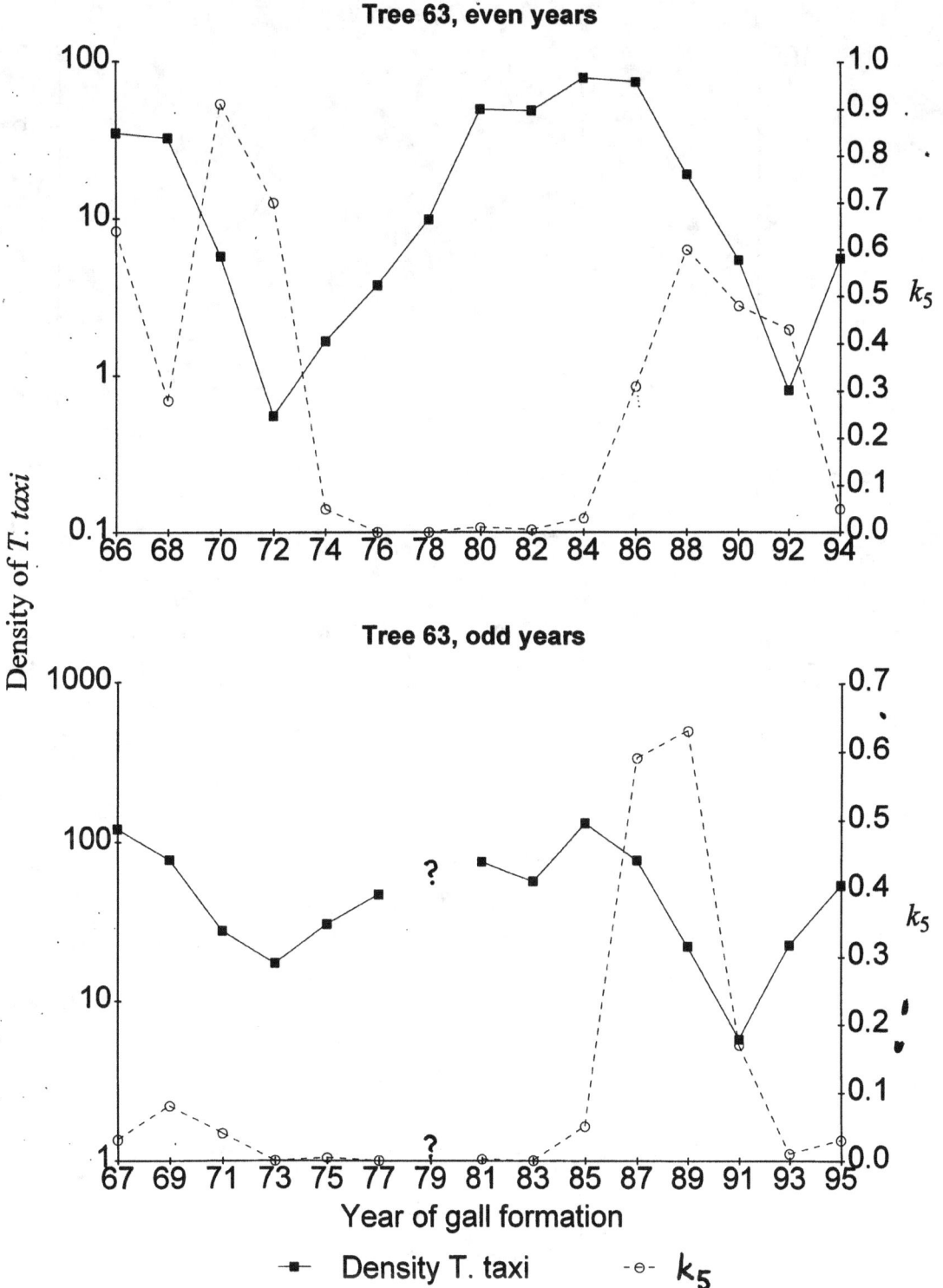

(*Figure 5 continued on next page*)

(*Figure 5 continued*)

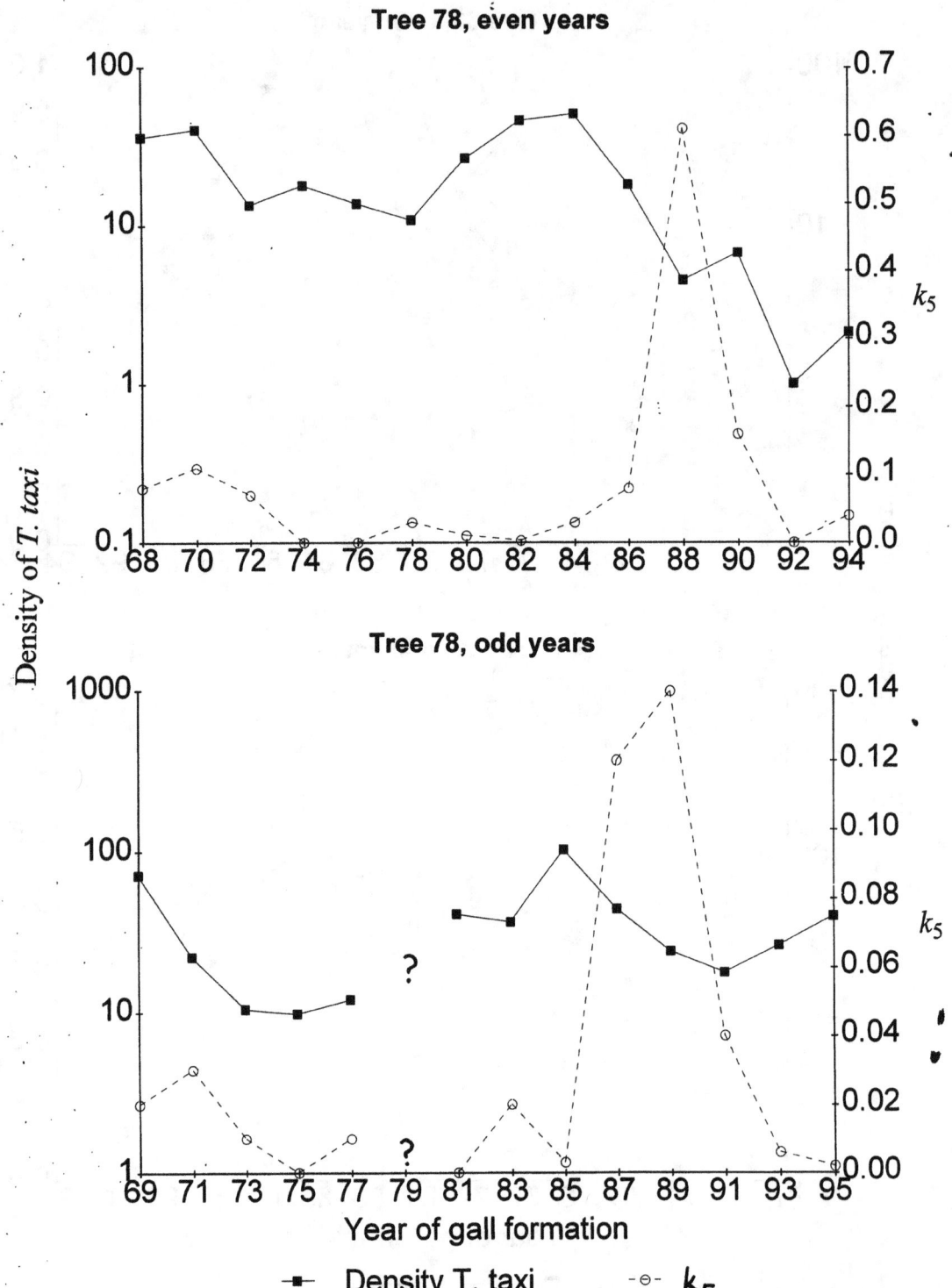

Table 6.—*Correlations (r) between k_5 and log. number of galls formed (N_1) in the next generation of* T. taxi *derived from the one on which k_5 acts, i.e., even year to next even, odd year to next odd. Asterisks as in previous tables; df, degrees of freedom.*

		Tree 5	Tree 63	Tree 78
Even				
	Mean of k_5	0.294	0.306	0.087
	r	-0.657*	-0.693**	-0.269
	df	11	12	11
Odd				
	Mean of k_5	0.074	0.117	0.031
	r	-0.829***	-0.709**	0.058
	df	11	12	11
All				
	Mean of k_5	0.184	0.210	0.062
	r	-0.724***	-0.629***	-0.276
	df	24	26	24

Table 7.—*Regressions of proportion of adult* T. taxi *emerging from 1-year galls against log. total numbers emerging each year. a, intercept; b, slope; df, degrees of freedom. Asterisks as in previous tables.*

	Tree 5	Tree 63	Tree 78
a	0.403	0.441	0.504
b	-0.203***	-0.257***	-0.304**
df	26	25	25

Thus, the proportion of *T. taxi* emerging from 1-year galls in any year is inversely proportional to total emergence. The double life cycle buffers the system. It allows rapid recovery in years in which 2-year cycle emergence is negligible due to heavy parasitism by *T. nigritarsus*, but enables much greater reproductive success overall than if 1-year cycles were involved alone, their density being tightly regulated at low values by *M. diffinis*.

The role of parasitoids in limiting host density is well illustrated by Briggs and Latto (1996) who excluded parasitoids of the cecidomyiid *Rhopalomyia californica* Felt (which causes galls on the perennial shrub *Baccharis pilularis*, Asteraceae) from experimental plots: densities of the hosts increased by three orders of magnitude.

Variable life cycle lengths are not unusual in cecidomyiids (Barnes 1956). In some species, this is achieved by variable diapause in the final instar. In others (as in *T. taxi*), earlier instars are involved (Takasu and Yukawa 1984). Relationships to the size and duration of the gall vary.

In the neolitsea leaf gall midge, *Pseudasphondylia neolitseae* Yukawa, two life cycles seem to have evolved in response to parasitism. As in *T. taxi*, one parasitoid is univoltine and monophagous, the other multivoltine and polyphagous. The pattern of parasitoid attack is different, however, and 1-year *P. neolitseae* are always more frequent than 2-year (Takasu and Yukawa 1984). In this case, unlike that of *T. taxi*, the galls pertaining to the different life cycles look similar. In other cases, it is quite common to find more than one kind of gall induced by one species, for example in cynipids (Redfern and Askew 1992) and pemphigids (Wool and Burstein 1991). These galls, however, are caused by successive generations.

The existence of density-dependent mortality factors is not in itself sufficient to ensure stability. In the 2-year cycles of *T. taxi*, the relationship between the host and *T. nigritarsus* goes beyond limit cycles into a zone of chaotic dynamics in which the host has periods of drastic fluctuations between even and odd years, and the parasitoid becomes locally extinct, at least to the limits of observation possible. The host may be protected from

Table 8.—*The cases of greatest increase from one 2-year generation (N_6) to the next (N_1) for each tree, and the effect of the contribution of the appropriate 1-year generation on each (see text). Densities in numbers per 1000 shoots. Implied k_0 based on the estimate of 107 eggs laid per adult emerging (Redfern and Cameron 1978, 1993).*

	Density (log.) of adults emerged (N_6)	Density (log.) of descendant galls formed (N_1)	Increase N_1 / N_6	Implied k_0
Tree 5				
2-year 1970	0.7	13.5	19.3	0.745
1-year 1971	0.8	1.9	2.37	1.655
1+2-year	1.5	15.4	10.3	1.019
2-year 1992	0.2	13.1	65.5	0.214
1-year 1993	1.8	1.7	0.94	2.055
1+2-year	2.0	14.8	7.4	1.161
Tree 63				
2-year 1972	0.1	4.0	40.0	0.429
1-year 1973	0.6	1.2	2.0	1.730
1+2-year	0.7	5.2	7.4	1.160
2-year 1992	0.3	9.4	31.3	0.534
1-year 1993	0.7	1.6	2.28	1.672
1+2-year	1.0	11.0	11.0	0.989
Tree 78				
2-year 1989	1.1	8.7	7.90	1.163
1-year 1990	2.6	2.3	0.88	2.083
1+2-year	3.7	11.0	2.97	1.556

local extinction by transfers from 1-year generations, and there is undoubtedly some metapopulation structure involved, even though dynamics in neighboring trees are positively correlated.

Such annual alternation of high and low densities has also been recorded by Wool (1990) in the pemphigid *Baizongia pistaceae* (L.), which induces galls on pistachio *Pistacia palaestina*. *B. pistaceae* has a 2-year life cycle and Wool suggests that two possible explanations are plausible: odd- and even-year generations, which do not interbreed, have genetically different capacities for population growth or colonization, or the load imposed by galling in one year adversely affects the suitability of the leaves of *Pistacia* for infestation in the next. "Leapfrogging" generations which are genetically isolated are unstable (Bulmer 1977). Wool did not study mortality or its causes in these populations, which might shed more light on their dynamics.

Ideally, our long-term study should have included more trees, so that spatial effects could be examined in more detail; Wool's (1990) study indicates the possibilities of detecting varying degrees of synchronicity in population fluctuations. Even on the data available, however, prolonged study coupled with analysis of mortality reveals complex patterns explicable only in terms of both ecological and evolutionary processes.

ACKNOWLEDGEMENT

We thank English Nature for permission to collect galls in Kingley Vale.

LITERATURE CITED

Barnes, H.F. 1956. Gall midges of economic importance. Vol. VII. London: Crosby Lockwood. 261 p.

Briggs, C.J.; Latto, J. 1996. The window of vulnerability and its effects on relative parasitoid abundance. Ecological Entomology. 21: 128-140.

Bulmer, M.G. 1977. Periodical insects. American Naturalist. 111: 1099-1117.

Cameron, R.A.D.; Redfern, M. 1978. Population dynamics of two hymenopteran parasites of the yew gall midge *Taxomyia taxi* (Inchbald). Ecological Entomology. 3: 265-272.

Hassell, M.P.; Latto, J.; May, R.M. 1989. Seeing the wood for the trees: detecting density dependence from existing life-table studies. Journal of Animal Ecology. 58: 883-892.

Manly, B.F.J. 1990. Stage-structured populations: sampling, analysis and simulation. London: Chapman and Hall.

Redfern, M. 1975. The life history and morphology of the early stages of the yew gall midge *Taxomyia taxi* (Inchbald) (Diptera: Cecidomyiidae). Journal of Natural History. 9: 513-533.

Redfern, M.; Askew, R.R. 1992. Plant galls. Naturalists' Handbooks 17. Slough, Richmond Publishing Co. 99 p.

Redfern, M.; Cameron, R.A.D. 1978. Population dynamics of the yew gall midge *Taxomyia taxi* (Inchbald) (Diptera: Cecidomyiidae). Ecological Entomology. 3: 251-263.

Redfern, M.; Cameron, R.A.D. 1993. Population dynamics of the yew gall midge *Taxomyia taxi* and its chalcid parasitoids: a 24-year study. Ecological Entomology. 18: 365-378.

Redfern, M.; Cameron, R.A.D. 1994. Risk of parasitism on *Taxomyia taxi* (Diptera: Cecidomyiidae) in relation to the size of its galls on yew *Taxus baccata*. In: Williams, M.A.P., ed. Plant galls: organisms, interactions, populations. Systematics Association Special Volume No. 49: 213-230.

Takasu, K.; Yukawa, J. 1984. Two-year life history of the neolitsea leaf gall midge, *Pseudasphondylia neolisteae* Yukawa (Diptera: Cecidomyiidae). Kontyû. 52: 596-604.

Wool, D. 1990. Regular alternation of high and low population size of gall-forming aphids: analysis of ten years of data. Oikos. 57: 73-79.

Wool, D.; Burstein, M. 1991. A galling aphid with extra life-cycle complexity: population ecology and evolutionary considerations. Research in Population Ecology. 33: 307-322.

ON THE BEGINNINGS OF CECIDOLOGY IN 19TH CENTURY NORTH AMERICA: BASSETT'S DISCOVERY OF HETEROGONY IN OAK GALL WASPS (HYMENOPTERA: CYNIPIDAE)

Alfred Wehrmaker

Biologisches Institut, Universität Stuttgart, GERMANY (retired)
Present address: Im Lehenbach 10, D-73650 Winterbach, GERMANY

Abstract.—The discovery of bivoltinism and heterogony in North American oak cynipids resulted from the mutually stimulating endeavors of R. Osten Sacken, B.D. Walsh, and H.F. Bassett. Osten Sacken and Walsh, though themselves not solving the problem completely, contributed valuable ideas to challenge Hartig's conjecture of permanent agamy. Thus the way was paved for Bassett's fundamental observation, on 28 June 1864, of *Cynips quercusoperator* females ovipositing, upon emergence from catkin galls, into half-grown acorns. This discovery, indicative of connection between voltinism, dimorphism and agamy, preceded Adler's breeding experiments in Europe.

INTRODUCTION

In Europe, the discovery of cyclical parthenogenesis or heterogony (alteration of asexual and sexual generations) in oak-galling cynipids is traditionally credited to Hermann Adler (1841-1921), a German physician and amateur entomologist who succeeded in demonstrating its occurrence in correlation with seasonal change in the type of gall induced, in gall wasps native to the northern regions of the continent. Adlers's thorough breeding experiments, however, were predated (as has frequently been overlooked) by field studies carried out in North America by three dedicated naturalists working at places far apart in the country: Osten Sacken in Washington DC, and New York, Walsh in Illinois, and Bassett in Connecticut. How these three men so extremely different in origin and station in life, tackled the cardinal problem of cynipid reproduction and their individual shares in solving it, will subsequently be outlined.

Carl Robert von der Osten Sacken (1828-1906)

When in early June 1856 the young Baron Osten Sacken embarked at Liverpool for the United States to take up the duties of his new position in the diplomatic service of the Russian Empire, he envisaged, as he relates in his autobiography (Osten Sacken 1903-1904), devoting his spare time to collecting and "working up" the Diptera of the Nearctic fauna, a task which, partly in co-operation with the German dipterist Hermann Loew, was accomplished in the course of the next 21 years (Alexander 1969). An enthusiastic collector since his boyhood, he had already acquired a good knowledge of the insects of the St. Petersburg area in Russia, and even published a first article on his lifelong favorites, the crane flies or Tipuloidea. Scion of a prosperous German-Baltic family with a landed estate in Estonia, he was in a position to pursue his hobby with remarkable vigor, soon entering into contacts with leading entomologists of his time. His multilingual training (in addition to German, his mother tongue, he spoke at least Russian, French, and English) enabled him to use the international literature to the fullest. Possibly the diversity of the oak galls he encountered in the environs of Washington moved him, in the autumn of 1860, to direct his attention, for nearly 10 years, to the Hymenoptera inducing them. Previous work on American gall wasps being almost negligible, his first concern, of course, was to describe and name the numerous new species he was confronted with. In this way he soon became intrigued by the well-pronounced selectivity of the oak-galling wasps with respect to the kind of oak they attack. His observations suggested that in eastern North America there occur two separate sets of cynipid species causing galls on *Quercus*: those of the first set confined to the "white oaks", those of the other, to the "red oaks" (Osten Sacken 1865). This rule (termed "Osten Sacken" rule, Wehrmaker 1990) has proved fundamental to understanding cynipid host-plant specificity in Nearctic America.

Furthermore, Osten Sacken (1861c) pointed out the necessity of voltinism studies in oak gall wasps. In North America, as in Europe, each oak gall appears only at a well-defined, and gall-specific, time of the year, thus suggesting that the gall-producer is single-brooded. Usually, a space of several or even many months elapse between the emergence of the adult wasp from the gall and the subsequent reappearance of this kind of gall on the plant. As the cynipid female, being extremely short-lived, oviposits soon after leaving its gall, Hartig (1840) took it for granted that the egg passed through a resting stage long enough to account for the apparent delay in the formation of the gall. Osten Sacken, by contrast, when observing that, from unmistakable galls on the very young leaves of the Pin Oak, the adults of "*Cynips querous palustris* OS." emerged already "very early in the season", instantly conjectured that this species may be double-brooded, with the second annual generation producing, on the more advanced leaves perhaps, a different kind of gall. This supposition was undoubtedly based (though he did not state it expressly) on the fact that, in the middle of May when the wasps were on the wing, the buds of next spring's growth were still too undeveloped to allow for a proper placement of the cynipid egg. Also for the bisexual cynipids emerging toward "midsummer" from vernal oak-apples, Osten Sacken (1862) felt inclined to consider bivoltinism.

The problem, however, that, from the very beginning of his studies, puzzled Osten Sacken most, arose from the insufficient knowledge of the reproductive habits of the oak-galling cynipids, and the uncertainty resulting therefrom in cynipid taxonomy. As one would expect, Osten Sacken's approach to cynipid classification was initially guided by the principles proposed by the German chief forester, forest botanist and forest entomologist Theodor Hartig (1805-1880) for the species of Central Europe (Hartig 1840, 1841, 1843). A peculiarity of Hartig's cynipid system was the extraordinary importance attributed to the mode of reproduction he believed to have ascertained. Noticing that from a given kind of oak gall either both males and females emerge (as is the case in bisexual reproduction), or exclusively females (which would always propagate agamically by parthenogenesis), Hartig concluded that in oak cynipids many purely agamic species do exist. Moreover, because all of the agamic species thus recognized also differ in structural traits from all of the bisexual species, it even appeared appropriate to him to distinguish wholly agamic "genera" from wholly bisexual ones. Among the latter, for example, he named the genus "*Andricus*", the etymology of which (from Greek adjective meaning "featuring males") still reminds us of this classificatory concept now long since obsolete.

Osten Sacken, to be sure, quickly realized how difficult it was to apply Hartig's taxonomic principles and definitions to the oak-galling cynipids of North America, the more as he was not really convinced that Hartig's inferences about the taxonomic relevance of agamy in gall wasps were free from error. So, when on 17 June 1861, a male wasp emerged from a spindle-shaped oak-gall that was closely similar to the supposedly agamic female he had obtained from the globular oak-apples of *Cynips confluens*, the idea immediately struck him that in the cases of supposed agamy the males were by no means lacking. On the contrary, they grew up simultaneously, but in galls distinctly differing in shape from those for the female sex. Hartig, unaware of this possibility, had presumably misidentified the males.

Osten Sacken hastened to communicate his observation, and the conclusion drawn from it, to the Academy of Natural Sciences in Philadelphia, to be reported at the Academy's meeting of 2 July 1861 (Osten Sacken 1861a, b, c). However his bold attempt at solving the "vexed question" of the sexes in oak-galling cynipids remained unsupported by conclusive evidence: In Europe, no kinds of oak gall had ever been found that produced male wasps only. In America, males were finally reared from globular oak-apples also. Consequently, he was compelled to drop his assertions a short while later (Osten Sacken 1862). Nonetheless, though he had not been the first to challenge Hartig's views, his temperamental way of approaching the problem gave fresh impetus to resuming the study of cynipid agamy.

Osten Sacken's manifold professional duties, at first as Secretary of the Russian Legation at Washington, then (from mid-1862) as Russian Consul-General at New York, prevented him from pursuing his hunches with the leisure required for such studies. Fortunately, however, two gifted American naturalists, encouraged by his writings and his letters, had joined him in his endeavors to unveil the secrets of cynipid life cycles. Thus, at the time when the Entomological Society of Philadelphia had just been founded to provide its "Proceedings" a ready means for publishing and disseminating new entomological knowledge, the stage was set for discovery.

Benjamin Dann Walsh (1808-1869)

Born in England and 20 years older than Osten Sacken, Walsh was a personality in many aspects quite unlike the German-Baltic baron with whom he so heartily co-operated. Educated at Cambridge University, he was a classmate of Darwin, who in a late edition of the "Origin of Species", called Walsh "a distinguished entomologist of the United States". Majoring in classics, he had obtained the M.A. degree, but refused to enter the ministry. He emigrated in 1838 to North America, became a farmer in Illinois, and turned to dealing in lumber. He was successful in this business and became interested in insect pests. His interest in insects in general led to his appointment in 1867 as the first professional "State Entomologist" of Illinois (Hagen 1870, Osten Sacken 1903-1904). Although working far away from academic institutions and almost excluded from direct personal contact with fellow entomologists, he contributed a number of pioneering studies to the knowledge of insect life in his adopted country. Perhaps his most remarkable paper (Walsh 1864) was devoted to the oak-galling cynipids of Illinois. In this article, he communicated the results of a 1-year field survey of the reproductive behavior of an oak-apple inducer named *Cynips aciculata* by Osten Sacken. From the galls of this wasp, Walsh had always bred only females, a fact suggesting that he had met with an American counterpart of Hartig's supposedly agamic species. Determined "to solve this mystery", he thus made a point of collecting, in 1863, representative samples of these galls at appropriate intervals spread over the whole year, a project feasible because in a small wood lot near Rock Island, on Black oak, *Quercus velutina*, oak-apples of this kind occurred abundantly.

But what Walsh did observe, under such favorable conditions, turned out to be rather perplexing. For, to his astonishment, many adult wasps, both males and females, emerged in June, all of them from galls freshly formed in May, and all of them evidently belonging to *C. spongifica*, a "species" considered different from *C. aciculata*. From many others of the oak-apples gathered in June, however, adult wasps emerged only during the following autumn and winter, this time all of them females, and exhibiting on face and thorax the dense little grooves ("aciculations") typical of *C. aciculata*. Since the galls issuing *spongifica* and those issuing *aciculata* were indistinguishable in shape and size, and had appeared on the same individual trees, Walsh arrived to the subsequent conclusions: (1) The two alleged species are by no means specifically distinct. (2) Rather, the *aciculata* female is a dimorphic form of the *spongifica* female, with the peculiarity of hatching at a later season and propagating agamically. (3) Obviously, *spongifica* and *aciculata* do not constitute different broods that alternately generate each other. (4) Presumably, the two types of females also differ in the kind of offspring, with the mated *spongifica* female generating females only, either *spongifica* or *aciculata*, the agamic *aciculata* female, only males.

This well-contrived interpretation, which Osten Sacken (1864) promptly communicated to entomologists in Germany, was bound to provoke strong criticism in that country, not only by Hartig (1864) himself, but also by Reinhard (1865), both stressing that Walsh's conjectures were totally inapplicable to the agamic oak-galling cynipids at least of Europe. Reinhard went so far as to place *spongifica* and *aciculata* in separate genera, for this purpose assigning *spongifica* to a new genus, *Amphibolips*, a name still in use. Yet, though clouded with much guesswork, Walsh's basic tenet that *spongifica* and *aciculata* are conspecific, was valid. It seems that the extremely early appearance of the *aciculata* gall (a timing quite exceptional for an "autumnal" gall!), and its perfect agreement in structure and size with the *spongifica* gall, had misled Walsh in his attempt to understand the elusive annual cycle of this species of wasp now properly called *Amphibolips confluens*. As Bassett (1873) surmised, "another generation may have intervened".

Homer Franklin Bassett (1826-1902)

So the credit of having discovered heterogony in oak-galling cynipids goes to H.F. Bassett, the only American-born of the three founders of American cecidology. Brought up in Ohio, and educated in Berea (Ohio) University and Oberlin College, he was headmaster of a private school in Waterbury, CT, until, in 1867, ill health forced him to discontinue teaching. Years of uncertainty followed but finally, in 1872, he was appointed librarian of the Brandon Library in Waterbury, a task in which he excelled (Anon. 1902, Osten Sacken 1903-1904).

To Bassett, who was acquainted with the vagaries of life only too well, watching gall wasps in the field or rearing them at home was, for 40 years, a matter nearest to his heart, and "always an unfailing source of purest happiness" (Bassett 1900). Motivated by Osten Sacken's advice to pay particular attention to the habits of these insects, he noticed as early as June 1863 that, in *Cynips quercusoperator* OS., copulation took place in his breeding boxes immediately upon adult emergence, suggesting to him that females would oviposit soon after, "probably" in the buds for next years' growth, which were already "fully formed" at the end of this month (Bassett 1863). Thus 1 year later it was a bit of a surprise to him when in the afternoon of 8 June 1864, "on visiting a shrub oak (*Q. ilicifolia*) thicket" he "found hundreds of *C.quercusoperator* with their ovipositor..." inserted the full length into the cups of young acorns." He instantly realized that the next generation of this wasp was destined to grow up, not in the beautiful plurilocular catkin galls derived from buds, but somewhere in the space between acorn cup and acorn. Bassett grew so excited about what he witnessed that on the very same day he communicated his observation to the president of the Entomological Society of Philadelphia. According to this letter, which was read at the Society's meeting on 11 July 1864 and also published (Bassett 1864a,b), he was at once conscious of the importance of the discovery "that will probably solve the mystery" of voltinism in vernal oak galling cynipids and "may possibly clear up the subjects of 'agamic species' and 'dimorphism'": "I shall await, impatiently, further developments."

Bassett was right. When, in the middle summer, pip-like galls resembling stunted chestnuts became visible between acorn and cup, no doubt was left that there are two generations in *C. quercusoperator*, one vernal and one autumnal, producing dissimilar galls, with those of the latter brood overwintering on the ground. And when in 1873, though with some delay, agamic females were obtained from such pip galls, and were also observed ovipositing in buds, the cycle was finally closed, and heterogony was established to be the regular mode of reproduction in this species of wasp (Bassett 1873, 1877, 1889, 1900; Riley 1873).

To pay proper tribute to their merits, one must acknowledge that Osten Sacken and Walsh, too, were on the right track when searching for explanations: Osten Sacken when proposing double-voltinism (a habit common in oak-gallers), or when considering the occurrence, in the same species of two kinds of gall (different kinds indeed occur but he did initially not imagine that apparently agamic species males might develop in galls of an alternate brood). Walsh, when suggesting dimorphism to exist in the female sex (there is dimorphism but, contrary to what he conjectured, the two forms of female alternate). Bassett, however provided more than a "first clue to the solution of the mystery." as he modestly wrote in 1889. Rather, his was the first demonstration of the event that is central to oak cynipid voltinism: the change from one kind of gall to an alternate one. His discovery of 1864, therefore, is a highlight of the early years of American cecidology.

Concluding Remarks

In North America, Bassett (1877, 1889, 1900), able to present further instances of cynipid heterogony, repeatedly returned to discussing this subject but unfortunately, compelled to earn his living until close to his death, he was frequently lacking the leisure required for extended field work. Walsh died prematurely in 1869, and Osten Sacken left the United States finally in 1877. Until recently, almost no American hymenopterist, including Ashmead, Beutenmuller, and Welsh has really cared to resume such studies on a large scale.

Kinsey, whose first efforts along this line (Kinsey 1920) seemed promising (he correctly recognized, in a species of *Zopheroteras*, the alternate generation *of C.*

quercuspalustris) later turned his attention to the habits of living things of quite another kind. Consequently, cynipid taxonomy in North America in still partly in a state of chaos.

In Europe, Bassett's original report (1864a) remained unknown, not only to Reinhard (1865), but also, except for Kieffer (1914), to later cecidologists. Having been published merely in the minutes section of a Society's Proceedings, without a title of its own and presumably without reprints available for distribution, the paper was easy to disregard, a deplorable situation that even Kinsey's detailed reference to it (1920) did not improve. When Adler, in 1874, began his famous experiments, he was totally ignorant of what Bassett had observed and recorded, with so much enthusiasm, on 28 June 1864 (Adler 1877). When he finally learnt of Bassett's pioneering studies, it was, not Bassett's letter of that date, but Bassett's article of 1873 of which he became aware (Adler 1881). In any case, the priority of Bassett's discovery is evident: Bassett was the "Mendel" of this branch of science; and Adler, only the "Correns, Tschermak, and deVries."

LITERATURE CITED

Adler, H. 1877. Beiträge zur Naturgeschichte der Cynipiden. Deutsche Entomologische Zeitschrift. 21: 209-247.

Adler, H. 1881. Über den Generationswechsel der Eichen-Gallwespen. Zeitschrift für wissenschaftliche Zoologie. 35: 151-246.

Alexander, C.P. 1969. Baron Osten Sacken and his influence on American dipterology. Annual Review of Entomology. 14: 1-18.

Anon. 1902. Obituary (H.F. Bassett). Entomological News. 13: 203-205.

Bassett, H.F. 1863. Descriptions of several supposed new species of *Cynips*, with remarks on the formation of certain galls. Proceedings of the Entomological Society of Philadelphia. 2: 323-333.

Bassett, H.F. 1864a. (Communication dated 28 June 1864). Proceedings of the Entomological Society of Philadelphia 3: 197-198.

Bassett, H.F. 1864b. Descriptions of several new species of *Cynips*, and a new species of *Diastrophus*. Proceedings of the Entomological Society of Philadelphia 3: 679-691.

Bassett, H.F. 1873. On the habits of certain gall insects of the genus *Cynips*. Canadian Entomologist. 5: 91-94.

Bassett, H.F. 1877. Agamic reproduction among the *Cynipidae*. Proceedings of the American Association for the Advancement of Science. 26: 302-306.

Bassett, H.F. 1889. A short chapter in the history of the cynipidous gall flies. Psyche. 5: 235-238.

Bassett, H.F. 1900. New species of North American Cynipidae. Transactions of the American Entomological Society. 26: 310-336.

Hagen, H. 1870. Benjamin D. Walsh. Entomologische Zeitung (Stettin). 31: 354-356.

Hartig, T, 1840. Über die Familie der Gallwespen. Zeitschrift für die Entomologie (German). 2: 176-209.

Hartig, T. 1841. Erster Nachtrag zur Naturgeschichte der Gallwespen. Zeitschrift für die Entomologie (German). 3: 322-358.

Hartig, T. 1843. Zweiter Nachtrag zur Naturgeshicte der Gallwespen. Zeitschrift für die Entomologie (German). 4: 395-422.

Hartig, T. 1864. (Review: Über die Parthenogenesis der Cynipiden). Berliner Entomologische Zeitschrift. 8: 405.

Kieffer, J.J. 1914. Die Gallwespen (Cynipiden) Mitteleuropas, inbesondere Deutschlands. In: Schroeder, C., ed. Die Insekten Mitteleuropas, inbesondere Deutschlands. Stuttgart, Germany: Franckh. 3(3): 1-94.

Kinsey, A.C. 1920. Life histories of American *Cynipidae*. Bulletin of the American Museum of Natural History. 42(6): 319-357.

Osten Sacken, R. 1861a. Communication on the sex of Cynipidae. Proceedings of the Academy of Natural Sciences of Philadelphia. 1861: 150-152.

Osten Sacken, R. 1861b. Über die Gallen und andere durch Insecten hervorgebrachte Pflanzendeformationen in Nord-America. Entomologische Zeitung (Stettin). 22: 405-423.

Osten Sacken, R. 1861c. On the Cynipidae of the North American oaks and their galls. Proceedings of the Entomological Society of Philadelphia. 1: 47-72.

Osten Sacken, R. 1862. Additions and corrections to the paper entitled: "On the Cynipidae of North American oaks and their galls". Proceedings of the Entomological Society of Philadelphia. 1: 241-259.

Osten Sacken, R. 1864. Über den wahrscheinlichen Dimorphismus der Cynipiden-Weibchen. Entomologische Zeitung (Stettin). 25: 409-413.

Osten Sacken, R. 1865. Contributions to the natural history of the Cynipidae of the United States and of their galls. 4th Proceedings of the Entomological Society of Philadelphia. 4: 331-380.

Osten Sacken, R. 1903-1904. Record of my life work in entomology. Facsimile reprint 1978. Series Classica Entomologica 2. Faringdon, England: Classey. 240 p.

Reinhard, H. 1865. Die Hypothesen über die Fortpflanzungsweise bei deneingeschlechtigen Gallwespen. Berliner Entomologische Zeitschrift 9: 1-13.

Riley, C.V. 1873. Controlling sex in butterflies. The American Naturalist. 7: 513-521.

Walsh, B.D. 1864. On dimorphism in the hymenopterous genus *Cynips*, with an appendix, containing hints for a new classification of *Cynipidae*, including descriptions of several new species, inhabiting the oak-galls of Illinois. Proceedings of the Entomological Society of Philadelphia. 2: 443-500.

Wehrmaker, A. 1990. Die Roteiche (*Quercus rubra*): für Naturschutz und Gallwespen kein Ersatz für die europäischen Eiche. (Mit Bemerkungen über die Cynipiden-Gallen von Nova Scotia). Umweltamt der Stadt Darmstadt: Schriftenreihe. 13(1): 40-49.

WOLBACHIA-INDUCED THELYTOKY IN CYNIPIDS

Olivier Plantard and Michel Solignac *

Equipe INRA-CNRS Biologie evolutive des hyménoptères parasites, UPR 9034, Laboratoire "Populations, Génétique et Evolution", CNRS, Bât. 13, Avenue de la Terrasse, 91198 Gif-sur-Yvette Cedex, FRANCE

Abstract.—*Diplolepis spinosissimae* (Giraud) was found to exhibit two different reproductive systems. In 10 out of 12 French populations studied, males were very rare and females were all homozygous at three microsatellite loci. This "obligate homozygous thelytoky" is due to the endosymbiotic bacteria, *Wolbachia*, whose presence we demonstrate here for the first time in Cynipids. In the two remaining populations, deprived of *Wolbachia*, *D. spinosissimae* reproduce by arrhenotoky as indicated by the higher frequency of males and heterozygosity of females. *Wolbachia* were found in 11 other thelytokous *Diplolepis* species and four species in the paraphyletic Aylacini. Phylogenetic analyses of partial *Wolbachia* ftsZ gene sequences indicate that Cynipids have been infected through several infection events.

INTRODUCTION

In Hymenoptera, different types of parthenogenesis have been defined according to the sex of the offspring of a virgin female. Most species exhibit arrhenotoky: unfertilized eggs produce haploid males, while fertilized eggs produce diploid females (haplo-diploid sex-determination) (Crozier 1975). Thelytokous species persist through asexual reproduction, virgin females producing only females. Cynipids are known to exhibit the two reproductive systems, but also a rarer type of parthenogenesis called cyclical parthenogenesis or heterogony, with alternation of two generations—one sexual (i.e., with functional males) and one asexual (Askew 1984). Various agents responsible for parthenogenesis have been described in insects (White 1964) but none of them have been identified to date in the family Cynipidae. Among these agents, *Wolbachia*, which are endosymbiotic bacteria of the alpha proteobacteria section, are responsible for various alterations of their host's reproduction. First discovered in the cells of *Culex pipiens* (Hertig and Wolbach 1924), these rickettsia-like micro-organisms have been found since in a large number of arthropods (mites, terrestrial crustaceans, 17 percent of insects species belonging to various orders according to Werren *et al.* (1995a). They are responsible for cytoplasmic incompatibility between uninfected males and infected females (e.g., *Culex pipiens* (Yen and Barr 1971)), various *Drosophila* species (Hoffmann *et al.* 1986)), feminization (in some terrestrial crustaceans—Rigaud *et al.* 1991) or thelytoky (in parasitoid Hymenoptera such as *Trichogramma* spp. (Stouthamer *et al.* 1990), *Encarsia* spp. (Zchori-Fein *et al.* 1992), *Aphytis* spp. (Zchori-Fein *et al.* 1995), *Muscidufurax* spp. (Stouthamer *et al.* 1993) or *Leptopilina* spp. (Werren *et al.* 1995b).

In this study, we investigated the reproductive system of species of the genus *Diplolepis* (Cynipidae : Rhoditini tribe) which induce galls on *Rosa*. Several species of this genus are considered to be thelytokous because males are very rare (Askew 1960, Eady and Quinlan 1963, Nieves-Aldrey 1980). According to the cytogenetic study performed by Stille and Dävring (1980) and Stille (1985a, b), *Diplolepis rosae* (L.) and *D. mayri* (Schlechtendal) reproduce by "obligate homozygous parthenogenesis" or "gamete duplication" and diploid females result from the fusion of mitotic products of the pronucleus. No heterozygote individuals were detected by the authors in their study of various populations of both species in Sweden using enzyme electrophoresis.

In *Diplolepis spinosissimae* (Giraud), a Eurasian species whose biology was poorly known, our samples revealed highly contrasting sex ratios among French populations. Because of its crucial consequences on the genetic structure of populations, we first determined the reproductive system in each population. Twelve populations of *D. spinosissimae* were studied using three methods: (i) analysis of sex ratio, (ii) analysis of heterozygosity using hypervariable genetic markers (microsatellite loci), (Estoup et al. 1993) and (iii) search for *Wolbachia*. Results were also obtained for other *Diplolepis* species from Europe, North America and Japan, and some European Aylacini species.

We also investigated phylogenetic relationships between the different microbial organisms found in these Cynipids to (i) evaluate their possible co-speciation with their host, (ii) test the relevance of reproduction system as a character in Cynipid phylogeny reconstruction and (iii) identify putative organisms responsible for horizontal transmission of infection in Cynipids.

MATERIAL AND METHODS

Biological Material

Diplolepis spinosissimae is a univoltine species which induces galls on the leaves, stems or fruits (cynorhodons) mainly of *Rosa pimpinellifolia* L. Eighty percent of the galls are unilocular, but the biggest ones can contain up to 10 larval cells (Plantard 1997). Females emerge from the gall in April or May in the studied sites and lay their eggs in the still unopened buds of rose bushes. Galls grow during spring and leaf galls fall on the ground in autumn while stem galls remain on the bushes during winter. Larvae hibernate in the gall and pupate in March before chewing an exit hole to emerge from the gall.

This Eurasian cynipid has a patchy distribution due to its association with *R. pimpinellifolia*, a plant with isolated populations. Galls of *D. spinosissimae* were collected in 12 populations in France (fig. 1) over 3 consecutive years (1994, 1995, and 1996). Nine populations were from coastal sites (Western France) where *R. pimpinellifolia* covers important areas in sand dunes. The three remaining populations, Argenton, Fontainebleau, and Queyras, are continental and the two first are highly isolated from the nearest populations. All cynipids studied were reared from galls collected in nature. Adults of *D. spinosissimae* and of other Cynipids collected in Europe were stored at -80°C, while *Diplolepis* from North America and Japan were stored in Ethanol 95 percent at 4°C before DNA extraction.

Molecular Studies

Microsatellite Isolation and Amplification

Enzyme polymorphism is generally low in Hymenoptera, possibly due to the haplo-diploid system of reproduction (Graur 1985) (but see Cook et al. 1998, Csóka et al. 1998). To get polymorphic markers for *D. spinosissimae*, we screened a partial genomic library to isolate microsatellite loci. Screening was performed as described in Estoup and Cornuet (1994), with $(TG)^{10}$, $(TC)^{10}$, $(CAC)^5$, $(CCT)^5$, $(ATCT)^5$, and $(TGTA)^6$ probes. Microsatellite loci were relatively rare in the two species used (*Diplolepis spinosissimae* and *D. rosae*). Eighteen positive clones were found out of 5,150 clones containing an insert of 400 to 600 bp (i.e., one locus every 140 kb). Primers were designed for 16 loci but only three of them gave clear amplification patterns (sequences of the primers and PCR conditions are given in table 1). PCR reactions were performed in a 12.5 µl volume. The products were labeled using gamma 33 P and separated electrophoretically in a denaturing polyacrylamide Gel (7 percent urea), and the gels exposed to an autoradiographic film.

DNA Extraction

Total DNA was prepared from single adults by SDS/proteinase K digestion and phenol-chloroform extraction according to Kocher et al. (1989). DNA was resuspended in 50 µl of ultra pure water. PCRs were performed with 1 µl of 1/10 dilution DNA.

Detection of the Endosymbiotic Bacteria *Wolbachia* using a PCR Assay

The presence of the endosymbiotic bacteria *Wolbachia* can be demonstrated by PCR assay using specific primers (unable to amplify other bacteria). We used primers amplifying the 16 S ribosomal DNA of *Wolbachia* designed by Rousset (1993) (Position 76-99 and 1012-994) to check for the presence of *Wolbachia* in individuals of *D. spinosissimae* and other Cynipid species. PCRs were performed in 12.5 µl, containing 1.25 µl 10X buffer, 1.0 µl $MgCl_2$ (50mM), 0.125 µl dNTP mix (100 mM), 0.25 µl BSA (10 mg/ml), 0.0625 µl of each primer (200mM), 0.0625 µl Taq DNA polymerase, 1µl of DNA solution and ddH_2O q.s.p. The mixture was covered with mineral oil before amplification in a Biometra thermal cycler programmed as follows: 5 min at 93°C followed by 37 cycles of 1 min at 93 °C (slope = 1 percent), 1 min at 47°C, 1 min at 72°C, and ended by 5 min at 72°C. After PCR, 5 µl of amplification product were separated electrophoretically on a 1.4 percent agarose gel and visualized by ethidium bromide staining and UV fluorescence. For each individual DNA, two PCRs were performed.

Sequencing the *Wolbachia ftsZ* Gene

To characterize more precisely the *Wolbachia* strains, we used primers specific to the *ftsZ* gene of *Wolbachia*, (base 471 to 493 for primer 1, and base 1242 to 1222 for primer 2). PCRs were performed in 50 µl (same concentrations for the mix components as in the previous PCRs, except $MgCl_2$ = 1.5 mM; same PCR program as the

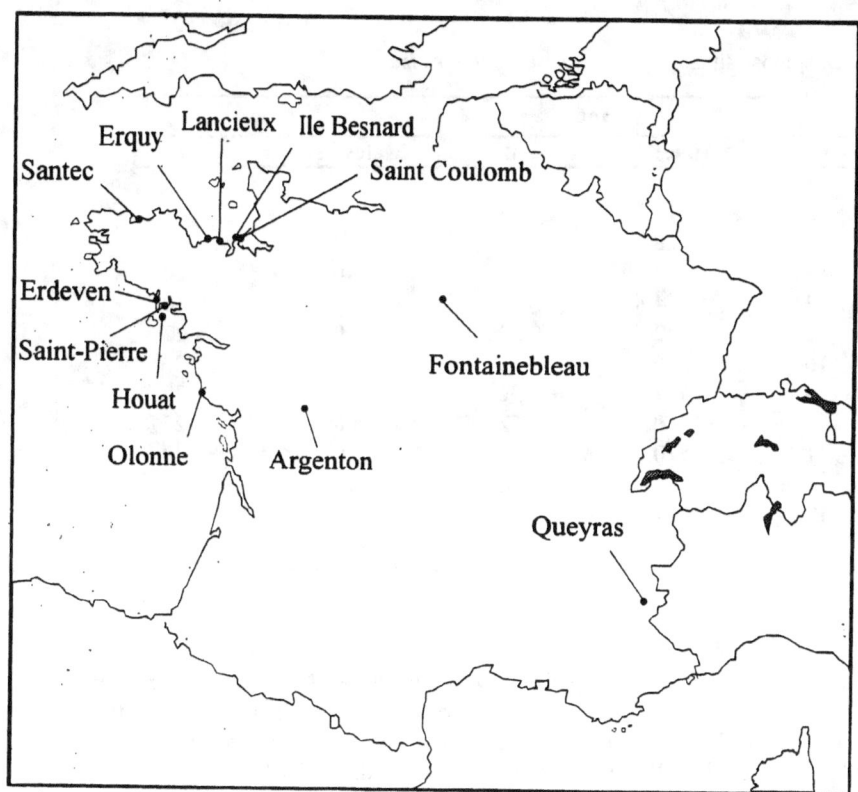

Figure 1.—*Localization of sampling sites of* Diplolepis spinosissimae *galls.*

Table 1.—*Microsatellite loci isolated and used in* Diplolepis spinosissimae. *Primer sequences and PCR conditions.*

Locus	Repeat	Primer	PCR Conditions	
			T/°C	$MgCl_2$ (mM)
DR 06	(TC) TT (TC) 7 TA (TC) (T) 26	5'-CTC ATC TCT TCT TCT TAT CTC AG-3' 5'-CCC AGG AGA GCA GAG G-3'	52	1.0
DS 24	(GA) 4 (TAGA) 7	5'-CAG AGC GTT GAT AAA CCG-3' 5'-CCT ATT CCA TTT TCC CAG-3'	50	1.2
DS 50	(TA) 9	5'-CAA AGA AGA AAT AAG AAA GGA-3' 5'-CAA GGC AAA TAT GAA AAA TG-3'	52	1.6

previous one, except that the annealing temperature is 52°C and the number of cycles, 35). PCR products were purified using a Qiaquick purification kit, and both strands of PCR products were sequenced using an ABI 373 automated sequencer using the dye terminator cycle sequencing kit (Perkin Elmer). Sequences were then aligned (612 b.p. = 204 amino-acid) and phylogenetic relationships were investigated through neighbor joining analysis using MEGA software (version 1.01, Kumar et al. 1993; 500 bootstrap replications). Sequences are available from the authors by request. The data set consists of 10 new sequences of *Wolbachia* strains associated with Cynipids and 29 sequences of *Wolbachia* from other arthropods (mainly from Werren et al. 1995b, with a few additional sequences from Genebank).

RESULTS

Sex Ratio of *Diplolepis Spinosissimae* Populations

Sex ratio, defined as the ratio of the number of males to the total number of individuals, is very different in coastal and continental populations. In all the coastal populations and also in the Argenton population (ca 120 km from the nearest coastal populations), males were very rare (= 9 males out of 591 individuals) and represented about 1.5 percent of the individuals (table 2). The frequencies of males in these populations are not significantly different among years (2 test). A very different situation was found in the other two continental populations (Fontainebleau and Queyras), where males represented between 19 and 28 percent of the individuals.

Table 2.—*Sex-ratio of the different* D. spinosissimae *populations (n = sample size).*

Population	1994		1995		1996	
	Males%	n	Males %	n	Males %	n
Fontainebleau	23.8	63	19.1	94	0	1
Queyras	-	-	28.9	38	-	-
Argenton	-	-	0	46	0	20
Saint Pierre	0	12	4.8	21	0	10
Erdeven	0	30	2.6	39	0	22
Erquy	6.25	16	0	53	0	29
Lancieux	-	-	2.2	92	0	37
Ile Besnard	-	-	0	52	8	25
Saint Coulomb	0	11	10.0	20	0	23
Olonne	-	-	0	6	-	-
Santec	0	17	-	-	-	-
Houat	0	10	-	-	-	-

Heterozygosity at Three Microsatellite Loci of *Diplolepis Spinosissimae*

The same contrast between the two continental populations of Fontainebleau and Queyras, and coastal populations is also observed in the frequency of heterozygous females for the three microsatellite markers. In coastal (+ Argenton) populations, all females but two out of 179 are homozygous for the three loci (table 3). The absence of heterozygotes is not due to the fixation of one single allele because several different genotypes are found in each population (table 4). Conversely, in the Fontainebleau and Queyras populations, 78 and 100 percent, respectively, of females are heterozygous at least for one locus. All the males of those populations tested (n = 15) carried a single allele.

Detection of the Endosymbiotic Bacteria *Wolbachia* in *D. Spinosissimae*

Wolbachia-specific primers revealed the presence of the symbiotic bacteria in all *D. spinosissimae* populations where males are extremely rare (table 5). Almost all females of the coastal populations and Argenton were infected by *Wolbachia* and only seven females (4.0 percent out of 174) were free of *Wolbachia*. One of the two heterozygous females (see above) was infected by *Wolbachia*. In contrast to the situation in the coastal populations, none of the 77 females tested from Fontainebleau or Queyras were infected by *Wolbachia*.

Distribution of *Wolbachia* in the *Diplolepis* Genus and in Some "Aylacini" Species

Among the six other European *Diplolepis* species, all populations of *D. rosae*, *D. mayri*, four out of five individuals from *D. fructuum* Rübsaamen and the only individual of *D. eglanteriae* (Hartig) tested were found to be infected by *Wolbachia* (table 6). Conversely, 10 females of *D. centifoliae* (Hartig) and two of *D. nervosa* (Curtis) were not infected. In North America, 6 out of 11 species were found to be infected. *Liebelia fukudae* Shinjii, a Japanese species belonging to the only other genus of the Rhoditini tribe, was uninfected (table 7). In the "Aylacini" tribe, four out of the eight species tested were found to be infected (table 8).

Phylogenetic Relationships Between *Wolbachia* Strains Found in Cynipid Species and in Other Arthropods

Sequence of the *ftsZ* gene were easily aligned with previously published sequence. The neighbor joining tree calculated with these sequences is given in the figure 2. *Wolbachia* strains associated with Cynipid belong to the two major groups of *Wolbachia*, namely the A and B groups (Werren *et al.* 1995b). The *ftsZ* gene sequences of the *Wolbachia* associated with the five infected European *Diplolepis* species were identical. Gene sequences of the bacterial strains associated with *D. nodulosa* and *D. bicolor* differed by only two transitions and one transversion. Those two groups differ from each other by only four transitions and one transversion out of 612 bp. The strain associated with *Liposthenes glechomae* belongs to a well-characterized sub-group (inside the B group). This sub-group, containing the two strains associated with *Gryllus pennsylvanicus* and *Tribolium confusum*, is characterized by the absence of a 9 b.p. deletion possessed by all B strains relative to A group. The *Wolbachia* associated with *D. spinosa* and *D. radicum* differ by four substitutions, while the strain associated to *D. californica* is less closely related.

Table 3.—*Frequencies of females heterozygous for at leat one microsatellite locus in the different D. spinosissimae populations (n = number of individuals tested).*

Population	1994		1995	
	Heterozygous females frequency	n	Heterozygous females frequency	n
Fontainebleau	0.78	50	-	-
Queyras	-	-	0.96	27
Argenton	-	-	0.05	40
Saint Pierre	0.00	13	0.00	9
Erdeven	0.00	29	-	-
Erquy	0.00	9	0.00	31
Saint Coulomb	0.00	7	0.00	14
Olonne	-	-	0.00	6
Santec	0.00	13	-	-
Houat	0.00	8	-	-

Table 4.—*Genotype frequencies of* Diplolepis spinosissimae *females in the coastal populations (loci 06 and 24 only, locus 50 being monomorphic in all these populations) (n = number of individuals tested). Genotypes found in Fontainebleau and Queyras populations are not reported in this table as respectively 17 and 15 different genotypes were represented there (n = 46 and 26, respectively).*

Population	n	Genotype (locus 06 then 24)	Frequency
Argenton	38	6161 8181	0.763
		6363 8181	0.184
		6361 8181	0.052
Olonne	5	0000 0101	1.000
Saint Pierre	18	6363 8181	0.888
		6363 7373	0.112
Houat	8	5959 8181	0.875
		6363 8181	0.125
Erdeven	23	5959 8181	0.478
		6363 8181	0.347
		6565 8181	0.173
Santec	8	5959 6969	0.500
		5959 8181	0.375
		5959 8585	0.125
Erquy	34	5959 8585	0.647
		5959 6969	0.352
Saint Coulomb	18	5959 8585	0.944
		5959 6969	0.055

Table 5.—*Frequency of females infected by* Wolbachia *in the different* Diplolepis spinosissimae *populations (n = number of individuals tested).*

Locality	1994		1995	
	Infected females %	n	Infected females %	n
Fontainebleau	0.00	50	-	-
Queyras	-	-	0.00	27
Argenton	-	-	0.95	38
Saint Pierre	1.00	12	1.00	9
Erdeven	0.97	29	-	-
Erquy	0.75	8	0.93	30
Saint Coulomb	1.00	7	1.00	13
Olonne	-	-	1.00	6
Santec	1.00	13	-	-
Houat	1.00	8	-	-

Table 6.—*Frequency of females infected by* Wolbachia *in different european* Diplolepis *species (n = number of individuals tested).*

Species	Populations (department)	Infected females frequency	n
D. spinosissimae	Queyras, Fontainebleau	0.00	77
D. spinosissimae	Coastal populations + Argenton	0.96	174
D. rosae	Gif-sur-Yvette (91)	0.71	14
D. rosae	Gap (05)	1.00	10
D. rosae	Redon (35)	0/1	1
D. rosae	Erquy	1.00	5
D. rosae	Chinon (37)	1.00	5
D. rosae	Argenton	1.00	5
D. mayri	Gap (05)	1.00	10
D. mayri	Uppsala (Sweden)	0/2	2
D. fructuum (Rübsaamen)	Taleghan (Iran)	0.80	5
D. eglanteriae (Hartig)	Ouzouère sur Loire (45)	1/1	1
D. centifoliae (Hartig)	Fontainebleau	0.00	2
D. centifoliae	Erquy	0.00	5
D. centifoliae	Erdeven	0.00	2
D. centifoliae	Saint Coulomb	0/1	1
D. centifoliae	Causse de la Selle	0/1	1
D. nervosa (Curtis)	Erdeven	0/1	1
D. nervosa	Erquy	0/1	1 male
D. nervosa	Causse de la Selle	0/1	1

Table 7.—*Frequency of females infected by* Wolbachia *in different nearctic and Japanese* Diplolepis *species (n = number of individuals tested).*

Species	Populations (country)	Frequency of infected females	n
Diplolepis bicolor (Harris)	Dominion Bay, Manitoulin I. (Ontario)	1.00	3
Diplolepis californica (Beutenmueller)	Cosumnes river preserve (California)	2/2	2
Diplolepis fusiformans (Ashmead)	Chelmsford (Ontario)	0.00	5
Diplolepis ignota (Osten Sacken)	SM, Lethbridge (Alberta)	0.00	5
Diplolepis nebulosa (Bassett)	Manitoulin I. (Ontario) Wiliams Lake Cranbrook (B.C.)	0.00	6
Diplolepis nodulosa (Beutenmueller)	Povidence Bay, Manitoulin I. (Ontario)	1.00	3
Diplolepis polita (Ashmead)	Chelmsford (Ontario)	1.00	5
Diplolepis radicum (Osten Sacken)	2 km SW TR, Manitoulin I. (Ontario)	2/2	2
Diplolepis rosaefolii (Cockerell)	Manitoulin I. (Ontario), L'Assoumption (Quebec)	0.00	5
Diplolepis spinosa (Ashmead)	Chelmsford, Sudbury, (Ontario) Fahler Kelowna, (B.C.) (Alberta), Fort St John's,	1.00	5
Diplolepis triforma Shorthouse and Ritchie	Chelmsford, Manitoulin I. (Ontario)	0.00	5
Diplolepis variabilis (Bassett)	Penticton (British Colombia)	0.00	5
Diplolepis sp. near *spinosissimae*	Aomori Apple Exp. Stn (Japan)	1.00	5
Diplolepis sp. near *centifoliae*	Aomori Apple Exp. Stn (Japan)	0.00	5
Liebelia fukudae Shinjii	Ajigasawa (Japan)	0.00	5

Table 8.—*Presence/Absence of* Wolbachia *in different Aylacini species (2 females tested per species).*

Species	Population (department)	*Wolbachia*
Aulacidea hieracii i	Pointe du Meinga (35)	-
Aylax papaveris	Rennes (35)	-
Timaspis lampsanae	Barbechat (44)	+
Phanacis hypochaeridis	Erquy (22)	+
Phanacis centaurea	Iffendic (35)	+
Xestophanes brevitarsis	Savigné sous le Lude (49)	-
Diastrophus rubi	Lancieux (22)	-
Liposthenes glechomae	Lancieux (22)	+

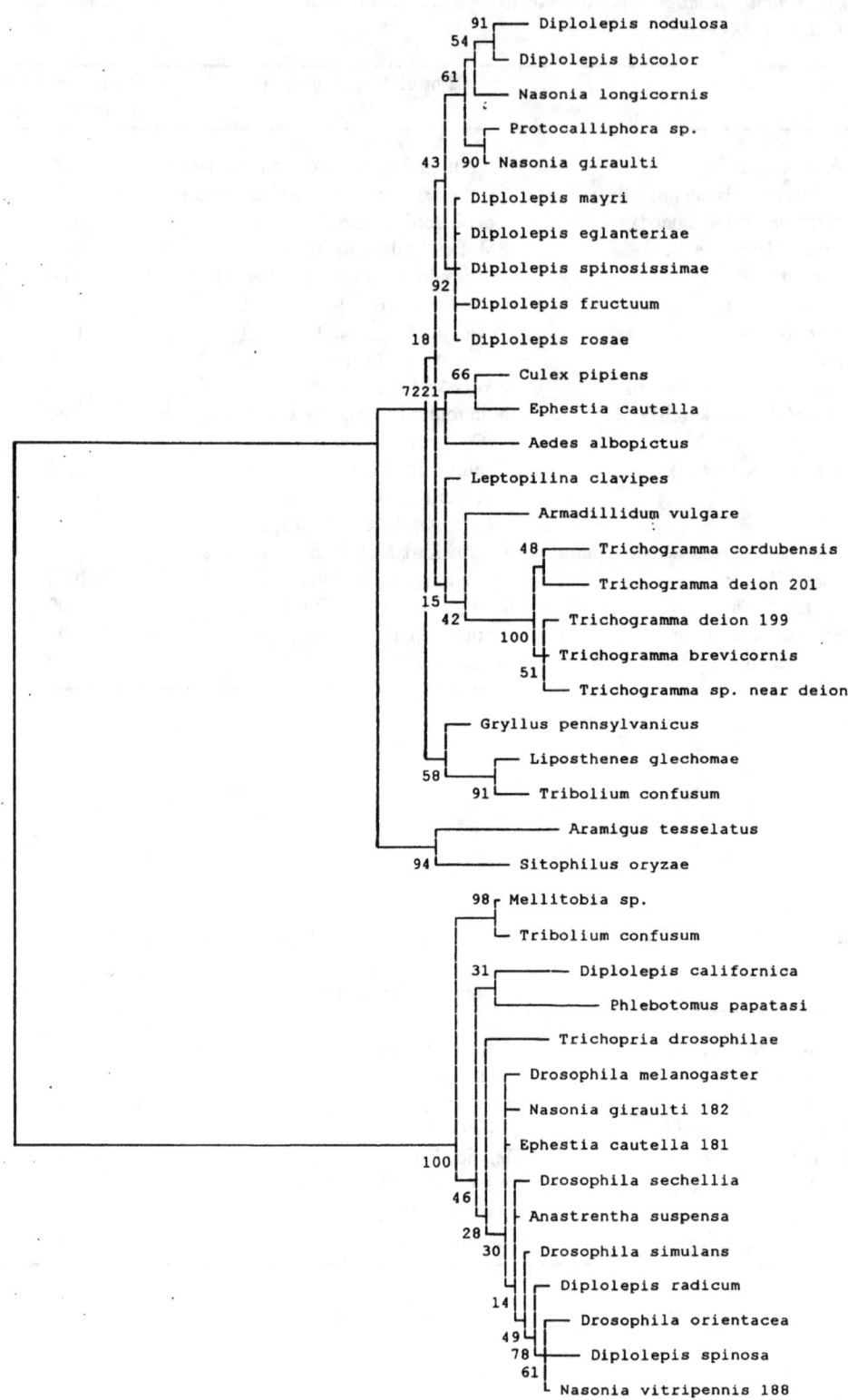

Figure 2.—*Phylogenetic tree of* Wolbachia *strains associated with various arthropods, based upon partial sequences of* ftsZ *gene (Neighbor joining). Sequences of strains associated with non-Cynipid species are from Werren* et al. *1995b.* Wolbachia *strains associated to Cynipids are shown in bold and underlined.*

DISCUSSION

Among Population Variation in the Reproductive System of *Diplolepis Spinosissimae*

The three data sets (sex-ratio, frequency of heterozygous females for at least one locus, and presence/absence of *Wolbachia*) allow two different types of populations to be distinguished. In the two most continental populations (Queyras and Fontainebleau), the percentage of males lies between 20 and 30 percent. Females are highly heterozygous at the three loci studied, while males present only one band (one allele). These *D. spinosissimae* populations are entirely *Wolbachia* free; they exhibit an haplo-diploid system, hemizygous males being produced by arrhenotoky, and diploid females by syngamy; their genetic structure is in agreement with sexual biparental reproduction. In all other populations (i.e., all the coastal populations and the continental population of Argenton), males are very rare (one or two individuals in each sample). Females are always homozygous (except two females from Argenton) and 90 to 100 percent are infected by the endosymbiotic bacterium *Wolbachia*. This microorganism is responsible for thelytoky of some parasitoid wasps (Stouthamer et al. 1990). In those populations, females reproduce by *Wolbachia*-induced thelytoky. New infected females arise from unfertilized but infected eggs. As demonstrated by Stouthamer and Kazmer (1994) for thelytokous *Trichogramma*, the endosymbiotic bacterium is probably responsible for homozygosity in *D. spinosissimae* too. *Wolbachia* modify the behavior of chromosomes in the first mitosis of unfertilized eggs, resulting in the fusion of the two sets of chromosomes. Hence, diploidy is restored but only homozygous females which possess the same genotype as their mother are produced. Females from these populations can thus be considered as independent clonal lines. We hypothesize that the more important spatial connectedness of littoral populations (roughly linearly distributed) could explain the fact that all populations are infected. Continental populations being more isolated would be less exposed to invasion by infected females and more prone to random loss of *Wolbachia* infection.

Distribution of *Wolbachia* in Other Cynipid Species

About half of the *Diplolepis* species tested were found to be infected by *Wolbachia*. Five out of the seven European *Diplolepis* species are infected. In the nearctic fauna, 6 out of the 13 species tested were also found to be infected. The five individuals of *Liebelia fukudae* tested, a species which belong to the second genus of the Rhoditini tribe and which also induces galls on *Rosa* spp., were found to be free of *Wolbachia*. Approximately the same proportion of infected species was found (four out of eight) in the Aylacini tribe. Our conclusions concerning the reproductive system of these species (arrhenotoky versus thelytoky) based on the presence/absence of *Wolbachia* are in complete agreement with the findings of Folliot (1964) who obtained galls in laboratory conditions, and determined the sex of the offspring of virgin females (table 9).

Phylogenetic Relationships Between the Different *Wolbachia* Strains Associated with Cynipids

Wolbachia strains associated with cynipids appear to be scattered over most of the branches of the phylogenetic tree of *Wolbachia* associated with arthropods (fig. 2) and consequently do not form a monophyletic clade. Moreover, the phylogenetic relationships between the hosts are not congruent with those of their parasites. The *Wolbachia* strain associated with the European *Diplolepis*

Table 9.—*Comparisons between Folliots' conclusions on the reproductive system of various Aylacini species with our results on presence/absence of* Wolbachia.

Species	Wolbachia	Parthenogenesis
Aulacidea hieracii (Bouché)	-	arrhenotoky
Aulacidea pilosellae (Kieffer)		thelytoky
Aylax minor or *papaveris* Hartig	-	arrhenotoky
Timaspis lampsanae Perris	+	
Phanacis hypochaeridis (Kieffer)	+	thelytoky
Phanacis centaurea Förster	+	
Xestophanes brevitarsis (Thompson)	-	arrhenotoky
Xestophanes potentillae (Retzius in DeGeer)		arrhenotoky
Diastrophus rubi (Bouché)	-	arrhenotoky
Liposthenes glechomae (Giraud)	+	thelytoky

is phylogenetically closer to *Wolbachia* strains associated with *Liposthenes glechomae* (Giraud), than with strains associated with some nearctic *Diplolepis* species.

D. nodulosa, despite being closely allied to *D. radicum* (Osten Sacken), *D. california* (Beutenmuller) and *D. spinosa* (Ashmead) (a nearctic species group well characterized by a flanged tibia) has a *Wolbachia* strain closely allied to European *Diplolepis* species, but very distant from the strains associated with the other flanged tibia *Diplolepis* species. Our results show that characters concerning the reproductive system of cynipids are not relevant for phylogeny reconstruction. Note also that strains associated with Cynipidae are relatively distant from those associated with *Leptopilina* spp. a parasitic member of the Cynipoidea in the family Eucoilidae (sequence determined by Werren *et al.* 1995b). Indeed, these results suggest that multiple independent infection events of cynipids by *Wolbachia* have occurred. Similar results where found for *Wolbachia* associated with *Trichogramma* (Schilthuizen and Stouthamer 1997). Although no transmission mechanism for *Wolbachia* other than vertical maternal transmission has been proven to date, parasitoids have already been advocated as possible agent for horizontal transmission (Breeuwer *et al.* 1992, Werren *et al.* 1995b). We propose that host feeding behavior of parasitoid (Askew 1971, Fulton 1933), existing even for concealed hosts such as gall makers through the construction of a feeding tube (O.P. pers. observ.), could create particularly favorable conditions for horizontal transmission.

ACKNOWLEDGEMENTS

We would like to thank all the people who provided biological material for the study of the distribution of *Wolbachia* in various *Diplolepis* species. We are particularly indebted to Joe Shorthouse (Laurentian University, Ontario). Thank you also to Doctor Norio Sekita (Japan), to the Modelling and Biometry laboratory of INRA - Montpellier (France), to Kathy Schick (California) and to Fredrik Ronquist (Sweden). We have also benefited from the valuable help in molecular work provided by Gwenaelle Genson, Isabelle Le Clainche (INRA, Gif-sur-Yvette) and Dominique Vautrin (CNRS, Gif-sur-Yvette). Jean-Yves Rasplus and Michael Hochberg provided useful help at various stages of this study. This work was supported in part by a grant from the French Ministère de l'Environnement (SRAE n° 95150).

LITERATURE CITED

Askew, R.R. 1960. Some observations on *Diplolepis rosae* (L.) (Hym., Cynipidae) and its parasites. Entomologist's Monthly Magazine. 95: 191-192.

Askew, R.R. 1971. Parasitic insects. New York: Elsevier.

Askew, R.R. 1984. The biology of gall wasps. In: Ananthakrishnan T.N., ed. Biology of gall insects. London: Edward Arnold: 223-271.

Breeuwer, J.A.J.; Stouthamer, R.; Barns, S.; Pelletier, D.; Weisburg, W.; Werren, J.H. 1992. Phylogeny of cytoplasmic incompatibility microorganisms in the parasitoid wasp genus *Nassonia* (Hymenoptera, Pteromalidae) based on 16S ribosomal DNA sequences. Insect Molecular Biology. 1: 25-36.

Cook, J.M.; Stone. G.N.; Rowe, A. 1998. Patterns in the evolution of gall structure and life cycles in oak 9911 wasps (hymenoptera: Cynipidae). In: Csóka, Gyuri; Mattson, William J.; Stone, Graham N.; Price, Peter W., eds. The biology of gall-inducing arthropods: proceedings of the international symposium; 1997 August 14-19; Mátraküred, Hungary. Gen. Tech. Rep. NC-199. St. Paul, MN: U.S. Department of Agriculture, Forest Service, North Central Research Station: 261-279.

Crozier, R.H. 1975. Hymenoptera. Animal Cytogenetics. Gebrüder Borntraeger, Berlin, Stuttgart: 1-95.

Csóka, G.; Stone, G.; Atkinson, R.; Schönrogge, K. 1998. The population genetics of postglacial invasions of northern Europe by cynipid gall wasps (Hymenoptera: Cynipidae). In: Csóka, Gyuri; Mattson, William J.; Stone, Graham N.; Price, Peter W., eds. The biology of gall-inducing arthropods: proceedings of the international symposium; 1997 August 14-19; Mátraküred, Hungary. Gen. Tech. Rep. NC-199. St. Paul, MN: U.S. Department of Agriculture, Forest Service, North Central Research Station: 280-294.

Eady, R.D.; Quinlan, J. 1963. Hymenoptera: Cynipoidea. Key to families and subfamilies and Cynipinae (including galls). London: Royal Entomological Society of London.

Estoup, A.; Cornuet, J.-M. 1994. Utilisation de sondes oligonucléotidiques marquées à la digoxigénine pour la recherche de microsatellites. Brin complémentaire, Le Journal des Biotechnologies. 10: 5-7.

Estoup, A.; Solignac, M.; Harry, M.; Cornuet, J.-M. 1993. Characterization of (GT)n and (CT)n microsatellites in two insect species: *Apis mellifera* and *Bombus terrestris*. Nucleic Acids Research. 21: 1427-1431.

Folliot, R. 1964. Contribution à l'étude de la biologie des Cynipides gallicoles (Hyménoptères : Cynipoidea). Annales des Sciences Naturelles, Zoologie. 12: 407-564.

Fulton, B.B. 1933. Notes on *Habrocytus cerealellae*, parasite of the angoumois grain moth. Annals Entomological Society of America. 26: 536-553.

Graur, D. 1985. Gene diversity in Hymenoptera. Evolution. 39: 190-199.

Hertig, M.; Wolbach, S.B. 1924. Studies on rickettsia-like micro-organisms in insects. Journal of Medical Research. 44: 329-374.

Hoffmann, A.A.; Turelli, M.; Simmons, G.M. 1986. Unidirectional incompatibility between populations of *Drosophila simulans*. Evolution. 40: 692-701.

Kocher, T.D.; Thomas, W.K.; Meyer, A.; Edwards, S.V.; Paabo, S.; Villablanca, F.X.; Wilson, A.C. 1989. Dynamics of mitochondrial DNA evolution in animals: amplification and sequencing with conserved primers. Proceedings of the National Academy of Sciences U.S.A. 86: 6196-6200.

Kumar, S.; Tamura, K.; Nei, M. 1993. MEGA: molecular evolutionary genetics analysis, version 1.01. Pennsylvania: Pennsylvania State University.

Nieves-Aldrey, J.L. 1980. Datos sobre *Diplolepis rosae* (L.) (Hym. Cynipidae) y sus himenopteros parasitos en Salamanca. Boletin de la Asociacion Espanola de Entomologia. 4 : 107-113.

Plantard, O. 1997. Ecologie des communautés de parasitoïdes associés aux Cynipidae galligènes (Hymenoptera): rôle des caractéristiques des galles, de la structure génétique des populations et de la phylogénie des hôtes sur leur cortège parasitaire. Paris: University of Paris VI. 286 p. Ph. D. dissertation.

Rigaud, T.; Souty-Grosset, C.; Raimond, R.; Mocquard, J.-P.; Juchault, P. 1991. Feminizing endocytobiosis in the terrestrial crustacean *Armadillidium vulgare* Latr. (Isopoda): recent acquisitions. Endocytobiosis and Cell Research. 7: 259-273.

Rousset, F. 1993. Les facteurs déterminant la distribution des *Wolbachia*, bactéries endosymbiotiques des arthropodes. Paris XI, Orsay. 144 p. Ph. D. dissertation.

Schilthuizen, M.; Stouthamer, R. 1997. Horizontal transmission of parthenogenesis-inducing microbes in *Trichogramma* wasps. Proceedings of the Royal Society of London (B). 264: 361-366.

Stille, B. 1985a. Host plant specificity and allozyme variation in the parthenogenetic gall wasp *Diplolepis mayri* and its relatedness to *D. rosae* (Hymenoptera: Cynipidae). Entomologia Generalis. 10: 87-93.

Stille, B. 1985b. Population genetics of the parthenogenetic gall wasp *Diplolepis rosae* (Hymenoptera, Cynipidae). Genetica. 27: 145-151.

Stille, B.; Dävring, L. 1980. Meiosis and reproductive strategy in the parthenogenetic gall wasp *Diplolepis rosae* (L.) (Hymenoptera: Cynipidae). Hereditas. 92: 353-362.

Stouthamer, R.; Breeuwer, J.A.J.; Luck, R.F.; Werren, J.H. 1993. Molecular identification of microorganisms associated with parthenogenesis. Nature. 361: 66-68.

Stouthamer, R.; Kazmer, D. 1994. Cytogenetics of microbe-associated parthenogenesis and its consequences for gene flow in *Trichogramma* wasps. Heredity. 73: 317-327.

Stouthamer, R.; Luck, R.F.; Hamilton, W.D. 1990. Antibiotics cause parthenogenetic *Trichogramma* (Hymenoptera/ Trichogrammatidae) to revert to sex. Proceedings of the National Academy of Sciences, USA. 87: 2424-2427.

Werren, J.H.; Windsor, D.W.; Guo, L.R. 1995a. Distribution of *Wolbachia* in neotropical arthropods. Proceedings of the Royal Society of London (Serie B). 262: 197-204.

Werren, J.H.; Zhang, W.; Rong Guo, L. 1995b. Evolution and phylogeny of *Wolbachia*: reproductive parasites of arthropods. Proceedings of the Royal Society of London, Série B. 261: 55-71.

White, M.J.D. 1964. Cytogenetic mechanisms in insect reproduction. In: Highnam, K.C., ed. Royal Entomological Society Symposium, No. 2: 1-12.

Yen, J.H.; Barr, A.R. 1971. New hypothesis on the cause of cytoplasmic incompatibility in *Culex pipiens* L. Nature. 232: 657-658.

Zchori-Fein, E.; Roush, R.; Hunter, M.S. 1992. Male production induced by antibiotic treatment in *Encarsia formosa*, an asexual species. Experentia. 48: 102-105.

Zchori-Fein, E.; Faktor, O.; Zeidan, M.; Gottlieb, Y.; Czosnek, H.; Rosen D. 1995. Parthenogenesis-inducing microorganisms in *Aphytis* (Hymenoptera: Aphelinidae). Insect Molecular Biology. 4(3): 173-178.

THE GALL-INHABITING WEEVIL (COLEOPTERA) COMMUNITY ON *GALENIA AFRICANA* (AIZOACEAE): CO-EXISTENCE OR COMPETITION?

Schalk Louw

Department of Zoology & Entomology, University of the Orange Free State, P. O. Box 339. Bloemfontein 9300, South Africa

Abstract.—The life-stages of *Urodontus scholtzi* Louw (Anthribidae) and an undescribed *Baris* sp. (Curculionidae) inhabit galls induced by the former species on *Galenia africana* in South Africa. Incidence of this weevil fauna in 141 galls sampled from four localities over 2 years, on average showed 39 percent of the galls harboring only the galler species, 9 percent harboring only the inquiline species and 52 percent containing both. Mortality through parasitism or other means was less than 4 percent, indicating that this scenario is most likely due to other factors exerting pressure on the system. This breakdown of species interaction most probably reveals evidence of interspecific competition as the actual driving force within the system.

INTRODUCTION

Albeit that insect herbivores are considered to be pivotal in trophic systems between autotrophic plants and higher trophic levels (Price 1997), the role of competition in such systems, although subtle and difficult to detect should not be ignored altogether (Strong et al. 1984). Competition in phytophagous insects, as an ecological phenomenon, has been caught in the crossfire of those authors who consider it to be an important, demonstrable driving force which structures insect herbivore communities (e.g., Damman 1992, Denno et al. 1995, Stewart 1996) and those who advocate that selection favors the avoidance of the high fitness costs of competition and that resource partitioning is the outcome of other ecological processes (e.g., Lawton and Strong 1981, Schoener 1983). An instance that is argued to favor the occurrence of interspecific competition is when a microhabitat, e.g., galls, not only provides plant tissue of high nutritional value, but also reduces the impact of natural enemies. In such cases of discrete resource availability, the value of these advantages overrides the disadvantages of facing competition (Zwölfer 1979), rendering competition a viable ecological force worthy of recognition.

The intention of this paper is to describe the patterns of co-existence of the Curculionoidea (Coleoptera) species that occupy the spatially restricted stem galls found on *Galenia africana* (Aizoaceae). Although the probable explanations relating to the functioning of this system have not yet been experimentally validated, the paper does provide basic information on the interaction between insect herbivores and their host, thereby addressing aspects of the ecology of a plant that is reported to have an influence on small-stock farming in its area of occurrence. It also supplements studies dealing with the effect of gall induction on architectural modification of the plant (Price and Louw 1996). Two questions were addressed: (1) what is the composition of the weevil community occupying *G. africana* galls? and (2) how do the different species in this system interact with one another?

METHODS

Galenia africana is an erect, low-growing woody shrub occurring throughout Namibia (excluding the extreme north-eastern parts) and central, western, and southern South Africa. The shrub is one of the first pioneers to regrow in heavily disturbed areas and as such is thus an important disturbance indicator. It can be commonly found in overgrazed areas, abandoned fields, around watering points and small-stock pens, and along roadsides.

Stem galls from *G. africana* were collected in Namaqualand, a winter rainfall region in western South Africa from Buffelsfontein (30.36S, 18.08E) in June 1995 (72 galls) and from Tweefontein (30.32S, 18.09E)

(26 galls), Nassau (30.27S, 18.08E) (12 galls) and Garies (30.35S, 18.10E) (31 galls) in June 1996. Since the number of active galls (= galls containing insect herbivores) present per shrub showed considerable patchiness, a sample consisted of 1 to 15 galls per shrub, but these were all pooled per locality for the purposes of this study. Galls were cut open and the larvae from each gall extracted, recorded and stored separately per species in an ethyl-alcohol based fixative. Larval separation was based on differentiation in larval frass texture from the feeding galleries within a gall. In this manner the occupants of an empty gallery where the adult of a species had already eclosed and left the gall could also be determined. The numbers of parasitized larvae, their parasites (if present) and any other mortalities from each gall were also listed. Visual calculation, in terms of gall size and percentage of gall consumption to the nearest 10 percent, was also carried out once all specimen material had been removed from a gall.

CHARACTERISTICS OF THE WEEVIL COMPLEX

Developing larvae of *Urodontus scholtzi* Louw (Coleoptera: Anthribidae) induce galls on *G. africana* stems after hatching from eggs deposited inside the soft tissue of vigorous new growth on the plant. Within these galls the individuals of this species develop to maturity. In terms of reproductive significance, gall induction within urodontine Anthribidae is regarded as an advanced trait (Louw 1994) and is encountered in only two species out of a total of ca. 80 (Louw 1993). Although a detailed analysis of the biological development of *U. scholtzi* is lacking, it was established by correlating weevil and gall development on the plant that this weevil has a 2 year life-cycle, while galls incited by it persist for at least 4 years (Price and Louw 1996).

An undescribed *Baris* sp. (Curculionidae) is a gall inquiline within this system. Where this *Baris* sp. oviposits on the plant and how it attains access to the gall has not been determined. However, based on biological knowledge of the closely related *Baris dodonis* Marshall which develops on *Amaranthus hybridus* (Amaranthaceae) (Louw, pers. observ.) and by careful dissection and investigation of galls and *G. africana* stems in all stages of plant growth, it is assumed that females oviposit directly into developing galls.

Larvae of the two species tunnelling in the galls excrete frass that differs in texture (i.e., fine in *U. scholtzi* and coarse in *Baris* sp.), which makes it possible to the determine activity patterns of each species within the gall without first having to differentiate the species by intricate morphological means. Adults of both species eclose in the feeding galleries, sclerotize and extend the galleries to directly below the gall surface from where they will leave the gall once environmental conditions become favorable.

RESULTS

Dissection of galls from the study area revealed three possible traits of gall occupation, namely those bearing only the galler, i.e., *U. scholtzi*; those bearing only the inquiline, i.e., *Baris* sp. and those bearing both. From a total of 141 galls, 39 percent, 9 percent, and 52 percent revealed these respective traits. Numerical comparison of gall occupation by the two weevil species between localities showed no pattern and called for a more detailed analysis. A subsequent detailed breakdown of data revealed two noteworthy facts. First, that on average more than a single weevil specimen occupied the galls, and second that the total mortality in the system was estimated at less than 4 percent (table 1). These two points, respectively, justify the posing of questions on

Table 1.—*Gall localities and weevil occurrence and mortality in* Galenia africana *galls. NAG = number of galls sampled; NS = total number of weevil specimens extracted from galls; Ave = average number of specimens per gall; U = galls with only* Urodontus scholtzi; *B = galls with only* Baris *sp.; U & B = galls with both* U. scholtzi *and* Baris *sp.; M = galls with mortalities (including parasitism).*

Locality	NAG	NS	Ave	U	B	U & B	M
Buffelsfontein (30.36S, 18.08E)	72	120	1.66	19	23	30	4
Tweefontein (30.32S, 18.09E)	26	44	1.69	7	9	10	3
Nassau (30.27S, 18.08E)	12	20	1.66	8	0	4	1
Garies (30.35S, 18.10E)	31	40	1.29	20	4	7	0

species interaction, and indicate that other factors than mortality (i.e., parasitism, fungal infection, and desiccation) exert pressure on the system when seeking answers for the species differentiation in gall occupation.

Worthy of mention with regard to fungal infection as a mortality factor, is that such *G. africana* galls were notably spongy in texture and the tissue streaked with black lines and blotches, strongly suggesting that infection is in the gall tissue. Mortality of the gall inducer in such cases is as a result of the gall tissue dying with the fungus thus acting as an inquiline (Wilson 1995).

Occurrence of one or the other species in galls in both cases show occupation of the gall center during larval development, whereas co-occurrence of the species in galls show *Baris* sp. life-stages occupying the gall center and *U. scholtzi* life-stages occurring in the extreme periphery of the gall just below the gall surface. Nutritional values of the different portions of gall tissue were not tested, but indications are that these are similar throughout the gall, since in all cases of occurrence *U. scholtzi* seemingly developed normally, ranging in size from 2.5-5.0 in length and 1.2-1.8 in elytral width. In all cases gall utilization never exceeded 50 percent of the gall volume and feeding was orderly with clearly defined feeding galleries. A clear majority (82 percent) of the feeding galleries of the two species were well-separated from one another, with direct contact only present in a few (18 percent). In the latter case frass accumulations of each of the species separated individuals from one another, indicating that direct contact, if at all, was only for brief periods during early development of larval instars.

Extraction of weevil material from galls often showed different instar larvae or different life-stages of the same species within the same gall. A dual explanation is provided for this. First, *U. scholtzi* presumably has an extended oviposition period during which oviposition, not necessarily by the same female, at the same site on a stem over a period of time is often possible. Second, since *U. scholtzi* has a biennial life-cycle and *Baris* sp. an annual life-cycle (presumed on knowledge of another species—see above), *Baris* sp. oviposition is probably later than that of *U. scholtzi* with *Baris* sp. larval development more accelerated and obviously not synchronized with that of *U. scholtzi*. In the light of this it is also likely that oviposition choice in the *Baris* sp. is based on selection of young developing galls.

An area in central Namaqualand, consisting of four localities in close proximity of one another (see above), provided unexpected information which subsequently initiated this study. In spite *U. scholtzi* having been dissected from *G. africana* galls from 22 other localities (Louw 1993), none of these galls revealed the galler-inquiline relationship described here. This phenomenon remains to be clarified.

DISCUSSION

Studies in gall biology within the larger insect-plant interactions arena have high relevancy and gall inducers have often been demonstrated as effective in the biocontrol of weeds, especially since they tend to have a narrow host range (Harris and Shorthouse 1996). Competition in gallers with narrow host plant ranges have also been reported to promote host plant shifts and subsequent speciation (Abrahamson *et al*. 1994). The plant studied here, *G. africana,* survives well during spells of drought and during such periods is reported to cause the disease ascites in small-stock (Le Roux *et al*. 1994), all of which increases the relevancy of this study beyond that of basic scientific investigation.

The system described here comprises a single pair-wise species interaction within stem galls of the host with gall occupation traits suggesting the occurrence of interspecific competition. Here the agent causing competitive interaction would be the plant, which confidently ties onto Denno *et al*. (1995) and Stewart (1996) who, in a layout of potential mediators in interspecific competition, show that plants account for 53 percent in 193 pair-wise species interactions studied. Breakdown of the interactions between the species pair considered in this study suggests that for both species the population structures are based on sequential evolution, albeit along different routes and not necessarily with the same polarities. Ultimately the adaptive value for both species could be clarified by the Gall Nutrition Hypothesis (Price *et al*. 1987). The galling action of *U. scholtzi* on *G. africana* stems results in growth dominance in galled stems and moribund growth or death in ungalled stems (Price and Louw 1996), with the result that a resource sink through the galled stems is presumably created. A second suggestion is thus that there is competitive interaction by the species pair for this sink (*sensu* Larson and Whitham 1997) that exists in the few large, galled stems and that the performance and ultimately the population composition of both the galler and the inquiline are determined by it.

As an alternative to this, and against the background that competitive interaction is considered as insignificant in structuring insect communities (e.g., Lawton and Strong 1981), the question whether this galler-inquiline interaction is a case of harmonious co-existence also has to be considered. It this context it is argued that a more natural process would be for strong selective pressures to function towards avoidance of the fitness costs associated with competition (Stewart 1996) and that instead current patterns of resource partitioning actually only reflect the result of historical competitive periods (Connell 1980) and that these may be quite short in evolutionary terms (Bernays and Chapman 1994). Proponents of this opinion go on to consider that, at most, competition is infrequent with only brief bouts of intensity, e.g., during population outbreaks, and that this sufficiently serves to reinforce existing patterns of niche partitioning (Stewart 1996). For reasons outlined above none of these hypotheses seem to explain the system described here.

In an attempt to explain the three traits of weevil community interaction in *G. africana* galls, the following scenario is debated. First, in cases where the *Baris* sp. inquiline is absent in galls, it is proposed that selection of oviposition sites by *Baris* sp. females is a random process and that the species is at least oligophagous and most probably polyphagous (the latter feeding strategy has already been established for closely related *Baris* species (Louw, pers. observ.)). Second, in cases where the *U. scholtzi* galler is absent in galls, it is proposed that this be considered as exploitative competitive interaction for nutritive resources during early larval development, with subsequent *Baris* sp. success and *U. scholtzi* mortality. Third, co-occurrence of both species is regarded as a case of unidirectional interspecific interference competition with *Baris* sp. displacing *U. scholtzi*. The latter two traits could thus be seen as examples of asymmetric interspecific competition (Lawton and Hassell 1981) or competitive dominance (Keddy 1989), whereby the competitive interactions are unequal and as such render the *Baris* sp. as the dominant species and *U. scholtzi* as the subordinant species. Furthermore, since competitive interactions comprise a sequence of effects and responses in the competitors (Goldberg 1987, Goldberg and Fleetwood 1987), application of these actions to this system will imply that the dominant *Baris* sp. has far-reaching effects on *U. scholtzi* and that the response of the latter to this, in cases of co-occurrence within the same galls, is movement away from the gall center to the gall periphery.

This system needs more study to evaluate interactions quantitatively and to document weevil behavior on *G. africana*. In this regard questions concerning timing and choice of *U. scholtzi* oviposition, *U. scholtzi* larval survival relative to oviposition choice and oviposition rate, and *Baris* sp. biology and host range need to be addressed.

ACKNOWLEDGEMENTS

Peter Price is thanked for both field work assistance and thought-provoking discussion during his 1995 visit, Annamé Wels is thanked for general assistance and the Foundation for Research Development and the University of the Orange Free State are acknowledged for research funding and research facilities, respectively. A special word of thanks also goes to Gyuri Nagy Bor Csóka, organizer of the Biology of Gall-inducing Arthropods symposium in Mátrafüred, Hungary for ensuring that we will never forget a truly memorable occasion.

LITERATURE CITED

Abrahamson, W.G.; Brown, J.M.; Roth, S.K.; Sumerford, D.V.; Horner, J.D.; Hess, M.D.; How, S.T.; Graig, T.P.; Packer, R.A.; Itami, J.K. 1994. Gallmaker speciation: an assessment of the roles of host-plant characters, phenology, gallmaker competition, and natural enemies. In: Price, P.W.; Mattson, W.J.; Baranchikov, Y.N., eds. The ecology and evolution of gall-forming insects. Gen. Tech. Rep. NC-174. St. Paul, MN: U.S. Department of Agriculture, Forest Service, North Central Forest Experiment Station: 208-222.

Bernays, E.A.; Chapman, R.F. 1994. Host-plant selection by phytophagous insects. New York, NY: Chapman and Hall. 312 p.

Connell, J.H. 1980. Diversity and the evolution of competitors, or the ghost of competition past. Oikos. 35: 131-138.

Damman, H. 1992. Patterns of interaction among herbivore species. In: Stamp, N.E.; Casey, T.M., eds. Caterpillars: ecological and evolutionary constraints on foraging. New York, NY: Chapman and Hall: 132-169.

Denno, R.F.; McClure, M.S.; Ott, J.R. 1995. Interspecific interactions in phytophagous insects: competition re-examined and resurrected. Annual Review of Entomology. 40: 297-331.

Goldberg, D.E. 1987. Neighbourhood competition in an old-field plant community. Ecology. 68: 1211-1223.

Goldberg, D.E.; Fleetwood, L. 1987. Competitive effect and response in four annual plants. Journal of Ecology. 66: 921-931.

Harris, P.; Shorthouse, J.D. 1996. Effectiveness of gall-inducers in weed biological control. The Canadian Entomologist. 128: 1021-1055.

Keddy, P.A. 1989. Competition. New York, NY: Chapman and Hall. 202 p.

Larson, K.C.; Whitham, T.G. 1997. Competition between gall aphids and natural plant sinks: plant architecture affects resistance to galling. Oecologia. 109: 575-582.

Lawton, J.H.; Hassel, M.P. 1981. Asymmetrical competition in insects. Nature. 289: 793-795.

Lawton, J.H.; Strong, D.R. 1981. Community patterns and competition in folivorous insects. American Naturalist. 118: 317-338.

Le Roux, P.M.; Kotzé, C.D.; Nel, G.P.; Glen, H.F. 1994. Bossieveld. Grazing plants of the Karoo and karoo-like areas. Bull. 428. Pretoria, South Africa: Department of Agriculture. 231 p.

Louw, S. 1993. Systematics of the Urodontidae (Coleoptera: Curculionoidea) of southern Africa. Entomology Memoir, Department of Agriculture, Republic of South Africa. 87: 1-92.

Louw, S. 1994. Seed-feeding Urodontidae weevils and the evolution of the galling habit. In: Price, P.W.; Mattson, W.J.; Baranchikov, Y.N., eds. The ecology and evolution of gall-forming insects. Gen. Tech. Rep. NC-174. St. Paul, MN: U.S. Department of Agriculture, Forest Service, North Central Forest Experiment Station: 186-193.

Price, P.W. 1997. Insect ecology. 3d ed. New York, NY: John Wiley & Sons, Inc. 874 p.

Price, P.W.; Louw, S. 1996. Resource manipulation through architectural modification of the host plant by a gall-forming weevil *Urodontus scholtzi* Louw (Coleoptera: Anthribidae). African Entomology. 4(2): 103-110.

Price, P.W.; Fernandes, G.W.; Waring, G.L. 1987. Adaptive nature of insect galls. Environmental Entomology. 16: 15-24.

Schoener, T.W. 1983. Field experiments on interspecific competition. American Naturalist. 122: 240-285.

Stewart, A.J.A. 1996. Interspecific competition reinstated as an important force structuring insect herbivore communities. Trends in Ecology and Evolution. 11(6): 233-234.

Strong, D.R.; Lawton, J.H.; Southwood, R. 1984. Insects on plants: community patterns and mechanisms. Oxford: Blackwell Scientific Publications. 313 p.

Wilson, D. 1995. Fungal endophytes which invade insect galls: Insect pathogens, benign saprophytes, or fungus inquilines? Oecologia 103: 255-260.

Zwölfer, H. 1979. Strategies and counterstrategies in insect population systems competing for space and food in flower heads and plant galls. Fortschrift der Zoologie. 25(2/3): 331-353.

GALL-FORMING APHIDS ON *PISTACIA*: A FIRST LOOK AT THE SUBTERRANEAN PART OF THEIR LIFE CYCLE

David Wool

Professor of Zoology, Department of Zoology, Tel Aviv University, Ramat Aviv, 69978 Israel

---

Abstract.—Fifteen species of aphids (Pemphigidae: Fordinae) induce galls on *Pistacia* trees (Anacardiaceae) in Israel. Their 2-year holocycles involve alternation between *Pistacia* and grass roots. The ecology of the gall stage of several species has been investigated in detail, but the biology and ecology of the root-inhabiting stages on the secondary hosts remain obscure. This is the first report of a detailed investigation of root-inhabiting colonies of several Fordinae species in the laboratory, facilitated by the construction of a simple root cage. Two species, *Baizongia pistaciae* L. and *Aploneura lentisci* Pass., have been reared for several generations in the root cages and reproduced well. Other species were much less successful in the cages. The rearing method shows promise but needs to be adapted to the biology and behavior of each species, which can only be discovered by further experimentation.

Gall-inducing aphids on *Pistacia* (Pemphigidae: Fordinae) are known from Mediterranean and Middle-Eastern countries from Spain to Iran (Bodenheimer and Swirski 1957, Wool 1984). Their holocycle involves host alternation between *Pistacia* and secondary hosts, generally the roots of grasses (Zwolfer 1958). Alate aphids (fall migrants) leave the galls in autumn before leaf abscission and reproduce parthonogenetically on the ground. Their offspring (here referred to as "crawlers") find their way to grass roots and live and reproduce there in winter without inducing galls. Alate sexuparae (spring migrants) appear in spring and fly back to *Pistacia* (Wool et al. 1994), where they produce the sexual forms which give rise to the overwintering eggs that start a new cycle.

About 15 species of Fordinae colonize three common wild *Pistacia* species in Israel (Koach and Wool 1977, Wool 1995). In the last 25 years, the biology and ecology of several species was investigated in detail (Wool 1990; Wool and Burstein 1991; Burstein and Wool 1993; Burstein et al. 1994; Wool and Manheim 1986, 1988; Wool and Bar-El 1995; Inbar and Wool 1995; Inbar et al. 1995), but these studies were limited to the galls on the primary hosts, and the ecology of the subterranean stages on the secondary hosts have not been studied in any detail to date.

The root forms of (related) Pemphigidae were studied in the USA by Moran and Whitham (1988), Royer et al. (1991), Campbell and Hutchinson (1993), and Moran et al. (1993a, b). We tried to use their methods for rearing the Fordinae on grass roots, but failed. Finally, we devised a simple root cage which enabled rearing some species for several months on wheat roots in the laboratory. Ecological information accumulated so far is reported here.

METHODS

Root Cages

Root cages (fig. 1) are constructed from standard 9 cm diameter disposable petri dishes, notched at one point. The deep part of the dish is lined with filter paper. We germinate grass seeds (mostly wheat) in the laboratory and three to five seedlings are introduced into the cages when their rootlets reach 5-10 cm in length, so that the leaves protrude from the notch and the roots are spread on the paper (fig. 1). A second circle of filter paper may be put over the roots to facilitate root development. The paper is wetted with Hoagland plant nutrient solution. A small lump of cotton wool is placed opposite the notch to absorb excess moisture, and the cages are placed vertically in a tray containing the same solution, so that the lowest 1 cm or less of the cage is in contact with the solution. The aphids are transferred to the roots with a thin brush.

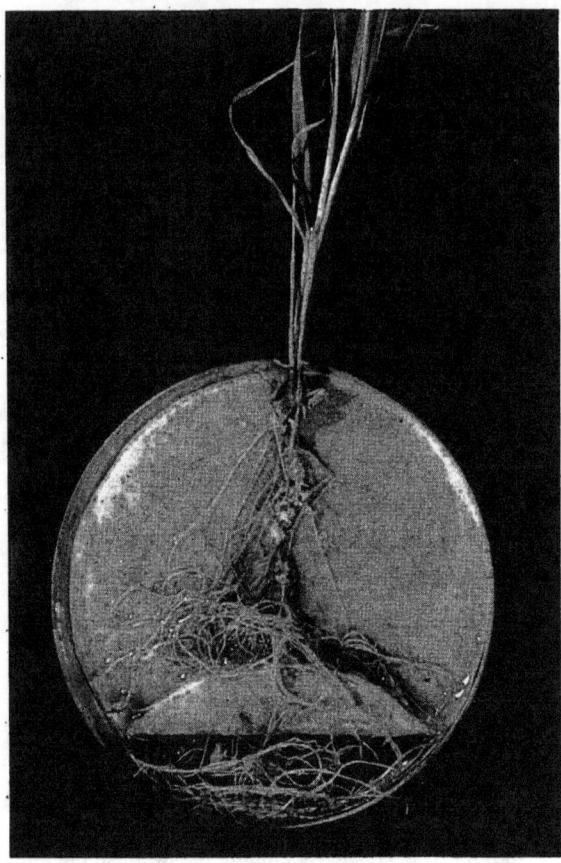

Figure 1.—*A photograph of a root cage. The white spots on the roots are accumulations of wax threads covering the aphids* Baizongia pistaciae.

Aphids and Handling

Galls of several species were collected in the autumn of 1995 and 1996. At this time of year the galls contain alates (fall migrants) ready to disperse. The galls were opened in the laboratory and the aphids were held in jars lined with filter paper. The alates readily gave rise to their offspring (crawlers) as they would have done in nature on the ground. In a cool and humid environment the crawlers may survive several weeks without food (Wertheim 1954). The jars were held at 20°C and samples of crawlers were periodically transferred to root cages. The cages were held at 20°C and 13 L:11 D light cycle or in an ambient temperature and continuous light.

RESULTS

A.—The first successful colonization of the roots was with crawlers of *Baizongia pistaciae* L., collected on *P. palaestina* in November 1995. The dark green crawlers molted once in the jars and commenced feeding immediately upon transfer to the roots. After another molt their color changed to light yellow. They tended to settle in groups along the roots. Feeding aphids produced long wax threads from 6 rows of dorsal wax pores. About 6 weeks after colonization the aphids began to reproduce parthogenetically. Despite two catastrophes (malfunctioning incubator), we succeeded in rearing the aphids for 19 successive generations on the roots, by transferring newborn offspring to new cages at 3-day intervals.

B.—Next came similar success with *Aploneura lentisci* Pass. Galls of this species were collected in November 1996 on its host plant *Pistacia lentiscus* and alates were held in jars. The crawlers of *A. lentisci* are light lemon-yellow, smaller and more elongated than all species of Fordinae examined so far. (This species is mentioned by Bodenheimer and Swirski (1957) as facultatively holocyclic; it belongs in the same tribe, Baizongiini, as the former species). They started feeding immediately after transfer to the cages, and soon produced dense colonies along the roots. They produce copious amounts of wax threads and honeydew droplets, which become coated with wax and look like small pearls. The reproductive rate of this species seems extremely high, the dense wax threads making it difficult to census this species in the cages. At the time of writing, we have the colonies going for 6 months and they seem to be thriving.

C.—Attempts to rear the crawlers of *Geoica* spp. were far less successful. We were particularly interested in rearing the winter generations of *Geoica* on roots in order to sort out the taxonomic problem of the "*G. utricularia*" species complex (Brown and Blackman 1994), because we seem to have at least two species (galling *P. palaestina* and *P. atlantica*, respectively) which were lumped together in previous work (Wertheim 1954 and earlier references there, Koach and Wool 1977, Wool 1977). Moreover, we expected to succeed with these species because anholocyclic populations of *Geoica* were described from North America and Northern Europe (Mordvilko 1928). Our efforts began in 1995, employing the methods of Campbell and Hutchinson (1995) and Royer (1991), but failed. In 1996, we began again, from galls collected in September and October on the two *Pistacia* hosts. Each gall was emptied into a separate jar and the alates readily reproduced. The crawlers were distinctly different in color: those from *P. palaestina* galls were orange, those from *P. atlantica* greenish. When introduced to the root cages after a first molt in the holding jars, the crawlers did not commence feeding. They remained small and most of them died. In a late attempt to induce feeding, we transferred two holding jars to a 4°C refrigerator for 2 weeks. The crawlers from these jars did start feeding when introduced to the cages, and some matured and reproduced. Keeping the root cages in a sheltered outdoor porch during the winter (January - March 1997) and taking extreme care in handling them, we managed to get F2 and some F3, which eventually died.

Although *Geoica* belongs in the Baizongiini, the root forms (from both hosts) do not secrete wax threads. Neither do they aggregate on the roots but tend to scatter and feed individually. The mature forms are pale, although those from *P. palaestina* (given the specific name *G. wertheimae* by Brown and Blackman 1994) tend to a pale orange tint. The reproductive rate of this form seems to be very low: they reproduce their young in batches of four or five, no more then three such batches in their lifetime. We hope to have better success when we try again next year.

The crawlers of two species of *Forda* behaved contrary to expectation. Galls of *Forda formicaria* von Heyden and *F. riccobonii* Stef. were collected in October and November 1996, and the alates produced crawlers in the holding jars. In the cages, the long-legged, greenish crawlers moved up the seedlings out of the cage through the notch and settled on the leaves. However, most of them were lost in a day or two. The few survivors were found near the seed where they grew and matured to produce F1 offspring, but no further reproduction was obtained.

Ants

Subterranean stages of Fordinae species are often reported as associated with ant nests. Shaposhnikov (1985) interpreted certain morphological structures in Fordinae as adaptations for these mutualistic interactions. In our laboratory and rearing rooms, "household" ants (*Monomorium pharoni*) are very abundant (to our dismay!). The ants invaded some of the rearing cages with *Baizongia pistaciae*. When observed under the stereomicroscope, these wild ants—which are not known to tend aphids—were seen, in many instances, to perform the exact kind of behavior that aphid-tending ants perform: they "tapped" the aphids with their antennae until a honeydew droplet was excreted, and devoured. "Milking" the aphids may not require specialized kinds of ant-aphid species associations.

DISCUSSION

The root cages, although still imperfect, seem suitable for the study of the ecology of subterranean stages of the Fordinae, which was previously unknown. The main unresolved problems are both technical (e.g., how to get rid of or prevent the development of fungi in the cages, which make observations difficult and may destroy the colonies) and biological (i.e., why only two of the species were successfully reared for long periods; solving the difficulties in rearing *Geoica* and *Forda* spp.).

The differences among species in adapting to the cages suggest that the method should be adjusted to the biology of each species. However, the biology is unknown, and only trial-and-error studies can break the vicious circle.

Until now, aphids of *A. lentisci* and *B. pistaciae* reproduced continuously on the roots and showed no evidence of becoming alate (producing sexuparae). In free-living aphids and in *Pemphigus*, short day length, low temperatures and crowding are known to induce sexuparae formation (Judge 1968, Moran et al. 1993a, b). In *Pemphigus*, sexuparae appear in the fall, while in the Fordinae they fly in spring, so it is unclear if the same factors are involved in sexuparae induction. Success with the use of the root cages should provide answers to these questions.

ACKNOWLEDGMENT

I benefitted from the experience of Dr. N.A. Moran, Dr. T.A. Royer, and particularly Dr. C.D. Campbell with rearing *Pemphigus* aphids on roots.

LITERATURE CITED

Bodenheimer, F.S.; Swirski, E. 1957. The Aphidoidea of the Middle East. Israel: The Weizmann Science Press.

Brown, P.A.; Blackman, R.L. 1994. Morphometric variation in the *Geoica ultricularia* (Homoptera: Aphididae) species group on *Pistacia* (Anacardiaceae), with descriptions of new species and a key to emigrant alatae. Systematic Entomology. 19: 119-132.

Burstein, M.; Wool, D. 1993. Gall aphids do not select optimal sites (*Smynthurodes betae*, Pemphigidae). Ecological Entomology. 18: 155-164.

Burstein, M.; Wool, D.; Eshel, A. 1994. Sink strength and clone size of sympatric gall-forming aphids. European Journal of Entomology. 91: 57-61.

Campbell, C.D.; Hutchinson, W.D. 1995. Rearing methods and demographic statistics for a subterranean morph of the sugarbeet root aphid (Homoptera, Aphididae). Canadian Entomologist. 127: 65-77.

Inbar, M.; Wool, D. 1995. Phloem-feeding specialists sharing a host tree: resource partitioning minimizes interference competition among galling aphid species. Oikos. 73: 109-119.

Inbar, M.; Eshel, A.; Wool, D. 1995. Interspecific competition among phloem-feeding insects mediated by induced hostplant sinks. Ecology. 76: 1506-1515.

Judge, F.D. 1968. Polymorphism in a subterranean aphid, *Pemphigus bursarius*. I. Factors affecting the development of sexuparae. Annals of the Entomological Society of America. 61: 819-827.

Koach, J.; Wool. D. 1977. Geographic distribution and host specificity of gall forming aphids (Homoptera, Fordinae) on *Pistacia* trees in Isreal. Marcellia. 40: 207-216.

Moran, N.A.; Whitham, T.G. 1988. Population fluctuations in complex life cycles: an example from *Pemphigus* aphids. Ecology. 69: 1214-1218.

Moran, N.; Seminoff, J.; Johnstone, L. 1993a. Genotypic variation in propensity for host alternation within a population of *Phemphigus betae* (Homoptera: Aphididae). Journal of Evolutionary Biology. 6: 691-705.

Moran, N.; Seminoff, J.; Johnstone, L. 1993b. Induction of winged sexuparae in root-inhabiting colonies of the aphid *Pemphigus betae*. Physiological Entomology. 18: 296-302.

Mordvilko, A. 1928. *Geoica* Hart and its anolocyclic forms. Compte Rendu. Academy of Science USSR. 1928: 525-528.

Royer, T.A.; Harris, M.K.; Edelson, J.W. 1991. A technique and apparatus for isolating and rearing subterranean aphids on herbaceous roots. Southwestern Entomologist. 16: 100-106.

Shaposhnikov, G.C. 1985. The main features of the evolution of aphids. In: Evolution and systematics of aphids. Proceedings, International Aphidological symposium at Jablona, 1981. Anonymous Wroclaw: Polsha Akademia Nauk: 19-99.

Wertheim, G. 1954. Studies on the biology and ecology of the gall-producing aphids of the tribe Fordini (Homoptera: Aphidoidea) in Israel. Transactions of the Royal Entomological Society of London. 105: 79-96.

Wool, D. 1977. Genetic and environmental components of morphological variation in gall-forming aphids (Homoptera, Aphididae, Fordinae) in relation to climate. Journal of Animal Ecology. 46: 875-889.

Wool, D. 1990. Regular alternation of high and low population size of gall-forming aphids: analysis of ten years of data. Oikos. 57: 73-79.

Wool, D. 1984. Gall-forming aphids. In: Ananthakrishnan, T.N., ed. Biology of gall insects. Oxford & IBH: 11-58.

Wool, D. 1995. Aphid-induced galls on *Pistacia* in the natural Mediterranean forest of Israel: which, where, and how many? Israel Journal of Zoology. 41: 591-600.

Wool, D.; Manheim, O. 1986. Population ecology of the gall-forming aphid, *Aploneura lentisci* in Israel. Research on Population Ecology. 28: 151-162.

Wool, D.; Manheim, O. 1988. The effect of host plant properties on gall density, gall weight and clone size in the aphid, *Aploneura lentisci* (Pass.) (Aphididae, Fordinae) in Israel. Researches on Population Ecology. 30: 227-234.

Wool, D.; Burstein, M. 1991. A galling aphid with extra life-cycle complexity: population ecology and evolutionary considerations. Researches on Population Ecology. 33: 307-322.

Wool, D.; Bar-El, N. 1995. Population ecology of the galling aphid *Forda formicaria* von Heyden in Israel: abundance, demography and gall structure. Israel Journal of Zoology. 41: 175-192.

Wool, D.; Manheim, O.; Burstein, M.; Levi, T. 1994. Dynamics of re-migration of sexuparae to their primary hosts in the gall-forming Fordinae (Aphidoidea: Pemphigidae). European Journal of Entomology. 91: 103-108.

Zwolfer, I. 1958. Zur Systematik, Biologie und Okologie unterirdisch lebender Aphiden (Homoptera, Aphidoidea) (Anoeciinae, Tetraneurini, Pemphigini und Fordinae). Zeitschrift fuer Angewandte Entomologie. 43: 1-52.

THE LIFE-CYCLE OF THE BLACKCURRANT GALL MITE, *CECIDOPHYOPSIS RIBIS* WESTW. (ACARI : ERIOPHYIDAE)

Dariusz Gajek[1] and Jan Boczek[2]

[1] Research Institute of Pomology and Floriculture, Department of Applied Entomology, Pomologiczna 18, 96-100 Skierniewice, Poland
[2] Warsaw Agricultural University, Department of Applied Entomology, Nowoursynowska 166, 02-787 Warszawa, Poland

Abstract.—Four to six generations of *C. ribis* develop each year with the highest population level reached just before dispersal to new buds. Chemical control of the pest is necessary and can be done only during its disperal period. Only a few blackcurrant, *Ribes nigrum* (L.), cultivars show resistance to the mite and can therefore be recommended in breeding work. The most common predators of the pest are phytoseiid mites and the wasp, *Tetrastichus eriophyes*.

INTRODUCTION

Fourteen species of eriophyid mites are known on plants of the genus *Ribes* by Amrine (1992). Csapo (1992) found that there are six species on blackcurrant, *Ribes nigrum* (L.) alone. Four species of *Cecidophyopsis* mites occur on currants (black, red, and golden) and on gooseberry (Amrine et al. 1994). The most important eriophyid mite on black currant is the blackcurrant gall mite, *Cecidophyopsis ribis* Westw. Other mite species are free living vagrants. The movement of infested nursery material has contributed to the wide distribution of *C. ribis* in Poland. There are many problems associated with this pest and most chemicals are ineffective against it.

C. ribis sucks the sap of leaf tissues and buds and in so doing, the introduction of its saliva triggers abnormal growth of plant cells leading to hypertrophy and hyperplasy. The consequence is that infested buds become markedly swollen to several times larger than uninfested ones.

DIRECT DAMAGE

Using light and polarization microscopy to study longitudinal sections of infested and uninfested buds (stained in a water solution of safranine and in an alcohol solution of permanent green), we found that in June, unifested buds had a vegetative tip which was differentiated into young leaves. By September this vegetative tip formed the germs of flowers, each surrounded by four leaves. By the next April, flower germs were very numerous. On the other hand, in infested June buds, the vegetative tips were differentiated into young flower germs. By September, infested buds had already become several times larger than those without *C. ribis*. During this time, 3-4 generations developed and mites were present throughout the entire bud between tightly arranged leaves. During the same period, damage to vegetative tips was observed, caused by the partial lysis of meristem cells and by strong necrotic changes. Feeding by *C. ribis* caused epidermal cells to undergo divisions, forming atypical meristematic galls. By April, the mites developed through 1-2 further generations resulting in the decay of whole germs of flowers (fig. 1A) and leaves (fig. 1B). Parenchymal cells of the leaf tissue were very often atypically lignified or formed meristematic galls on the epidermis. Population levels of the mite were still higher in spring and were present in all bud parts. When the plants started to grow, the onset of bud break was distinctly delayed. Most buds then dried out completely. The result was a smaller photosynthetic area which typically leads to a decreased resistance to frost and a lower fruit yields. The viability of such crop plants is reduced. According to Brennan (1992), if more than 5 percent of buds are infested, a currant plantation becomes unprofitable.

INDIRECT DAMAGE

C. ribis is the only known vector of a serious disease of blackcurrants, known as reversion. Both nymphs and adult mites are able to transmit this disease. However, for a plant to develop symptoms, the mites have to feed for at least 3 hours. So far, two types of reversion are

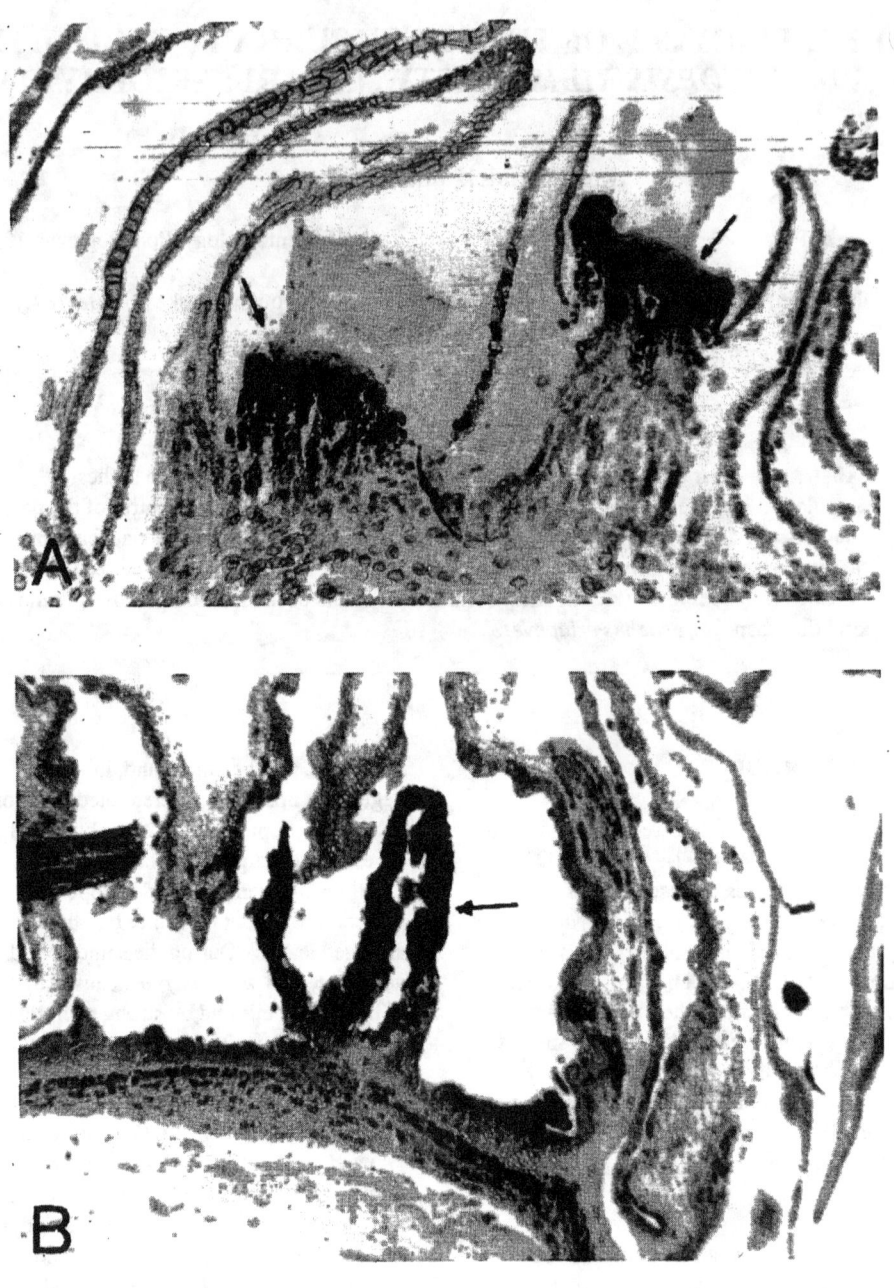

Figure 1.—*The decay of germs of flowers (A) and leaves (B) caused by feeding of* C. ribis.

known: (1) Common form (E), which is distributed throughout Europe, part of Great Britain and New Zealand. Symptoms include a reduction in leaf flaps, smaller leaves with deep incisions and with fewer serrations. Some cultivars show a chlorotic discoloration and take on the shape of oak leaves.

(2) Severe form (R) is found mainly in Eastern Europe and sometimes in Scandinavia. Symptoms are as in Common form (E). In addition, infested plants form sterile flowers without the onset of fruits. Although reversion disease has been known from the beginning of this century, the causal agent remains unknown.

DISPERSAL OF *C. RIBIS*

C. ribis mites remain hidden inside buds for most of the year. Their control is therefore difficult and can only be carried out during their dispersal period. In Poland dispersal begins when the spring day-time temperature averages 11°C. The average temperature for the previous 20 days before dispersal is usually around 7°C. In such conditions, mites leave the old buds and actively colonize other plant parts. During this time, they are transferred by wind, rain, insects, and other animals.

Besides temperature, dispersal is related to the development of the host plant during its flowering stage. In reaching this phenological stage, particular varieties may take several days. This is an important factor for the timing of chemical applications. In 1994 and 1995, observations of consecutive stages of eriophyid dispersal on varieties with different blooming times were carried out (Gajek 1997). Varieties included the cultivars Bona (blooming early) and Titania (blooming later).

Blooming Period

The differences in the blooming periods of the Bona and Titania cultivars were more obvious in 1994 than in 1995 (table 1). In 1994 the initiation of blooming in cultivar Bona was 9 days earlier than in Titania. Bona reached full blooming 5 days sooner than Titania. Likewise, end of blooming occurred 9 days earlier in cultivar Bona, the differences in blooming stages were only 2 to 3 days apart. In that year the start of the growing season was at

Table 1.—*Blooming period of studied cultivars*

Cultivar	Beginning of blooming		Full blooming		End of blooming	
	1994	1995	1994	1995	1994	1995
Bona	4/16	4/24	4/24	5/2	5/1	5/9
Titania	4/25	4/27	4/29	5/4	5/10	5/12

first delayed, but development rapidly accelerated later due to an increase in temperature.

Onset of Dispersal

On 12 April, 1994, the first dispersing mites were observed on cultivar Bona, 4 days before blooming (table 2). The average daily temperature was 11.1°C with a maximum of 13.7°C. Mites on cultivar Titania started dispersing 2 days later, on 14 April, 11 days before blooming. The average daily temperature was 7.1°C with a maximum of 10.1°C. In 1995, the beginning of dispersal coincided exactly with the timing of blooming. On cultivar Bona, dispersal began at the same time as blooming, on 24 April. On cultivar Titania this occurred 2 days later, coincident with the beginning of blooming.

Table 2.—*Periods of C. ribis dispersal on two currant cultivars in 2 years*

Dispersal phases	Bona		Titania	
	1994	1995	1994	1995
Beginning of migration	4/12	4/24	4/14	4/26
Inhabitation of new buds	5/2	5/4	5/2	5/4
Height of migration	5/4	5/4	5/4	5/4
End of migration	6/15	6/4	6/15	6/6
Time (days)	59	42	57	42

Occupation of New Buds

On 25 April, 1994, in both cultivars, single mites were found in the leaf axils, earlier than shown in table 2. At that stage, new buds were still unformed. Table 2 shows young buds inhabited by mites a few days later, on 2 May. In 1995 as a result of faster plant growth, these differences were less obvious. The mites were found on May 4, both in the leaf axils and buds of both cultivars.

Peaks and Termination of Dispersal

The dispersal peaks of the two varieties studied in 1994 and 1995 did not differ. In 1994 in cultivar Bona it was 4 days earlier than in Titania but after this period, the temperature decreased rapidly which lowered the intensity of dispersal. When the temperature started to increase, on 4 May, there was a mass evacuation of buds by mites in both cultivars. In 1995, the same type of evacuation was observed on 4 May. This occurred directly after the lower temperature.

There were no differences found in the end of the dispersal phases. In 1995, mites were observed until 4

June and 6 June for cultivar Bona and Titania, respectively. The average period of dispersal lasted for about 50 days.

Although the periods of dispersal for both varieties were similar, the courses taken by the mites were slightly different. The differences were due to the way in which infested buds developed. Bud scales of cultivar Bona were not as tight as Titania. The first mites were hence, more common in the buds of the Bona variety. This difference was even more prominent later. At the end of blooming, the buds of Bona were like loose rose buds. Such morphological changes caused an intensive departure of mites from the buds. Loose buds took on the appearance of fully developed roses and dried out quickly, which led to decay of the buds and the death of the mites. The dates given above to mark the end of dispersal dealt with single, living specimens existing on some incompletely decayed buds. These buds were usually found towards the center of bushes.

Morphological changes within infested buds of cultivar Titania were less distinct and the consecutive stages of dispersal were shorter. During this whole period, buds remained fairly tight resembling small cabbage heads. Such buds showed no signs of decay. The dispersal from these buds was less dynamic, making the point at which it reached its maximum peak difficult to discern. In contrast to cultivar Bona, the last migrants were found on the most viable buds which were most numerous. After this period, buds started to wilt and did not open. Some differences in the course of dispersal were also observed within and between bushes of the same variety. The dispersal period was shorter on bushes which had greater infestations. It seems that dispersal was slightly faster on peripheral than on the interior branches of a bush. This is probably related to a greater penetration of sunlight. If there are a large number of galled buds on bushes, there is better penetration of light and as a result plant physiological processes and mite dispersal are faster.

On the basis of these observations, it is difficult to precisely determine the best timing for the application of chemical treatments. According to actual recommendations for 1997/1998 the first spraying should be at the beginning of blooming, the second at the moment of full blooming and third, at the end of blooming. It is essential that the first treatment is made before first mites reach the axils of leaf petals. However, it was found that this period does not always coincide with the beginning of flowering in black currants. An example is cultivar Titania which began to bloom 25 April, 1994, at the time when mites were already present in the axils of leaf petals. The second treatment performed during full blooming and did not always coincide with the height of dispersal. In 1994, the height of dispersal in cultivars Bona and Titania was observed a few days after blooming. The third treatment after the end of blooming is pointless, and furthermore can lead to an accumulation of pesticide residue in fruits. There is a slight possibility that mites which disperse later in the season may inhabit new buds. The active dispersal of such mites towards young, uncovered buds is difficult as these buds are further removed from old buds during stem growth. In this situation only passive dispersal is possible. Mites are washed out by rain to the lower parts of bushes where they can easily find new buds to inhabit.

VARIATIONS IN INFESTATION OF DIFFERENT CULTIVARS

Control of *C. ribis* is difficult. In 1993-1995, more than a hundred cultivars were selected to learn if any showed resistance to this pest (Gajek 1997). Such cultivars could then be recommended when setting up new plantations and used in future plant breeding work. This evaluation included both cultivars with the best utilization traits and all others in our collections. Results showed that the degree of resistance to *C. ribis* and degree of infestation, were not permanent cultivar traits but rather were influenced by environmental factors. For example, cultivars Ben Tirran and Titania in Dabrowice showed moderate levels of infestation (table 3), whereas in Brzezna, Ben Tirran was not attacked at all and Titania showed the highest level of infestation (table 4). Infestation levels were also very much related to the locality of a plantation. For example in Milobadz (table 5), infestation levels were several times higher than in other localities. Another factor which affects the level of infestation is the application of chemicals for the control of the pest. Chemical application varies between localities and sometimes is not carried out at all. Experimental plots of differing age and sanitary condition were studied, showing that older plots were inhabited by a variety of pests and that the development of *C. ribis* was slower.

Only a few black currant genotypes showed some level of resistance to *C. ribis*. High levels of resistance were observed in varieties Titania Z, Bogatyr, and Russian cultivars such as Aranka, Korelskaja, Nariadnaja and in clones Ri 1650 and X23/25. In other studies it was shown that the main mechanism for resistance to *C. ribis* was induced antibiosis (Gajek 1997). Biochemical and cytochemical studies of the resistant cultivar Ceres under the influence of mite piercing showed an increase in the activity of oxidizing enzymes. This led to the production of oxidized phenols. Necroses, which were the result, reduced the mites access to food.

NATURAL ENEMIES OF *C. RIBIS*

Observations of natural enemies were carried out in 1994-1996 with special consideration given to phytoseiid

Table 3.—*Average number of buds per currant bush infested with* C. ribis; *Dabrowice*

Cultivar	Date of observation		
	13 April 1993	21 March 1994	25 March 1995
Ojebyn	0.16 abc[1]	0.24 abc	0.04 ab
Ben Alder	0.66 cd	3.58 e	3.16 d
Ben Lomond	0.16 abc	0.52 abcd	0.06 ab
Ben Sarek	0.10 abc	0.02 a	0.10 abc
Ben Tirran	0.98 d	1.50 cde	0.92 c
Stor Klass	0.12 abc	0.72 abcd	0.16 abc
Titania	0.30 abcd	1.26 bcd	0.78 bc
Ri 740206	0.58 cd	2.20 de	0.76 bc

[1] Means with dates (columns) followed by the same letter are not significantly different (P = 0.05), Duncan's multiple range test.

Table 4—*Average number of buds per bush infested with* C. ribis; *Brzezna*

Cultivar	Date of observation 5 April 1994
Ojebyn	0.05 a[1]
Ben Lomond	0.08 a
Ben Tirran	0 a
Titania	0.38 b
Tsema	0 a
Ri 740209	0.05 a

[1] Means followed by the same letter are not significantly different (P = 0.05), Duncan's multiple range test.

Table 5.—*Average number of buds per bush infested with* C. ribis; *Brzezna*

Cultivar	Date of observation	
	10 March 1994	28 March 1995
Ojebyn	0.16 a[1]	2.66 a
Ben Lomond	0.10 a	3.94 a
Ben Sarek	0.44 a	1.06 a
Ben Tirran	5.90 b	29.66 b
Titania	0.44 a	0.96 a
Ri 740209	3.38 ab	14.80 ab

[1] Means followed by the same letter are not significantly different (P = 0.05), Duncan's multiple range test.

mites and the predatory wasp *Tetrastichus eriophyes*, whose larvae feed on *C. ribis*. The presence of *T. eriophyes* was observed in from 0 to 40 percent of buds infested by the pest. This wasp develops one generation a year in the climatic conditions of Poland. Females predominantly lay eggs in June but can also deposit single eggs to infested buds later. The occurrence of more than one larva per bud was rare. The first wasp larvae were seen at the beginning of September and their development extended to the period of mass mite dispersal the following spring. During the second half of May, larvae pupated and adults started to appear in early June. Buds infested by *C. ribis* and larvae of *T. eriophyes* did not differ in appearance to buds that were infested by mites only, except for the fact that they seemed to be smaller. Mite dispersal from buds that were inhabited by the wasp was still high. This suggests that *T. eriophyes* is not effective in the population regulation of *C. ribis*.

Two species of phytoseiid mites were found on black currant bushes in experimental plots. These were Amblyseius andersoni (Chant) and *Typhlodromus pyri* (Schr.). *A. andersoni* was more numerous. Further observations on these mites are required to determine their potential as predators of *C. ribis*.

LITERATURE CITED

Amrine, J.M. 1992. Eriophyid mites on Ribes. In: Proceedings for the Ribes risk assessment workshop; 1992 August 17-18; Corvallis, OR: USA: 17-22.

Amrine, J.M.; Jones, A.T.; Gordon, S.; Roberts, I.M. 1994. Cecidophyopsis mites (Acari: Eriophyidae) on Ribes spp. (Grossulariaceae). International Journal of Acarology. 20(30): 139-168.

Brennan, R.M. 1992. Ribes breeding in the United Kingdom. In: Proceedings for the Ribes risk assessment workshop; 1992 August 17-18; Corvallis, OR: USA: 33-35.

Csapo, Z. 1992. Eriophyid mites (Acarina Eriophyoidea) on currants: morphology, taxonomy and ecology. Department of Applied Entomology of the Warsaw Agricultural University. Warsaw, Poland. Ph.D. dissertation.

Gajek, D. 1997. Resistance mechanisms of black currant cultivars to the gall (Cecidophyopsis ribis Westw.) and biological aspects of its control. Skierniewice, Poland: Department of Applied Entomology of Research Institute of Pomology and Floriculture. Ph.D. dissertation.

CAN INDUCING RESISTANCE COST AS MUCH AS BEING SUSCEPTIBLE?

Carolyn Glynn[1], Stig Larsson[1] and Urban Gullberg[2]

[1] Department of Entomology Swedish University of Agricultural Sciences Box 7044, S-750 07 Uppsala, Sweden
[2] Department of Plant Breeding Swedish University of Agricultural Sciences Box 7003, S-750 07 Uppsala, Sweden
Correspondence to Carolyn Glynn: e-mail Carolyn.Glynn@entom.slu.se

Abstract.—Susceptible *Salix viminalis* genotypes respond to *Dasineura marginemtorquens* feeding attempts by forming a gall. Resistant willow genotypes respond by killing the larvae before they can cause galls to form. We found that exposure to feeding gall midge larvae has a negative impact on both willow types. Resistant genotypes surprisingly lost nearly as much growth potential as did infested susceptible genotypes, even though midges died before galls were formed, and plants showed no visible damage. Resistant genotypes showed no trade off in allocation between defense and growth; when grown in a midge-free environment both willow types achieved higher biomass. The lower biomass of the resistant genotypes exposed to midges implies the presence of an induced defense.

INTRODUCTION

There are only a few ways of coping with enemies: either flee, fight, or weather the attack (Berryman 1988). Plants can flee by growing rapidly and thereby limiting the amount of time in which vulnerable tissue is exposed to potential herbivores (Aide 1988, Janzen 1975, Rhoades and Cates 1976). Plants can also escape by being less apparent (Feeny 1976). Resistant plants fight and defend themselves by being unsuitable as food for the insect's development or survival ('antibiosis') (Renwick 1983); plants can also be resistant by means of characters that make insects prefer them less ('antixenosis') (Kogan and Ortman 1978). Plants may, in addition, cope with herbivory through tolerance: the ability to grow and reproduce in spite of herbivory (Rosenthal and Kotanen 1994). These strategies are not exclusive and plants may utilize several at the same time or different ones at different phases in their development.

Exploration of the range of interspecific variation in plant resistance to herbivores has been the subject of recent interest (Fineblum and Rausher 1995, Simms and Triplett 1994, Trumble et al. 1993, van der Meijden et al. 1988, Weis 1994). Empirical studies abound in the agricultural literature (see reviews by Kennedy and Barbour 1992, Maxwell and Jennings 1980). Natural plant populations have been the focus of fewer studies, but those that searched for evidence of genetically-based variation in resistance have generally found it (Anderson et al. 1989, Christiansen et al. 1987, Haukioja and Hanhimaki 1985, Maddox and Root 1987, Raffa 1991).

We studied the variation in resistance within basket willow, *Salix viminalis,* to herbivory by a monophagous gall midge, *Dasineura marginemtorquens*. Genotypes of *S. viminalis* respond in different ways to gall initiation attempts by *D. marginemtorquens*. Strong et al. (1993) found that 240 genotypes displayed a normal distribution when classified by number of galled leaves per shoot, with the exception of about 10 percent of the genotypes which proved to be very resistant outliers. Larval survival on these genotypes was less than 5 percent whereas on the most susceptible genotypes it was over 90 percent (Larsson and Strong 1992).

Models of the evolution of defense strategies assume that plants are constrained by finite resources (Mooney 1972) and that defenses are costly (Coley et al. 1985, Feeny 1976, McKey 1979). We expected, therefore, that any resistance to *D. marginemtorquens* that would evolve within *S. viminalis* would prove to be one of little cost since otherwise the alternative strategy of tolerating or compensating for damage would be employed by this species with such high growth potential (Coley et al. 1985).

Our prediction of a low cost of resistance was further based on estimates of stem growth production from an earlier investigation. Strong *et al.* (1993) found that resistant *S. viminalis* genotypes produced almost as much as susceptible ones, 122 and 126 g dry weight respectively, (averaged over years and nutrient treatments), despite an order of magnitude difference in insect damage on the two willow types (unpubl. data). Their interpretation was that resistance was not costly. That experiment, however, included no control without damage and thus an alternative explanation could be that both infested susceptible genotypes and attacked resistant plants suffered from reduced growth.

The present study was designed to test these two alternatives. Clones of resistant and susceptible willow genotypes were grown in field cages with and without midges. With this experimental design it was also possible to examine whether the resistance was of an induced type. The experiment carried on for two consecutive years and encompassed the effects of six generations of gall midge attack.

S. viminalis was first introduced into Sweden for basket weaving material in the late 1600's (Hylander 1971). Naturalized stands can be found along waterways in central and southern Sweden. *S. viminalis* grows extremely rapidly and has a remarkable capacity for regrowth following damage (see Sennerby-Forsse *et al.* 1992). Vigorous growth and its clonal habit make *S. viminalis* an important species for plantation biomass production (Mitchell *et al.* 1992). *D. marginemtorquens* has become a pest in willow plantations where the feeding activity of midge larvae causes the margins of willow leaves to roll under and swell. Like most gall-makers, *D. marginemtorquens* can initiate galls only on meristematic tissue but the indeterminate growth pattern of *S. viminalis* allows the midge to complete three to four generations per year. Midge population numbers can be very high in cultivated willow plantations, and in some years as many as 85 percent of all leaves in a plantation can be heavily infested (Larsson *et al.* 1995).

MATERIALS AND METHODS

Plant Material

Willows used in this experiment originated from crosses of eight seed parents and eight pollen parents of 15 wild plants collected in Sweden and one from Holland (Gullberg 1989). Six full-sib groups were made for each crossing and the resulting 240 seedlings were raised to small plants, cloned by planting stem cuttings, and then tested in a common garden experiment for vulnerability to *D. marginemtorquens* attack. In previous studies, clones were ranked with respect to resistance based on the number of galled leaves per shoot (Larsson and Strong 1992, Strong *et al.* 1993). On 3 May 1991, 5 cm stem cuttings were made of ten of the most resistant and ten of the most susceptible frozen-stored ($-5^{\circ}C$) genotypes. The most resistant genotypes were represented by five families: two families (#38 and #46) each contributed three full siblings (Swedish University of Agricultural Sciences willow clone identification numbers 3810, 3813, 3814 and 4608, 4609, 4610, respectively) to the experiment and three other families (#14, #22, and #30) each contributed one or two genotypes (clones 1414, 2211, 2212, 3006). Susceptible genotypes were represented by eight families: one genotype each came from families #4, #21, #22, #28, #50, #51 (clones 409, 2110, 2214, 2813, 5010, 5107), and two genotypes each came from families #23 and #29 (2308, 2309, 2912, 2913). Plants were planted in sandy soil in pots (15 cm diameter) in the greenhouse where they were watered and fertilized daily. They were acclimatized in early June in an open unheated glasshouse for several weeks before the experiment started.

Experimental Design and Treatments

Plants were transferred to 12 blocks (4.4 m wide x 3.2 m deep and 3 m high) on June 26, 1991. Each block had a wooden frame covered by netting that allowed for 87 percent light transparency and effectively contained or excluded midges. Cages were spaced 2 m from each other and were situated in a long row in the middle of a newly plowed field of the heavy clay soil that is typical for the Uppsala area. No fertilization occurred during the experiment. Ten resistant clones alternated with 10 susceptible clones to form a randomly-assigned checkerboard pattern unique for each of the 12 blocks. Each block consisted of four rows of five experimental plants that were spaced 40 cm x 40 cm between individuals and rows. A row of border plants of randomly selected resistant and susceptible genotypes surrounded each set of experimental plants.

Cages were randomly assigned to one of two treatments: with or without midges. *D. marginemtorquens* individuals used in the experiment originated from an outbreak population in a commercial willow plantation of susceptible clone 78021 near Uppsala. Galled leaves with pupating larvae were collected from field plants and placed in small net cages in the laboratory on 1 and 4 July. Wrapping petioles in moist cotton wool kept the leaves fresh for the few days it took until adults eclosed from the galls. Most midges mated in the eclosion cages soon after emerging from the galls. At mid-morning midges were collected and transferred to vials for the 300 m transport to the field cages. The close contact that the midges had in the vials offered unmated individuals opportunity to find partners. Vials were opened inside the field cages and midges were allowed to fly out. Such inoculations were carried out daily from 1 to 6 July, resulting in a total of about 600 females and as many males in each of the six treatment cages. This resulted in about 14 ovipositing females per willow plant, a number high enough to ensure a heavy infestation. Edge plants, three resistant and three susceptible genotypes, were randomly selected from each of the inoculated cages and sampled on 7 July, to check for proportion of eggs laid on the respective willow types. Terminal leaf bundles were dissected and number of eggs was recorded.

In late August 1991 and 1992 the number of galled leaves as well as the number of galls per leaf on the tallest shoot per plant was recorded. The netting was removed from each cage in late autumn of 1991 after all leaves had fallen from the plants and replaced in April 1992. In January 1992 and 1993 the previous season's growth from each experimental plant was harvested 10 cm from the ground. All stems on every plant were dried for 3 days at 80°C and then weighed.

Because the experiment was carried out for two successive growing seasons, no new inoculations were necessary during the second year of the experiment. Midges that overwintered in the soil or in the leaf litter in the inoculated cages, emerged in spring and re-infested the plants.

Data Analysis

The effects of gall midge infestation on resistant and susceptible willow genotypes were investigated using an analysis of variance performed with the general linear model (GLM) procedure from the SAS statistical package using type-III sums of squares (SAS Institute Inc. 1985). The factorial design had fixed effects: willow type (10 resistant genotypes analyzed as "resistant type" and 10 susceptible genotypes as "susceptible types") and infestation level (with or without midges) as well as a random effect: replicate (six replicates of cages with midges and six without midges) (Zar 1984). Variation in stem dry weights in cages within each treatment were analyzed by the GLM procedure sorted by infestation level. Dry weights for each plant were correlated with the number of galled leaves and total number of galls on the tallest shoot in each plant (REG procedure SAS).

RESULTS

Dasineura marginemtorquens infestation reduced the stem dry weight production of both resistant and susceptible *Salix viminalis* genotypes (table 1). A two factor analysis of variance showed that gall midge infestation affected the stem weights of susceptible and resistant willow types in the second but not in the first year of the experiment (table 2). Stem biomass of susceptible and resistant willow types did not differ significantly from each other. The interaction between willow type and infestation was not statistically significant in either year (table 2), which indicates that susceptible and resistant types were affected similarly by gall midge infestation. In 1992, susceptible willows without midges attained higher mean weights than susceptible willows with midges, a difference of 36.5 percent. Likewise, resistant willows with midges had on average 34 percent less stem biomass than resistant willows without midges.

Egg deposition was about the same on resistant and susceptible willows. In 1991 a few days after the inoculations, resistant plants had an average of 35.4 (sd 5.68, n = 18) eggs per terminal leaf bundle and susceptible plants had 37.3 (sd 6.25, n = 18). No estimate was performed for 1992 as we had no reason to assume that oviposition behavior would be different to the preceding year.

Number of galled leaves differed on the willow types. In 1991, susceptible willows had on average 15.3 (sd 7.25, n = 60) galled leaves per shoot, and 126.9 (sd 73.76, n = 60) galls per shoot. In 1992, susceptible plants with midge infestation had on average 26.8 (sd 6.55, n = 60) galled leaves per shoot and 531.3 (sd 143.79, n = 60) galls per shoot. Resistant willows had no galled leaves. In some instances necrotic spots and unsuccessful galls were noted on the leaves of the resistant plants, indicating that midge larvae had attempted, but were not successful at gall initiation.

On susceptible plants in 1992, stem biomass was inversely related to number of galled leaves per shoot, as well as number of galls per shoot (no. galled leaves: $r^2 = 0.736$, y = 75.57 - 1.74x, $p < 0.001$; no. galls: $r^2 = 0.295$, y = 55.67 - 0.051x, $p < 0.001$, n = 36 plants). Regression coefficients for 1991 were not significant (no. galled leaves: $r^2 = 0.003$, y = 6.32 - 0.139x, p = 0.549, n = 36; no. galls: $r^2 = 0.046$, y = 6.08 + 0.002x, p = 0.314, n = 36 plants).

DISCUSSION

Dasineura marginemtorquens infestation significantly reduced stem growth in *Salix viminalis*. The reduction in stem biomass in genotypes that supported galls in the second year of the study was 36.5 percent. Surprisingly, there was also a considerable reduction in biomass in the resistant genotypes (34 percent), despite the fact that these plants had no galls and supported no feeding larvae. When grown without insects the two willow types had about the same stem biomass, indicating that there is no constitutive cost associated with resistance. The great differences in stem biomass in resistant genotypes exposed to midges compared with the same genotypes without midges, indicates the presence of an induced resistance.

Table 1.—*Mean dry stem weight (g) and standard deviation of susceptible and resistant willows with and without midge infestation. Each number represents the average of 6 replicates of 10 resistant or 10 susceptible willow genotypes.*

Year	Plant type	Treatment	
		With midges	Without midges
1991	Resistant	5.6 (±1.65)	6.9 (±1.47)
	Susceptible	5.1 (±1.09)	7.3 (±1.39)
1992	Resistant	33.46 (±9.91)	50.5 (±9.56)
	Susceptible	30.7 (±1.09)	48.3 (±6.49)

Table 2—*Analysis of variance for effects of infestation on stem dry weight of* Salix viminalis *willow types (10 resistant and 10 susceptible genotypes), and infestation (with or without D. marginemtorquens).*

Year	Source	MS	df	F	p
1991	Replicate	17.187	5	2.47	0.0337
	Willow type	0.05922	1	0.0027	0.9266
	Infestation	174.88	1	16.04	0.1558
	Willow type*infest.	10.90	1	1.565	0.2123
	Genotypes	17.68	18	2.5388	0.0008
	Error	6.9645	26		
1992	Replicate	1435.76	5	6.386	0.0001
	Willow type	745.967	1	4.49	0.115
	Infestation	17334.7	1	78.05	0.0232
	Willow type*infest.	23.106	1	0.1028	0.7488
	Genotypes	367.656	18	1.6354	0.0537
	Error	224.816	26		

These findings are in accordance with data from the study by Strong *et al.* (1993). They compared damage by *D. marginemtorquens* on 240 genotypes of *S. viminalis* growing in two different nutrient environments; the focus of their report was on genetic differences in resistance and included no data on plant production. We have analyzed stem biomass production from the Strong *et al.* (1993) study for the 10 resistant and 10 susceptible genotypes in the present study. For 1988, the year of highest level of *D. marginemtorquens* infestation (see Strong *et al.* 1993), stem production, averaged over high and low nutrient environments, amounted to 36.5 and 38 g dry weight for resistant and susceptible genotypes, respectively. In 1990, when there was very little damage, resistant and susceptible genotypes produced 187 and 184 g dry weight (the higher figure for 1990 is partly an effect of higher production of a more established plant and does not only reflect the release from insect attack). Thus, as in the present study, stem biomass production was about the same irrespective of insect density.

The effect of galling on stem growth in genotypes that supported galls was strong. The galls of *D. marginemtorquens* prevent the leaf edges from unfurling and we estimate a 40-50 percent leaf area reduction due to heavy galling in the treatment cages (C. Glynn, pers. observ.). Loss of photosynthetic potential can, in itself, negatively affect the growth of the attacked plant (e.g., Kulman 1971). This, in addition to possible sink effects of the galling process (Abrahamson and McCrea 1986) as well as energy losses to feeding larvae, most likely explains the biomass reduction caused by the gall makers.

The strong effect on stem biomass in the resistant genotypes cannot, however, be explained by a reduction in leaf area; leaves suffered no visible damage. Rather, the effects measure the indirect costs of resistance (see Baldwin 1994). It is not known, however, what proportion of the biomass reduction can be attributed to lost growth potential due to physiological changes incurred as a result of unsuccessful gall initiation attempts on the resistant plants (cf. Welter 1989).

The willows in this experiment faced the classic plant dilemma: to grow or defend (Herms and Mattson 1992). While forced to defend themselves against attack by a large number of gall midge larvae, they had to establish roots and attain competitive status over nearby conspecifics. Plant defense theory centers on the assumption that plants are constrained in their ability to simultaneously support different processes (Coley et al. 1985, Fagerström et al. 1987, Herms and Mattson 1992, Loomis et al. 1990, McKey 1979) and thus, the dilemma. Does this mean that the resistant willow types in our experiment had a limited ability to secure enough carbon and nitrogen to meet demands for defense as well as for biomass acquisition? If so, then it seems that the resistant genotypes of S. viminalis invested in defense rather than in growth.

In a situation of intense attack and high plant densities, all individuals in the near-plant community will be exposed to infestation. In our experiment, there was a tendency for growth reduction in resistant plants to be less than that of the willows with many galls. This small percentage may be enough for even costly resistance to have a selective advantage (Samson and Werk 1986). Competitive interactions have proved to be important in other plant/herbivore systems. Defoliation of 75 percent had no effect on the reproduction of the annual Abutilon theophrasti when the plant was grown at low densities. The same level of defoliation at high plant densities, however, reduced seed production by 50 percent (Lee and Bazzaz 1980).

Our results indicate that defense is as costly as sustaining an infestation. It is intuitively difficult to understand why such a costly defense is employed in a plant species that has such high capacity for regeneration. The fact that resistant genotypes had no visible damage and yet acquired so much less biomass than did the same genotypes without exposure to midges is even more counter-intuitive.

Data from another experiment involving field-grown resistant and susceptible S. viminalis do not corroborate the results from the present experiment. Resistant willows grown on plantations in 1994 and 1995 attained higher biomasses than did their heavily infested susceptible neighbors (S. Larsson, unpubl. data). This discrepancy between data sets may be due to environmental differences. There are a number of differences between the field-grown willow experiment and the present experiment. Field willows were commercial genotypes (87195-resistant and 87021-susceptible), not the experimental clones included in the present study. Field willows were planted in double rows, with 1 m between double rows and a distance of 0.5 m between plants. Experimental willows were grown with 0.4 m between plants on all sides. Data on biomass accumulation during heavy- versus low-infestation years was collected from well-established willows which attained heights of about 2.5-3 m and were more than 5 years old. Experimental willows were in their first and second year of growth, grown under netting, and were considerably less vigorous in stature than were the field-grown willows. Any of these factors alone or in combination with each other are potential factors that could explain our results.

The alternative explanation for such remarkable results is that there is some inadvertent flaw in the experiment. The 12 experimental cages were erected in a single row. The cages assigned to be the treatment were randomly selected, as it happened five of six were positioned on the southern half of the row. A hypothetical gradient of decreasing nutritional quality from north to south could produce the exact results that we attribute to the influence that midge herbivory has on its host plant, and to an extra-ordinarily costly induced resistance. We see no reason for why such a gradient should exist.

This ambiguity, however, demands a follow-up experiment to confirm or reject our results. Controlled experimental conditions with less potential environmental variation will either establish or refute: (1) that D marginemtorquens attack has a negative impact on both susceptible and resistant S. viminalis genotypes, and that (2) the existence of induced resistance in certain genotypes of S. viminalis. In our view of willow resistance, based on the present study's results, S. viminalis seems not to make much fitness gains by being resistant to D. marginemtorquens. As such, our data do not qualify the putative induced resistance to be termed as a defense (see Karban and Myers 1989). Even though our data indicate a high cost of resistance, they support to the "no cost" side of the debate on cost of resistance (Simms and Rausher 1987). In this environment in which herbivory was unusually high, the defended plants were at no competitive disadvantage to the undefended plants.

Regardless of whether our results reflect a true induced and costly resistance or not, the variation in resistance within S. viminalis still remains. The question therefore stands; does the variation in resistance displayed by *S. viminalis* reflect an evolutionary stable strategy for coping with damage caused by attacking organisms like *D. marginemtorquens*? Gall midges are a special type of herbivore in that they are intimately associated with their host plant. The gall initiation attempts by gall midge larvae cause damage to the leaf more like that of a pathogen than, for example, that of a chewing larva. It may be that the signals perceived by a gall midge infested *S. viminalis* may initiate a general response that is meant to resist attack from pathogens which can have very serious consequences (cf. Hatcher 1995). For example, rust infestations by *Melampsora* can, in combination with environmental factors such as frost, cause considerable damage to willow plants (Verwijst 1990). We suggest that the two willow types in this polymorphism compose a strategy for avoiding the impacts of infestation of this type, regardless if they are caused by herbivory of gall midge type or by pathogens.

The variation in resistance among genotypes of *S. viminalis* may well be a transient situation, and not a system of stable polymorphism. One reason for this may be the domestication of willows. The very resistant genotypes with high costs of defense may have been eliminated if *S. viminalis* had not been under artificial selection by humans interested in good willow crops for basket weaving. *S. viminalis* is naturalized in Sweden (Larsson and Bremer 1991) and the genotypes in our experiment were taken from this population that is not in genetic equilibrium. This new ecological situation between insect and plant may be another explanation for a transient occurrence of the polymorphism.

ACKNOWLEDGMENTS

We thank I. Åhman, I. Baldwin, C. Björklund, H. Bylund, B. Ekbom, R. Fritz, D. Herms, P. Johnson, G. Nordlander, A. Rönnberg-Wästljung, C. Solbreck, A. Wicklund Glynn for discussions around and comments on this paper. A. Glynn, E. Glynn, S. Höglund, A. LeGrand, U. Pettersson, E. Winkler provided excellent field assistance. The Swedish Natural Science Research Council (NFR) financed this work.

REFERENCES

Abrahamson, W.G.; McCrea K.D. 1986. Nutrient and biomass allocation in *Solidago altissima*: effects of two stem gallmakers, fertilization, and ramet isolation. Oecologia. 68: 174-180.

Aide, T.M. 1988. Herbivory as a selective agent on the timing of leaf production in a tropical understory community. Nature. 336: 574-575.

Anderson, S.S.; McCrea, K.D.; Abrahamson, W.G.; Hartzel, L.M. 1989. Host genotype choice by the ball gallmaker *Eurosta Solidaginis* (Diptera: Tephritidae). Ecology. 70: 1048-1054.

Baldwin, I.T. 994. Chemical changes rapidly induced by folivory. In: Bernays, E.A., ed. Insect-plant interactions. Boca Raton, FL: CRC Press: 5: 1-23.

Berryman, A.A. 1988. Towards a unified theory of plant defense. In: Mattson W.J.; Levieux, J.; Bernard-Dagan, C., eds. Mechanisms of woody plant defenses against insects. New York: Springer-Verlag: 39-55.

Christiansen, E.; Waring R.H.; Berryman, A.A. 1987. Resistance of conifers to bark beetles attack: searching for general relationships. Forest Ecology and Management. 22: 89-106.

Coley, P.D.; Bryant, J.P.; Chapin, F.S., III. 1985. Resource availability and plant anti herbivore defense. Science. 230: 895-899.

Fagerström, T.; Larsson, S.; Tenow, O. 1987. On optimal defence in plants. Functional Ecology. 1: 73-81.

Feeny P. 1976. Plant apparency and chemical defense. Recent Advances in Phytochemistry. 10: 1-40.

Fineblum, W.L.; Rausher, M.D. 1995. Tradeoff between resistance and tolerance to herbivore damage in a morning glory. Nature. 377: 517-520.

Glynn, C.; Larsson, S. 1994. Gall initiation success and fecundity of *Dasineura marginemtorquens* on variable *Salix viminalis* host plants. Entomologia Experimentalis et Applicata. 73: 11-17.

Gullberg, U. 1989. Växtförädling av Salix 1986-1989. Res. Notes 42. Uppsala: Department of Forest Genetics.

Hatcher, P.E. 1995. Three-way interactions between plant pathogenic fungi, herbivorous insects and their host plants. Biological Reviews. 70: 639-694.

Haukioja, E.; Hanhimaki, S. 1985. Rapid wound-induced resistance in white birch (*Betula pubescens*) foliage to the geometrid *Epirrita autumnata*, a comparison of trees and moths within and outside the outbreak range of the moth. Oecologia. 65: 223-228.

Herms, D.A.; Mattson, W.J. 1992. The dilemma of plants: to grow or defend. Quarterly Review of Biology. 67: 283-335.

Hylander, N. 1971. Första litteraturuppgifter för Sveriges vildväxande kärlväxter jämte uppgifter om första fynd. Svensk Botanisk Tidskrift. 64: 1-322.

Janzen, D.H. 1975. Ecology of plants in the tropics. Studies in Biology 58. London: Edward Arnold.

Karban, R.; Myers, J.H. 1989. Induced plant responses to herbivory. Annual Review of Ecology and Systematics. 20: 331-348.

Kennedy, G.G.; Barbour, J.D. 1992. Resistance variation in natural and managed systems. In: Fritz, R.S.; Simms, E.L., eds. Plant resistance to herbivores and pathogens: ecology, evolution, and genetics. Chicago, IL: The University of Chicago Press: 13-42.

Kogan, M.; Ortman, E.F. 1978. Antixenosis–a new term proposed to replace Painter's "nonpreference" modality of resistance. Bulletin of the Entomological Society of America. 24: 175-176.

Kulman, H.M. 1971. Effects of insect defoliation on growth and mortality of trees. Annual Review of Entomology. 16: 289-324.

Larsson, G.; Bremer, B. 1991. Korgviden-nyttoväxter förr och nu. Svensk Botanisk Tidskrift. 85: 185-200.

Larsson, S.; Strong, D.R. 1992. Oviposition choice and larval survival of *Dasineura marginemtorquens* (Diptera: Cecidomyiidae) on resistant and susceptible *Salix viminalis*. Ecological Entomology. 17: 227-232.

Larsson, S.; Glynn, C.; Höglund, S. 1995. High oviposition rate of *Dasineura marginemtorquens* on *Salix viminalis* genotypes unsuitable for offspring survival. Entomologia Experimentalis et Applicata. 77: 263-270.

Lee, T.D.; Bazzaz, F.A. 1980. Effects of defoliation and competition on growth and reproduction in the annual plant *Abutilon theophrasti*. Journal of Ecology. 68: 813-821.

Loomis, R.S.; Lou, Y.; Kooman, P.L. 1990. Integration of activity in the higher plant. In: Rabinge, R.; Goudriaan, J.; van Keulen, H.; Penning de Vries, F.W.T.; van Laar, H.H., eds. Theoretical production ecology: reflections and prospects. Wageningen: Pudoc: 105-124.

Maddox, G.D.; Root, R.B. 1987. Resistance to 16 diverse species of herbivorous insects within a population of goldenrod, *Solidago altissima*: genetic variation and heritability. Oecologia. 72: 8-14.

Maxwell, F.G.; Jennings, P.R. 1980. Breeding plants resistant to insects. New York: John Wiley and Sons.

McKey, D. 1979. The distribution of secondary compounds within plants. In: Rosenthal, G.A.; Janzen, D.H., eds. Herbivores: their interaction with secondary plant metabolites. New York: Academic Press: 55-133.

Mitchell, C.P.; Ford-Robertson, J.B.; Hinckley, T.; Sennerby-Forsse, L. 1992. Ecophysiology of short rotation forest crops. Essex, England: Elsevier Science Publishers Ltd.

Mooney, H.A. 1972. The carbon balance of plants. Annual Review of Ecology and Systematics. 3: 315-346.

Raffa, K.F. 1991. Induced defensive reactions in conifer-bark beetle systems. In: Tallamy, D.; Raupp, M.J.; Wiley, M.; eds. Phytochemical induction by herbivores. New York: 245-276.

Renwick, J.A.A. 1983. Nonpreference mechanisms: plant characteristics influencing insect behavior. In: Hedin, P.A., ed. Plant resistance to insects. Washington, DC: American Chemical Society: 199-213.

Rhoades, D.F.; Cates, R.G. 1976. Towards a general theory of plant antiherbivore chemistry. Recent Advances in Phytochemistry. 10: 168-213.

Rosenthal, J.P.; Kotanen, P.M. 1994. Terrestrial plant tolerance to herbivory. Trends in Ecology and Evolution. 9: 145-148.

SAS Institute Inc. 1985. SAS,, user's guide: statistics, version 5 edition. Cary, NC.

Samson, D.A.; Werk, K.S. 1986. Size-dependent effects in the analysis of reproductive effort in plants. American Naturalist. 127: 667-680.

Sennerby-Forsse, L.; Ferm, A.; Kauppi, A. 1992. Coppicing ability and sustainability. In: Mitchell, C.P.; Ford-Robertson, J.B.; Hinckley, T.; Sennerby-Forsse, L., eds. Ecophysiology of short rotation forest crops. Essex, England: Elsevier Science Publishers Ltd.: 146-184.

Simms, E.L.; Rausher, M.D. 1987. Costs and benefits of plant defense to herbivory. American Naturalist. 130: 570-581.

Simms, E.L.; Triplett, J. 1994. Costs and benefits of plant responses to disease: resistance and tolerance. Evolution. 48: 1973-1985.

Strong, D.R.; Larsson, S.; Gullberg, U. 1993. Heritability of host plant resistance to herbivory changes with gallmidge density during an outbreak on willow. Evolution. 47: 291-300.

Trumble, J.T.; Kolodny-Hirsch, D.M.; Ting, I.P. 1993. Plant compensation for arthropod herbivory. Annual Review of Entomology. 38: 93-119.

van der Meijden, E.; Marijke, W.; Verkaar, H.J. 1988. Defense and regrowth alternative plant strategies in the struggle against herbivores. Oikos. 51: 355-363.

Verwijst, T. 1990. Clonal differences in the structure of a mixed stand of *Salix viminalis* in response to *Melampsora* and frost. Canadian Journal of Forest Research. 20: 602-605.

Weis, A.E. 1994. What can gallmakers tell us about natural selection on the components of plant defense? In: Price, P.W.; Mattson, W.J.; Baranchikov, Y.N., eds. The ecology and evolution of gall-forming insects. Gen. Tech. Rep. NC-174. St. Paul, MN: U.S. Department of Agriculture, Forest Service, North Central Forest Expeirment Station: 157-171.

Welter, S.C. 1989. Arthropod impact on plant gas exchange. In: Bernays, E.A., ed. Insect-plant interactions. Boca Raton, FL: CRC Press: 1: 135-150.

Zar, J.H. 1984. Biostatistical analysis. Englewood Cliffs, NJ: Prentice Hall.

PLANT HYPERSENSITIVITY AGAINST TISSUE INVASIVE INSECTS: *BAUHINIA BREVIPES* AND *CONTARINIA* SP. INTERACTION

Tatiana G. Cornelissen and G. Wilson Fernandes

Ecologia Evolutiva de Herbívoros Tropicais/ DBG, ICB/Universidade Federal de Minas Gerais
CP 486, Belo Horizonte MG 30161-970, Brazil

Abstract.—*Bauhinia brevipes* induces a hypersensitive reaction (HR) to the galling insect, *Contarinia* sp. HR is observed as a necrotic tissue around the gall induction site. It was the most important mortality factor against *Contarinia* sp. in two consecutive years but was not influenced by plant genotype, shoot length, or year. Density of attack by *Contarinia* sp. was significatively higher in 1995 than in 1996. Most plants showed high levels of resistance (> 90 percent HR) to *Contarinia* sp. There was an overall increased level of resistance in susceptible plants from 1995 to 1996 lending some support to the existence of plant memory in this system.

INTRODUCTION

Studies on plant defenses against herbivores are centered on constitutive defenses (e.g., Rosenthal and Janzen 1983, Fritz and Simms 1992). Nevertheless, over the past decade much attention has been directed towards the study of induced defenses (e.g., Agrios 1988, Karban and Baldwin 1997). While several studies have documented the importance of induced responses against herbivores (Karban and Myers 1989, and reviews by Tallamy and Raupp 1991, Karban and Baldwin 1997), more detailed and long term experiments have shown an enormous variability in the defense by the hosts (e.g., Karban 1987, Coleman and Jones 1991).

The Hypersensitive Reaction (HR) is an important mechanism of induced defense against pathogens, but has rarely been studied as a source of plant's resistance to insect herbivores, except for bark beetles where the defense is directed against the mutualistic fungi associated with the beetles (e.g., Berryman 1972, Raffa and Berryman 1983, Berryman and Ferrell 1988, see also Fernandes 1990). Hypersensitivity is an induced response which encompasses morphological, histological and biochemical changes that result in the death of the attacked tissue. Increasing levels of metabolic compounds such as phytoalexins and other toxic metabolites lead to the localization, containment, inactivation, and death of the invasive agent. This induced reaction of the host is the primary event in resistance to fungal parasites, bacteria and viruses (Maclean *et al.* 1974, Agrios 1988). It leads to a disruption of nutrient, water and oxygen supplies that result in the cessation of the microorganism's growth (Bayley and Mansfield 1982) hence decreasing the probability of establishment and success by the invading organism (Wong and Berryman 1977). HR culminates in localized necrosis, due to a disturbance in the balance between oxidative and reductive processes resulting in an excessive oxidation of polyphenolic compounds and a breakdown of cellular and subcellular structures (Király 1980, Beckman and Ingram 1994).

For the past 6 years we have been studying the population dynamics of insect herbivores associated with *Bauhinia brevipes* Vogel (Leguminosae) (Cornelissen *et al.* 1997, Fernandes 1998, see also Fernandes and Price 1992). One of the most conspicuous herbivores on *B. brevipes* is *Contarinia* sp. (Diptera: Cecidomyiidae) which induces galls on the leaves of the host plant. The attack by *Contarinia* sp. on *B. brevipes* is concentrated on the longest shoots, although these represent the smallest proportion of the available shoot population (Cornelissen *et al.* 1997). No relationship has been found between female preference for longer, more vigorous shoots and larval performance and progeny of females that attack the longest shoots do not experience higher reproductive success (Fernandes 1998). Perhaps, this lack of correlation between preference and performance in the system is due to plant hypersensitivity, which may occurs indiscriminately and independently of module growth.

The HR exhibited by the *B. brevipes* to the galling larvae is easily observed as a brown circular spot around the site of tissue penetration by the larva of *Contarinia* sp. The interactions between *B. brevipes* and *Contarinia* sp. provide an excellent system to study the effects of induced plant defense on herbivore population dynamics. The HR is easy to identify while *Contarinia* sp. has all the characteristics of galling organisms, which make this system well suited for population dynamics studies.

The goal of this study was to answer the following questions: (a) What is the relative importance of HR against *Contarinia* sp. galls?; (b) What is the influence of plant genotype and module growth on the HR of *Bauhinia brevipes* against *Contarinia* sp.?; (c) Do *B. brevipes* show memory after the attack by *Contarinia* sp.?

METHODS

System and Area of Study

The study was performed in the Estação Ecológica de Pirapitinga (IBAMA) in Três Marias (MG), southeastern Brazil between 1995 and 1996. The biological station is a man-made island, created in 1965 in the Três Marias Reservoir (18° 23'S and 45° 20'W) at an altitude of 560 m above sea level. The vegetation of the island is primarily cerrado (savanna) (Azevedo *et al.* 1987) and *B. brevipes* is a leguminous shrub up to 3 meters high and abundant in the island (Cornelissen *et al.* 1997).

Samples of galls were taken in October of both 1995 and 1996 by the end of the galling period. *Contarinia* sp. induces galls on the adaxial leaf surface of *B. brevipes*. Galls are spherical, with long red hairs covering their external walls and contain a single chamber, with only one larva per chamber (Fernandes 1998, see also Fernandes *et al.* 1988, Gagné 1994).

Relative Importance of HR Against *Contarinia* sp. Galls

To answer the question "what is the relative importance of HR against *Contarinia* sp. galls", 25 shoots were haphazardly collected around the canopy of 170 arbitrarily selected plants in three subpopulations of *B. brevipes*. The subpopulations were located in three adjacent trails (subpopulation a = 50 plants, subpopulation b = 60 plants, subpopulation c = 60 plants). We measured the length (to the nearest mm), total number of leaves, and the number of galled leaves on each of the 8,216 sampled shoots. Shoot length was recorded and then transformed into shoot length classes of 3 cm (n = 24 classes), (see Price 1991). To evaluate the mortality factors associated with *Contarinia* sp., galls were dissected for observation of their contents, and factors associated with mortality recorded and categorized as parasitism, predation, pathogen attack, hypersensitivity reaction and unknown factors (see Fernandes and Price 1992). Galls killed by the plant's hypersensitivity could not be opened because the necrotic reaction fuses the gall walls. Data were then transformed into percent mortality and survivorship.

Because the subpopulations were not clearly distanced from each other, we asked whether they would belong to a single population. We tested this hypothesis by verifying if the number of shoots and leaves per shoot length class differed among subpopulations. The frequency distribution of shoots and mean number of leaves per shoot length class were similar in the three subpopulations studied (fig. 1). Short shoots were abundant while long shoots were rare, and larger shoot length classes supported larger number of leaves in all populations (One-way Anova, $p < 0.0001$). Since there was no difference in the pattern of distribution of shoots and leaves per shoot length class among the subpopulations studied, we pooled the data into a single data set.

Influence of Plant Genotype and Module Length on HR

To answer the question "what is the influence of plant genotype and module growth on the HR of *Bauhinia brevipes* against *Contarinia* sp.?" we analyzed the proportion of galls killed by natural enemies, plant resistance, galls successfully formed, and HR reaction through an ANCOVA test, which allowed us to control for shoot length (continuous variable), while testing for differences between plants and year (categorical variables) (Zar 1996). Before data analysis with an ANCOVA, we observed whether there was no significant interaction between the covariates and independent variables (test of the homogeneity of slopes) (see SYSTAT 1992). Number of galls and percent HR were used as grouping factors, while shoot length (n = 8,216) was used as the covariate. Data on percent HR were arcsine transformed (Zar 1996).

Plant Memory

To answer the question "Do *B. brevipes* show memory after the attack by *Contarinia* sp.?", we divided the studied plants into two arbitrarily chosen categories that indicate the level of induced resistance to *Contarinia* sp. attack: Resistant plants (> 90 percent HR), Susceptible plants (< 75 percent HR). Plants that felt into the intermediate level of HR were not considered in the analysis. We then verified the number of plants in each category in 1995 and 1996 and the number of plants that moved from one category to another from 1995 to 1996.

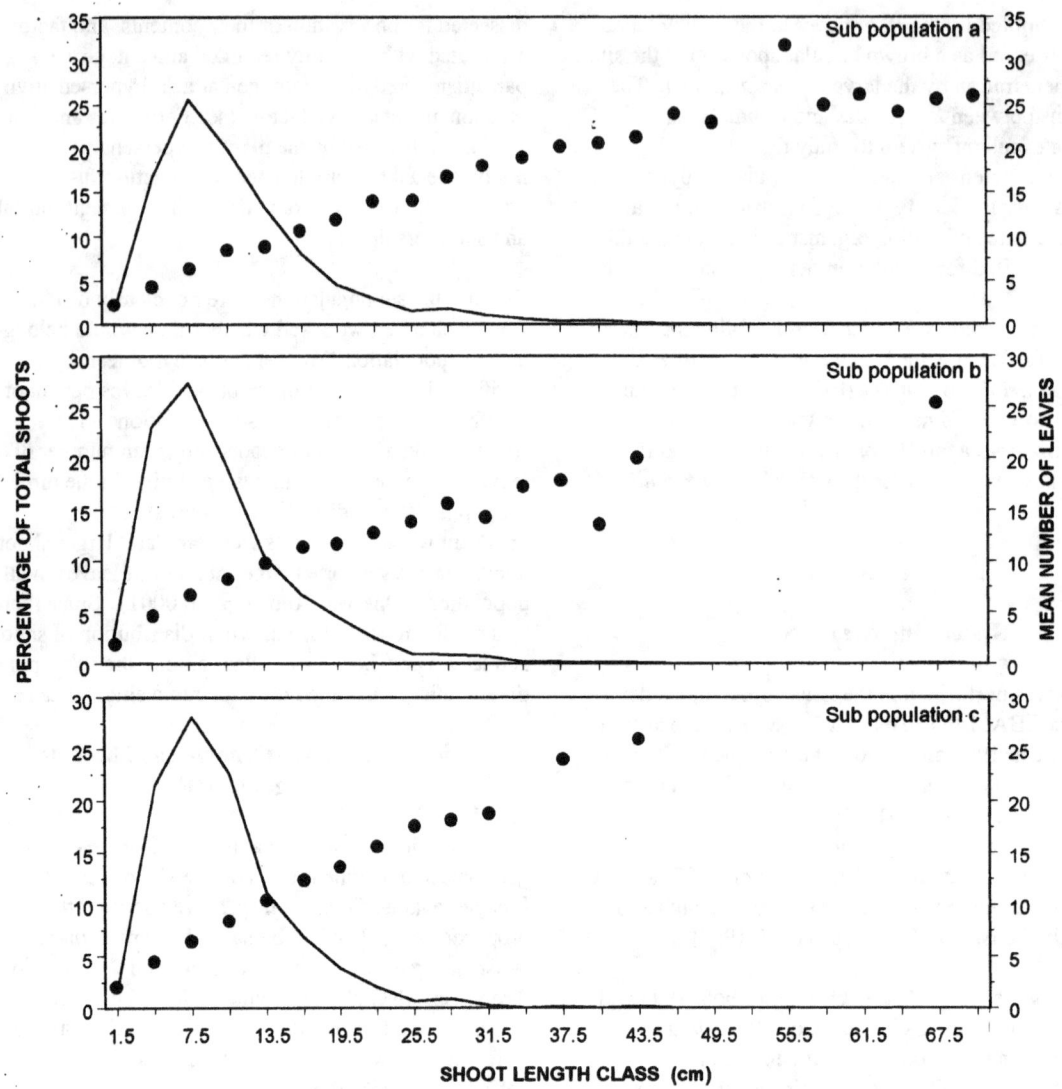

Figure 1.—*Pattern of distribution of shoots (solid lines) and mean number of leaves (solid circles) on three sub populations of* Bauhinia brevipes. *Note the high frequency of short shoots in the lower length classes.*

Differences between the number of plants that moved between categories were tested by a Chi-Square test. We also verified the similar density of attack by *Contarinia* sp. between years by the Mann-Whitney test (Zar 1996).

RESULTS AND DISCUSSION

Importance of HR against *Contarinia* sp. and Occurrence in Other Plants

HR exerted the strongest pressure upon *Contarinia* sp. in both 1995 and 1996. More than 92 percent of the larvae did not trigger gall development because of this plant's resistance mechanism (table 1). Therefore, less than 8 percent of the potential galls were left to be attacked by parasitic hymenopterans, predators that feed on gall walls, and pathogens. Only 0.1 percent of the *Contarinia* sp. survived in both years. These data corroborate previous data obtained in different plant individuals in 1993 and 1994 (see Fernandes 1998).

Table 1.—*Relative importance of HR against* Contarinia *sp. galls on* Bauhinia brevipes.

Mortality factor	Year	
	1995	1996
HR	95.5% ± 1.2	92.4% ± 0.6
Parasitism	1.5% ± 0.2	2.4% ± 0.3
Predation	2.7% ± 0.3	1.9% ± 0.3
Pathogens	0.7% ± 0.1	0.7% ± 0.1
Unknown factors	1.2% ± 0.2	2.1% ± 0.3
Survivorship	0.1% ± 0.0	0.1% ± 0.0

HR is an important mechanism whereby *B. brevipes* resist the induction of the gall by *Contarinia* sp. During four consecutive years the level of defense was higher than 90 percent (see also Fernandes 1998). HR constitutes the primary line of defense of several unrelated plant taxa against their major herbivores [e.g., wheat (Shukle et al. 1992), rice (Bentur and Kalode 1996), *Solidago* (Anderson et al. 1989), pines (Raffa and Berryman 1983, Berryman and Ferrel 1988), and possibly *Sorghum* (G.W. Fernandes, unpubl. data)].

In a field survey in the cerrado vegetation of two different localities in Brazil we observed that plants of very different and unrelated taxa were resisting gall induction on leaves by eliciting a HR (table 2). These reactions resembled that of *B. brevipes* against *Contarinia* sp., i.e., were formed by necrotic, brownish tissue, generally circular and always around the plant tissue that was penetrated by the first instar larvae of the galling cecidomyiid. Interestingly, all galls in which such reaction was noted were induced by this family of insect. In *Chrysothamnus nauseosus hololeucus* (Asteraceae) in Arizona, the HR may be the most important mechanism against a galling cecidomyiid (*Rhopalomyia chrysothamni*) (Fernandes and Price, in prep.).

Influence of Plant Genotype and Module Length on galls and HR

The abundance of galls formed by *Contarinia* sp. on *B. brevipes* was influenced by plant (genotype) ($p < 0.0001$), and shoot length ($p < 0.0001$), while we did not find any effect caused by year (table 3). Strong interactions were also observed between all variables studied (table 3). Many other studies have shown that herbivory is strongly influenced by plant genotype (e.g., Fritz and Simms 1992, Fritz et al. 1994) and module growth (e.g., Price et al. 1990, Price 1991, Prezler and Price 1995, Kimberling and Price 1996). The lack of a relationship between attack and year may have been caused by the weak relationship between the number of galls per plant in 1995 and 1996. The number of galls (all galls, including plant HR and galls killed by natural enemies) in 1995 was a poor predictor of the number of galls induced in 1996 ($r^2 = 0.05$, $y = 23.019 + 0.268 x$, $F_{1,161} = 8.913$, $p < 0.005$, fig. 2a). The same trend was observed when we analysed the relationship between the number of attacked leaves in 1995 and 1996 ($r^2 = 0.09$, $y = 0.075 + 0.328 x$, $F_{1,161} = 16.506$, $p < 0.001$, fig. 2b).

Rates of HR were not influenced either by genotype (= plant effect), shoot length, or year (ANCOVA, $p > 0.05$, table 4). This trend was also reflected in the relationship between percent HR per plant in 1995 and 1996. Although statistically significant, the relationship between percent HR in *B. brevipes* against *Contarinia* sp. between years was very weak ($r^2 = 0.05$, $y = 1.012 + 0.203 x$, $F_{1,161} = 8.078$, $p < 0.005$, fig. 2c).

Altogether these results may suggest that *Contarinia* sp. selects plant genotype and modules consistently (table 3), while HR occurs indiscriminately among genotype and module within plants (table 4). These results also corroborate the hypothesis on the efficiency of HR in locating and killing the gall inducing larvae and the lack

Table 2.—*Host plants where HR was elicited against gall-forming insects in cerrado vegetation.*

Host Plant	Plant Family	Galling Herbivore
Annona coriacea	Annonaceae	Cecidomyiidae
Aspidosperma tomentosum	Apocynaceae	Cecidomyiidae
Baccharis dracunculifolia	Asteraceae	Cecidomyiidae
Tabebuia ochracea	Bignoniaceae	Cecidomyiidae
Kielmeyeria coriacea	Clusiaceae	Cecidomyiidae
K. cf. rubiflora	Clusiaceae	Cecidomyiidae
Hyptis cana	Dilleniaceae	Cecidomyiidae
Acosmium dasicarpum	Leguminosae	Cecidomyiidae
Andira sp.	Leguminosae	Cecidomyiidae
Hymenea stignocarpa	Leguminosae	Cecidomyiidae
Tibouchina sp.	Melastomataceae	Cecidomyiidae
Ouratea sp.	Ochnaceae	Cecidomyiidae
Pouteria torta	Sapotaceae	Cecidomyiidae
Qualea grandiflora	Vochysiaceae	Cecidomyiidae
Qualea multiflora	Vochysiaceae	Cecidomyiidae

Table 3.—*Analysis of covariance on the influence of plant (genotype), shoot length, and year on the number of galls induced by* Contarinia *sp. on* B. brevipes.

Variable	Sum-of-Squares	DF	Mean-Square	F-Ratio	P
Plant	97.981	1	97.981	20.800	0.000
Shoot Length	595.612	1	595.612	126.443	0.000
Year	66.537	1	66.537	1.388	0.239
Plant* Shoot Length	483.229	1	483.229	102.585	0.000
Plant* Year	33.490	1	33.490	7.110	0.008
Shoot Length* Year	65.498	1	65.498	13.905	0.000
Plant*Shoot Length*Year	222.127	1	222.127	47.155	0.000
Error	386635.807	8202	4.711		

Table 4.—*Analysis of covariance on the influence of plant (genotype), shoot length, and year on the frequency of HR in* B. brevipes.

Variable	Sum-of-Squares	DF	Mean-Square	F-Ratio	P
Plant	0.294	1	0.294	1.819	0.177
Shoot Length	0.219	1	0.219	1.352	0.245
Year	0.182	1	0.182	1.125	0.289
Error	669.740	4142	0.162		

of female preference and larval performance in the system (see Fernandes 1998). Nevertheless, it raises the question as to whether plants maintain similar levels of defense through the years, in other words, whether *B. brevipes* has "memory".

Preliminary Observations on Plant Memory

Karban and Niiho (1995) stressed that "nothing is known about whether induced resistance in plants against herbivores has a memory in an immunological sense, i.e., whether responses are greater or faster following a second bout of damage, independent of the amount of damage" (see also Harvell 1990a, b). We adopted the immunological definition stated in Karban and Niiho (1995) and postulated that *B. brevipes* memory to attack by *Contarinia* sp. would mean that the HR would be stronger (= more effective) in plants that had experienced a certain level of attack in 1995.

Density of attack (number of galls per plant) by *Contarinia* sp. was significantly higher in 1995 than in 1996 (Mann-Whitney U test = 17505.5, p < 0.001). During the 2 years of study most of plants showed high levels of resistance (> 90 percent HR) to *Contarinia* sp. (1995 = 74.8 percent; 1996 = 66.4 percent) (fig. 3). Plants that showed lower resistance (< 75 percent HR) to *Contarinia* sp. decreased from 9.2 percent of the population in 1995 to 6.1 percent in 1996. On the other hand, plants with intermediary levels of HR (between 75 percent and 90 percent HR) increased from 16.0 percent in 1995 to 23.5 percent in 1996. These data corroborate our previous observation on the efficiency of HR against this galling herbivore. This general analysis canot, however, indicate the fate of each individual plant. Hence, a better analysis is to observe the proportion of plants in each category that moved among resistance levels.

Approximately 78.8 percent of the plants that were resistant in 1995 (n = 122) remained resistant in 1996 (n = 95) while only 4.0 percent (n = 5) became susceptible to attack by *Contarinia* sp. in 1996 (fig. 4). Of the 26 plants placed into the category of intermediate levels of HR in 1995, 69.2 percent (n = 18) became resistant in 1996. From the small number of plants that were susceptible to galling by *Contarinia* sp. in 1995 (n = 15) only 3 plants (20 percent) remained susceptible in the following year, while 4 plants (27 percent) became resistant in 1996. These data differed statistically between resistant and susceptible categories (χ^2 = 23.463, p < 0.001), hence showing an overall increased level of resistance of formerly susceptible plants as stated in the plant memory hypothesis. We also observed that 19.6

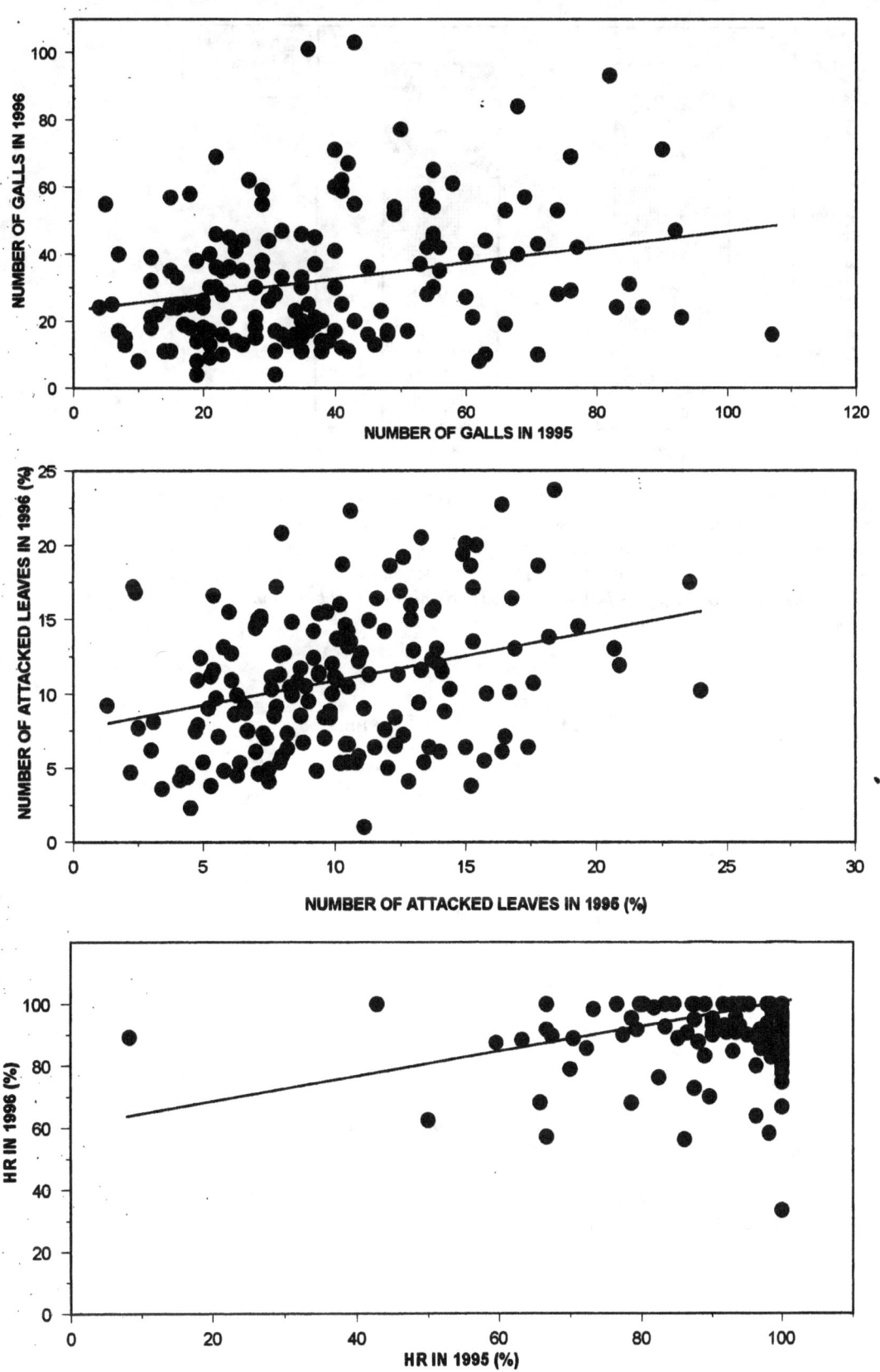

Figure 2.—*Relationship between total number of galls per plant (a), percent of attacked leaves (b), and frequency of HR per plant (c) in 1995 and 1996.*

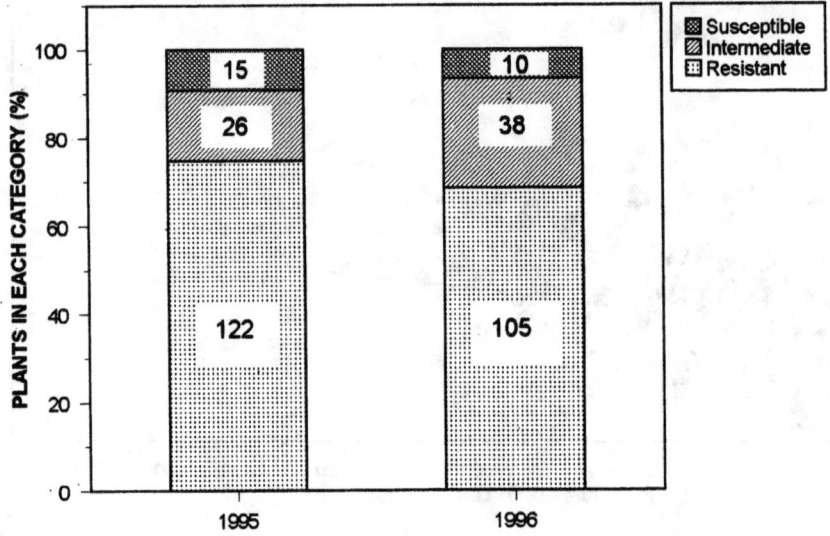

Figure 3.—*Percentage of individuals of* Bauhinia brevipes *in three categories of hypersensitive reaction [resitant > 90 percent; susceptible < 75 percent, intermediate > 75 percent and < 90 percent) against* Contarinia *sp. in 1995 and 1996. Note that the total number of plants decreased from 1995 to 1996 due to mortality caused by biotic and/or environmental factors (n_{1995} = 170, n_{1996} = 163). Numbers inside bars represent the number of plants in each category.*

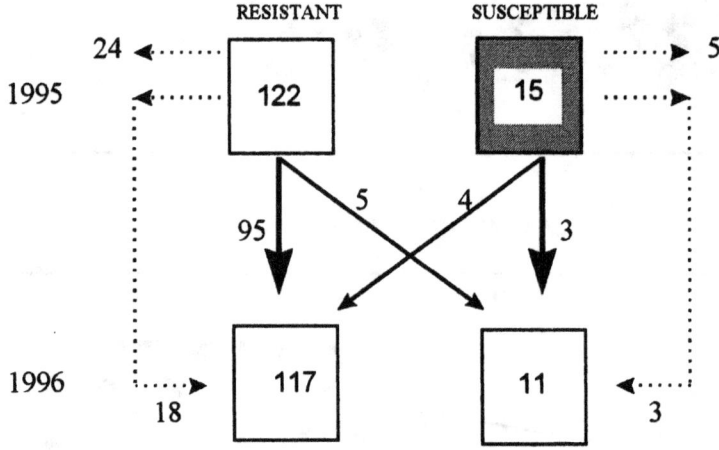

Figure 4.—*Number and relative proportion of plants of* Bauhinia brevipes *that exhibited HR against* Contarinia *sp. in 1995 and 1996. Plants were categorized into resistant (> 90 percent HR), and susceptible (< 75 percent HR). Solid arrows indicate the number and proportion of plants that moved among resistant and susceptible categories and broken arrows indicate plants of intermediate level of resistance that entered or left the system.*

percent (n = 24) of the plants that were resistant in 1995 had their resistance lowered to an intermediate level in 1996. Overall, these data indicate the existence of memory in this system, as they show that resistance levels may have increased as stated in the definition of memory (Karban and Niiho 1995) as well as they show certain variability in defense. Although the density of attack decreased from 1995 to 1996, most plants remained resistant to attack by *Contarinia* sp. and the number of susceptible plants decreased almost by one third. Future studies shall focus on the fate of these plants in attempt to observe the consistency of the trends reported here and perform experimental studies on the levels of attack by the galling cecidomyiid.

ACKNOWLEDGMENTS

We are very grateful to D. Yanega, S.P. Ribeiro, M.L. Faria, and K. Schonrogge for their reviews on the manuscript. Logistical support was provided by Estação Ecológica de Pirapitinga - Instituto Brasileiro do Meio Ambiente e Recursos Naturais Renováveis. The project was supported by the Conselho Nacional de Desenvolvimento Científico e Tecnológico (CNPq) (52.1772/95-8), Fundação do Amparo à Pesquisa de Minas Gerais (1950/95), and by the Graduate Program in Ecologia, Conservação e Manejo de Vida Silvestre/ Universidade Federal de Minas Gerais.

LITERATURE CITED

Agrios, G.N. 1988. Plant pathology. 3d ed. New York, NY: Academic Press.

Anderson, S.S.; McCrea, K.D.; Abrahamson, W.G.; Hartzel, L.M. 1989. Host choice by the ball gallmaker *Eurosta solidaginis* (Diptera: Tephritidae). Ecology. 70: 1048-1054.

Azevedo, L.G.; Barbosa, A.A.A.; Oliveira, A.L.C.; Gorgonio, A.S.; Bedretchuk, A.C.; Siqueira, F.B.; Rizzo, H.G.; Silva, I.S.; Moura, L.C.; Araújo Filho, M.; Santos, R.V. 1987. Ensaio metodológico de identificação e avaliação de unidades ambientais - A Estação Ecológica de Pipapitinga, MG. Secretaria Especial do Meio Ambiente, Embrapa-CPAC, Brasília.

Bayley, J.A.; Mansfield J.W. 1982. Phytoalexins. New York, NY: John Wiley & Sons.

Beckman, K.B.; Ingram, D.S. 1994. Physiology of Molecular Plant Pathology. 45, 229

Bentur, J.S.; Kalode, M.B. 1996. Hypersensitivity reaction and induced resistance in rice against the Asian rice gall midge *Orseolia oryzae*. Entomologia Experimentalis et Applicata. 78: 77-81.

Berryman, A.A. 1972. Resistance of conifers to invasion by bark beetle-fungus association. BioScience. 22: 598-602.

Berryman, A.A.; Ferrell, G.T. 1988. The fir engraver beetle in western states. In: Berryman, A.A., ed. Dynamics of forest insect populations: patterns, causes, implications. New York, NY: Plenum Press: 555-577.

Coleman, J.S.; Jones C.G. 1991. A phytocentric perspective of phytochemical induction by herbivores. In: Tallamy, D.W.; Raupp, M.J., eds. Phytochemical induction by herbivores. New York, NY: John Wiley & Sons.

Cornelissen, T.G.; Madeira, B.G.; Allain, L.R.; Lara, A.C.F.; Araújo, L.M.; Fernandes G.W. 1997. Multiple responses of insect herbivores to plant vigor. Ciência and Cultura. 49: 285-288.

Fernandes, G.W. 1990. Hypersensitivity: a neglected plant resistance mechanism against insect herbivores. Environmetal Entomology. 19: 1173-1182.

Fernandes, G.W. 1998. Hypersensitivity as a phenotypic basis of plant induced resistance against a galling insect (Diptera: Cecidomyiidae). Environmental Entomology. (in press).

Fernandes, G.W.; Price, P.W. 1992. The adaptive significance of insect gall distribution: survivorship of species in xeric and mesic habitats. Oecologia. 90: 14-20.

Fernandes, G.W.; Tameirão Neto, E.; Martins, R.P. 1988. Ocorrência e caracterização de galhas entomógenas na vegetação do Campus-Pampulha, UFMG, Belo Horizonte-MG. Revista Brasileira de Zoologia. 5: 11-29.

Fritz, R.S.; Simms, E.L. 1992. Plant resistance to herbivores and pathogens. Chicago, IL: Chicago University Press.

Fritz, R.S.; Orians, C.M.N.; Brunsfield, S.J. 1994. Interspecific hybridization of plants and resistance to herbivores: hypotheses, genetics, and variable responses in a diverse herbivore community. Oecologia. 97: 106-117.

Gagné, R.J. 1994. The gall midges of the Neotropical Region. Ithaca, NY: Comstock: 352.

Harvell, C.D. 1990a. The ecology and evolution of inducible defenses. Quaterly Review of Biology. 65: 323-340.

Harvell, C.D. 1990b. The evolution of inducible defence. Parasitology. 100: S53-S61.

Karban, R. 1987. Environmental conditions affecting the strength of induced resistance against mites in cotton. Oecologia. 73: 414-419.

Karban, R.; Baldwin, I.T. 1997. Induced responses to herbivory. Chicago, IL: University of Chicago Press. (in press).

Karban, R.; Myers, J.H. 1989. Induced plant responses to herbivory. Annual Review of Ecology and Systematics. 20: 331-348.

Karban, R.; Niiho, C. 1995. Induced resistance and susceptibility to herbivory: plant memory and altered plant development. Ecology. 76: 1220-1225.

Kimberling, D.N.; Price, P.W. 1996. Variability in grape phylloxera preference and performance on canyon grape (*Vitis arizonica*). Oecologia. 107: 553-559.

Király, Z. 1980. Defenses triggered by the invader: hypersensitivity. In: Horsfall, J.; Cowling, E.B., eds. Plant Diseases. 5: 201-225.

Maclean, D.J.; Sargent, J.A.; Tommerup, I.C.; Ingram, D.S. 1974. Hypersensitivity as the primary event in resistance to fungal parasites. Nature. 249: 186-187.

Preszler, R.W.; Price, P.W. 1995. A test of plant vigor, plant stress, and plant-genotype effects on leaf-miner oviposition and performance. Oikos. 74: 485-492.

Price, P.W. 1991. Patterns in comunities along latitudinal gradients. In: Price, P.W.; Lewinsohn, T.M.; Fernandes, G.W.; Benson, W.W., eds. Plant-animal interactions: evolutionary ecology in tropical and temperate regions. New York, NY: John Wiley & Sons: 51-69.

Price, P.W.; Cobb, N.; Craig, T.P.; G.W. Fernandes; Itami, J.; Mopper, S.; Preszler, R.W. 1990. Insect herbivore population dynamics on trees and shrubs: new approaches relevant to latent and eruptive species and life table development. In: Bernays, E., ed. Focus on insect-plant interactions. Boca Raton, FL: CRC Books. 3: 1-38.

Raffa, K.F.; Berryman, A.A. 1983. The role of host plant resistance in the colonization behavior and ecology of bark beetles (Coleoptera: Scolytidae). Ecological Monographs. 53: 27-49.

Rosenthal, G.A.; Janzen, D.H. 1983. Avoidance of nonprotein aminoacid incorporation into protein by the seed predator *Caryedes brasiliensis* (Bruchidae). Journal of Chemical Ecology. 9: 1353-1361.

Shukle, R.H.; Grover, P.B., Jr.; Mocelin, G. 1992. Responses of susceptible and resistant wheat associated with Hessian fly (Diptera: Cecidomyiidae) infestation. Environmental Entomology. 21: 845-853.

SYSTAT. 1992. Systat for Windows. Evanston, IL: Systat Inc.

Tallamy, D.W.; Raupp, M.J. 1991. Phytochemical induction by herbivores. New York, NY: John Wiley & Sons.

Wong, B.L.; Berryman, A.A. 1977. Host resistance to the fir engraver beetle. Lesion development and containment of infection by resistant *Abies grandis* inoculated with *Trichosporium symbioticum*. Canadian Journal of Botany. 55: 2358-2365.

Zar, J.H. 1996. Biostatistical analysis. 3d ed. Englewood Cliffs, NJ: Prentice-Hall.

REPROGRAMMING PLANT DEVELOPMENT: TWO APPROACHES TO STUDY THE MOLECULAR MECHANISM OF GALL FORMATION

K. Schönrogge[†], L. J. Harper[†], S. E. Brooks[‡], J. D. Shorthouse[‡] and
C. P. Lichtenstein[†]

[†] Queen Mary and Westfield College, School of Biological Sciences, Mile End Rd, London, E1 4NS, Great Britain
[‡] Department of Biology, Laurentian University, Sudbury, Ontario, P3E 2C6, Canada

Abstract.—Each species of cynipid gall wasps induce the growth of unique galls which are both structurally distinct and anatomically novel structures to their host plant. Although much has been written on the anatomy of cynipid galls, little is known about the molecular mechanisms responsible for gall initiation and growth. Presumably the gall wasps send signals to the host plant to bring about gall formation. Here we present the first results of two approaches to identify these signals.

First, we are trying to identify molecular markers to distinguish between gall and non-gall plant tissue. If such markers can be established, it should also be possible to use marker expression in a bioassay to find and characterise the compounds which serve as signal molecules. We have identified two proteins, expressed in galls of *Diplolepis spinosa* on *Rosa rugosa*, which could prove useful as molecular markers. One protein (90kDa) was found to be exceptionally more abundant in the inner gall tissue, compared to non-gall tissues, such as leaf and stem tissues. The second protein (60kDa) was not detected in the non galled tissues, and appears to be differentially produced in the inner gall tissue, making it an excellent molecular marker.

Nod-factors are signal molecules involved in the interaction/communication between nitrogen fixing bacteria, *Rhizobium* spp., and their legume hosts. To address whether gall wasps use similar signals we are also searching for homologues to nodC, a gene known to be involved in the synthesis of all Nod-factors, in gall wasp genomic DNA. A homologue to nodC in gall wasp DNA suggests that the signal molecules used by cynipids to induce galls might be of a Nod-factor nature. In a polymerase-chain-reaction we used oligo-nucleotide primers designed from NodC and DG42 (a nodC homologue from *Xenopus laevis*) amino acid sequences on gall wasp template DNA. Both sets of primers amplified a fragment of 400bp length from the gall wasp DNA. However, the primers designed from DG42, also amplified two more fragments of 440bp and 1kb length.

INTRODUCTION

Gall formation is arguably the most intimate relationship between herbivorous insects and their host plants, and galls induced by cynipid gall wasps might well be the most sophisticated structures of induced plant growth. Cynipids are able to change the natural growth patterns of their host plants to such an extent that the galls have been described as new plant organs. However, very little is know about the actual mechanisms employed by gall wasps to control the growth of such novel structures.

The shapes and morphologies of cynipid galls vary widely not only between the species of gall-former, but also between generations, where the gall former has more than one generation per year (Dregger-Jauffret and Shorthouse 1992). Mature cynipid galls can be as small as 2-3 millimetres or as large as 10 cm in diameter. Galls of some species house only a single larva whereas others are inhabited by several hundred. Cynipids induce galls on virtually all plant organs and many of them have impressive surface structures, such as spines of variable shapes or glands, which can secrete sugary or otherwise sticky compounds. Although the gross morphology of cynipid galls may vary, the inner organization of tissues is similar (Rohfritsch 1992). Tissues found in all cynipid galls include a layer of cytoplasmically dense nutritive cells which line the larval chamber, followed by a layer of vacuolate parenchyma, a layer of sclerenchyma, a layer of parenchymatous gall cortex and epidermis (Rohfritsch 1992). Nutritive cells are unique to insect galls and serve the inducers as the sole source of food. Adjoining parenchyma are converted to nutritive cells as the inducer feeds (Bronner 1976). All nutritive and vacuolate parenchyma are considered inner gall tissue in the current study.

Cynipid wasps also control the physiology of gall tissues (Bagatto et al. 1996, Harris and Shorthouse 1996). Not only do galls serve as physiological sinks for nutrients and assimilates, but tannins and phenolic compounds, thought to serve as feeding inhbitors for herbivorous insects, are concentrated in the peripheral parenchyma while inner gall tissues have none (Berland and Bernard 1951, Bronner 1976). Starch and glycogen are broken down to their constituents in the inner gall tissue, which results in a measurable gradient from the center to the outside of the gall (Bronner 1976). Transcription and translation rates are increased also in the inner gall tissue, and the cytoplasm of the cells in this tissue are known to be rich in ribosomes and contain large amounts of protein. While the cytology and physiology of cynipid galls has been much studied, the mechanism, i.e., the putative signal molecules or "morphogens", used by the gall wasp to cause all the described changes in the plant tissue is still a mystery.

A number of hypotheses have been put forward concerning the nature of the morphogens, which archive the reprogramming of plant development. These hypotheses suggest the involvement of symbiotic viruses or virus-like-particles, plant hormones, and "other" signal molecules (Cornell 1983, Hori 1992). However, little experimental evidence exists to support any of the hypotheses.

One of the best studied interactions and communication between plants and another organism, is that of *Rhizobium* bacteria and their legume hosts (Denarie et al. 1996). Van Brussel et al. (1986) showed the significance of Nod-factors for the induction of root-nodules. Chemically Nod-factors are lipo-chito-oligosaccharides, which cause the host plant to develop an infection-thread, through which the rhizobia enter the host plant, and ultimately to develop root nodules in which the bacteria reside. While the Nod-factors produced by different *Rhizobia* species differ in variations of the residual substitutes, they all share the 3-6 monomer long oligosaccharide backbone. NodC was identified as the β-glycosyl-transferase to establish the β-linkages between the glucosamin sugar residues. The fact that NodC produces only oligomeres of a defined length makes it unusual compared to other carbohydrate-synthase-enzymes (Denarie et al. 1996). Since the identification of the amino acid sequence for NodC, a number of homologous proteins have been found in a variety of different organisms, e.g., mice, yeast and the frog *Xenopus* laevis (Bulawa 1992, Atkinson and Strong 1992, Spicer et al. 1996). It is intriguing that the gene DG42 is only expressed between the midblastula and the neurulation stage of the embryonal development (Semino and Robbins 1995). Like NodC, it appears that the DG42 protein is involved with development. Thus our first approach is to identify homologues to nodC, the gene which codes for the NodC enzyme, in the genomic DNA of *Andricus quercuscalicis* (Burgsdorf).

A more general approach makes no assumptions about the nature of the morphogens. An essential tool to use in the identification of the proposed morphogens would be a bioassay to test extracts made from wasp larvae for gall forming activity. However, since the control of gall-formation could be a complex mechanism, it might be unlikely that it will be possible to induce a gall artificially. It is therefore desirable to have molecular markers in the form of RNA molecules or proteins, which indicate that gall initiation has taken place. Since the cells of the inner gall tissues have been reported to have a dense cytoplasm with high concentrations of proteins, we compare protein extracts from galled and non-galled tissues to identify proteins, which are potential molecular markers. In this second approach we studied the protein contents of inner-gall, outer-gall, and non-galled tissues using galls of *Diplolepis spinosa* (Ashmead) on *Rosa rugosa* Thunb.

The many questions that could be asked in the context of gall formation include: what are the morphogens, how many are there, where do they originate, and what is their effect on the host plant. Here we present the first results from the two approaches outlined above.

METHODS

Protein Extracts From Galled and Non-galled Tissues

Female *D. spinosa* reared from galls collected near Sudbury, Ontario, Canada during April 1995, were exposed to 40 potted *R. rugosa* (Hansa variety) cultivated in growth cabinets at Laurentian University. Galls were harvested and dissected 6 weeks (±1 week) after oviposition. Larvae were removed and inner gall tissue (nutritive tissue and nutritive parenchyma) and outer gall tissue were harvested separately and snap frozen in liquid nitrogen. Simultaneously, non-galled tissues, here leaf tissue and pieces of stem from just below the gall, were harvested and snap frozen. Protein extracts were made by grinding the tissues under liquid nitrogen using mortar and pestle and subsequent boiling for 5 minutes in SDS-PAGE extraction buffer. Any solid debris was removed by centrifugation. Protein contents were established using the Biorad assay based on Bradford (1976). 20µg total protein of each sample were loaded on a 10 percent SDS-PAGE gel (Laemmli 1970) together with the wide range molecular weight markers from Sigma. The gel was subsequently stained with Coomassie Blue after Dunn (1993).

Polymerase-Chain-Reactions (PCR) in the Search for NodC Homologues

The primers for the PCR reactions were designed as degenerate primers based on the consensus of the aminoacid sequence of 12 known NodC enzymes from different *Rhizobia* species. To achieve a higher specificity with the degenerate primers, the PCR was carried out in two steps, i.e., a nested reaction. In a nested PCR, two different pairs of primers are used whereby the sequences of one pair (the outer primers) are located on the amino acid sequence outside the second pair (the inner primers; table 1). After using the outer primers with template DNA, a second PCR was carried out on the amplified products of the first reaction. Similarly, another set of primers was designed from the consensus sequences between NodC and the DG42 protein, isolated from *X. laevis*. To further increase the efficiency of the PCR, 5 permissive cycles where run at a lower annealing temperature of 42°C followed by 35 cycles with the more stringent annealing temperature of 48°C. The complete PCR program is described in table 2.

The template for the PCR reactions were the cDNAs of nodC and DG42 as controls and genomic DNA of the gall wasp *A. quercuscalicis*. To obtain the gall wasp

Table 1.—*Description of the primers used including the amino acid sequences from NodC and DG42, which were the templates for the design of the primers*

Primer location	Primer name	Amino acid sequence of the template
NodC		
5' end - outer	B5out	YVVDDG
5' end - inner	B5in	NVDSDT
3' end - inner	B3in	MCCCGP
3' end - outer	B3out	QQLRWA
DG42		
5' end - outer	V5out	QVCDSD
5' end - inner	V5in	EMVKLV
3' end - inner	V3in	DDRHLT
3' end - outer	V3out	NQQTRW

Table 2.—*PCR programming steps in the PCR for nodC homologues*

Number of cycles	Temperature	Duration
1	94°C	4min
5	94°C	60sec
	42°C	60sec
	72°C	60sec
35	94°C	60sec
	48°C	60sec
	72°C	60sec
1	72°C	5min

DNA, larvae were dissected from the autumn galls. The guts of the larvae were removed before the genomic DNA was extracted after a protocol provided by Dr. J. Cook.

RESULTS

Identifying Proteins in the Inner Gall Tissue

Extracts of total protein, made from leaf-(L), stem-(ST), outer gall-(OG) and inner gall tissues-(IG), show a variety of abundant structural proteins (fig. 1). By comparing the bands in all four lanes, we found that two of the proteins in the extract made from inner gall tissue appear to be particular to this tissue (see arrow heads; fig. 1). At a molecular weight of 60 kDa, the band found in the IG-lane does not appear in any of the other samples and might be specific to inner gall tissues. The protein at 90 kDa is much more abundant in IG, although it is present in all four samples shown here (fig. 1).

Searching for Homologues to NodC and DG42 in Gall Wasps

Degenerate primers, as used here, generally produce some non-specific amplification, so that the PCR-product appears as a smear. However, where homology exists, annealing is more likely and the amplified product will produce a brighter band within the smear. After the first reactions, using the "outer" primers, the PCR products were then used as the template with the appropriate "inner" primers in the second reaction of the nested PCR (fig. 2). The lane marked M shows a 100bp ladders as a size marker, where the top band has the size of 1kb, the next of 900bp and the smallest band visible 200bp. Lane 1 shows the PCR products of the first control reaction with nodC primers and nodC as the template, and lane 3 the second control reaction using DG42 primers and DG42 as the template. In lane 2 and 4 the PCR products of the nodC primers (lane 2) and the DG42 primers (lane 4) are shown, used with the genomic DNA of the gall wasp *A. quercuscalicis* as the template.

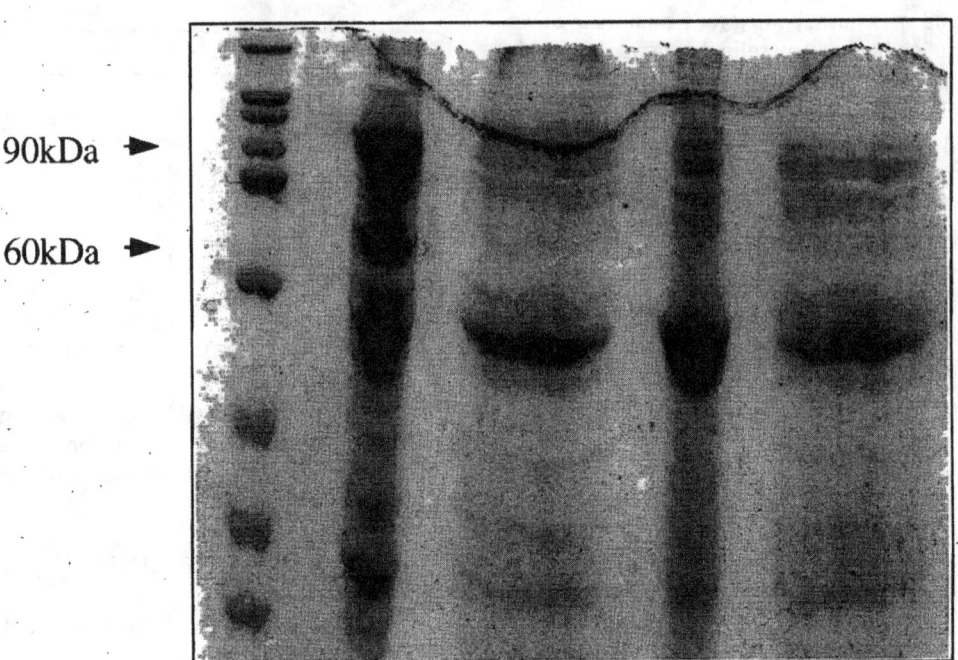

Figure 1.—*SDS-PAGE gel of total protein extracts of inner gall tissue (IG), outer gall tissue (OG), leaf tissue (L) and stem tissue (St). The lane marked M shows the wide range molecular weight marker and the arrow heads point out the two proteins at 60 and 90 kDa, which are candidates to be molecular markers for inner gall tissue.*

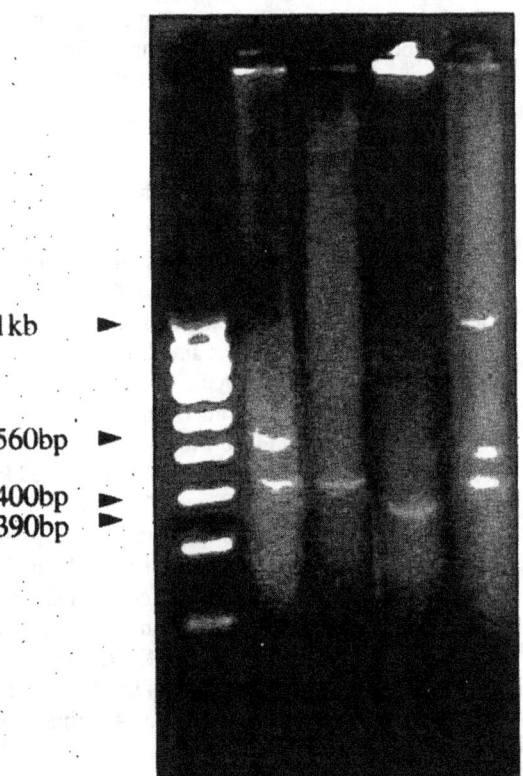

Figure 2.—*Products of the second reaction of the nested PCR. Lane M shows a 100bp ladder as a size marker. The amplified product of nodC primers using nodC cDNA as template is shown in lane 1, nodC primers on gallwasp genomic DNA template in lane 2, DG42 primers on DG42 cDNA template in lane 3 and DG42 primers on gallwasp genomic DNA template in lane 4.*

Table 3.—*Expected and observed fragment sizes from the nested PCR reactions shown in figure 2*

Combinations of primers and template DNA	Expected size of the fragment to be amplified	Size of the bands observed
nodC primers on nodC cDNA template	198bp	440bp, 560bp
DG42 primers on DG42 cDNA template	397bp	390bp
nodC primers on gall wasp genomic DNA template	unknown	400bp
DG42 primers on gall wasp genomic DNA template	unknown	400bp, 440bp, 1kb

The DG42 control reaction produced one band at 390bp, which is the expected size for the DG42 gene (fig. 2; table 3). The expected size for the nodC fragment would have been 198bp (table 3), which we could not detect. It is possible that the nodC fragment was amplified, but was below a detectable concentration against the background of unspecific amplification. Using the gall wasp DNA as the template, both sets of primers amplified a 400bp fragment. The DG42 primers also amplified two larger fragments at 440bp and 1kb. The 400bp fragment, if identical in both reactions, might be the best candidate for a nodC homologue. However, we do not have an expected size for the fragments amplified from the gall wasps genomic DNA, because it might contain introns, unlike the cDNAs, which served as controls.

DISCUSSION

Mechanisms of the initiation and control of organogenesis are currently some of the biggest challenges for developmental biologists. Because during gall formation the morphogens originate outside the developing organism, the interaction between host plant and ceciduous insects will perhaps provide an excellent model system for plant development. However, little is known about these interactions on a molecular level. Here we have shown the early results of two different approaches to gain insight into what the morphogens are and how they affect the host plant.

We were able to amplify DNA fragments from gall wasp genomic DNA using degenerate primers designed from amino acid sequences of NodC and DG42. Sequence analysis will reveal whether the amplified fragments share sequence identity with nodC, which could suggest that the morphogens might be oligosaccharides. Apart from Nod-factors, derivatives of xyloglucan, a polysaccharide which makes up the matrix of cell-walls together with cellulose, have been found to be signalling molecules, which affect cell-cyle and proliferation (Guillen *et al.* 1995). Intriguingly, the biological active derivatives of xyloglucan have the same length as the Nod-factors (3-6 monomers). Longer fragments are not recognized by the cells as signal molecules (Guillen *et al.* 1995). The xyloglucan derivatives represent a plant internal signalling mechanism and it is conceivable that *Rhizobia*, and possibly also cynipid wasps, make use of the plant cell receptors designed for plant internal signalling with oligosaccharides as signal molecules.

A prerequisite to identify and study such morphogens is a bioassay for gall formation. Such a bioassay could be based on morphological changes (Higton and Mabberly 1994). However, gall formation is potentially complex making it unlikely to assemble all components to induce a gall in the absence of a gall former. Here we present the first results of a different approach. By studying the differences of the protein contents between galled and non-galled tissues it is possible to establish gall formation activity without necessarily having to observe morphological changes. The 90 kDa and 60 kDa proteins we identified are good candidates for being molecular markers: (a) they are abundant and in case of the 60 kDa protein seems specific to the inner gall tissue and (b) the inner gall tissue is the first to differentiate during gall formation. The process of tissue differentiation during gall formation has been studied for a variety of different cynipid galls as well as galls formed by other inducers (Hough 1953a, b; Rey 1992; Rohfritsch 1992).

The two proteins offer a number of possibilities to proceed in future studies. We are currently determining the end-sequences of the two proteins, which would allow us to do a database search for homologous proteins with known function. Further research might also involve raising antibodies against the proteins, which could be used to screen tissues of the host plant to investigate where and when in the normal development of the plant these proteins might be produced. Further screening of the inner gall tissues of other cynipid galls would be of interest to establish whether or not the two proteins are generally produced during gall formation or if they are specific to the interaction between *D. spinosa* and *R. rugosa*. It would also be possible to determine the cDNA sequences for the proteins, which could then be used as probes to identify gall formation activity in RNA expression studies.

Insect-Plant Interaction During Gall Formation

We propose here that oligosaccharide molecules, similar to Nod-factors, might be the type of morphogens used by cynipid wasps to reprogram plant development. It is important to note that the *Rhizobia*—host interaction involves a whole chain of events. The plant produces flavenoids at its root tips, which the bacteria use to orientate themselves towards the hosts. The *Rhizobia* then begin the production of Nod-factors, causing the host-plant to develop a so called infection thread, by which the bacteria can enter the roots. Ultimately the plant will develop nodule primodia into fully developed nodules, where the bacteria will reside. Thus, the recognition of the Nod-factor starts a whole cascade of developmental events. Gene regulation during organogenesis is known to be a complex process. Generally a series of regulatory genes are expressed before any structural genes. Therefore it seems likely that cynipids make use of developmental cascades in the same way that Nod-factors stimulate nodule formation. The inner structure of galls can be compared to that of seeds and it will be interesting to see whether or not the proteins we isolated show any relation to seed tissue. If so, the result would strongly suggest that a "seed developmental cascade" is at least involved in the gall formation process. What these cascades are and how many might be involved in the formation of a gall will be subject of future research.

While the notion that Nod-factor like molecules might be involved in the gall formation process is new to the discussion about active substances, a variety of other compounds have been proposed and are discussed in detail by Hori (1992). Plant hormones, such as auxins and cytokinins, have most often been suspected as morphogens. Other compounds in question are those involved with the regulation of plant hormones, such as the phenol-polyphenol system or indol-acetic-acid (IAA) oxidase. Auxins (IAA, etc.) were found in larvae of *Cynips quercusfolii*, and larvae kept in a solution of tryptophan (a precursor of IAA) seemed to produce IAA (Kaldewey 1965). Kaldewey, however, concluded that the IAA originated in the diet of the larvae. Matsui and Torikata (1970) found only low concentrations of IAA in the larvae of *Dryocosmus kuriphilus*, but high concentrations of tryptophan, suggesting that the secretion of the precursor of the plant growth factor is important to gall formation. Later Matsui *et al.* (1975), studying again extracts of the same cynipid species, suggested cytokinins were the important compounds, while Yokota *et al.* (1973) claimed it was not the plant hormones, although he found an increase of activity by indigenous plant hormones. Thus the role of plant hormones in the process of gall formation is far from clear. In the context here, the question is whether or not plant hormones are morphogens themselves and therefore it would be crucial to show that cynipid larvae are able to produce them. In fast proliferating tissues like those inside a developing gall, plant hormones should play an important role and it would be easy to imagine that a larva would ingest plant produced hormones and even be able to concentrate them in some way in its salivary glands to secret them back to the plant tissue. If this would be the case, however, plant hormones would aid gall formation, but there would still be another signal necessary to initially induce them. Weidner (1957) suggested that cynipid galls are too complex in form, and that cecidogenesis could not simply be explained by plant hormone activity. In fact, we know that plant hormones play the most important role in the formation of "crown galls", induced by *Agrobacterium tumifascians*, and they lack all the morphological complexity so characteristic of cynipid galls.

Other substances suggested as morphogens include RNA (Taylor 1949). RNAs are generally short lived molecules within a cell. To move RNA through intercellular space and into a host plant cell suggests a highly sophisticated system of chaperoning to keep it safe from degradation and RNAase activity. A possible way would be the involvement of a symbiotic virus as suggested by Cornell (1983). Unlike Taylor, who found high concentration of RNA in the saliva of the wasp larvae, Cornell proposed a transfer of viruses with a fluid deposited by the ovipositing female. This would be an intriguing parallel to parasitoid relatives of cynipid wasps, which are known to suppress the immune system of their insect hosts via "viroid particles" which are injected during oviposition (Edson *et al.* 1981, Vinson 1990). Cornell (1983) himself pointed out that gall development stops, if the gall wasp larva is killed, makes it necessary to envisage a mechanism by which the "virus" is only active in the presence of a factor emanating from the gall wasp larva (Magnus 1914, Rohfritsch 1975, Bronner 1976). This scenario would make ecological sense, as it would forge a permanent link in a mutualistic relationship between "viroid" and gall wasps (Cornell 1983). However, there might be another paradox in this kind of relationship. The fact that gall morphologies are species- or even generation-specific would mean that the speciation rates of gall wasp and "viroid" should be synchronized and we are not aware of any proposed mechanism allowing this to occur. Alternatively, one could imagine a situation where the function of a "viroid" would be to facilitate gall formation rather than being the main factor. Thus, while it is not possible to discard the presence of a mutualistic virus-like organism, we would still suggest that the wasp larvae are the most important source for morphogens.

Rohfritsch (1992) describes the presence of a "cocktail" that bathe the cells which line the larval chamber. This "cocktail" containing salivary enzymes, such as amylases and proteases, from the larva, as well as hydrolases, amino acids, soluble sugars, various metabolites and cell wall fragments from the consumed plant cells. Whether or not the larva actually produces Nod-factor like molecules cannot be decided as yet, but with the presence of the insects chitin and the cell wall fragments swimming in a digestive cocktail there might well be the 3-6 monomer long oligomers present, similar to Nod-factors or the derivatives of xyloglucan (Guillen et al. 1995). Hori (1992) pointed out that the microfeeding behavior of the gall wasp larva might determine the ultimate morphology of a gall. While this implies it might be extremely difficult to induce a gall artificially, i.e., without the gall inducing insect, it also suggests that the development of bioassays, as the one proposed here, is all the more important to identify the morphogens, which allow a reprogramming of plant development.

ACKNOWLEDGEMENTS

For this project CPL received funds from the BBSRC and JDS from NSERC and LURF. We would like to thank A. Downie and Jim Smith who provided us with nodC- and DG42 cDNA.

LITERATURE CITED

Atkinson, E.M.; Strong, S.R. 1992. Homology of *Rhizobium meliloti* NodC to polysaccharide polymerizing enzymes. Molecular Plant Microbe Interactions. 5: 439-442.

Bagatto, G.; Paquette, L.C.; Shorthouse, J.D. 1996. Influence of galls of *Phanacis taraxaci* on carbon partitioning within common dandilion, *Taraxacum officinale*. Entomologia Experimentalis et Applicata. 79: 111-117.

Berland, L.; Bernard, F. 1951. Ordre de Hymenoptères. In: Grassé, P., ed. Traité de Zoologie. 10: 771-1276.

Bradford, M. 1976. A rapid and sensitive method for the quantification of microgram quantities of protein utilizing the principle of protein-dye binding. Anales of Biochemistry. 72: 248.

Bronner, R. 1976. Contribution à l'étude histochimique des tissus nouriciers des Zoocécidies. Marcellia. 40: 1-136.

Bulawa, C.E. 1992. CSD2, CSD3, and CSD4, genes required for chitin synthesis in *Saccharomyces cerevisiae*: the CSD2 gene product is related to chitin synthases and to developmentally regulated proteins in Rhizobium species and Xenopus laevis. Molecular and Cell Biology. 12: 1764-1776.

Cornell, H.V. 1983. The secondary chemistry and complex morphology of galls formed by the Cynipinae (Hymenoptera): why and how? American Midland Naturalist. 110: 223-234.

Dénarié, J.; Debellé, F.; Promé, J-C. 1996. Rhizobium lipo-chitooligosaccharide nodulation factors: signaling molecules mediating recognition and morphogenesis. Annual Revue of Biochemistry. 65: 503-535.

Dreger-Jauffret, F.; Shorthouse, J.D. 1992. Diversity of gall-inducing insects and their galls. In: Shorthouse, J.D.; Rohfritsch, O., eds. Biology of insect induced galls. New York: Oxford University Press: 8-34.

Dunn, M.J. 1993. Gel electrophoresis: proteins. Oxford: Bios Scientific Publishers.

Edson, K.M.; Vinson, S.B.; Stoltz, D.B.; Summers, M.D. 1981. Virus in a parasitoid wasp: suppression of the cellular immune response in the parasitoid's host. Science. 211: 582-583.

Guillen, R.; York, W.S.; Pauly, M.; An, J.; Impallomeni, G.; Albersheim, P.; Darvill, A.G. 1995. Metabolism of xyloglucan generates xylose-deficient oligosaccharidesubunits of this polysaccharide in etiolated peas. Carbohydrate Research. 277: 291-311.

Harris, P.; Shorthouse, J.D. 1996. Effectiveness of gall inducers in weed biological control. Canadian Entomologist. 128: 1021-1055.

Higton, R.N.; Mabberly, D.J. 1994. A willow gall from the galler's point of view. In: Williams, M.A.J., ed. Plant galls: organisms, interactions, populations. Oxford: Clarendon Press: 301-312.

Hori, K. 1992. Insect secretions and their effect on plant growth, with special reference to Hemipterans. In: Shorthouse, J.D.; Rohfritsch, O., eds. Biology of insect induced galls. New York: Oxford University Press: 157-170.

Hough, J.S. 1953a. Studies on the common spangle gall of oak I. the developmental history. New Phytology. 52: 149-177.

Hough, J.S. 1953b. Studies on the common spangle gall of oak III. the importance of the stage in laminar extension of the host leaf. New Phytology. 52: 229-237.

Kaldewey, H. 1965. Wachstumsregulatoren aus Pflanzengallen und Larven der Gallenbewohner. Berichte der Deutschen Botanischen Gesellschaft. 78: 73-84.

Laemmli, U.K. 1970. Most commonly used discontinous buffer system for SDS electrophoresis. Nature. 227: 680.

Magnus, W. 1914. Die Entstehung der Pflanzengallen verursacht von Hymenopteren. G. Fischer Verlag, Jena.

Matsui, S.; Torikata, H. 1970. Studies on the resistance of chestnut trees to chestnut gall wasps. III. Plant growth regulators contained in chestnut gall wasps and host gall tissues. Journal of the Japanese Society of Horticultural Science. 39: 115-123.

Matsui, S.; Torikata, H.; Munakata, K. 1975. Studies on the resistance of chestnut trees (*Castanea* spp.) to chestnut gall wasps (*Dryocosmus kuriphilus* Yasumatsu). V. Cytokinin activity in leaves of gall wasps and callus formation of chestnut stem sections by larval activity. Journal of the Japanese Society of Horticultural Science. 43: 415-422.

Rey, L.A. 1992. Developmental morphology of two types of hymenopterous galls. In: Shorthouse, J.D.; Rohfritsch, O., eds. Biology of insect induced galls. New York: Oxford University Press: 87-101.

Rohfritsch, O. 1975. Etude comparative de cellules du tissue nourricier de la jeune galle de l' *Aulax glechomae* L. et de cellules du tissu nourricier abandonné par le parasite. Marcellia. 38: 185-196.

Rohfritsch O. 1992. Patterns in gall development. In: Shorthouse, J.D.; Rohfritsch, O., eds. Biology of insect induced galls. New York: Oxford University Press: 60-86.

Semino, C.E.; Robbins, P.W. 1995. Synthesis of "Nod"-like chitin oligosaccharides by the Xenopus developmental protein DG42. Proceedings of the National Academy of Science USA, 92: 3498-3501.

Spicer, A.P.; Augustine, M.L.; McDonald, J.A. 1996. Molecular cloning and characterization of a putative mouse hyaluronan synthase. Journal of Biological Chemistry. 271: 23400-23406.

Taylor. S.H. 1949. Initiation and development of the gall of *Aylax glechomae* on *Nepeta hederacea*. American Journal of Botany. 36: 83-86.

Van Brussel, A.A.; Zaat, S.A.; Cremers, H.C.; Wijffelman, C.A.; Pees, E.; Tak, T.; Lugtenberg, B.J. 1986. Role of plant root exudate and Sym plasmid-localized nodulation genes in the synthesis by *Rhizobium leguminosarum* of Tsr factor, which causes thick and short roots on common vetch. Journal of Bacteriology. 165: 517-522.

Vinson, S.B. 1990. How parasitoids deal with the immune system of their host: an overview. Archives of Insect Biochemistry and Physiology. 13: 3-27.

Weidner, H. 1957. Neuere Anschauungen über die Entstehung der Gallen durch die Einwirkung von Insekten. Zeitschrift für Pflanzenkrankheiten, Pflanzenpathologie und Pflanzenschutz. 64: 287-309.

Yokota, T.; Okabayashi, M.; Takahashi, N.; Shimura, I.; Umeya, K. 1973. Plant growth regulators in chestnut gall tissue and wasps. Proceedings of the 8th International Conference on Plant Growth Substances. Tokyo, Japan: 28-38.

PLANT HORMONES AND GALL FORMATION BY *EUROSTA SOLIDAGINIS* ON *SOLIDAGO ALTISSIMA*

Carol C. Mapes[1] and Peter J. Davies[2]

[1] Biology Department, Kutztown University of Pennsylvania, Kutztown, PA 19530 USA
[2] Section of Plant Biology, Cornell University, Ithaca, NY 14853 USA

Abstract.—The levels of cytokinins and of indole-3-acetic acid in the gall-forming larvae of *Eurosta solidaginis* have been determined in order to gain insight into the mechanism of ball gall formation on *Solidago altissima*. First instar larvae of *E. solidaginis* contained 350 ng/g f.wt. isopentenyladenine, 43 ng/g f.wt. isopentenyladenosine, 12 ng/g f.wt. zeatin riboside, and 8 ng/g f.wt. zeatin. Third instar larvae removed from fully formed galls exhibiting a "green island effect" contained zeatin riboside, isopentenyladenine, and isopentenyladenosine. First instar larvae were also shown to contain high levels of indole-3-acetic acid. The high concentrations of cytokinins and indole-3-acetic acid associated with the larvae are suggestive of a role for these hormones in ball gall development.

As is the case for insect-induced plant galls in general the mechanism of ball gall formation by the larva of the tephritid fly, *Eurosta solidaginis* Fitch, on goldenrod, *Solidago altissima* L., has not yet been determined. Extracts and secretions of various species of gall-inducing insects have been shown to cause growth promotion and in some cases structures resembling galls (Boysen Jensen 1948, Leatherdale 1955, Martin 1942, McCalla et al. 1962, Plumb 1953). The growth promoting activity has, in some cases been determined to be associated with salivary or accessory glands (Hovanitz 1959, McCalla et al. 1962, Plumb 1953) but it remains to be determined which chemical or chemicals are responsible for insect mediated gall induction and development.

LIFE HISTORY OF *EUROSTA SOLIDAGINIS*

The ball gall of *Solidago altissima* is a stem gall that forms in response to the activities of a single larva of *Eurosta solidaginis*. Adults emerge from overwintering galls in late May or early June in the Ithaca, NY area and live for an average of 9-11 days. After mating, adult females lay eggs in the folded leaves of the terminal bud of *S. altissima*. A larva, approximately 0.2 by 0.6 mm hatches from an egg 7-12 days after oviposition (Uhler 1951). It travels down through the folded leaves of the bud and burrows into stem tissue generally right below the apical dome. Gall formation results from cell division, enlargement, and differentiation (Beck 1947). Gall growth continues over a 3-4 week period during which time a 3 mm diameter stem will enlarge to form a gall up to 30 mm in diameter. The average width of a mature gall is 22 mm and the length 26.9 mm (Uhler 1951).

During the time that the gall grows, the larva does not increase greatly in size. Larvae are only about 1.0 by 1.5 mm by the time the galls have reached maximal size. Larvae continue to grow after gall growth has ceased and they reach maximal size 3 1/2 months after hatching. During their development, larvae spend approximately 4 weeks as first instars; 3 weeks as second instars; and 7 1/2 to 8 months as third instars. In September, fully grown third instar larvae prepare exit holes by scraping away the pithified tissue of the gall to form a tunnel from the center of the gall to the epidermis which is left intact. Larvae overwinter as third instars and form puparia and pupate in the spring (Uhler 1951).

BALL GALL DEVELOPMENTAL ANATOMY

The initial reaction to the presence of an *E. solidaginis* larva in the stem of *Solidago altissima* is the initiation of cell division in several layers of cells in the cortex. As the larva burrows through the region of the vascular bundles and into the pith, cell division is stimulated in the cell layers around the passageway of the larva. Some of the cells enlarge considerably and exhibit spiral thickenings causing them to resemble vessel cells. After the larva stops its downward travel through the pith at a position generally within several millimeters of the apical dome, it burrows up and down forming a very small central cavity. A circle of cell division appears to radiate out from the center and eventually includes the entire pith region which exhibits active cell division during the growth of the gall. The pith cells are thin walled, hexagonally or radially elongated toward the center of the gall, and much smaller than those of normal stem pith (Beck 1947). The cells of the pith adjacent to the larval chamber, termed the nutritive-pith by Weis *et al.* (1989) are cytoplasmically rich during the early stages of gall development and frequently contain irregularly shaped nuclei and chloroplasts (Bross *et al.* 1992).

Bands of meristematic tissue initially form in the pith and then split into strands of meristematic tissue which are separated by less actively dividing parenchyma cells. Some meristematic activity is also induced in the outer regions of the pith. The meristematic strands radiate from the central pith region surrounding the larval chamber outward to the fasicular cambia of the vascular bundles. As the gall matures, the number of meristematic cells in the strands decreases to only a few and the older parenchyma cells become quite large, sometimes equaling the size of those in the pith of normal stems (Beck 1947). Bross *et al.* (1992) observed that cell divisions in the meristematic strands are most evident in the early stages of gall development.

In a mature gall, some of the cells of the meristematic strands show reticular thickenings which stain more deeply and resemble the thickenings of the walls of some vessel cells (Beck 1947). Cosens (1912) and Ross (1936) referred to the strands which radiate outward from the pith region surrounding the larval chamber as "vascular strands". Blum (1952) presented support for the vascular nature of the strands by indicating that the strands stain with eosin and by noting that the strands originate adjacent to the primary xylem. He also noted that some of the cells in the strands mature into scalariform-reticulate elements.

The cortex of the gall becomes considerably wider than that of the stem. There are approximately twice as many cells present and they are 2-4 times wider than stem cortical cells. The resin ducts of the cortex become enlarged as do those of the pith. The hypodermis contains more cell layers, and epidermal cells increase in number in response to the gall stimulus (Beck 1947).

STUDIES ON THE NATURE OF THE BALL GALL FORMING STIMULUS

Beck (1947) attempted to determine the chemical nature of the ball gall forming stimulus by injecting goldenrod stems with a variety of compounds using a hypodermic syringe and glass micropipettes. In some cases, stem enlargement occurred, but gall formation did not occur in response to any of the treatments. Stem splitting was a recurring problem in the study. Applications of 0.01 percent ammonium hydroxide, 0.01 percent ammonium carbonate, trypsin, bacto-peptone, bacto-tryptone, and tryptophan caused cell division, and induced the formation of spirally or reticularly thickened cells resembling vessel cells. The protein digest bacto-tryptone caused the initiation of "meristematic strands" from the edge of the fasicular cambium in two cases. Injections of 0.01 percent IAA and 0.1 percent NAA caused cell division without differentiation. Beck also reported that tryptophan mixed with "gall juice" for 24 hours resulted in the production of IAA detected with the nitrite HCl test. Distilled water extracts of the larvae and distilled water in which larvae had been kept for 3 or more days resulted in some cell division, while crude organic extracts of young galls and of larvae injected into goldenrod stem tips, in water, yielded no positive results. Beck concluded that the secretion of proteolytic enzymes and the elevated pH of the gall region surrounding the larva (pH 6.0) are important in keeping conditions in the larval infected region similar to conditions in the growing point of the stem. He also suggested that formation of IAA from tryptophan is a second important factor in ball gall formation. However, his evidence for enzyme activity capable of converting tryptophan to IAA is questionable as he did not test the "gall juice" for IAA, an essential control, and did not consider microbial conversion of tryptophan to IAA by contaminants as a possibility.

Mills (1969) did not observe gall formation following injections of goldenrod plants with IAA, parachlorophenoxyacetic acid, kinetin, *N*-6-benzyladenine, ecdysterone, farnesyl methyl ether, or saline extracts of second or third instar larvae of *E. solidaginis*. He reported that butanol extracts of larvae resulted in galls in 3 out of 28 injections, but provided no information regarding the size, morphology, and/or anatomy of the galls that resulted. As Mills did not provide sufficient details on the manner or location of the chemical injections, factors which are very important when attempting to simulate insect-induced gall formation, one cannot assess the significance of the mainly negative results.

As a follow up to Beck's (1947) findings with respect to the effects of protein digests and tryptophan, Uhler et al. (1971) and Heady et al. (1982) studied changes in the free amino acid levels in *Eurosta solidaginis*, in the ball gall, and in normal *Solidago* stem tissues over time. The amino acids were separated from 80 percent ethanol tissue extracts by two-dimensional paper chromatography and semi-quantified after spraying with ninhydrin by visual comparison of color intensity (Heady et al. 1982) or by spectrophotometry of eluted spots (Uhler et al. 1971). Tryptophan and indoles were reacted with Ehrlich's reagent (Uhler et al. 1971). Uhler et al. (1971) detected tryptophan only in third instar larvae, at the end of August, after gall growth had ceased. Uhler concluded that the absence of significant concentrations of tryptophan or indoles argued against the production of an auxin by the gall-producing first instar. Heady et al. (1982) detected tryptophan in *E. solidaginis* larvae earlier in the season, in July, but at a point when most of the gall growth had ceased. However as both groups did not indicate the amounts of tissues extracted, did not present values on a per gram tissue basis, and did not account for losses or indicate the sensitivity of their assays, quantitative comparisons cannot be made and the reported absence of a particular amino acid at any stage cannot be considered conclusive.

Carango et al. (1988) detected hyperinduction of a native 58 kilodalton protein in developing ball gall tissues. Induction of the protein was most evident during the peak period of gall growth. They suggested that ball gall formation is the result of an alteration of existing plant growth mechanisms in response to secretions of the larva. Abrahamson et al. (1991) detected high levels of phenolics in developing ball galls and suggested that phenolics may play a role in ball gall formation.

STUDIES OF CYTOKININS IN INSECT GALLS

Leitch (1994) using radioimmunoassay following high pressure liquid chromatography (HPLC), found that the levels of isopentenyladenine, isopentenyladenosine, isopentenyladenine ribotide, and isopentenyladenine-9-glucoside were higher on a per gram fresh weight basis in *Pontania proxima* Lep. induced gall tissue during the first two weeks of gall development compared to leaf tissue in the same stage of development. McDermott et al. (1996) found that levels of isopentenyladenosine were 50-fold higher in hackberry, *Celtis occidentalis* L., gall tissues than in uninfested leaf tissues when detected by radioimmunoassay following immunoaffinity chromatography and HPLC. In contrast, in a study of the cytokinin bioactivity of galls of *Mikiola fagi* Htg. assessed by the tobacco callus bioassay, Engelbrecht (1971) found that levels in galls were not elevated when compared to healthy leaves. Similarly, Van Staden and Davey (1978) in a study of leaf galls formed by a chalcid wasp on *Erythrina latissima* E. Mey., reported that cytokinin concentrations in gall tissues were consistently lower than in surrounding leaf tissues.

In a study of galls caused by several species of *Phylloxera* on pecan, *Carya illinoensis* Koch, Hedin et al. (1985) reported the detection of the cytokinins zeatin, zeatin riboside, and kinetin, as well as gibberellic acid, abscisic acid, and IAA in leaves, stems, leaf galls and stem galls by absorption at 254 nm following HPLC. They reported confirmation of zeatin, kinetin, gibberellic acid, abscisic acid and IAA by mass spectrometry of derivatized samples. Levels of all of the compounds, assessed by integration of peak areas following HPLC, were reported as lower in leaf galls than in leaves and lower in stem galls than in stems. It should be noted that detection of all of the reported plant hormones at 254 nm is unacceptable as a general method of detection because impurities absorbing at this wavelength far exceed the compounds in question in any absorbancy peak, and that "purported" confirmation by direct-injection mass spectrometry in the absence of gas chromatography is also extremely unreliable for the same reason. In addition, it is well established that kinetin is not a naturally occurring plant hormone, but rather an "artefactual rearrangement product of heated DNA" (McGaw and Burch 1995).

A few studies have focused on assessing levels of cytokinins in gall-inducing insects. McCalla et al. (1962), associated growth promotion with two unidentified adenine derivatives in female accessory glands of *Pontania pacifica* Marlatt, a sawfly which forms leaf galls on willow, but did not detect significant cytokinin bioactivity of gland extracts in the *Xanthium* leaf disc assay. Wood and Payne (1987), Ohkawa (1974), and Van Staden and Davey (1978) presented evidence for cytokinins in the larvae of insect gall formers, with detection by enzyme immunoassays with monoclonal antibodies; by bioassay and coincidence with standards following LH_{20} column chromatography, gas liquid chromatography, and thin layer chromatography; and by bioassay following paper chromatography and fractionation on an LH_{20} column, respectively.

CYTOKININS AND BALL GALL FORMATION BY *EUROSTA SOLIDAGINIS*

Cytokinins that were extracted from *Eurosta solidaginis* larvae, from ball galls containing larvae, and from normal stem tissues were positively identified by gas chromatography/mass spectrometry with selected ion monitoring GC-MS (SIM). The extraction procedure followed was essentially the procedure of Davies et al. (1986). [$^2H_5$]-trans zeatin and [$^2H_5$]-trans zeatin riboside (98 percent [$^2H_5$], Apex Organics Ltd., Oxford, UK) were added as internal standards in quantitative studies, but not in qualitative studies. Extracts were purified by preparative and analytical high pressure liquid chromatography (HPLC) before and after permethylation. In qualitative studies, aliquots of fractions collected following reverse phase HPLC were screened for cytokinin bioactivity in the *Amaranthus* betacyanin bioassay (Biddington and Thomas 1973). Cytokinins in permethylated samples were detected using GC-MS with selected ion monitoring (SIM) and quantified by use of isotope ratios. Details of methods are described in Mapes (1991).

The concentrations of the cytokinins zeatin (Z), zeatin riboside (ZR), isopentenyladenine (iP), and isopentenyladenosine (iPA) were lower on a weight/weight basis in developing galls (12-15mm in diameter) as compared to the top 3.5 cm of stem where actively growing galls of this diameter tend to be found (table 1). However, when the concentrations were expressed on a weight/stem length basis, the levels of the cytokinins were about six times higher in the gall tissues compared to the stem tissues, indicating that the presence of the larva of *Eurosta solidaginis* results in higher cytokinin amounts in a given length of stem than are found in its absence. In galls, the greater width of the gall negates the increased cytokinin levels when recorded on a weight/weight basis. The most abundant cytokinin in galls and in stem tissues was ZR, followed by iP, and iPA, with Z present at the lowest concentrations in gall and stem tissues.

Four cytokinins were detected in extracts of first instar larvae of *E. solidaginis* by GC-MS (SIM) (table 2). Isopentenyladenine was the most abundant cytokinin, present in larvae at 349.61 ng/g fresh weight or 0.073 ng/larva. The second most abundant cytokinin was isopentenyladenosine at 43.44 ng/g fresh weight or 0.009 ng/larva. Similar amounts of zeatin riboside and zeatin were present at 12.22 ng/g fresh weight and 8.36 ng/g fresh weight or 0.003 ng/larva and 0.002 ng/larva, respectively. The concentrations of each of the cytokinins expressed on a weight/weight basis in the larvae, were much higher than the weight/weight concentrations of the corresponding cytokinins in gall or stem tissues. The concentration of iP was 53 times greater in larvae than in stems and 117 times greater in larvae than in whole galls, while the concentration of iPA was 16 times greater in larvae than in stems and 33 times greater in larvae than in whole galls.

Following extraction of 127 third instar *E. solidaginis* larvae isolated from fully formed galls exhibiting a "green island effect", found in the field in October, three cytokinins, zeatin riboside, isopentenyladenine, and isopentenyladenosine, were detected and quantified by GC-MS (SIM) (table 3). The most abundant cytokinin was isopentenyladenine at 4.75 ng/g fresh weight or 0.273 ng/larva. Zeatin riboside was present at 1.98 ng/g fresh weight or 0.114 ng/larva and isopentenyladenosine was present at 1.62 ng/g fresh weight or 0.093 ng/larva (table 3).

Bioassay data from partially purified extracts of third instar larvae provided evidence for the existence of cytokinins in overwintering larvae as well. In an extract of 460 third instar larvae from overwintering galls, a small amount of bioactivity in the *Amaranthus* betacyanin test of one-tenth of each fraction eluting after preparatory HPLC (fig. 1), ran parallel with ZR and a relatively large zone of bioactivity ran at the retention times of iPA (elution time 36.63) and iP (elution time 37.49). When the bioactive fractions which ran parallel with iP and iPA were run on analytical HPLC, one-tenth aliquots of fractions 14 and 15 were shown to contain

Table 1.—*Levels of extractable cytokinins from developing ball galls and goldenrod stem tissues. Zeatin, zeatin riboside, isopentenyladenine, and isopentenyladenosine from purified extracts of developing galls (12-15 mm diameter, 2.0 cm in length) and stem tissue (top 3.5 cm below apex) detected and quantified by GC-MS (SIM) with [$^2H_5$]-trans zeatin and [$^2H_5$]-trans zeatin riboside as internal standards to correct for losses during extraction. Results are reported as ng/gram and ng/cm. For replicated values, results are presented as means ± standard errors.*

Extracted Cytokinin:	Gall [Cytokinin]	Gall [Cytokinin]	Stem [Cytokinin]	Stem [Cytokinin]
	(ng/gram)	(ng/cm)	(ng/gram)	(ng/cm)
Zeatin	0.45 ± 0.09	0.42 ± 0.08	0.87 ± 0.15	0.07 ± 0.02
Zeatin riboside	3.79 ± 0.61	3.60 ± 0.81	9.35 ± 3.00	0.67 ± 0.18
Isopentenyladenine	2.98	2.82	6.58	0.49
Isopentenyladenosine	1.31 ± 0.11	1.25 ± 0.11	2.78	0.21

Table 2.—*Levels of extractable cytokinins from first instar larvae of Eurosta solidaginis removed from developing galls. Zeatin, zeatin riboside, isopentenyladenine, and isopentenyladenosine from a purified extract of 340 larvae removed from developing galls (7-21 mm diameter) detected and quantified by GC-MS (SIM) with[$^2H_4$]-trans zeatin and [$^2H_5$]-trans zeatin riboside as internal standards to correct for losses during extraction. Results are presented as ng/gram and ng/larva.*

Extracted cytokinin:	[Cytokinin] (ng/gram)	[Cytokinin] (ng/larva)
Zeatin	8.36	0.002
Zeatin riboside	12.22	0.003
Isopentenyladenine	349.61	0.073
Isopentenyladenosine	43.44	0.009

Table 3.—*Levels of extractable cytokinins from third instar larvae of Eurosta solidaginis removed from galls exhibiting a "green island effect". Zeatin riboside, isopentenyladenine, and isopentenyladenosine from a purified extract of 340 third instar larvae removed from galls exhibiting a "green island effect", detected and quantified by GC-MS (SIM) with[$^2H_5$]-trans zeatin and [$^2H_5$]-trans zeatin riboside as internal standards to correct for losses during extraction. Results are presented as ng/gram and ng/larva.*

Extracted cytokinin:	[Cytokinin] (ng/gram)	[Cytokinin] (ng/larva)
Zeatin riboside	1.98	0.114
Isopentenyladenine	4.75	0.273
Isopentenyladenosine	1.62	0.093

bioactivity, coinciding with the elution time of iP at 14.72 (fig. 2). Bioactivity did not run parallel with iPA which had an elution time of 12.54. The amount of putative iP in the sample was estimated at 110 ng (5.56 ng/g f.wt. or 0.239 ng/larva). After the bioactive fractions were permethylated and run on normal phase HPLC, no peaks were found with the retention times of Me-iP or Me-iPA. The permethylation efficiency of the sample was not tested, and as subsequent problems occurred with the permethylation of samples, it is possible that the compounds were not permethylated. The initial bioassay of fractions after preparatory HPLC was repeated with an extract of 491 third instar larvae, yielding similar zones of bioactivity (data not shown).

Comparison of Ball Gall Cytokinin Levels with Reports of Cytokinins in Other Galls and Gall Formers

The results presented here are similar to the results of cytokinin analyses in insect-induced gall tissues by Engelbrecht (1971) and Van Staden and Davey (1978), in that the nongalled tissues were shown to contain higher levels of cytokinin activity and cytokinins, expressed on a fresh weight basis, compared to the gall tissues. However, these results differ from those of Leitch (1994) and McDermott et al. (1996) who reported higher concentrations of cytokinins in insect-induced gall tissues compared to uninfested leaf tissues. Ohkawa (1974), Wood and Payne (1987), and Van Staden and Davey (1978) have previously presented evidence for cytokinins in the larvae of insect gall formers. However none of these studies utilized GC-MS for the detection of the cytokinins.

Ohkawa (1974) detected zeatin in larvae of the oriental chestnut gall wasp, *Dryocosmus kuriphilus* Yasumatsu, but presented no quantitative data. Wood and Payne (1987) found that levels of zeatin, zeatin riboside, and dihydrozeatin were similar in developing galls of Chinese chestnut, *Castanea mollissima* Bl., to the levels in normal shoots. However dihydrozeatin riboside-like substances were much higher in developing galls compared to normal shoots. The larvae of *Dryocosmus kuriphilus* contained high levels of dihydrozeatin only. Wood and Payne suggested that the levels of cytokinins in the larvae were not high enough to be of significance to gall formation.

Van Staden and Davey (1978) found that the larvae of the chalcid wasp which forms leaf galls on *Erythrina latissima* had very high concentrations of cytokinins, containing 90 percent of the cytokinins found in the gall tissue, but the larvae did not contain cytokinin activity when they were found in senesced galls. They suggested that the larvae accumulate cytokinins from the surrounding gall tissue causing nutrient mobilization by the larvae and a resultant decrease in cytokinin concentration in the surrounding gall tissue.

Evidence presented in this study shows a comparatively high level of only one of the cytokinins in the first instar larvae, iP, which is found in a weight/weight concentration that is 53 times more than the amount found in the stem tissues. The elevated levels of iP in first instar larvae of *E. solidaginis* in comparison to the other cytokinins, were not reflective of the relative amounts of cytokinins in the gall or stem tissues. ZR was the cytokinin that was found in highest concentrations in gall and stem tissues. In addition, as iP is the precursor of ZR in plant tissue (McGaw and Burch 1995), the iP in the larva is unlikely to derive from the ZR in the plant.

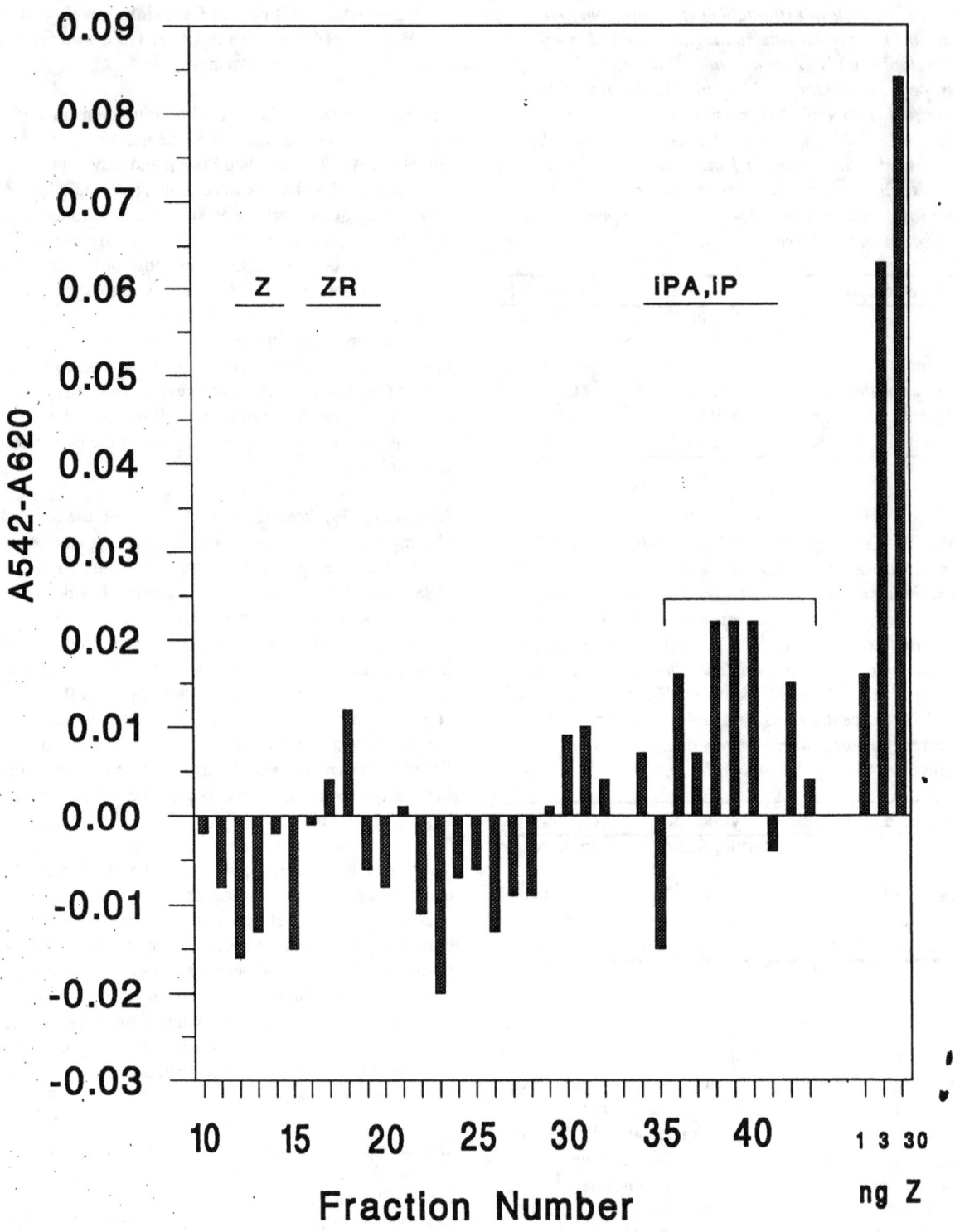

Figure 1.—*Bioactivity of preparative HPLC column fractions in the Amaranthus betacyanin bioassay of an extract of third instar larvae following separation on HPLC. The results of the bioassay were measured as increases in absorbancy at 542 nm, with the background at 620 nm subtracted (A542-A620), indicating levels of betacyanin in Amaranthus seedlings in response to 1/10 aliquots of fractions, with the bioactivity of 1 ng, 3 ng, and 30 ng zeatin standards indicated at the right. The retention times of standard zeatin (Z), zeatin riboside (ZR), isopentenyladenosine (iPA), and isopentenyladenine (iP) are indicated, and the bioactive fractions collected for further purification are shown in brackets.*

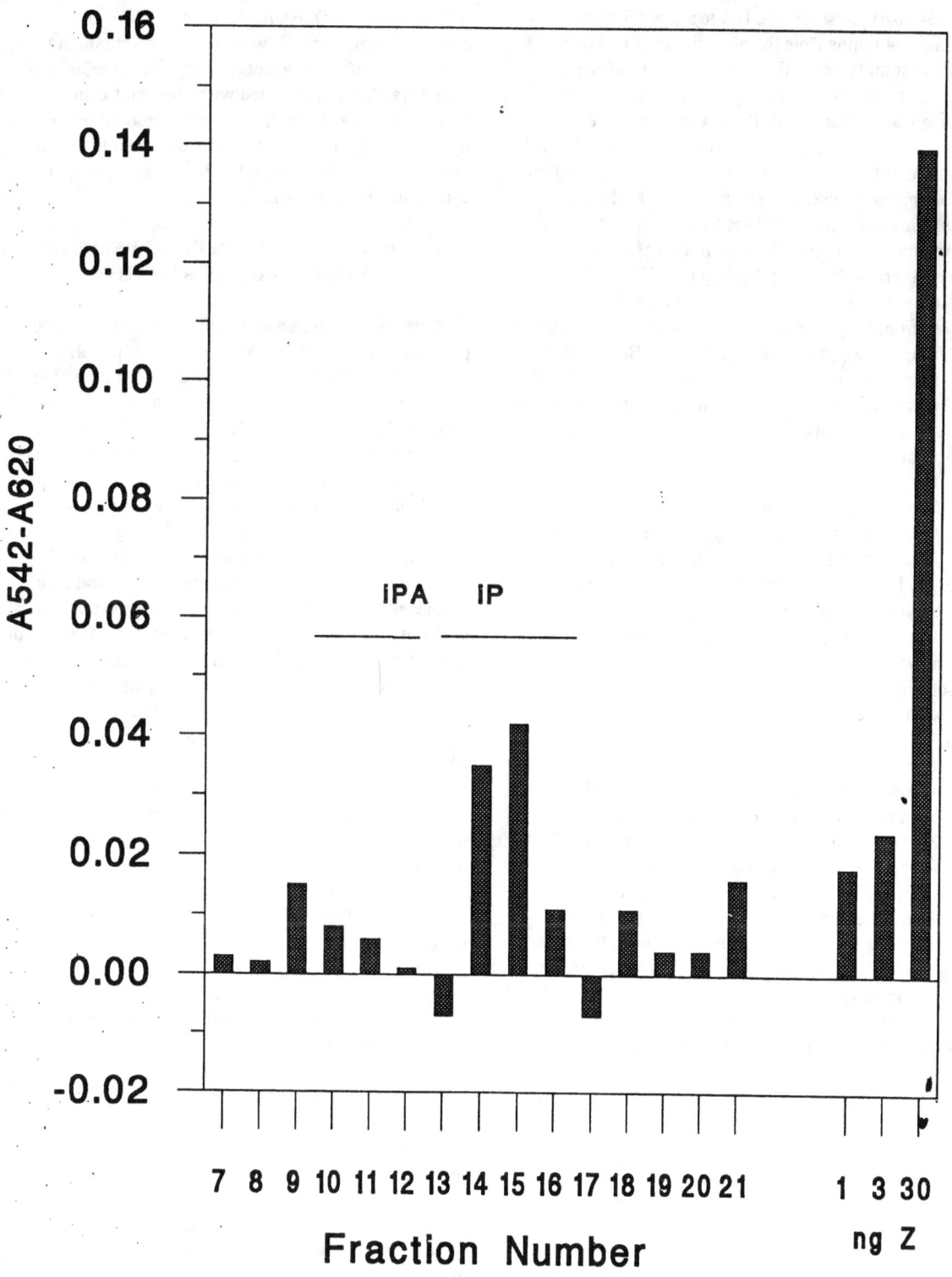

Figure 2.—Bioactivity of analytical HPLC column fractions in the Amaranthus betacyanin bioassay of an extract of third instar larvae following separation on HPLC. The results of the bioassay were measured as increases in absorbancy at 542 nm, with the background at 620 nm subtracted (A542-A620), indicating levels of betacyanin in Amaranthus seedlings in response to 1/10 aliquots of fractions, with the bioactivity of 1 ng, 3 ng, and 30 ng zeatin standards indicated at the right. The retention times of standard isopentenyladenosine (iPA), and isopentenyladenine (iP) are indicated.

These results suggest that the larvae are not just accumulating the cytokinins from the plant tissue. In addition, specific accumulation of iP from the plant tissues and storage in the fat body of *E. solidaginis*, is not likely given the water solubility of iP. The results of the bioassay of overwintering larvae suggest that a high level of iP is present at a time when the larvae are not consuming plant materials and are surrounded by dead plant tissues, a result supportive of the larvae as a source of iP. The amount of iP indicated by the results of the bioassay of overwintering larvae at 5.56 ng/g or 0.239 ng/larva is very similar to the amount found in the third instar larvae removed from fully formed galls in October at 4.75 ng/g or 0.273 ng/larva (quantified by GC-MS). Both of these values represent an increase in the amount of iP on a per larva basis compared to the 0.073 ng/larva value of the first instar larvae, and a decrease on a per gram fresh weight basis.

As preferential accumulation of iP by the larvae of *E. solidaginis* seems unlikely, the high concentration of iP in the larvae (350 ng/g f.wt.) could be a result of larval iP synthesis. It is interesting to note that the ribotide of isopentenyladenine has been detected in the nongall-forming insect, *Locusta migratoria* L., and has been shown to be conjugated to the insect hormone ecdysone in newly laid eggs of *L. migratoria* with evidence obtained by NMR and GC-MS (Tsoupras *et al.* 1983).

No previous studies have provided evidence for iP or iPA in gall-forming arthropods. However the cytokinins which have been shown to be associated with other plant tumor forming organisms including *Frankia*, an actinomycete which forms nitrogen fixing nodules on *Alnus glutinosa* L. (Stevens and Berry 1988), and the root nodule forming bacterium *Rhizobium* strain LPR1105 (Wang *et al.* 1982), have been iP or its riboside, iPA. It has been shown that one of the genes on the T-DNA of the crown gall inducing bacterium, *Agrobacterium tumefaciens* (Smith and Townsend) Conn, codes for isopentenyl transferase which is responsible for the formation of isopentenyladenine metabolites (Barry *et al.* 1984).

Evidence for a Role for Cytokinins in the Green Island Effect Induced by Larvae

The high levels of iP, iPA, and ZR in the larvae from fully formed galls (table 3) may be partially responsible for the "green island effect" that is evidenced by the persistence of green tissue in the immediate vicinity of the larvae of *E. solidaginis* in a region of otherwise senesced plant tissues. Engelbrecht (1971) found high levels of cytokinin bioactivity associated with the "green islands" surrounding *Hartigola annulipes* Htg. galls and surrounding leaf mines on beech, birch, and poplar. Leaf mining larvae of *Stigmella argentipedella* Z. and of *Stigmella argyropeza* Z. were shown to contain high levels of zeatin ribonucleotide bioactivity, the bulk of which was found associated with the labial glands. Engelbrecht concluded that the large amount of bioactivity associated with the larvae was more likely due to larval synthesis than a result of the consumption of cytokinins in plant tissues.

EVIDENCE FOR AUXIN INVOLVEMENT IN INSECT GALL FORMATION

Early studies suggested that applications of auxins to plant tissues resulted in structures similar to galls (Guiscafré-Arrillaga 1949, Hamner and Kraus 1937), while others indicated unsuccessful attempts to produce galls artificially (Mandl 1957, Plumb 1953). Applications of auxins to plant tissues have not proven to be successful in causing the formation of galls of complex morphologies (Hough 1953) and have not resulted in formation of structures that exhibit the degree of hyperplasia found in most galls (Mani 1964). Link *et al.* (1940) determined that ether extracts of several aphid species resulted in positive *Avena* coleoptile curvature tests but indicated they were unable to tell if this was due to production of auxin by the insects or accumulation from the surrounding plant tissues. Nysterakis (1948) and Hopp (1955, cited in Mani (1964)) detected IAA in extracts of aphid saliva by bioassays and concluded that IAA was the active cecidogenetic (gall-forming) factor in the aphid species studied. However according to Mani (1964), the aphid saliva was contaminated by honey-dew, invalidating the results. Others have reported the presence of IAA in the saliva and salivary glands of some homopterans (reviewed in Hori 1992, Hori and Endo 1977). All of these studies were done using bioassays, which are of limited validity.

Few studies have considered the levels of IAA in insect-induced gall tissues. Kaldeway (1965) found that oak apple gall tissues of *Quercus robur* L. had twice as much auxin activity as normal leaf tissues, and the three gall inhabitants contained compounds with auxin activity in the *Avena* coleoptile test. The gall former, *Cynips quercusfolii* L. was shown to secrete approximately 3 ng of auxin active compounds per larva, which was five times the amount of extractable "IAA" active compounds (0.5 ng/larva). The chalcid inquiline larvae and the parasite, *Synergus* sp., secreted much lower levels of auxin active compounds and had much lower levels of extractable auxin compared to the gall forming larvae. The main source of the excreted auxin of the chalcid larvae was their frass, while the *Synergus* sp. larvae gave off IAA during the process of pupation. Kaldeway hypothesized that these larvae give off tryptophan which could be converted to IAA by symbiotic microorganisms. He suggested that the proposed microbial activity probably occurs in the intact gall, as freshly removed frass also contains auxin active compounds. Kaldeway concluded that the gall former and the two coinhabitants, are able to produce auxin active compounds, but only the gall former gives off the compounds continuously, the factor which results in its ability to form galls.

In another study using bioassays, Byers et al. (1976) found that pinyon, *Pinus edulis* Engelm., needles with basal galls induced by larvae of the midge, *Janetiella* sp. near *J. coloradensis*, contained 3.7 times higher concentrations of auxin bioactivity in *Avena* coleoptile tests compared to needles lacking galls on a fresh tissue weight basis, and 17 times more auxin activity per needle. The highest levels of auxin activity per unit volume were found in the youngest galls. Levels dropped with developmental age of the galls. No auxin activity was detected in an extract of 200 of the gall inducing larvae.

Indole-3-Acetic Acid and Ball Gall Formation by *Eurosta Solidaginis*

IAA was isolated from extracts of larvae of *E. solidaginis*, ball galls containing larvae, and normal stem tissues and was purified by HPLC before and after methylation. $[^{13}C_6]$-IAA (99 percent ^{13}C, Cambridge Isotope Laboratories, Woburn, MA) was added as an internal standard, and 1-$[^{14}C]$-IAA (Amersham, specific activity 55 •Ci •mol^{-1}) was added to allow detection of IAA during the extraction procedure. IAA was detected on GC-MS by the coincidence of ions at m/z 130 (endogenous IAA) and 136 ($[^{13}C_6]$-IAA internal standard) at the retention time for MeIAA. The amount of IAA was calculated by isotope ratios. Details of methods are described in Mapes (1991).

The concentration of IAA on a weight/weight basis was lower in developing galls (6-9 mm in diameter) compared to that found in stem tissues removed from the top one centimeter of stem just below the apex, the region where actively growing galls of this diameter tend to be found (table 4). However, when the concentrations were expressed on a weight/length basis, the level of IAA was elevated in the gall tissue compared to the stem tissue, indicating that the presence of the larva results in higher IAA amounts in a given length of stem than are found in its absence. First instar larvae removed from galls up to 9 mm in diameter, were shown to contain 9139 ng/g f.wt. or 1.21 ng/larva (table 4) as detected by GC-MS with $[^{13}C_6]$-IAA as an internal standard to account for losses during extraction. Preliminary evidence suggested the presence of IAA in an extract of overwintering third instar larvae. When a portion of the third instar larval extract was run on a thin layer chromatogram and treated with Ehrlich's reagent, a positive reaction was obtained at an Rf of 0.83 compared to the Rf of 0.84 for standard IAA. When the sample was run on HPLC with detection by spectrofluorimetry (excitation at 280 nm, emission at 350 nm; slit width 20 nm) a peak was obtained at the retention time of IAA.

The first instar larvae of *E. solidaginis* were shown to contain very high levels of IAA. The concentration of IAA on a weight/weight basis in the larvae was 33 times greater than the concentration found in stem tissues removed from the top 1 centimeter of stem below the apex and 198 times greater than the concentration found in whole galls (6-9 mm in diameter). These results suggest that a high concentration of IAA may exist in the center of the gall due to the presence of the larva, a factor that could play an important role in ball gall development. Evidence for the continued presence of IAA in

Table 4.—*Levels of IAA from developing ball galls, stem tissue, and larvae of Eurosta solidaginis. IAA from purified extracts of developing galls (6-9 mm diameter, 1 cm in length), stem tissue (top 1 cm below apex), and from a purified extract of first instar larvae removed from developing galls, detected and quantified by GC-MS (SIM) with $[^{13}C_6]$-IAA as an internal standard to correct for losses during extraction. For replicated values, results are presented as means ± standard errors.*

Extracted tissue:	[IAA] (ng/gram)	[IAA] (ng/unit)
Stem	275 ± 24.0 ng/gram	11.6 ± 1.6 ng/cm
Gall	46.1 ± 2.7 ng/gram	14.2 ± 0.6 ng/cm[1]
Larvae	9139 ng/gram	1.21 ng/larva

[1] *The ng/cm value for the gall extract is equivalent to ng/gall as the galls were 1 cm in length.*

larvae of *E. solidaginis* removed from senesced overwintering galls suggests that the IAA may be synthesized by the larvae, and not just accumulated from the tissues. In support of this interpretation, studies on the fate of ^{14}C-IAA ingested by *Creontiades dilutus* Stal and the bug, *Lygus disponsi* Lin. have shown that almost all of the ingested IAA was decomposed or metabolized in the gut and excreted (Hori and Endo 1977, Miles and Hori 1977).

The results of the IAA analyses in goldenrod stem and gall tissues were similar to the results found for cytokinins, in that the nongalled tissues were shown to contain a higher concentration of IAA expressed on a fresh weight basis than the gall tissues. However the higher concentration of IAA in gall tissues compared to stem tissues when the results were expressed on a tissue length basis are consistent with a high concentration of IAA in the larvae. Given the well established role of auxin in vascular tissue differentiation (Aloni 1995), the vascular tissue pattern of the gall is also suggestive of the larvae as a central source of auxin, as vascular tissue radiates out from the nutritive-pith tissue surrounding the larva and connects with the stem vascular column in the periphery of the gall (Weis *et al.* 1989). As vascular strand formation has been shown to be triggered along the path of auxin flow from a source of auxin (Jacobs 1952), the vascular tissue pattern in the ball gall may reflect diffusion of IAA from the larvae. The lack of an increase in IAA concentrations in whole gall extracts could be reflective of the diffusion of IAA out of the gall. IAA may not be accumulating in the tissues but diffusing out, causing vascular tissue differentiation along the paths of diffusion. As IAA transport typically occurs at a rate of 5 to 20 mm/hour (Rubery 1987), tissue levels may not be increasing.

Alternatively the IAA may be rapidly metabolized and/or conjugated, without a concomitant increase in free IAA levels. Continuous uptake of exogenously applied radiolabelled IAA by light grown stems of bean and pea has been shown to result in increasing tissue levels of free IAA primarily during the first 6 hours of uptake only, with free IAA representing only 10-15 percent of radioactivity in the tissues after 24 hours of uptake, and evidence that most of the IAA was present as conjugates at 24 hours (Davies 1973). Others have found that increased auxin biosynthetic activity as a result of transformation by the Ti plasmid of *Agrobacterium tumefaciens* has not resulted in an increase in the levels of free IAA, (Ishikawa *et al.* 1988, Wyndaele *et al.* 1985). Sitbon *et al.* (1991) found that transgenic tobacco plants which contained the two IAA biosynthesis genes of the Ti plasmid had only slightly higher concentrations of free IAA but contained significantly higher levels of IAA conjugates compared to wild type plants. These studies provide evidence that the free level of IAA may be controlled by conjugation and may not increase despite greater biosynthetic activities in the tissue.

CONCLUSIONS

Zeatin, zeatin riboside, isopentenyladenine, isopentenyladenosine, and indole-3-acetic acid have been positively identified in ball galls, goldenrod stem tissue and larvae of *Eurosta solidaginis*. When concentrations were expressed on a weight/weight basis they were not higher in developing ball galls compared to concentrations in stems, but they were elevated on a weight/stem length basis. These findings are consistent with a localized high concentration of cytokinins and IAA existing in the center of the gall. The high levels of iP and IAA found in larvae point to the larvae as potential sources of iP and IAA. The evidence for the presence of iP and IAA in overwintering larvae supports the larvae as the source of these hormones. The vascular tissue pattern of the gall is indicative of a central source of auxin and the "green island effect" that is evident in the tissue surrounding the larva is supportive of a high concentration of cytokinins associated with the larvae. Larvae of *Eurosta solidaginis* may act as point sources of cytokinins and of IAA in developing ball galls.

ACKNOWLEDGEMENTS

This work was partially funded by Grants-In-Aid of Research from Sigma Xi and from the Cornell Chapter of Sigma Xi. The authors would also like to thank Angelika Behrooz, Angela Kell, and Silvia Mapes for their assistance with manuscript translations, as well as Dan Coffman and Maureen Kelly for their assistance in the preparation of the figures.

LITERATURE CITED

Abrahamson, W.G.; McCrea, K.D.; Whitwell, A.J.; Vernieri, L.A. 1991. The role of phenolics in goldenrod ball gall resistance and formation. Biochemical Systematics and Ecology. 19(8): 615-622.

Aloni, R. 1995. The induction of vascular tissues by auxin and cytokinin. In: Davies, P.J., ed. Plant hormones: physiology, biochemistry, and molecular biology. Dordrecht, The Netherlands: Kluwer Academic Publishers: 531-547.

Barry, G.F.; Rogers, S.G.; Fealey. R.T.; Brand, L. 1984. Identification of a cloned cytokinin biosynthetic gene. Proceedings of the National Academy of Science. 81: 4776-4780.

Beck, E.G. 1947. Some studies on the *Solidago* gall caused by *Eurosta solidaginis* (Fitch). Ann Arbor, MI: University of Michigan. 103 p. Ph.D. dissertation.

Biddington, N.L.; Thomas, T.H. 1973. A modified *Amaranthus* betacyanin bioassay for the rapid determination of cytokinins in plant extracts. Planta. 111: 183-186.

Blum, J.L. 1952. Vascular development in three common goldenrod galls. Papers of the Michigan Academy of Science, Arts and Letters. 38: 23-34.

Boysen Jensen, P. 1948. Formation of galls by *Mikiola fagi*. Physiologia Plantarum. 1: 95-108.

Bross, L.S.; Weis, A.E.; Hanzely, L. 1992. Ultrastructure of cells of the goldenrod (*Solidago altissima*) ball gall induced by *Eurosta solidaginis*. Cytobios. 71: 51-65.

Byers, J.A.; Brewer, J.W.; Denna, D.W. 1976. Plant growth hormones in *Pinyon* insect galls. Marcellia. 39: 125-143.

Carango, P.; McCrea, K.D.; Abrahamson, W.G.; Chernin, M.I. 1988. Induction of a 58,000 dalton protein during goldenrod gall formation. Biochemical and Biophysical Research Communications. 152(3): 1348-1352.

Cosens, A. 1912. A contribution to the morphology and biology of insect galls. Transactions of the Canadian Institute. 9: 297-387.

Davies, P.J. 1973. The uptake and fractional distribution of differentially labeled indoleacetic acid in light grown stems. Physiologia Plantarum. 28: 95-100.

Davies, P.J.; Horgan, R.; Heald, J.K.; McGaw, B. 1986. Endogenous cytokinins in vegetative shoots of peas. Plant Growth Regulation. 4: 311-323.

Engelbrecht, L. 1971. Cytokinin activity in larval infected leaves. Biochemie Physiologie der Pflanzen. 162: 9-27.

Guiscafré-Arrillaga, J. 1949. Formation of galls in stems and leaves of sugar cane in response to injections of growth regulating substances. Phytopathology. 39: 489-493.

Hamner, K.C.; Kraus, E.J. 1937. Histological reactions of bean plants to growth-promoting substances. Botanical Gazette. 98: 735-807.

Heady, S.E.; Lambert, R.G.; Covell, C.V. 1982. Determination of free amino acids in larval insect and gall tissues of the goldenrod, *Solidago canadensis* L. Comparative Biochemistry and Physiology. 73B: 641-644.

Hedin, P.A.; Neel, W.W.; Burks, M.L.; Grimley, E. 1985. Evaluation of plant constituents associated with pecan *Phylloxera* gall formation. Journal of Chemical Ecology. 11: 473-484.

Hopp, H.H. 1955. Wirkungen von Blattreblausspeichel auf Pflanzengewebe. Weinbau, Wissenschaftliche Beihefte. 9: 9-23.

Hori, K. 1992. Insect secretions and their effect on plant growth, with special reference to hemipterans. In: Shorthouse, J.D.; Rohfritsch, O., eds. Biology of insect-induced galls. New York, NY: Oxford University Press: 157-170.

Hori, K.; Endo, M. 1977. Metabolism of ingested auxins in the bug *Lygus disponsi*: conversion of indole-3-acetic acid and gibberellin. Journal of Insect Physiology. 23: 1075-1080.

Hough, J.S. 1953. Studies on the common spangle gall of oak. II. A general consideration of past work on gall induction. The New Phytologist. 52: 218-228.

Hovanitz, W. 1959. Insects and plant galls. Scientific American. 201: 151-162.

Ishikawa, K.; Kamada, H.; Harada, H. 1988. Morphology and hormone levels of tobacco and carrot tissues transformed by *Agrobacterium tumefaciens*. I. Auxin and cytokinin contents of cultured tissues transformed with wild type and mutant Ti plasmids. Plant and Cell Physiology. 29: 461-466.

Jacobs, W.P. 1952. The role of auxin in differentiation of xylem around a wound. American Journal of Botany. 39: 301-309.

Kaldeway, H. 1965. Wachstumregulatoren aus Pflanzengallen und Larven der Gallenbewohner. Berichte der Deutschen Botanischen Gesellschaft. 78: 73-84.

Leitch, I.J. 1994. Induction and development of the bean gall caused by *Pontania proxima*. In: Williams, M.A.J., ed. Plant galls: organisms, interactions, populations. Oxford, UK: Clarendon Press: 283-300.

Leatherdale, D. 1955. Plant hyperplasia induced with a cell-free insect extract. Nature. 175: 553-554.

Link, G.K.; Eggers, V.; Moulton, J.E. 1940. *Avena* coleoptile assay of ether extract of aphids and their hosts. Botanical Gazette. 101: 928-939.

Mandl, L. 1957. Wachstum von Pflanzengallen durch Synthetische Wuchsstoffe. Osterreichische Botanische Zeitschrift. 104: 185-208.

Mani, J. 1964. Ecology of plant galls. The Hague, Junk. 434 p.

Mapes, C.C. 1991. The regulation of ball gall growth and development on *Solidago altissima*. Ithaca, NY: Cornell University. 162 p. Ph.D. dissertation.

Martin, J.P. 1942. Stem galls of sugar-cane induced with insect extracts. Science. 96: 39.

McCalla, D.R.; Genthe, M.; Hovanitz. W. 1962. Chemical nature of an insect gall growth-factor. Plant Physiology. 37: 98-103.

McDermott, J.; Meilan, R.; Thornburg, R. 1996. Plant-insect interactions: the hackberry nipple gall. World Wide Web Journal of Biology. 2: 7 p.

McGaw, B.A.; Burch, L.R. 1995. Cytokinin biosynthesis and metabolism. In: Davies, P.J., ed. Plant hormones: physiology, biochemistry, and molecular biology. Dordrecht, The Netherlands: Kluwer Academic Publishers: 98-118.

Miles, P.W.; Hori, K. 1977. Fate of ingested ß-indolyl acetic acid in *Creontiades dilutus*. Journal of Insect Physiology. 23: 221-226.

Mills, R.R. 1969. Effect of plant and insect hormones on the formation of the goldenrod gall. National Cancer Institute Monograph. 31: 487-491.

Nysterakis, F. 1948. Autres preuves sur la sécrétion d'auxines par certain insectes. Un nouveau test très sensible pour le dosage des substances de croissance. Comptes Rendus Hebdomadaires des Seances. Academie des Sciences (Paris). 226: 1917-1919.

Ohkawa, M. 1974. Isolation of zeatin from larvae of *Dryocosmus kuriphilus* Yasumatsu. HortScience. 9: 458-459.

Plumb, G.H. 1953. The formation and development of the Norway Spruce gall caused by *Adelges abietes* L. New Haven, CT: Agricultural Experiment Station Bulletin. 557: 2-77.

Ross, F.W. 1936. *Solidago* cecidology. Ithaca, NY: Cornell University. 129 p. Ph.D. dissertation.

Rubery, P.H. 1987. Auxin transport. In: Davies, P.J., ed. Plant hormones and their role in plant growth and development. Dordrecht, The Netherlands: Martinus Nijhoff: 341-362.

Sitbon, F.; Sundberg, B.; Olsson, O.; Sandberg, G. 1991. Free and conjugated indoleacetic acid (IAA) contents in transgenic tobacco plants expressing the iaaM and iaaH IAA biosynthesis genes from *Agrobacterium tumefaciens*. Plant Physiology. 95: 480-485.

Stevens, G.A.; Berry, A.M. 1988. Cytokinin secretion by *Frankia* sp. HFPArI3 in defined medium. Plant Physiology. 87: 15-16.

Tsoupras, G. 1983. A cytokinin (isopentenyl-adenosyl-mononucleotide) linked to ecdysone in newly laid eggs of *Locusta migratoria*. Science. 220: 507-509.

Uhler, L.D. 1951. Biology and ecology of the goldenrod gall fly, *Eurosta solidaginis*. Cornell University Agricultural Experiment Station Memoir. 300: 3-51.

Uhler, L.D.; Crispen, C.R., Jr.; McCormick, D.B. 1971. Free amino acid patterns during development of *Eurosta solidaginis* (Fitch). Comparative Biochemistry and Physiology. 38B: 87-91.

Van Staden, J.; Davey, J.E. 1978. Endogenous cytokinins in the laminae and galls of *Erythrina latissima* leaves. Botanical Gazette. 139: 36-41.

Wang, T.L.; Wood, E.A.; Brewin, N.J. 1982. Growth regulators, *Rhizobium* and nodulation in peas. Planta. 155: 345-349.

Weis, A.E; Wolfe, C.L.; Gorman, W.L. 1989. Genotypic variation and integration in histological features of the goldenrod ball gall. American Journal of Botany. 76: 1541-1550.

Wood, B.W.; Payne, J.A. 1987. Phytohormones in Chinese Chestnut as influenced by the Oriental Chestnut gall wasp. In: Proceedings of the 14th annual Plant Growth Regulator Society of America meeting; 1987 August 2-6. Plant Growth Regulator Society of America: 278.

Wyndaele, R.; Van Onckelen, H.; Christansen, J.; Rudelsheim, P.; Hermans, R.; De Greef, J. 1985. Dynamics of endogenous IAA and cytokinins during the growth cycle of soybean crown gall and untransformed callus. Plant and Cell Physiology. 26: 1147-1154.

THE EFFECTS OF GALL FORMATION BY *LIPARA LUCENS* (DIPTERA, CHLOROPIDAE) ON ITS HOST *PHRAGMITES AUSTRALIS* (POACEAE)

L. De Bruyn, I. Vandevyvere, D. Jaminé and E. Prinsen

Department of Biology, University of Antwerp (RUCA-UIA), Belgium: Groenenborgerlaan 171, B- 2020

Abstract.—The flies of the genus *Lipara* (Diptera, Chloropidae) are monophagous herbivores of common reed, *Phragmites australis*, on which they induce characteristic cigar-like galls. We investigated the morphological and hormonal changes in the reed shoots during gall development. To this end, *Lipara lucens* galls of different ages, cultivated in a greenhouse and collected in the field, were examined morphologically, histochemically and biochemically.

Shoot elongation is strongly reduced. Ungalled shoots are longer and bear more leaves than galled shoots. Total shoot biomass does not differ between galled and ungalled shoots. Biomass allocated to gall tissues reduces biomass allocated to the rest of the stem and the leaves.

Internally, the gall consists of a central marrow zone of rich nutritive cells surrounded by a thickened and lignified tissue cylinder. At the transition zone between the marrow parenchyma and the tissue cylinder, a layer of longitudinal and radial sclerenchyma cells arises during gall development. Vascular bundles from the tissue cylinder branch off to this zone of sclerenchyma cells, connecting the larval chamber with the vascular bundles.

In the apical zone of parasitized shoots, the concentration of free indole-3-acetic acid (IAA) decreases in conjunction with an increase of conjugated IAA. At the same time, parasitized stems show a reduced basipetal gradient of free IAA. The altered free IAA/conjugated IAA balance probably plays an important role in the inhibition of shoot elongation. The abscissic acid (ABA) concentration in the apical zone of the gall increases. Probably the latter is due to the presence of the chewing larvae between the young enwrapped leaves above the gall apical region.

INTRODUCTION

Over evolutionary time, gall forming insects have developed the ability to manipulate the growth and differentiation of plant cells in the vicinity of their offspring, providing them with an optimal place to complete maturation (Shorthouse 1986). Most species are very species- and even organ-specific, only able to induce galls at a specific site on the host plant (Dreger-Jauffret and Shorthouse 1992). As a rule, galls possess a high content of mineral nutrients, carbon and energy because they act as a resource sink (Inbar et al. 1995, Larson and Whitham 1991, McCrea et al. 1985). As a consequence, plants carrying galls usually show reduced growth and reproduction (Fay and Hartnett 1991, Hartnett and Abrahamson 1979).

In contrast to gall morphology, anatomy, and ecology, the process of gall initiation (or cecidogenesis) is only poorly understood. Bronner (1977) and Shorthouse and Lalonde (1988) point to the necessity of the presence of the gall inhabitant to maintain gall development. Experiments with extracts from gall forming insects have lead to the hypothesis that chemical components should be involved. Insect saliva sometimes contains products such as amino acids, phenols or plant hormones (or products with an analogous action) (Hori 1992). On the other hand, several studies have shown an abnormal phytohormone balance in gall tissues (Byers et al. 1976, Van Staden and Bennett 1991, Wood and Payne 1988).

The host plants attacked by gall insects belong to nearly all branches of the plant kingdom (Buhr 1965, Mani 1964). In contrast to galls induced by other groups of cecidizoans, dipterous galls are also found on monocotyledons, and especially Poaceae (Dreger-Jauffret and Shorthouse 1992). Compared to more complex plants such as trees, grasses only possess simple plant architecture, relatively low protein concentration, and a low diversity of secondary compounds (Bernays and Barbehenn 1987, Strong and Levin 1979).

The common reed, *Phragmites australis* (Cav.) Trin. Ex Steud., a member of the family Poaceae, is the host of many herbivorous species. Most of the species are monophagous (Tscharntke 1994) and comprise a wide range of feeding habits such as leaf chewers and miners, sap suckers, stem miners, and a number of gall formers (e.g., De Bruyn 1985; Skuhravy 1978, 1981; Tscharntke 1989).

All species of the genus *Lipara* are strictly monophagous herbivores of the common reed. They all form a typical terminal shoot gall on their host plant. In the Western Palaearctic region, five species can be found (Beschovski 1984). In Belgium, three *Lipara* species are encountered frequently: *L. lucens* Meigen, *L. pullitarsis* Doskoil and Chvála, and *L. rufitarsis* (Loew) (De Bruyn 1985). One species reaches the western boundary of its distribution and was only recently discovered in Belgium (Beschovski 1984, De Bruyn 1988). The structural complexity of the galls is very variable. *L. similis* and *L. pullitarsis* hardly alter host plant growth while *L. lucens* and *L. rufitarsis* induce more complex galls. Gall complexity in turn strongly influences the life history of inhabiting species (De Bruyn 1994a, b; Tscharntke 1994).

Here we present knowledge on how the galling fly, *Lipara lucens*, affects its host plant, *Phragmites australis*. We discuss the influence of galling on biomass allocation to the different constituent parts of the host plant. We then describe changes in various aspects of morphology that occur during cecidogenesis. Finally we report on our preliminary findings concerning the involvement of phytohormones in the system.

NATURAL HISTORY

The Host plant

The common reed is a perennial rhizomatous grass that produces fresh shoots every year in spring (April-May) (Björk 1967, Van der Toorn 1972). A shoot consists of a stem and leaves. The stem is made up of internodes separated by nodes. The erect culms (0.5-3 m high) bear a leaf at each node. A leaf consists of a sheath and a lanceolate blade. Above the growing point, only the newly formed enwrapped leaves are present. These leaves point upwards in line with the shoot. Below the soil surface, the different shoots of a reed clone are connected via a rhizome. The vegetational phase lasts until the end of June, after which an ear is formed (the start of the reproductive phase). In autumn, the shoots dry up and the above ground parts of the reed clone die.

Shoots emerging from thicker buds (measured by basal shoot diameter) are more vigorous. They grow faster, reach a greater final height, carry more leaves and have a higher probability of producing an ear (Haslam 1969). Furthermore, several life history aspects, such as female oviposition choice and larval survival, of *L. lucens* are closely tied to plant vigor (De Bruyn 1994a, 1995; Mook 1967). As a consequence, the galls of this species are particularly found on thin shoots growing in dry and nutrient poor reed beds (De Bruyn 1987, 1996; Mook 1967).

Life History of *Lipara Lucens*

Adult *L. lucens* emerge in spring from the end of May until early July depending on climatological conditions. There is only one annual generation. The newly emerged females carry many mature eggs and can mate immediately. On average, one female can deposit 54 eggs with a maximum of 84 under laboratory conditions (De Bruyn 1994b). Egg deposition is very specific. Females carefully select thinner reed shoots on which larval survival is also higher than on thicker shoots (De Bruyn 1994a, Mook 1967). The females stick their eggs on the surface of the reed shoot. Under natural conditions in the field, most reed shoots carry only one egg, while practically all eggs are situated in the top 20 cm of the shoot. The eggs of *L. lucens* are very distinctive. They are long and rather slender, light yellow in color. The length to width ratio is about 5:1 (length 1.37-1.69 mm; width 0.25-0.36 mm; Chvála et al. 1974).

Eggs hatch after about 9 days. After hatching, the first instar larvae have to enter the reed shoot. According to Ruppolt (1957) they crawl up the shoot, enter under the edge of a leaf sheath and gnaw their way down through the enwrapped leaves until they reach the growing point. Here they feed on the young enwrapped leaves above the growing point. Due to presence of the larva, the newly formed internodes do not elongate any more and the species-specific gall forms (Chvála et al. 1974). After a few weeks, when gall formation is completed, the larvae gnaw through the growing point and enter the gall chamber wherein they continue feeding. *L. lucens* passes three larval instars (Mook 1967, Ruppolt 1957). The ivory white mature larva is large and stout (length 8-12 mm; width 2-3 mm; Chvála *et al.* 1974). Only the anterior segments are dorsally sclerotised. The anterior spiracles carry 11 to 12 buds. Only one *Lipara* larva can develop per shoot. At the end of August, the last instar larva stops feeding, turns its head upwards and goes into diapause. Pupation takes place the following spring. The new flies leave the gall by crawling up between the enwrapped leaves above the growing point.

Under the influence of a feeding larva between the enwrapped leaves above the growing point, the extension of the newly formed internodes is inhibited and a terminal shoot gall develops (Chvála *et al.* 1974, Reijnvaan and Docters van Leeuwen 1906). Usually 9 to 12 internodes are involved. Due to the shortening and widening of the shoot's internodes, the surrounding leaf sheaths can no longer grow parallel to the shoot axis and are pushed outwards. The typical cigar-like gall is formed (fig. 1-1).

DEVELOPMENTAL MORPHOLOGY

To understand the processes involved in gall induction and development in the *Phragmites-L. lucens* system, we studied field collected and laboratory grown galls and uninfected shoots of different ages. Reed rhizomes were dug out at the end of the winter in a reed bed where *L. lucens* is common, and potted in boxes (35 x 33 x 41 cm) filled with black earth and placed in a climatic chamber under an L:D photoperiod regime of 14:10, with a temperature and relative humidity of 21°C and 80 percent during the day and 14°C and 99 percent during the night. *L. lucens* flies were reared from field collected galls. After mating, females were forced to oviposit on the young reed shoot under a cylindrical plastic cage with a muslin top (circumference = 100 cm; height = 62 cm).

Comparison of overall shoot growth between galled and ungalled shoots was studied under controlled greenhouse conditions. Specific developmental changes in the gall were observed with microscope sections under the light microscope and specific histochemical stains. Methodological techniques are described in detail in Vandevyvere (1995).

Figure 1-1 to 1-3.—*Cigar-shaped gall induced by* Lipara lucens *on the common reed,* Phragmites australis. *Figure 1-1.—external morphology.* Figure 1-2.—*transverse section through the gall.* Figure 1-3.—*a gall where the larva died prematurely and normal shoot growth resumed.*

External Changes of Galled and Ungalled Shoots

Most aerial shoots of *P. australis* start to grow during a 1-3 month emergence period in spring (Haslam 1969). Shoots can be considered to be composed of nodal units, each consisting of a node bearing an internode, surrounded by a leaf sheath, and a leaf. Above the growing point only leaf blades and sheaths are present. The outer sheaths enclose the inner ones partially or wholly. The newest leaves, which are the smallest, are entirely concealed. With growth, the new leaf blades emerge telescopically from the outer sheaths. After the base of a blade becomes free, it unfolds and stands out from the stem. Below the shoot apex there are 2(-3) growing internodes on thin shoots, 3(-4) on wide, rapidly growing shoots. The blade completes its development first, the sheath second, and the stem last. In each, the upper part matures while the lower part is still forming, since most growth is from the intercalary meristem. The mature internode is often longer than the surrounding sheath in the lower part of the stem, and is always shorter in the upper part.

According to Ruppolt (1957), egg development takes about nine days. In the climatic room, eggs on the infested shoots hatched after 10 days (mean ± SD = 10.1 ± 1.4). About 27 days (27.41 ± 3.06; minimum = 22) after the oviposition date, gall initiation becomes visible externally. The second youngest leaf unfolds its leaf blade while it is still enwrapped by the next older leaf. A few centimeters lower, a node forms the border between the normal stem and the developing gall. The region above this node thickens so that gall initiation becomes evident. Gall widening follows a sigmoid pattern (fig. 2) In the early phase, the differences between shoot and gall width increase slowly, later accelerating and then decreasing again at the end when gall width exceeds stem diameter by about 3-3.5 times (measured on live galls with leaf sheets present). Although thicker reed shoots carry thicker galls, gall widening is more pronounced in thinner shoots (fig. 3).

Ungalled shoots reached a height of nearly 90 cm (mean ± SE = 87.07 ± 6.47 cm) and carried on average 15 leaves (15.60 ± 1.08) on the reed stem (table 1). As expected, bud width, measured as basal shoot diameter, strongly influences shoot development. Thicker shoots grow taller (length = 7.60 + 24.99 diameter, R^2 = 0.55; $p < 0.001$; min = 35 cm; max = 131 cm) and tend to carry more leaves (# leaves = 7.54 + 1.65 diameter; R^2 = 0.22; $p = 0.07$; min = 6; max = 20) than thinner shoots. Because plant vigor influences final shoot appearance, differences between galled and ungalled shoots were assessed by analysis of covariance using basal shoot diameter as the covariate.

Galled shoots were significantly smaller (mean ± SE = 47.20 ± 3.03 cm). This is caused by a strongly reduced elongation (maximum ± 1 cm long) of the internodes above the point of infestation. The resulting gall is 6-15 cm long. The shoots also produced significantly fewer leaves (11.44 ± 0.30) than ungalled shoots. Although a normally developed gall consists of 9 to 12 shortened internodes, only 5 (4.92 ± 0.13) leaf blades extrude from the gall.

Although ungalled shoots are significantly longer than galled shoots, total shoot biomass does not differ between both groups (table 2). Compared to ungalled shoots, galling reduced total biomass of stems and leaf blades, but increased total sheath biomass. Proportionally, infested shoots allocated 34 percent of the total

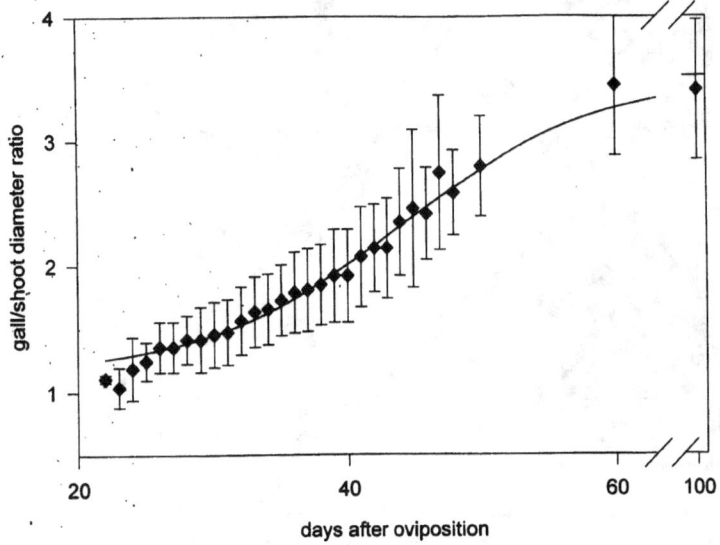

Figure 2.—*Changes in gall to shoot diameter ratio of* L. lucens *galls during development (data are mean ± SE).*

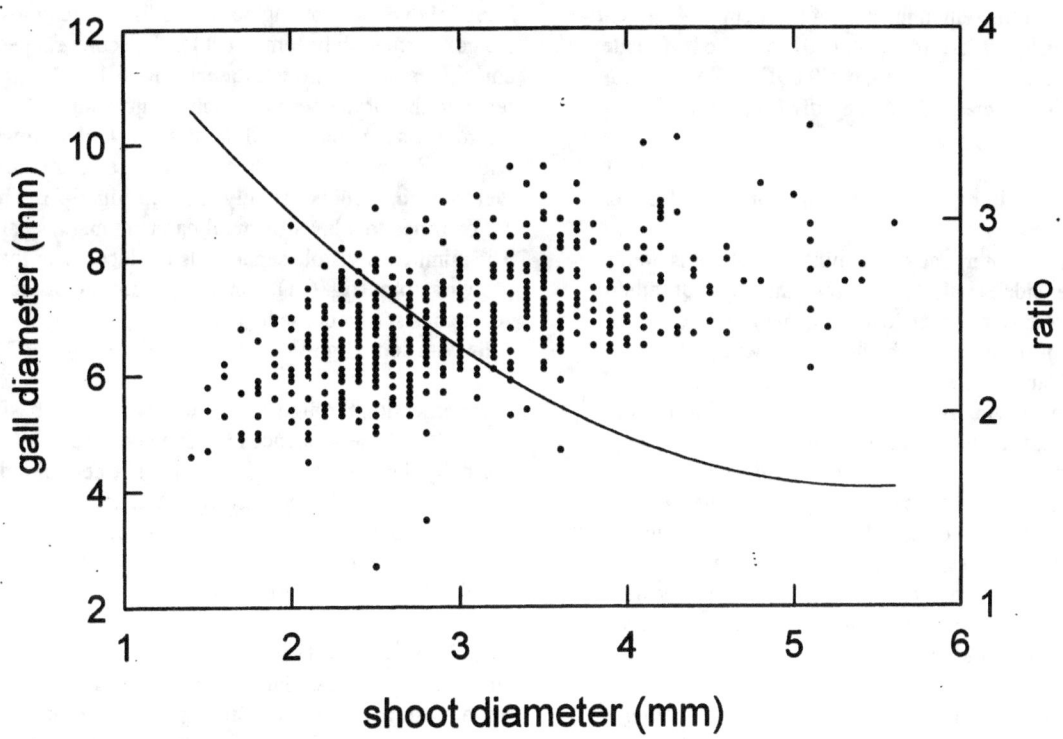

Figure 3.—*Relationship between shoot diameter and gall diameter. Scatter diagram = shoot versus gall diameter; line = proportional size increase as a function of shoot diameter.*

Table 1.—*Measures of final shoot length and leaf production from experimentally galled (n = 41) and ungalled (n = 15)* Phragmites australis *shoots. F = ANCOVA with basal shoot diameter as covariate.*

Plant trait	Ungalled shoot Mean	± SE	Galled shoot Mean	± SE	$F_{1,53}$	p
Shoot length (cm)	87.07	± 6.47	47.20	± 0.03	66.615	<0.001
# Leaves/stem	15.60	± 1.08	6.63	± 0.38		
# Leaves/gall	-	-	4.92	± 0.13		
Total # leaves	15.60	± 1.08	11.44	± 0.30	17.932	<0.001

Table 2.—*End-of-season dry mass (g) of constituent organs of galled (n = 32) and ungalled (n = 36)* Phragmites australis *shoots. F = ANCOVA with basal shoot diameter as covariate.*

Plant part(s)	Ungalled shoot Mean	± SE	Galled shoot Mean	± SE	$F_{1,60}$	p
Gall	—	—	263.41	± 141.21		
Stem	444.44	± 335.99	336.94	± 283.91	15.225	<0.001
Sheath	245.36	± 178.30	512.00	± 334.51	20.754	<0.001
Leaf	555.83	± 370.66	454.09	± 312.70	12.902	<0.001
Total	1260.81	± 871.76	1552.63	± 1019.02	0.235	0.630

biomass to gall tissues (fig. 4), mainly at the expense of allocation to the remaining part of the stem (-14 percent; ANCOVA $F_{1,60} = 131.149$, $p < 0.001$) and the leaf blades (-17 percent; $F_{1,60} = 109.938$, $p < 0.001$). Allocation to sheaths only decreased very slightly (-3 percent; $F_{1,60} = 4.257$, $p = 0.043$).

Internal Changes of Galls and Uninfested Shoots

A full-grown internode of an uninfected shoot is hollow. The inner side is only covered by a thin layer of rather loosely connected empty cells (fig. 5-1). These are the remains of the central pith which degenerates during internode maturation. In youngest internodes, right under the growing point (fig. 5-4), the pith consists of more or less rounded cells with small intercellular gaps. Lower in the shoot, larger cavities arise between the shrinking cells, until the pith completely degenerates. At a node a septum intersects the marrow cavity. This septum consists of xylem, phloem, small parenchyma cells and a few sclerenchyma cells (fig. 6-3). The actual tissue cylinder of a stem is bound on the outside by a single layered epidermis (fig. 5-1), followed by the subepidermal sclerenchyma and the subepidermal parenchyma. More centrally we find a second, outer sclerenchymatous layer. The inside consists of a thick layer of transversely rounded to polygonal, longitudinal cylindrical cells, the ground parenchyma. In the ground parenchyma run the collateral vascular bundles, surrounded by a perivascular sclerenchyma. At the height of the subepidermal parenchyma, the tissue cylinder can be provided with aerenchyma strands. These are particularly well developed in submerged shoots and rhizomes (Armstrong and Armstrong 1988, Rodewald-Rudescu 1974).

When the young larvae still feeds on the enwrapped leaves above the growing point, the first indications of the gall formation become visible. The cone-shaped apical meristem is more flattened (fig. 6-1). During feeding, the larva does not touch the growing point. The marrow tissue expands and the tissue cylinder grows wider. We found that septa are still present, although their development is strongly reduced (fig. 6-2). They are confined to a layer of small parenchymatous cells. The number of visible septa is also reduced near the growing point (fig. 6-1). Only at the bottom of the gall, septa consist of a dense layer of small parenchymatous cells with some lignified cells.

A cross section through a mature *L. lucens* gall basically reveals the same sequence of tissue layers as in a normally developed stem (fig. 5-6). The central pith however, is completely filled with a dense mass of a parenchymatous tissue, consisting of large round cells. Only at the bottom of the gall chamber is the marrow provided with some air channels.

In addition to the differences in the central marrow, the surrounding tissue cylinder also changes slightly. The larval chamber is coated with a special layer of sclerenchyma cells on the inside of the tissue cylinder, not found in an uninfested stem (fig. 5-6). This sclerenchymatous sheath consists of two distinct layers, each composed of several cell layers. The sclerenchyma cells of the inner layer are orientated longitudinally, parallel to the stem axis. These cells are orientated circular around the stem axis in the outer layer. This zone is perforated regularly by horizontally orientated vascular bundles which run around the larval chamber, perpendicular to the normal orientation. They run from the vascular bundles in the

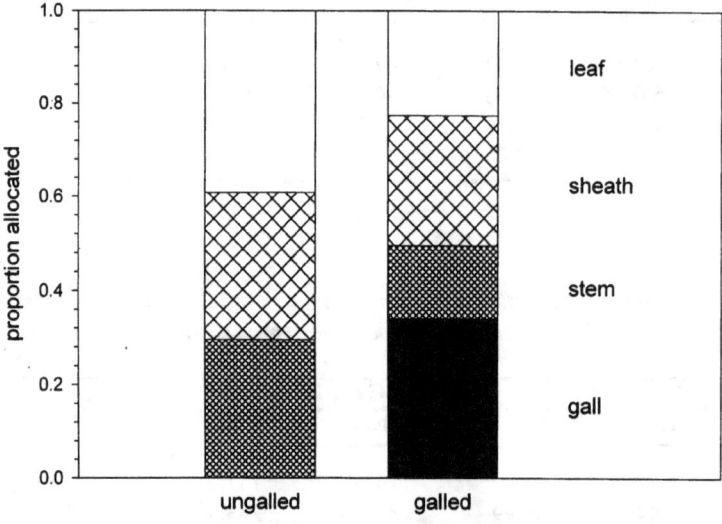

Figure 4.—*Proportion of biomass allocated to constituent organs in galled and ungalled shoots of Phragmites australis. The largest standard error was ± 0.030 for allocation to sheath.*

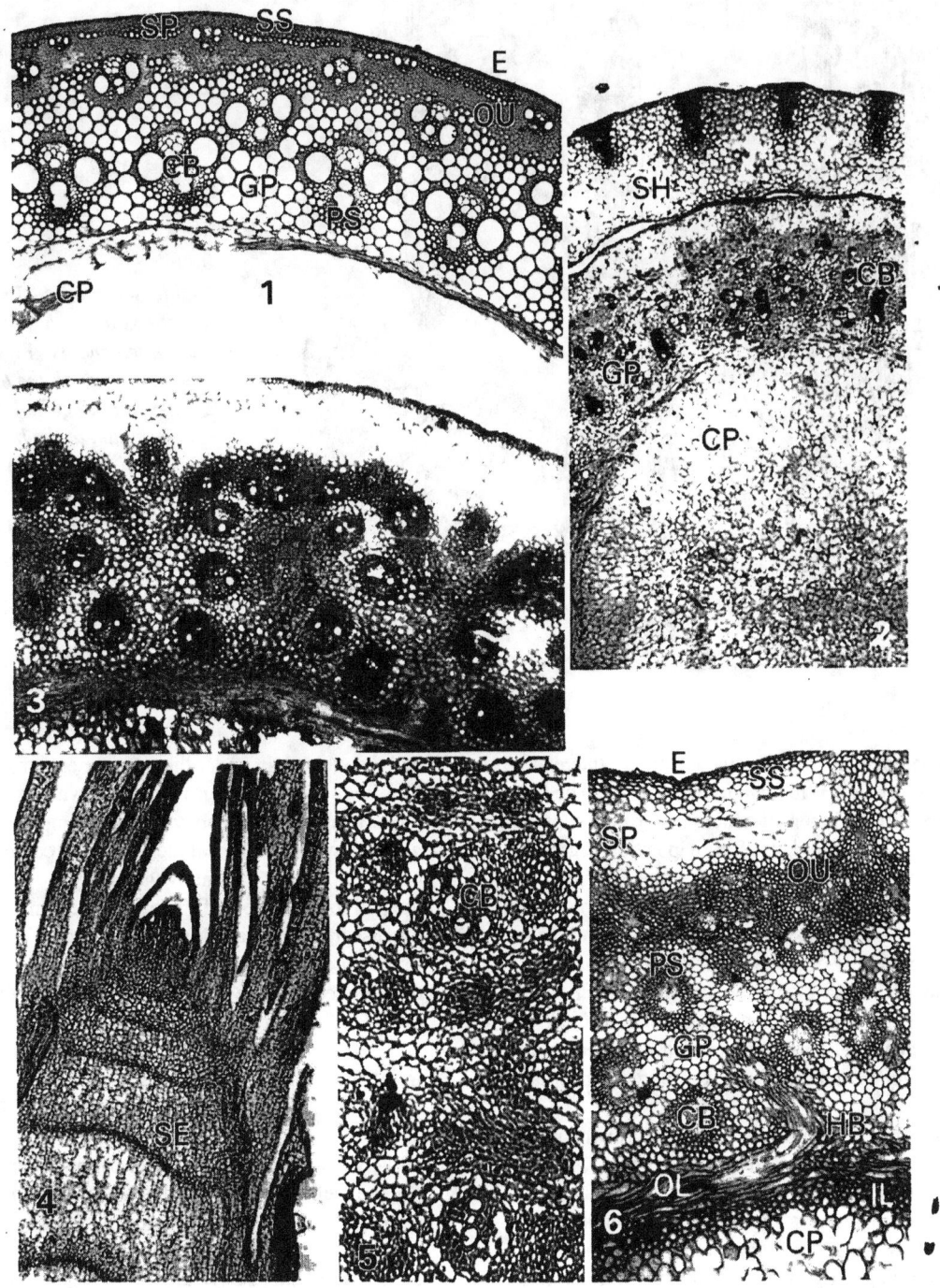

Figure 5-1 to 5-6.—*detailed structures of galled and ungalled shoots. Figure 5-1.—Transverse section through a full-grown stem of an ungalled shoot. Figure 5-2.—Transverse section through the top of a young gall. Figure 5-3.—small vascular bundles grouped in a sickle-shape. Figure 5-4.—Longitudinal section through the growing point of an uninfected shoot. Figure 5-5.—Detail of new forming vascular units a few millimeters under the gall apex. Figure 5-6.—Transverse section through a mature gall. CB, collateral vascular bundle; CP, central pith parenchyma; E, epidermis; GP, ground parenchyma; HB, horizontal connective vascular bundle; IL, inner layer special sclerenchyma; OL, outer layer special sclerenchyma; OU, outer sclerenchyma; PS, perivascular sclerenchyma; SE, septum; SH, leaf sheath; SP, subepidermal parenchyma; SS, subepidermal sclerenchyma.*

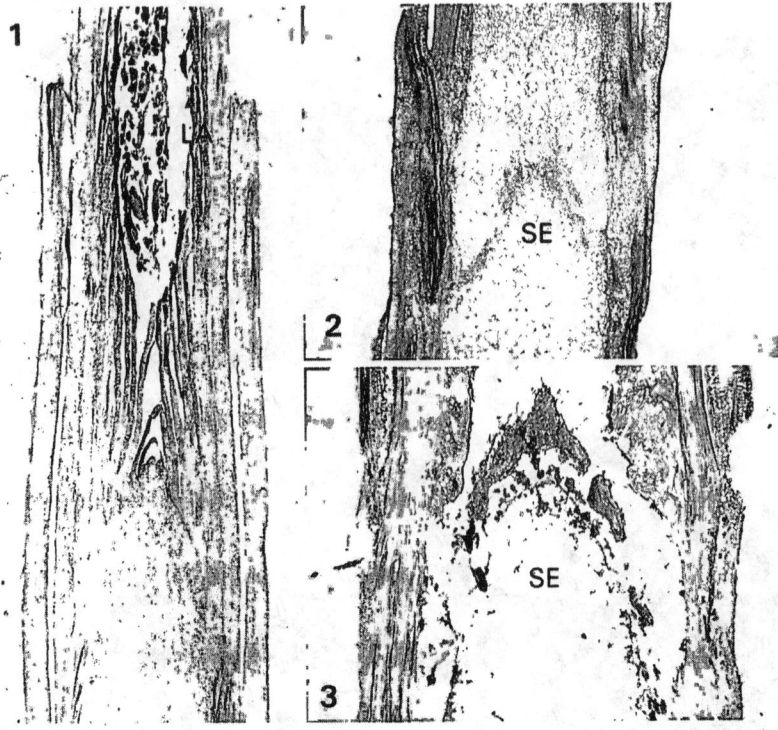

Figure 6-1 to 6-3.—*Transverse sections through galled and ungalled shoots.* Figure 6-1.—*Section through the top part of a developing gall.* Figure 6-2.—*septum at the bottom part of a gall.* Figure 6-3.—*Septum in a normally developed shoot. LA, larva feeding on unwrapped leaves; SE, septum.*

ground parenchyma to the larval chamber. Interestingly, the xylem is external relative to the phloem. The first indications of the inner sclerenchyma layers become visible after about 30 days after oviposition (fig. 5-2). The first cells of the outer layer become visible at the inner side of the ground parenchyma, in groups of small, and on cross-sections strongly lengthened parenchyma cells. The number of this outer layer-cell groups continuously increase until, about 10 days later, a complete band around the gall chamber is formed. Between this outer layer and the marrow parenchyma smaller parenchyma cells arise, forming the inner layer.

At the top, a few millimeters under the growing point, the vascular bundles start to shape (fig. 5-2). A number of vascular units are scattered in the tissue cylinder. These units consist of a central vascular bundle, surrounded by cells that start to differentiate into new vascular bundles (fig. 5-5). In a normal, uninfested stem these newly formed vascular bundles segregate and end up as singular bundles. In the gall, small vascular bundles can be grouped into a kind of sickle-shape, usually around a central, larger vascular bundle (fig. 5-3). The outer sclerenchymatous layer merges with the perivascular sclerenchyma. Between the vascular units run additional singular vascular bundles.

Gall differentiation continues for about 22 days. During this developmental period, the surrounding tissue cylinder is still quite soft. Only the bottom of the gall and the stem right under the gall is more strongly lignified. During gall growth, the larva was feeding on the enwrapped leaves above the growing point. Now it gnaws through the apical meristem and enters the gall chamber. This is externally visible by the browning of the gall top and top leaf. The sclerenchymatisation process in the tissue cylinder accelerates. The bands, immediately bordering the gall chamber. Lignification then proceeds to the ground parenchyma of the tissue cylinder, finally resulting in an extremely hardened gall.

In the mean time, the larva feeds on the nutritive cells of the central pith. This central pith tissue and the young enwrapped leaves above the growing point were quantitatively analyzed for water (wet-oven dry mass), protein (by the method of Markwell *et al.* 1978) and non structural carbohydrates (glucose, fructose, sucrose) and starch (by the method of Hendrix 1993) content in young developing galls, before the larva has entered the gall chamber. Water, protein, glucose, fructose and starch content were all significantly higher in the central pith tissue than in the enwrapped leaves of the corresponding gall (table 3). Compared to glucose and fructose however, the starch content was very low. The sucrose content was so low both in pith and leaves that it could not be quantified with the applied method. At the end of the feeding period the larval chamber is completely empty. At this time the larva turns the head capsule upwards and starts diapause.

Table 3.—*Nutritional composition of the parenchymatous pith tissue in the gall chamber and the enwrapped leaves above the growing point.*

Component	Enwrapped leaves		Parenchymatous pith		Paired-t	df	p
	Mean	± SE	Mean	± SE			
Water (mg/g wet mass)	882.47	± 6.06	933.52	± 4.36	9.167	24/23	<0.001
Protein (mg/g dry mass)	252.76	± 22.26	407.36	± 34.66	4.917	23/22	<0.001
Glucose (mg/g dry mass)	62.88	± 25.42	219.43	± 51.65	3.859	16/15	0.002
Fructose (mg/g dry mass)	39.87	± 6.23	199.06	± 41.42	3.612	15/14	0.003
Starch (mg/g dry mass)	5.46	± 0.80	18.24	± 3.77	2.545	15/14	0.023

When measured at the moment just before the larva enters the gall chamber, the central marrow channel in a gall (mean ± SE: 4805 ± 292 µm) is twice as wide as in a normal developing stem of the same age at the same height (2392 ± 166µm) (fig. 7). In the mean time, the surrounding tissue cylinder also increases its width up to 3.6 times (gall: 1303 ± 96µm; stem: 424 ±19µm). Overall gall width (7904 ± 513µm) is about 2.5 times larger than the width of a normally developed stem (3267 ± 192µm). Because the surrounding tissue cylinder expands faster than the central marrow, the marrow takes up 61 percent of the total gall width and 72 percent of a normal developed stem. The tissue cylinder takes up 39 and 28 percent, respectively, of the total width in a gall and a normal stem. In a full-grown gall, when the larva has eaten the gall chamber and is in diapause, the proportional differences between gall and stem disappear. Now the central marrow takes up 70 percent and the surrounding cylinder takes up 30 percent of the total width in both cases. This indicates that the tissue cylinder seemingly expands faster in the early stages of gall formation.

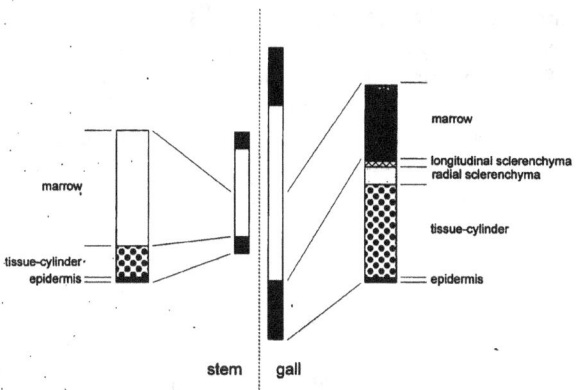

Figure 7.—*Diagrammatic representation of a transverse section through a L. lucens gall and a normally developed stem. Central = real size ratios, outside = different tissue layers.*

It is of interest that at the height of the nodes, mainly on the bottom part of the gall, a well defined group of parenchymatous cells, surrounded by vascular bundles, can develop. These will give rise to new side shoots.

THE INVOLVEMENT OF PLANT GROWTH SUBSTANCES

To understand how the gall inhabiting herbivores change the composition of the plant growth hormones, an initial survey was carried out of the auxin indole-3-acetic acid (IAA), abscissic acid (ABA), cytokinins and gibberellins content of the involved tissues. Both field collected and laboratory grown galls and uninfected shoots of different ages were analyzed. The material was obtained in the same was as for the morphological analyses. The plant material was transferred to liquid nitrogen immediately after harvesting. In the laboratory, it was stored in a freezer (-80°C) until processed. Plant hormones were extracted from the gall apical region (± 5 mM includes the growing point), the gall chamber (10-15 mm of the central zone) and the gall base, the zone directly under the gall chamber (± 15 mm). As a control, the same plant hormones were extracted from ungalled shoots of the same age and same diameter on the corresponding position on the shoot. Methods of phytohormone extraction, purification and quantification using liquid chromatography tandem mass spectrometry (LC-MS) and gas chromatography-coupled mass spectrometry (GC-MS) (Bialek and Cohen 1989; Pilet and Saugi 1985; Prinsen et al. 1995a, b) were evaluated and slightly adapted. Details are described in Jaminé (1996). Reliable differences between galled and ungalled *P. australis* shoots were obtained for IAA and ABA and are reported here. The results for cytokinins and gibberellins were too fragmentary to make interpretation possible.

The free IAA concentration in a *L. lucens* gall is clearly lower than in an ungalled shoot (fig. 8). In both treatments, the amount of free IAA decreases from the apical region to the base of the gall. Also conjugated IAA is predominantly present in the apical region of both the

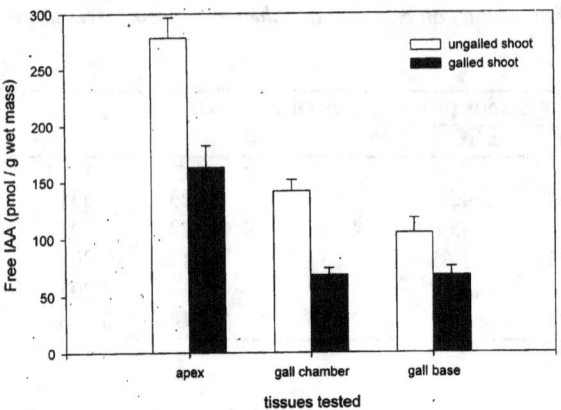

Figure 8.—*Endogenous concentrations of free IAA in the different constituents of galled (n > 5) and ungalled (n > 5) shoots. Data are means ± SE.*

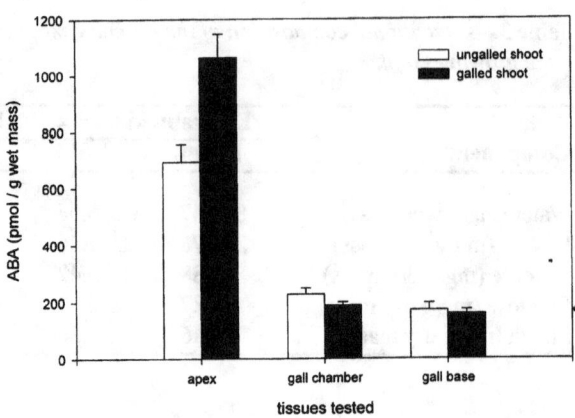

Figure 10.—*Endogenous concentrations of ABA in the different constituents of galled (n > 6) and ungalled (n > 7) shoots. Data are means ± SE.*

galled and ungalled shoots. In contrast to the free IAA concentration, the concentration of conjugated IAA is definitely higher (± 3 times) in the gall apical region than in the apical region on an uninfected shoot (fig. 9). This difference disappears lower in the gall. No difference is found between the two treatments at the height of the gall chamber and the gall base.

For ABA a comparable pattern is found as for conjugated IAA (fig. 10). Again the concentrations are higher in the apical region, with a significant excess in the galled shoots. Concentrations decrease lower in the gall. Because the ABA concentration drops faster in a gall than in an ungalled shoot, the difference disappears in the gall chamber and at the base of the gall.

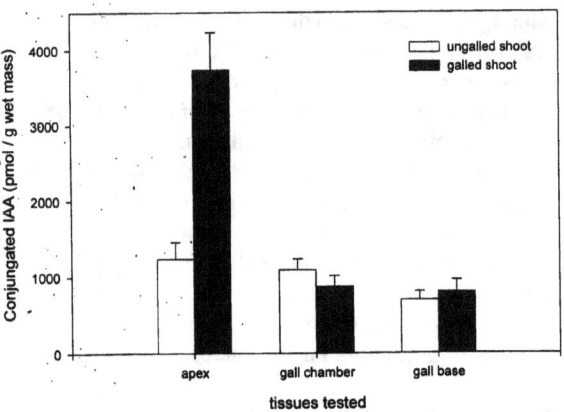

Figure 9.—*Endogenous concentrations of conjugated IAA in the different constituents of galled (n > 3) and ungalled (n > 5) shoots. Data are means ± SE.*

STAGES IN GALL DEVELOPMENT

In general, four phases can be distinguished during gall development: gall initiation, the growth phase, the maturation phase, and finally the dehiscence phase (Lalonde and Shorthouse 1984, Rohfritsch 1992). Gall initiation is the first and usually the most crucial step.

Gall Initiation

In internally feeding insects, the females have to select the host plant and carefully deposit their eggs at the right moment, at the correct spot on the host plant to insure that the young larva will develop successfully (e.g., Thompson 1988, Valladares and Lawton 1991, Weis *et al.* 1988). The insect takes over the control of plant organ development (Dawkins 1982, Weis and Abrahamson 1986). Cell conditioning which increases the sensitivity of the neighboring cells to the activities of the feeding herbivore larva should be a first step (Hori 1992). The stimulus which initiates development in the galls of *L. lucens* remains unknown. Four main hypotheses for gall initiation have been formulated in the past: (a) injection of a stimulant fluid during oviposition, (b) mechanical injury by a foreign body, (c) secretion of active saliva, or (d) excretion of metabolic products (Carter 1962).

In contrast with many other gall-forming insects where gall development starts due to adult feeding (i.e., mites, psyllids, thrips) and/or where the ovipositing female injects a specific stimulant fluid (i.e., sawflies, cynipids) (Leitch 1994, McCalla *et al.* 1962, Rohfritsch 1992), the female *L. lucens* plays no part in gall initiation. She only sticks the eggs to the outside of the reed shoot. Because gall formation only starts when the larva has reached the growing point, it is the latter that has to provide the necessary stimulus. The larvae of *L. lucens* have a primitive mode of feeding as in other cyclorrhaphous flies. They feed by slashing host tissues and ingesting the liquid cell contents. They can cause severe damage to the tissues. Whether this physical destruction is the primary cause for gall formation is not clear. However, other researchers working with other gallmaker systems have not supported the mechanical hypothesis and have suggested it to be a very improbable mechanism (Carter 1962, Hori 1992). Recent studies based on natural and artificial galls induced by insects or by injected insect extracts increasingly support the chemical hypotheses. The saliva of some investigated galling insects has proved to contain plant-growth substances or other active components that are able to change plant organ development (Hori 1992, Leitch 1994). A preliminary analysis of the phytohormone content of total homogenate of two field collected larvae showed the presence of IAA and some cytokinins and cytokinin precursors (unpubl. data). This may indicate that gall induction by *L. lucens* depends on chemical agents. However, more elaborate analyses will be necessary to confirm or disprove this hypothesis.

The Gall Growth Phase

During the growth phase the biomass of the gall increases by cellular hyperplasy and hypertrophy. In general this is characterized by rapidly dividing parenchyma cells around the larvae, and the development of vascular strands connecting the gall with the rest of the host plant (Bronner 1977, Meyer 1969, Shorthouse 1986).

The most striking external visible change occurring in the *L. lucens* galls is the strongly reduced internode elongation and increased widening which gives rise to the typical cigar-like shape of the gall. In addition, this internode shortening halves the full shoot length. The morphological changes are accompanied by a change in the concentrations of free and conjugated IAA (figs. 8 and 9). Free IAA is the principal active auxin in most plants and is mainly present in leaf primordia, young leaves and developing seeds. It plays an important role in a number of physiological and morphogenetic processes in plants such as the stimulation of cell elongation and stem growth, suppression of side shoots (apical dominance) and stimulation of cell division (Davies 1995). Possible functions of conjugated IAA are IAA transport, to protect IAA against peroxidase activity or simply to serve as a source of free IAA (Bandurski *et al.* 1995, Cohen and Bandurski 1978, Hangarter and Good 1981). A possible explanation for the drop in free IAA and an increase in conjugated IAA at the gall apical region is the (partial) destruction of an auxin source by the larva which feeds above the growing point. In this case however we expect a more even change in the free and conjugated IAA concentrations. Another possible explanation is that the feeding larva secretes certain active chemicals which alter the IAA concentrations, i.e., through conjugation. Several authors have indeed already reported on the secretion of phytohormones in other gall forming insects (e.g., Dieleman 1969, Hori 1975, Matsui *et al.* 1975, Maxwell *et al.* 1962, Ohkawa 1974). Moreover, the reduction in free IAA in the apical region might induce the reduction in apical dominance shown in *Lipara* galls.

A second major change occurs inside the gall. The central marrow cavity is completely filled with the dense parenchymatous tissue. The latter is rich in nutritional components. The amount of water, soluble carbohydrates and proteins is significantly higher than in the enwrapped leaves above the growing point. In the past, two possible hypotheses were formulated to explain this nutritive enrichment. The first postulates that the gall works as a nutrient block, filtering all the necessary components from a passing nutrient stream (Whitham 1978). According to the second hypothesis, a gall can act as a nutrient sink, actively attracting nutrients which would otherwise be directed to other plant modules (Harris 1980, Larson and Whitham 1991, McCrea *et al.* 1985, Weis and Kapelinski 1984). The localization at the top of a reed stem, right under the growing point, makes both hypotheses reasonable. The gall can intercept the nutrient flow to the growing point, and/or attract products from other parts of the plant.

Our results show no difference in total biomass between galled and ungalled shoots (table 2). On the other hand, 34 percent of the galled shoot biomass is allocated to the gall (fig. 4). Although the remaining part of the stem under the gall is 14 percent lighter, the total stem biomass is still 20 percent heavier in galled than ungalled shoots. As a consequence, gall weight mainly increases at the cost of biomass allocation to the leaves (-17 percent). The reduction of biomass allocation to leaves in turn can be explained by reduced leaf production at the height of the gall. A gall is composed of 9-12 internodes. Under normal conditions, these internodes elongate and carry a leaf. On the gall, on average only about five leaves develop (table 1). These data suggest that the *L. lucens* gall acts as a nutrient block, accumulating nutrients normally destined for shoot growth and leaf expansion.

During gall development, we measured an increased ABA concentration in the apical region of the galled shoot (fig. 10). ABA accumulation commonly occurs in response to tissue damage and wounding (Peña-Cortés and Willmitzer 1995). Probably the increased ABA concentration in the gall apical region is correlated with the mechanical stress caused by the feeding activity of the young larva in the near vicinity of the apex. We also want to mention that ABA inhibits stem growth (Davies 1995).

All major changes in gall morphology occur while the young *L. lucens* larva is feeding on the enwrapped leaves above the growing point, outside the actual developing gall. Nevertheless, the presence of the feeding larva is crucial. When it dies prematurely, normal reed shoot development resumes (fig. 1-3). Only traces of the old gall, some shortened internodes, remain visible. If infestation occurs late in spring and the flower primordia have established, a normal ear develops later in the season (Reijnvaan and Docters van Leeuwen, 1906).

The Gall Maturation and Dehiscence Phase

The maturation phase starts when the *L. lucens* larva gnaws through the growing point and enters the gall chamber. The insect enters the last larval stage. This is normally the period in the life of the herbivore in which it has to consume large amounts of food (Strong *et al.* 1984). The *L. lucens* larva now starts to feed on the mass of rich nutritive tissues that has accumulated in the gall chamber. The sclerenchymatisation process in the tissue cylinder accelerates. The first cells to lignify are those of the inner and the outer layer of the specially formed bands, immediately bordering the gall chamber. Lignification then proceeds to the ground parenchyma of the tissue cylinder, finally resulting in an extremely hardened gall.

In most galls, gall dehiscence or opening occurs towards the end of the maturation period (Rohfritsch 1992). This ripening phase facilitates the escape of the inhabiting larvae. The flow of sap and water to the gall stops. For *L. lucens* the gall gradually turns brown. No special escape channels are formed. The larva pupates in the gall chamber and will leave the gall in spring by crawling up through the enwrapped leaves on the top of the gall.

Gall structure and development in *L. lucens* galls are rather simple, and follow the main lines of normal shoot development. The gall can more or less be described as a strongly concentrated stem with reduced leaf growth. Only slight deviations occur: the central marrow cavity is completely filled with a dense, nutrient-rich parenchymatous tissue, and on the inner side of the surrounding tissue cylinder a special, double sclerenchymatous band arises.

Our data also suggest that plant hormones are involved in gall initiation and development. The data presented here however, do not provide us with a definite answer. IAA and ABA may be involved in altering growth patterns in the reed shoot, although other active plant growth substances may well also be involved.

ACKNOWLEDGMENTS

We want to thank F. Neefs for help with the sectioning and J.P. Timmermans for helpful advice with the histochemical analysis and for providing ultratome sections. This study was supported by grants of the "Rosa Blanckaert" foundation and of the Fund for Scientific Research - Flanders (Belgium) (F.W.O.-2.0128.84) to LDB. EP is a research associate of the Fund for Scientific Research - Flanders (Belgium).

LITERATURE CITED

Armstrong, J.; Armstrong, W. 1988. *Phragmites australis*—a preliminary study of soil-oxidizing sites and internal gas transport pathways. New Phytologist. 108: 373-382.

Bandurski, R.S.; Cohen, J.D.; Slovin, J.; Reinecke, D.M. 1995. Auxin biosynthesis and metabolism. In: Davies, P.J., ed. Plant hormones: physiology, biochemistry and molecular biology. Dordrecht, The Netherlands: Kluwer Academic Publishers: 39-65.

Bernays, E.A.; Barbehenn, R. 1987. Nutritional ecology of grass foliage-chewing insects. In: Slansky, F.; Rodriguez, J.G., eds. Nutritional ecology of insects, mites, spiders, and related invertebrates. New York, NY: John Wiley and Sons: 147-175.

Beschovski, V.L. 1984. A zoogeographic review of endemic Palaearctic genera of Chloropidae (Diptera) in view of origin and formation. Acta Zoologica Bulgarica. 24: 3-26.

Bialek, K.; Cohen, J.D. 1989. Quantification of indoleacetic acid conjugates in bean seeds by direct tissue hydrolysis. Plant Physiology. 90: 398-400.

Björk, S. 1967. Ecological investigations of *Phragmites communis*. Studies in theoretic and applied limnology. Folia Limnologica Scandinavica. 14: 1-248.

Bronner, R. 1977. Contribution a l'étude histochemique des tissus nourriciers des zoocécidies. Marcellia. 40: 1-134.

Buhr, H. 1965. Bestimmungstabellen der Gallen (Zoo- und Phytocecidien) an Pflanzen Mittel- und Nordeuropas. VEB Gustaf Fischer Verlag, Jena.. 2 vols.

Byers, J.A.; Brewer, J.W.; Denna, D.W. 1976. Plant growth hormones in pinyon insect galls. Marcellia. 39: 125-134.

Carter, W. 1962. Insects in relation to plant disease. New York, NY: Interscience.

Chvála, M.; Dosko il, J.; Mook, J.H.; Pokorny, V. 1974. The genus *Lipara* Meigen (Diptera, Chloropidae), systematics, morphology, behaviour and ecology. Tijdschrift voor Entomologie. 117: 1-25.

Cohen, J.D.; Bandurski, R.S. 1978. The bound auxins: protection of indole-3-acetic acid from peroxidase-catalysed oxidation. Planta. 139: 203-208.

Davies, P.J. 1995. The plant hormones: their nature, occurrence, and function. In: Davies, P.J., ed. Plant hormones: physiology, biochemistry and molecular biology. Dordrecht, The Netherlands: Kluwer Academic Publishers: 1-12.

Dawkins, R. 1982. The extended phenotype: the gene as the unit of selection. Oxford: Oxford University Press.

De Bruyn, L. 1985. The flies living in *Lipara* galls (Diptera: Chloropidae) on *Phragmites australis* (Cav.) Trin. ex Steud. Bulletin et Annales de la Société royale belge d'Entomologie. 121: 485-488.

De Bruyn, L. 1987. Habitat utilisation of three West-European *Lipara* species (Diptera Chloropidae), a pest of the Common Reed, *Phragmites australis*. Mededelingen van de Faculteit Landbouwwetenschappen, Rijksuniversiteit Gent. 52(2a): 267-271.

De Bruyn, L. 1988. *Lipara similis* (Diptera, Chloropidae), a fourth *Lipara* species for the Belgian fauna, with notes on its habitat selection. Bulletin et Annales de la Société royale belge d'Entomologie. 124: 317-320.

De Bruyn, L. 1994a. Life history strategies of three gall-forming flies tied to natural variation in growth of *Phragmites australis*. In: Price, P.W.; Mattson, W.J.; Baranchikov, N., eds. The ecology and evolution of gall-forming insects. Gen. Tech. Rep. NC-174. St. Paul, MN: U.S. Department of Agriculture, Forest Service, North Central Forest Experiment Station: 56-72.

De Bruyn, L. 1994b. Life cycle strategies in a guild of dipteran gall formers on the common reed. In: Williams, M.A.J., ed. Plant galls: organisms interactions populations. Oxford: Systematics Association, Clarendon Press. 49: 259-281.

De Bruyn, L. 1995. Plant stress and larval performance of a dipterous gall former. Oecologia (Berlin). 101: 461-466.

De Bruyn, L. 1996. Habitat preference and community structure in the genus *Lipara* (Diptera, Chloropidae). In: Walters, K.F.A.; Kidd, N.A.C., eds. Populations and patterns in Biology, I PPP, Silwood Park, UK: 32-37.

Dieleman, F.L. 1969. Effects of gall midge infestation on plant growth and growth regulating substances. Entomologica Experimentalis et Applicata. 12: 745-749.

Dreger-Jauffret, F.; Shorthouse, J.D. 1992. Diversity of gall-inducing insects and their galls. In: Shorthouse, J.D.; Rohfritsch, O., eds. Biology of insect-induced galls. New York: Oxford University Press: 8-33.

Fay, P.A.; Hartnett, D.C. 1991. Constraints on growth and allocation patterns of *Silphium integrifolium* (Asteraceae) caused by a cynipid gall wasp. Oecologia (Berlin). 88(2): 243-250.

Hangarter, R.P.; Good, N.E. 1981. Evidence that IAA conjugates are slow-release sources of free IAA in plant tissues. Plant Physiology. 68: 1424-427.

Harris, P. 1980. Effects of Urophora affinis Frfld. and U. quadrifasciata (Meig.)(Diptera: Tephritidae on Centaurea diffusa Lam. and C. maculosa Lam. (Compositae). Zeitschrift für Angewandte Entomologie. 90: 190-201.

Hartnett, D.C.; Abrahamson, W.G. 1979. The effects of stem gall insects on life history patterns in *Solidago canadensis*. Ecology. 60(5): 910-917.

Haslam, S.M. 1969. The development of shoots in *Phragmites communis* Trin. Annals of Botany. 33: 695-709.

Hendrix, D.L. 1993. Rapid extraction and analysis of nonstructural carbohydrates in plant tissues. Crop Science. 33: 1306-1311.

Hori, K. 1975. Pectinase and plant growth-promoting factors in the salivary glands of the larva of the bug *Lygus disponsi*. Journal of Insect Physiology. 21: 1271-1274.

Hori, K. 1992. Insect secretions and their effect on plant growth, with special reference to hemipterans. In: Shorthouse, J.D.; Rohfritsch, O., eds. Biology of insect-induced galls. New York: Oxford University Press: 157-170.

Inbar, M.; Eshel, A.; Wool, D. 1995. Interspecific competition among phloem-feeding insects mediated by induced host-plant sinks. Ecology. 76(5): 1506-1515.

Jaminé, D. 1996. Gall induction and gall development in the fly genus *Lipara* on reed (in Dutch). Licenthiat thesis, University of Antwerp.

Lalonde, R.G.; Shorthouse, J.D. 1984. Developmental morphology of the gall of *Urophora cardui* (Diptera: Tephritidae) in stems of Canada thistle (*Cirsium arvense*). Canadian Journal of Botany. 62: 1372-1384.

Larson, V.C.; Whitham, T.G. 1991. Manipulation of food resources by a gall-forming aphid: the physiology of sink-source interactions. Oecologia (Berlin). 88: 15-21.

Leitch, I.J. 1994. Induction and development of the bean gall caused by *Pontania proxima*. In: Williams, M.A.J., ed. Plant galls: organisms, interactions, populations. Oxford: Clarendon Press: 283-300.

Mani, M.S. 1964. Ecology of plant galls. Dr. W. Junk, Den Haag.

Markwell, M.K.; Haas, S.M.; Bieber, L.L.; Tolbert, N.E. 1978. A modification of the Lowry procedure to simplify protein determination in membrane and lipoprotein samples. Analytical Biochemistry. 87: 206-210.

Matsui, S.; Torikata, H.; Munakata, K. 1975. Studies on the resistance of chestnut trees (*Castanea* spp.) to chestnut gall wasps (*Dryocosmus kuriphilus* Yasumatsu). V. Cytokinin activity in larvae of gall wasps and callus formation of chestnut stem sections by larval extracts. Japanese Society of Horticultural Sciences. 43: 415-422.

Maxwell, F.G.; Painter, R.H. 1962. Plant growth hormones in ether extracts of the greenbug, *Toxoptera graminum*, and the pea aphid, *Macrosiphum pisi*, fed on selected tolerant and susceptible host plants. Journal of Economic Entomology. 55: 57-62.

McCalla, D.R.; Genthe, M.K.; Hovanitz, W. 1962. Chemical nature of an insect gall growth factor. Plant Physiology. 37: 98-103.

McCrea, K.D.; Abrahamson, W.G.; Weis, A.E. 1985. Goldenrod ball gall effects on *Solidago altissima*: 14C translocation and growth. Ecology. 66(6): 1902-1907.

Meyer, J. 1969. Irrigation vasculaire dans les galles. Bulletin de la Société Botanique de France. 75-97.

Mook, J.H. 1967. Habitat selection by *Lipara lucens* Mg. (Diptera, Chloropidae) and its survival value. Archives néerlandaises de Zoologie. 17: 469-549.

Ohkawa, M. 1974. Isolation of Zeatin from larvae of *Dryocosmus kuriphilus* Yasumatsu. HortScience. 9: 458-459.

Peña-Cortés, H.; Willmitzer, L. 1995. The role of hormones in gene activation in response to wounding. In: Davies, P.J., ed. Plant hormones: physiology, biochemistry and molecular biology. Dordrecht, The Netherlands: Kluwer Academic Publishers: 395-414.

Pilet and Saugi. 1985. Effect of applied and endogenous indole-3-acetic acid on maize root growth. Planta. 164: 254-258.

Prinsen, E.; Redig, P.; Strnad, M.; Galis, I.; Van Dongen, W.; Van Onckelen, H. 1995a. Quantifying phytohormones in transformed plants. In: Gartland, K.M.A.; Davey, M.R., eds. Methods in Molecular Biology. 44. Humana Press, Totowa: 245-262.

Prinsen, E.; Redig, P.; Van Dongen, W.; Esmans, E.; Van Onckelen, H. 1995b. Quantitative analysis of cytokinins by electrospray tandem mass spectrometry. Rapid Communications in Mass Spectrometry. 9: 948-953.

Reijnvaan, J.; Docters van Leeuwen, W.M. 1906. Die entwicklung der galle von *Lipara lucens*. Recueil de Travaux Botanique Néerlandaises. 12: 235-261.

Rodewald-Rudescu, L. 1974. Das Schilfrohr. Die Binnengewässer, E.Schweizerbart'sche Verlagbuchhandlung, Stuttgart.

Rohfritsch, O. 1992. Patterns in gall development. In: Shorthouse, J.D.; Rohfritsch, O., eds. Biology of insect-induced galls. New York: Oxford University Press: 60-86.

Ruppolt, W. 1957. Zur Biologie der Cecidogenen Diptere *Lipara lucens* Meigen (Chloropidae). Wissenschaftliche Zeitschrift der Ernst-Moritz-Arndt-Universität Greifswald. 6: 279-291.

Shorthouse, J.D. 1986. Significance of nutritive cells in insect galls. Proceedings of the Entomological Society of Washington. 88: 368-375.

Shorthouse, J.D.; Lalonde, R.G. 1988. Role of *Urophora cardui* (L.) (Diptera, Tephritidae) in growth and development of its gall on stems of Canada Thistle. The Canadian Entomologist. 120(7): 639-646.

Skuhravy, V. 1978. Invertebrates: destroyer of common reed. In: Dykyjová, D.; Kvet, J., eds. Ecological Studies. 28: 376-388.

Skuhravy, V. 1981. Invertebrates and vertebrates attacking common reed stands *(Phragmites communis)* in Czechoslovakia. Ceskoslovenská Akademie Ved., Praha.

Strong, D.R.; Lawton, J.A.; Southwood, T.R.E. 1984. Insects on plants, community patterns and mechanisms. Oxford: Blackwell Scientific Publications.

Strong D.R., Jr.; Levin, D.A. 1979. Species richness of plant parasites and growth form of their hosts. The American Naturalist. 114: 1-22.

Thompson, J.N. 1988. Evolutionary ecology of the relationship between oviposition preference and performance of offspring in phytophagous insects. Entomologica Experimentalis et Applicata. 47(1): 3-14.

Tscharntke, T. 1994. Tritrophic interactions in gallmaker communities on *Phragmites australis*: testing ecological hypothesis. In: Price, P.W.; Mattson, W.J.; Baranchikov, N., eds. The ecology and evolution of gall-forming insects. Gen. Tech. Rep. NC-174. St. Paul, MN: U.S. Department of Agriculture, Forest Service, North Central Forest Experiment Station: 73-92.

Valladares, G.; Lawton, J.H. 1991. Host-plant selection in the holly leaf-miner: does mother know best? Journal of Animal Ecology. 60(1): 227-240.

Van der Toorn, J. 1972. Variability of *Phragmites australis* (Cav) Trin ex Steudel in relation to the environment. Van Zee tot Land. 48: 1-122.

Vandevyvere, I. 1995. Morphological and chemical analyses of *Lipara lucens* (Diptera, Chloropidae) galls on *Phragmites australis* (in Dutch). Licenthiat thesis, University of Antwerp.

Van Staden, J.; Bennett, P.H. 1991. Gall formation in crofton weed: differences between normal stem tissue and gall tissue with respect to cytokinin levels and requirements for in vitro culture. Suider-Afrikaanse Tydskrift vir Plantkunde. 57(5): 246-248.

Weis, A.E.; Abrahamson, W.G. 1986. Evolution of host-plant manipulation by gall-makers: ecological and genetic factors in the *Solidago-Eurosta* system. The American Naturalist. 127: 681-695.

Weis, A.E.; Kapelinski, A. 1984. Manipulation of host plant development by the gall-midge *Rhabdophaga strobiloides*. Ecological Entomology. 9: 457-465.

Weis, A.E.; Walton, R.; Crego, C.L. 1988. Reactive plant tissue sites and the population biology of gall makers. Annual Review of Entomology. 33: 467-486.

Whitham, T.G. 1978. Habitat selection by *Pemphigus* aphids in response to resource limitation and competition. Ecology. 59(6): 1164-1176.

Wood, B.W.; Payne, J.A. 1988. Growth regulators in chestnut shoot galls infected with oriental chestnut gall wasp, *Dryocosmus kuriphilus* (Hymenoptera, Cynipidae). Environmental Entomology. 17(6): 915-920.

ANALYSIS OF PIGMENT-PROTEIN COMPLEXES IN TWO CECIDOMYIID GALLS

Chi-Ming Yang[1], Ming-Horn Yin[1], Kuo-Wei Chang[1],
Ching-Jeng Tsai[1], Shu-Men Huang[1] and Man-Miao Yang[2]

[1]Associate Research Fellow and Research Assistants of the Institute of Botany, Academia Sinica, Nankang, Taipei, Taiwan 115, Republic of China
[2]Associate Curator of Entomology, Division of Zoology, National Museum of Natural Science, Taichung, Taiwan 404, Republic of China

Abstract.—We examined the content of chlorophyll, carotenoid, flavonoid, anthocyanin and tannin, chlorophyll biosynthetic capacity, chlorophyll degradation products, and pigment-protein complexes in the chloroplast of two cecidomyiid galls, both infecting the leaves of *Machilus thunbergii* Sieb & Zucc. (Lauraceae). The galling insects drastically increase the content of anthocyanin and tannin, and the mole percentage of protoporphyrin IX, but decrease the content of flavonoid, chlorophyll, and carotenoid, the mole percentage of protochlorophyllide, and chlorophyll degradation products of infected leaves. A unique pigment-protein complex pattern was discovered in the thylakoid of the two insect galls. While the infected leaves possess CP1 and CP2 pigment-protein complexes fractionated by the Thornber system, or A1, AB1, AB2, AB3 pigment-protein complexes fractionated by the MARS system, the two insect galls (infected area of leaf) contain just CP2 or AB3. Electron microscopy demonstrated that the infected leaf has spindle-shaped chloroplast with abnormal grana thylakoid. It is still unknown how the cecidomyiid insects cause the deficiency of some pigment-protein complexes.

INTRODUCTION

Although all plant organs are subject to insect galling, about 80 percent of insect galls are on leaves (Dreger-Jauffret and Shorthouse 1992). While much attention has been focused on the morphology and anatomy of insect galls in the past (Dreger-Jauffret and Shorthouse 1992, Meyer 1987, Williams 1994), relatively little work has been done on the chloroplast of infected leaves. The limited information available about gall chloroplasts mostly concerns the agrana thylakoid membrane distributed in the stroma (Rey 1973, 1974, 1992). It seems that no researcher has studied the biochemical features of the agrana thylakoid membrane in the gall chloroplast. In this study, we therefore analyze the biochemical composition of pigment-protein complexes of the thylakoid membrane isolated from two cecidomyiid gall chloroplasts of *M. thunbergii* Sieb & Zucc. leaves. A unique pattern of pigment-protein complex different from normal chloroplast was discovered in the gall chloroplast.

MATERIALS AND METHODS

Plant Galls

Two galls on the leaves of *Machilus thunbergii* (Lauraceae) were collected from Chung-cheng mountain of Yang-Ming Shan National Park, in northern Taiwan. Mature galls were sampled from the infected mature leaf and surrounding healthy leaf tissue was trimmed off from the galls to avoid contamination.

Pigment Analysis

Following extraction of liquid-nitrogen frozen mature leaf with 80 percent acetone, the concentrations of chlorophyll, carotenoid, and three porphyrins (protoporphyrin IX, Mg-protoporphyrin IX, and protochlorophyllide) were determined according to the methods of Porra et al. (1989), Jasper (1965), and Kahn et al. (1976), respectively. Chlorophyllide and pheophytin were determined as described by Holden (1961). Absorbance, room temperature absorption spectra, and the first derivative spectra of pigment-protein complexes were obtained with a Hitachi U2000 UV-visible spectrophotometer.

Secondary Metabolite Analysis

The content of anthocyanin, flavonoid, and tannin were determined as previously described by Geisman (1955) and Hagerman and Butler (1978), respectively. Their contents are expressed in terms of optical density per gram of fresh leaf.

Pigment-Protein Complexes

Thylakoid membranes isolated from both leaf and detached galls were analyzed for constituent pigment-protein complexes by solubilization with SDS and electrophoresis on Thornber and MARS fractionation gel systems. Pigment-protein complexes were excised from the gel and their absorption spectra in the gel slices were determined (Markwell 1986).

Electron Microscopy

The outer part of galls and central part of the leaf were collected and cut into small cubes in fixation buffer containing 2.5 percent glutaraldehyde. After incubation at 4°C for 2 h in 0.1 M cacodylate buffer (pH 7.0) containing 2.5 percent glutaraldehyde, the samples were washed three times in plain buffer, postfixed in 1 percent osmium tetraoxide for 2 h, dehydrated through an ethanol series, infiltrated and embedded in Spurr's resin (Spurr 1969), and then polymerized at 70°C for 8 h. Gold sections were collected by ultramicroscopy and stained with ethanol uranyl acetate and lead citrate. The thylakoid morphology was examined with a Philips CM 100 transmission electron microscope at 75kV.

RESULTS AND DISCUSSION

Photosynthetic Pigment

Compared with normal leaf, all pigments relevant to chlorophyll biosynthesis in the gall, such as chlorophyll, protoporphyrin IX, Mg-protoporphyrin IX, and protochlorophyllide drastically decreased 10–50 fold.

Chlorophyll a/b ratios suggest that the two galls synthesize relatively more chlorophyll b than chlorophyll a in the late stage, therefore causing the decrease of the chlorophyll a/b ratio. Oppositely, even though both chlorophyll and carotenoid content decreased, the carotenoid/chlorophyll ratio suggests that the two galls synthesize relatively more carotenoid than chlorophyll in the late stage. While the mole percentage of total porphyrins (protoporphyrin IX + Mg-protoporphyrin IX + protochlorophyllide) extracted from M. thunbergii leaf is as normal as other plants, vis., 49, 33, and 18 percent, respectively, those of the two galls are about 64, 31, and 5 percent, respectively. The mole percentage of total porphyrins implies that the chlorophyll biosynthetic capacity of the gall is much different from that of the healthy leaf. It seems that most of the protochlorophyllide synthesized in the two galls is very quickly transformed into chlorophyll, causing the high percentage of protoporphyrin IX and low percentage of protochlorophyllide. Relatively, the two insect galls contain 3-12 percent of the carotenoid of the surrounding leaf (tables 1 and 2). Pheophytin and chlorophyllide are the catalytic products of Mg-dechelatase and chlorophyllase, respectively, using chlorophyll as substrate. The chlorophyll degradation products in the two galls decrease, as do the chlorophyll biosynthetic intermediates (table 3).

Secondary Metabolites

Many secondary metabolites are involved in the defense system of plants (Harborne 1988, Mazza and Miniati 1993). Mature leaves of M. thunbergii contain high amounts of flavonoid, no anthocyanin, and very low amounts of tannin. However, the two insect galls contain much less flavonoid and much more anthocyanin and tannin (table 4). The (anthocyanin+tannin)/flavonoid ratio of mature leaves is 0.007, whereas those of the two insect galls is 0.286 and 0.270, respectively. The data suggest that the two galls may induce the infected leaf to synthesize less photosynthetic pigment to save energy in order to produce more compounds involved with the plant defense system for protecting the insects in the gall.

Table 1.—*Chlorophyll and carotenoid content of two cecidomyiid galls and the leaf of M. thunbergii*[1]

Galls	Chlorophyll (μg/g)	Chlorophyll a/b ratio	Carotenoid (μg/g)	Carotenoid/chlorophyll ratio
Leaf	2977 (100)	2.78	442 (100)	0.148
Oval-pointed gall	60 (2.0)	2.20	16 (3.6)	0.258
Obovate gall	197 (6.6)	2.24	53 (12.0)	0.271

[1] The results were the average of three determinations and standard deviation is not shown. Number in parenthesis is the relative percentage.

Table 2.—*Porphyrin and mole percentage of porphyrin of two cecidomyiid galls and the leaf of* M. thunbergii[1]

Galls	(nmol/g)	Mole Percentage of Porphyrin (%)		
		Protoporphyrin IX	Mg-protoporphyrin IX	Protochlorophyllide
Leaf	3695 (100)	49.1	33.4	17.5
Oval-pointed gall	111 (3.0)	64.0	31.6	4.4
Obovate gall	366 (9.9)	64.6	29.4	6.0

[1] The results were the average of three determinations and standard deviation is not shown. Number in parenthesis is the relative percentage.

Table 3.—*Chlorophyll degradation products of two cecidomyiid galls and the leaf of* M. thunbergii[1]

Galls	Pheophytin a+b (μg/g)	Chlorophyllide a+b (μg/g)
Leaf	3336 (100)	80.8 (100)
Oval-pointed gall	74 (2.2)	4.8 (5.9)
Obovate gall	232 (7.0)	24.0 (29.7)

[1] The results were the average of three determinations and standard deviation is not shown. Number in parenthesis is the relative percentage.

Table 4.—*Second metabolites of two cecidomyiid galls and the leaf of* M. thunbergii[1]

Galls	Flavonoid (A_{540}/g)	Anthocyanin [(A_{530}−0.333A_{657})/g]	Tannin (A_{510}/g)	(Anthocyanin+tannin) /flavonoid ratio
Leaf	40.8 (100)	0	0.20 (100)	0.007
Oval-pointed gall	15.0 (36.8)	2.56 (∞)	1.74 (870)	0.286
Obovate gall	7.2 (17.5)	0.92 (∞)	1.01 (505)	0.270

[1] The results were the average of three determinations and standard deviation is not shown. Number in parenthesis is the relative percentage.

Pigment-Protein Complexes

The results of analysis of pigment-protein complexes of thylakoid membranes is consistent with the absorption spectra (data not shown). The Thornber electrophoretic system shows that the leaf of *M. thunbergii* contains both the CPI and CPII pigment-protein complexes commonly found in higher plants, whereas the two insect galls contain only CPII. CPI contains only chlorophyll a and is derived from PSI, whereas CPII contains both chlorophyll a and b and is derived from PSII. By using the MARS electrophoretic system, only one (AB3) of the three chlorophyll b-containing pigment-protein complexes (AB1, AB2, and AB3) present in the normal thylakoid membranes of leaf is detectable in the insect gall, which is also deficient in A1. The pigment-protein complex pattern was further confirmed by western blotting of LHCII apoproteins. The pigment-protein complex pattern of the insect gall is the same as that of mungbean testa, not found in any normal chloroplast (Yang et al. 1995). SDS-PAGE loaded with equal amount of chlorophyll demonstrated that the two insect galls contain almost the same high amount of LHCII apoproteins (LHCIIa, b, and c), CP1, CP43, and CP47. It is widely accepted that the LHCII complex regulates the formation of grana in the thylakoid membrane (Allen 1992, Bennett 1991); that is, no normal grana are formed if no LHCII complex is assembled in the chloroplast. However, the following electron microscopy showed that this is not the case.

Thylakoid Morphology

Ultrastructural studies showed that the chloroplast of *M. thunbergii* leaf has normal grana and is the same as that of other higher plants, whereas the chloroplast of the two insect galls is agrana or has abnormal grana with few layers of thylakoid membranes, similar to that of the chlorophyll-deficient mutants of higher plants (fig. 1). Rey (1973, 1974, 1992) reported that chloroplasts in galls of *Pontania proxima* on infected willow leaves never contain starch, but a bundle of tubules appears in their stroma which is very often isolated in a stretched lobe. Many chlorophyll-deficient mutants have been reported in barley, pea, maize, wheat, sweetclover, rice, soybean, sugar beet, *Arabidopsis thaliana*, *Chlamydomonas*, and other plants (Yang *et al.* 1993). Except for three cases (Nakatani and Baliga 1985, Quijja *et al.* 1988, Yang and Chen 1996), all chlorophyll-deficient mutants possess abnormal thylakoid morphology. As described above, the two insect galls have abundant LHCII apoproteins. However, their chloroplasts contain no grana or abnormal grana. Therefore, factors other than LHCII may be involved in the grana stacking.

CONCLUSION

In this report, we have examined several biochemical characteristics of two insect galls collected from the leaf of *M. thunbergii* and have shown that (1) the insect galls lack pigment-protein complexes CPI and are totally deficient in pigment-protein complexes A1, AB1, and AB2; (2) the insect galls are unable to operate the chlorophyll biosynthetic machine as fast and efficiently as the leaf does; (3) the insect galls may induce the infected leaf to produce more anthocyanin and tannin to protect the galling insects; (4) the insect galls contain normal amounts of LHCII complex, but contain agrana or abnormal thylakoid membranes; and (5) it should be carefully re-evaluated that LHCII is the only factor in regulation of grana stacking in higher plant chloroplasts.

It is still unknown (1) how commonly the deficiency phenomena of pigment-protein complexes occur in other insect galls; (2) how the galling insects shut down or slow down the chlorophyll biosynthetic machine; (3) how the galling insects induce the absence of some pigment-protein complexes; (4) how the galling insects trigger or initiate the synthesis of anthocyanin and tannin; and (5) what the physiology of the deficiency of some pigment-protein complexes is.

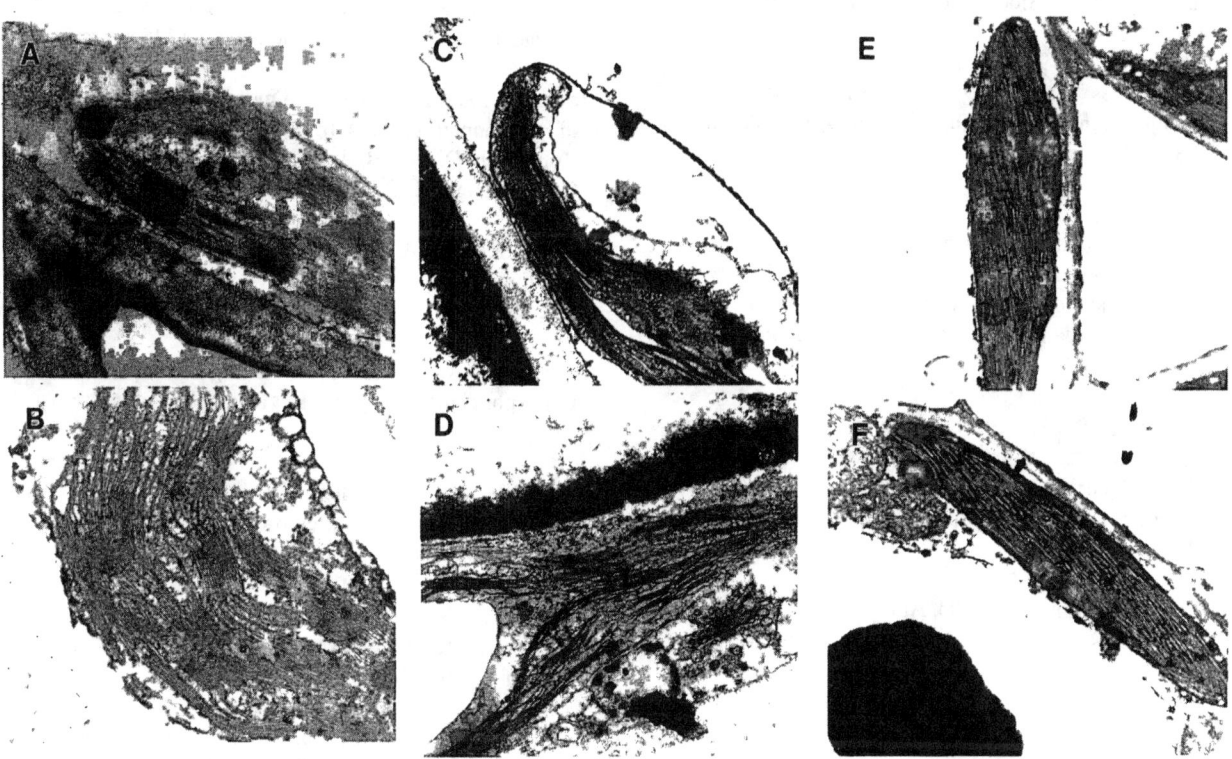

Figure 1.—*Ultrastructure of thylakoid membrane of two cecidomyiid galls and the leaf of* M. thunbergii. *A and B, mature leaf; C and D, oval-pointed gall; E and F, obovate gall.*

REFERENCES

Allen, J.F. 1992. Protein phosphorylation in regulation of photosynthesis. Biochimica et Biophysica Acta. 1098: 275-335.

Bennett, J. 1991. Protein phosphorylation in green plant chloroplasts. Annual Review of Plant Physiology and Plant Molecular Biology. 42: 281-311.

Dreger-Jauffret, F.; Shorthouse, J.D. 1992. Diversity of gall-inducing insects and their galls. In: Shorthouse, J.D.; Rohfritsch, O., eds. Biology of insect-induced galls. Oxford, England: Oxford University Press: 8-33.

Geisman, T.A. 1955. Modern methods of plant analysis. Berlin, Springer-Verlag: 420-433.

Hagerman, A.E.; Butler, L.G. 1978. Protein precipitation method for the quantitative determination of tannins. Journal of Agriculture and Food Chemistry. 26: 809-812.

Harborne, J.B. 1988. The flavonoids: recent advances. In: Goodwin, T.W., ed. Plant pigments. London, England: Academic Press: 299-343.

Holden, M. 1961. The breakdown of chlorophyll by chlorophyllase. Biochemical Journal. 78: 259-264.

Jasper, E.M.W. 1965. Pigmentation of tobacco crown gall tissues cultured in vitro in dependence of the composition of the medium. Physiologia Plantarum. 18: 933-940.

Kahn, V.M.; Avivi-Bieise, N.; von Wettstein, D. 1976. Genetic regulation of chlorophyll synthesis analyzed with double mutant in barley. In: Dhuchler, T., ed. Genetics and diagenesis of chloroplasts and mitochondria. Amsterdam: Elsevier/North-Holland Biomedical Press: 119-131.

Markwell, J.P. 1986. Electrophoretic analysis of photosynthetic pigment-protein complexes. In: Hipkins, M.F.; Baker, N. R., eds. Photosynthesis energy transduction: a practical approach. Oxford, England: IRL Press: 27-49.

Mazza, G.; Miniati, E. 1993. Anthocyanins in fruits, vegetables, and grains. Florida: CRC Press. 362 p.

Meyer, J. 1987. Plant galls and gall inducers. Gebruder Borntraeger, Berlin, Stuttgart, 291p 285p c of Chinaision of Zoology.

Nakatani, H.Y.; Baliga, V. 1985. A clover mutant lacking chlorophylls a- and b-containing protein antenna complexes. Biochimica et Biophysica Research Committee. 131: 182-189.

Porra, R.J.; Thompson, W.A.; Kriedelmann, P.E. 1989. Determination of accurate extractions and silmultaneous equations for assaying chlorophylls a and b extracted with four different solvents: verification of the concentrations of chlorophyll standards by atomic absorption spectroscopy. Biochimica et Biophysica Acta. 975: 384-394.

Quijja, A.; Farineau, N.; Cantrel, C.; Guillot-Salomon, T. 1988. Biochemical analysis and photosynthetic activity of chloroplasts and photosystem II particles from a barley mutant lacking chlorophyll b. Biochimica et Biophysica Acta. 932: 97-106.

Rey, L.A. 1973. Ultrastructure des chloroplastes au cours de leur evolution pathologique dans le tissue central de la jeune galle de Pontania proxima Lep. Comptes rendus hebdomadaires des Seances de l'Academie des Sciences. Serie D 276: 1157-1160.

Rey, L.A. 1974. Modification ultrastructurales pathologiques presentees par les chloroplastes de la galle de Pontania proxima Lep. En fin de croissance. Comptes rendus hebdomadaires des Seances de l'Academie des Sciences. Serie D 278: 1345-1348.

Rey, L.A. 1992. Developmental morphology of two types of hymenopterous galls. In: Shorthouse, J.D.; Rohfritsch, O., eds. Biology of insect-induced galls. Oxford, England: Oxford University Press: 87-101.

Spurr, A.R. 1969. A low viscosity epoxy resin embedding medium for electromicroscopy. Journal of Ultrastructure Research. 26: 31-43.

Williams, M.A.J. 1994. Plant galls: organisms, interactions, populations. Oxford, England: Clarendon Press. 488 p.

Yang, C.M.; Hsu, J.C.; Chen, Y.R. 1993. Light- and temperature-sensitivity of chlorophyll-deficient and virescent mutants. Taiwania. 38: 49-56.

Yang, C.M.; Hsu, J.C.; Chen, Y.R. 1995. Analysis of pigment-protein complexes in mungbean testa. Plant Physiology and Biochemistry. 33: 135-140.

Yang, C.M.; Chen, H.Y. 1996. Grana stacking is normal in a chlorophyll-deficient LT8 mutant of rice. Botanical Bulletin of Academia Sinica. 37: 31-34.

THE DEVELOPMENT OF AN OVAL-SHAPED PSYLLID GALL ON *CINNAMOMUM OSMOPHLOEUM* KANEH. (LAURACEAE)

Gene Sheng Tung[1], Man-Miao Yang[2], and Ping-Shih Yang[3]

[1] Graduate Student, Department of Forestry, National Taiwan University, Taipei, Taiwan, Republic of China
[2] Associate Curator of Entomology, Division of Zoology, National Museum of Natural Science, Taichung, Taiwan 404, Republic of China
[3] Professor of Entomology, Department of Plant Pathology and Entomology, National Taiwan University, Taipei, Taiwan, Republic of China

---

Based on our survey between 1995 and 1996, Lauraceae is considered the most important host family of galling insects in Taiwan (Yang and Tung 1998). In the Lauraceae, 26 species of *Cinnamomum* stand for nearly a quarter (23.2 percent) of the total number of galling host species. *Cinnamomum osmophloeum* is endemic to Taiwan, and contains cinnamaldehyde, which has been extracted for industrial and medical uses. On its peduncle, pedicel, and stem, we found a new oval-shaped psyllid gall formed by a new *Trioza* species. Several histological studies of psyllid galls have been documented: Wells (1916, 1920), Weiss (1921), Walton (1960), Dundon (1962), Lewis and Walton (1964), Raman (1987, 1991), and Beisler and Baker (1992). Traditional sectioning or scanning electronic microscopy (SEM) techniques were applied for the above studies on galls of *Celtis* and Myrtaceae leaves, but not Lauraceae. This study attempts to elucidate the life history of the gall maker and the development of the gall tissue of this oval-shaped psyllid gall.

Figure 1 summarizes our observations on the life history and tissue development. It took a year for the gall maker to complete its life cycle. Gall initiation started in early May and development stopped while dehiscence took place the next January (fig. 1, n = 45). Like most psyllid species, the nymph has five instars. According to the development of parenchyma and sclerenchyma tissues, the gall could be regarded as a prosoplasmatic gall sensu Küster (1911). The changes of gall tissue described below are based on the four major stages in gall development and follow the system of Lalonde and Shorthouse (1984) and Rohfritsch (1992).

1. INITIATION PERIOD (fig. 1A)—The first instar was responsible for gall initiation. Its stylets reached phloem and the plant was malformed into a pit at the beginning. The cortex covered the nymph gradually, and finally became oval shaped like a mung bean. Modification of the cortex was the major change of the plant tissue that followed an active anticlinal division.

2. GROWTH AND DEVELOPMENT PERIOD (fig. 1B and 1C)—Active gall enlargement prevailed during the growth period from the end of May; the insect was nearly completely enclosed by the gall, and the gall chamber was formed. Sporadic growth of short trichomes could be seen near the enclosed area at the beginning. The epidermis at the inner part of the gall formed a row of nipple-shaped cells while the cells at the outer part remained in normal shape. Another obvious change at this period was the mass growth of trichomes, which became longer with thickened cell walls. The insect eventually grew into the fourth instar by the end of November.

3. MATURATION PERIOD (fig. 1D)—In December, the gall reached the maturation period. The psyllid became a fifth instar and was ready to molt. The intermediate layers between pith axis and gall chamber, usually under the psyllid body, were inhibited. The cortex parenchyma along the wall of the gall became sclerenchymatous but the part near the gall chamber remained parenchymatous.

4. DEHISCENCE PERIOD (fig. 1E)—In the next January the insect emerged through the opening area. Trichomes on the opening protruded outside the gall and the cell walls of trichomes reached their thickest period, which was nearly equal to half of the radius. The trichomes pushed each other and easily fell off. In addition to their general role in protection, trichomes in this system seem to help in the mechanism of gall dehiscence.

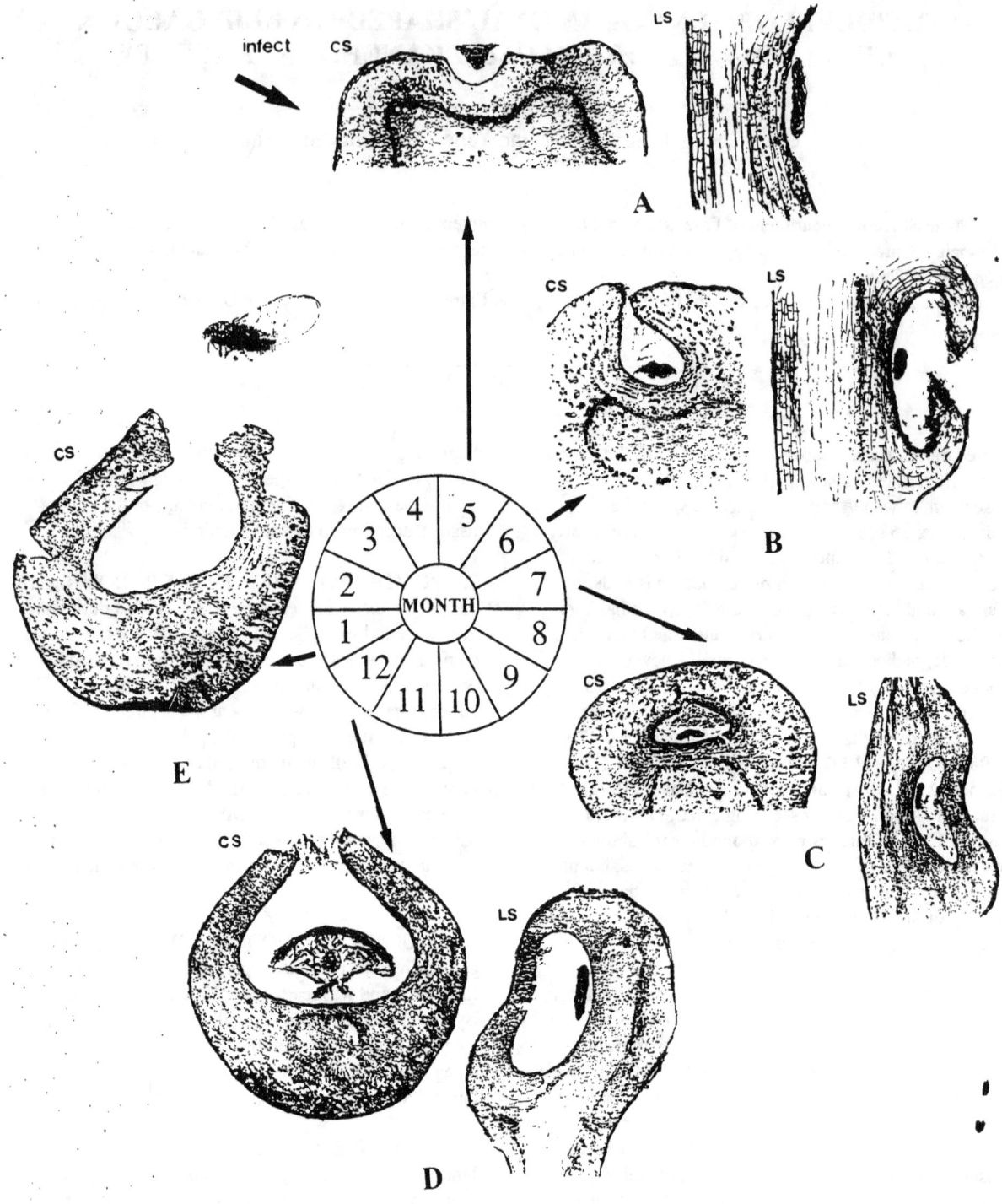

Figure 1.—*A summary showing the development of an oval-shaped psyllid gall on* Cinnamomum osmophloeum *Kaneh (Lauraceae), including the life history of the psyllid gall maker,* Trioza *sp., and the anatomy of its induced gall.* **A**. *During the initiation period in May, the first instar nymph infected the flower pedicel and initiated a pit on the surface.* **B & C**. *The stage of growth and development period started as the host plant was actively responding to the insect stimulation and completely enclosed the psyllid nymph into the gall. After the gall chamber was formed, the second instar nymph appeared in June and continued its development into the fourth instar.* **D**. *In December, the gall reached the maturation period and the psyllid nymph became a fifth instar.* **E**. *The dehiscence period occurred in January.*

LITERATURE CITED

Beisler, J.M.; Baker, G.T. 1992. *Pachypsylla celtidismamma* (Fletcher) (Homoptera: Psyllidae): morphology and histology of its adult and nymphal sensilla. Mississippi Entomology Museum. 3: 1-27.

Dundon, T.R. 1962. Multinucleate giant cell formation in a *Pachypsylla* gall on *Celtis*. American Journal of Botany. 49: 800-805.

Küster, E. 1911. Die gallen der pflanzen. Hirzel, Leipzig.

Lalonde, R.G.; Shorthouse, J.D. 1984. Developmental morphology of the gall of *Urophora cardui* (Diptera, Tephritidae) in the stems of Canada thistle (*Cirsium arvense*). Canadian Journal of Botany. 62: 1372-1383.

Lewis, I.F.; Walton, L. 1964. Gall formation on leaves of *Celtis occidentalis* L. resulting from material injected by *Pachypsylla* sp. Transactions of the American Microscopy Society. 83: 62-78.

Raman, A. 1987. On the cecidogenesis and nutritive tissues of the leaf galls of *Garuga pinnata* Roxburgh induced by *Phacopteron lentiginosum* Buckton (Pauropsyllinae: Psyllidae: Homoptera). Phytophaga. 1: 121-140.

Raman, A. 1991. Cecidogenesis of leaf galls on *Syzygium cumini* (L.) Skeels (Myrtaceae) induced by *Trioza jambolanae* Crawford (Homoptera: Psylloidea). Journal of Natural History. 25: 653-663.

Rohfritsch, O. 1992. Pattern in gall development. In: Shorthouse, J.D.; Rohfritsch, O., eds. Biology of insect-induced galls. Oxford: Oxford University Press: 60-85.

Walton, B.C.J. 1960. The life cycle of the hackberry gall-former, *Pachypsylla celtidis-gemma* (Homoptera: Psyllidae). Annals of the Entomological Society of America. 53: 265-277.

Weiss, H.B. 1921. Notes on the life-history of *Pachypsylla celtidis-gemma* Riley. Canadian Entomologist. 53: 19-21.

Wells, B.W. 1916. The comparative morphology of the zoocecidia of *Celtis occidentalis*. Ohio Journal of Science. 7: 249-298.

Wells, B.W. 1920. Early stages in the development of certain *Pachypsylla* galls on *Celtis*. American Journal of Botany. 7: 275-285.

Yang, M.M.; Tung, G.S. 1998. The diversity of insect-induced galls on vascular plants in Taiwan: a preliminary report. In: Csóka, Gyuri; Mattson, William J.; Stone, Graham N.; Price, Peter W., eds. The biology of gall-inducing arthropods: proceedings of the international symposium; 1997 August 14-19; Mátraküred, Hungary. Gen. Tech. Rep. NC-199. St. Paul, MN: U.S. Department of Agriculture, Forest Service, North Central Research Station: 44-53.

ADAPTIVE RADIATION OF GALL-INDUCING SAWFLIES IN RELATION TO ARCHITECTURE & GEOGRAPHIC RANGE OF WILLOW HOST PLANTS

Peter Price,[1] Heikki Roininen[2] and Alexei Zinovjev[3]

[1]Department of Biological Sciences, Northern Arizona University, Flagstaff, Arizona 86011-5640, U.S.A.
[2]Department of Biology, University of Joensuu, P.O. Box 111, SF-80101 Joensuu, Finland
[3]Laboratory of Insect Taxonomy, Zoological Institute, Russian Academy of Sciences, St. Petersburg 199034, Russia

Abstract.—We tested the hypothesis that gall-inducing sawflies (Hymenoptera: Tenthredinidae) have undergone more adaptive radiation on shrubs than on trees because shrubs lack apical dominance and generally provide juvenile shoots, whereas trees have apical dominance with inevitable ontogenetic and physiological aging. The hypothesis was based on detailed comparative studies on four *Euura* sawflies: *Euura lasiolepis* and *E. mucronata* on shrubs and *E. amerinae* and *E. atra* on trees. The test involved records of gall-inducing sawflies on 137 *Salix* species from the Palearctic Region ranging in architecture from dwarf shrubs to tall trees. We concluded that small and large shrubs are at least as easy to colonize as trees, a result which runs counter to the hypothesis on insect herbivore species richness in relation to plant architecture.

GENERAL PATTERNS OF INSECT HERBIVORE SPECIES RICHNESS IN RELATION TO PLANT ARCHITECTURE

A general relationship between plant structure and the number of insect herbivore species per plant species has been recognized for 20 years (Lawton and Schroder 1977, Lawton 1983, Lawton and MacGarvin 1986). Species richness increases with increasing architectural complexity from herbs to shrubs to trees. Lawton (1983) proposed two major hypotheses which may account for this pattern: the **size per se hypothesis** and the **resource diversity hypothesis**. Increasing size of a plant species correlates with increasing age and probability of encounter by insect species—large plant species are more apparent in time and space (Feeny 1975). However, as size increases so does resource diversity as architectural complexity increases. Resources may include bark, cambium, wood, twigs, leaves, flowers, etc., as food, and a similar range of oviposition sites, escape from enemies may change with architecture, and large plants provide more microhabitats in terms of zonation, shading, and modular types such as juvenile, sapling, sucker, and mature tree, and additional factors treated by Lawton (1983). As Lawton notes, the correlation between plant size and diversity of resources makes the investigation of the mechanistic processes challenging.

The pattern of increasing herbivore species richness with increased plant size and complexity is observed even when all plant species compared are very common with an extensive geographic range (e.g., Lawton and Schroder 1977), and when individual taxonomic groups are considered such as leafminers (Agromyzidae and microlepidoptera) and microlepidoptera in general (Lawton and Price 1979, Price 1980). The patterns are consistent, general, and intuitively clear enough to have raised little debate or to have stimulated more detailed investigation on mechanisms or responses of specific insect taxa.

Two discoveries, one concerning gall-inducing insects in general and the other involving gall-inducing sawflies, have provided evidence that gall formers may not fit these general patterns relating to plant architecture. We discuss these data next.

PATTERNS IN GALL-INDUCING INSECTS

On a global scale, gall-inducing insects in general increase in local species richness dramatically in the warm temperate latitudes and their equivalents, with decreasing richness toward the equator and the poles (Price et al. 1998a). Broadly speaking, on a global scale, insect species richness is associated most strongly with scleromorphic, shrubby vegetation, such as campina, caatinga, chaparral, cerrado, fynbos, Kwongan, and matorral. Some of these types are in the tropics but scleromorphic plants on poor soils are characteristic. In general, with a 1-hour census of local galling species richness, many more galls can be found on scleromorphic shrub vegetation than on tropical forest trees, even when the full canopy of trees is sampled (Price et al. 1998a). This pattern is driven especially by the speciose cecidomyiid gall inducers. Fernandes and Price (1991) developed a hypothesis to account for the high gall richness on scleromorphic plants involving a flow of influences from low nutrient soils to scleromorphy, to safe sites for gall-forming insects because of low abscission of leaves, leaf longevity, high carbon-based defenses in outer layers of galls, and low mortality imposed by fungal growth and parasitoid attack. We would now add the benefits of shrubs as sites for gall formers discussed next and the importance of fire as a disturbance factor which prunes back old growth and stimulates rapid growth of juvenile modules suitable for attack.

The second pattern involves gall-forming sawflies in the genus *Euura*, which utilize as hosts the willows (*Salix*, Salicaceae). We have noted that shrubs support sawfly populations almost indefinitely because juvenile ramets develop from the base of the plant with no apical dominance. Juvenile shoots are the most favorable modules for gall formation and larval survival. These patterns have been well studied for the stem-galler, *Euura lasiolepis*, on *Salix lasiolepis* in Arizona, U.S.A., and for the bud-galler, *Euura mucronata*, on *Salix cinerea* in North Karelia, Finland, but the pattern is general for many species (Price and Roininen 1993; Price et al. 1995, 1998b). *Euura* species on trees also attack long and juvenile shoot modules, but populations are not persistent in a natural vegetation because trees age, becoming resistant to attack, and a local population of sawflies goes extinct in a few years. We have documented this type of dynamics for the stem-galling *Euura amerinae* on *Salix pentandra* in Finland (Roininen et al. 1993). *Euura amerinae* galls cannot usually be found on trees older than about 12 years. In a similar way, the stem galls of *Euura atra* can be found in trees such as *Salix alba* in Finland, only on modules growing very vigorously on young trees, or after severe breakage of limbs or when trees are heavily pruned in parks and roadsides (Price et al. 1997). The inevitable aging of trees, characterized by apical dominance, results in ontogenetic and physiological aging with an evident increase in resistance to *Euura* species.

Trees appear to offer a challenge to colonization by *Euura* species, with resources available relatively briefly compared to long-lived shrubs. We call this the reduced **apparency hypothesis**, which probably applies most strongly to *Euura* species, and with weakening effects in the order of the *Phyllocolpa* leaf-edge gallers and the *Pontania* leaf-lamina gallers. We predicted that there would be fewer gall-forming sawflies on trees than on shrubs. [Note that our use of three genera for three types of gall formation follows Smith (1979). However, it is now apparent that so-called "open" galls, including leaf folds and rolls, are induced by members of the genus *Phyllocolpa* and members of one species group of *Pontania*. Also, *Phyllocolpa* may be regarded as a subgenus of *Pontania* (Zinovjev 1985, 1993a, 1998). For simplicity we have retained the earlier classification by Smith.]

The two patterns in gall-inducing insects related to local global richness of gallers and dynamics on shrubs and trees suggested that increasing architectural complexity of plants should not be correlated with an increase in gall-inducing sawfly species richness for plant species. Therefore, we developed a test of the reduced apparency hypothesis described next.

TEST OF THE REDUCED APPARENCY HYPOTHESIS

It is most unusual to find a single genus, like *Salix*, with such a wide architectural range, from dwarf shrubs through small and large shrubs to large trees. All species are woody, with all the basic resources available on such plants, which permits a focus on the plant size per se hypothesis emphasizing plant apparency in time and space. Concentrating on gall-forming sawflies also reinforces the testing of the plant size hypothesis because the sawflies utilize only leaves, midribs, petioles, buds, and stems as resources, all available on every willow species. In addition, the lower the architectural stature of the willow, the more allocation there is to the clonal habit, such that the clonal species may be represented by individuals and populations even older than the largest trees. In effect, apparency through time in the clonal species may compensate for apparency in size for the large tree species, with a full gradient between the extremes. The arena for the adaptive radiation of gall-forming sawflies across the willow species appears to be rather flat in terms of the hypotheses relating to insect richness and plant architectural complexity, with the possibility that speciation onto willow species may have been relatively independent of willow architecture.

In one scenario the adaptive radiation of the gall-inducing sawflies started with a free-feeding progenitor, moved to the leaf-edge galling habit and the genus *Phyllocolpa*, then to leaf gallers using the leaf lamina, in the genus *Pontania*, and finally to the galling of leaf midribs, petioles, buds, and stems, all in the genus *Euura* (Nyman et al. 1988). An alternative hypothesis was developed by Zinovjev (1982) with an endophagous feeding habit preceding both the free-feeding and gall-inducing species. The gall types and genera seem to have evolved early in the phylogeny, with subsequent radiation across willow species. Given that the genera *Phyllocolpa* and *Pontania* use leaves exclusively as the plant resource, but *Euura* species utilize four distinct resources, based on the resource diversity hypothesis we should predict a greater radiation of *Euura* species than in either of the other two genera.

To test the hypothesis we used data on the Palearctic willows from Skvortsov (1968) and a few other sources, including records of 137 species of *Salix* and the number of gall-forming sawfly species in each genus per plant species derived from personal observations by two authors (Zinovjev 1993b and pers. observ. by Zinovjev and Roininen) and many published records. Almost all willow species were included whether or not sawfly species had been recorded. Variables in willow species characters included plant height and architecture, geographic factors such as range, patchiness, and locality, and phylogenetic relationships relating to the sections in the genus *Salix*. When categorical data were used, such as dwarf shrub, small shrub, large shrub and tree, logistic regression was employed. Simple linear regression was used when continuous variables such as plant height were available. When investigating effects of plant architecture on insect herbivore species richness, the effect of geographic range of host plant has been removed first in most studies because of its strong influence (e.g., Lawton and Schroder 1977). Data available enabled use of three geographic range categories—large, medium, and small—so initially we investigated plant architectural effects within each of these categories.

Plants with a large geographic range include dwarf shrubs of 10 cm in height or less to tall trees up to 37 m, but there was no significant effect of plant height on species richness of gall-inducing sawflies (fig. 1). In

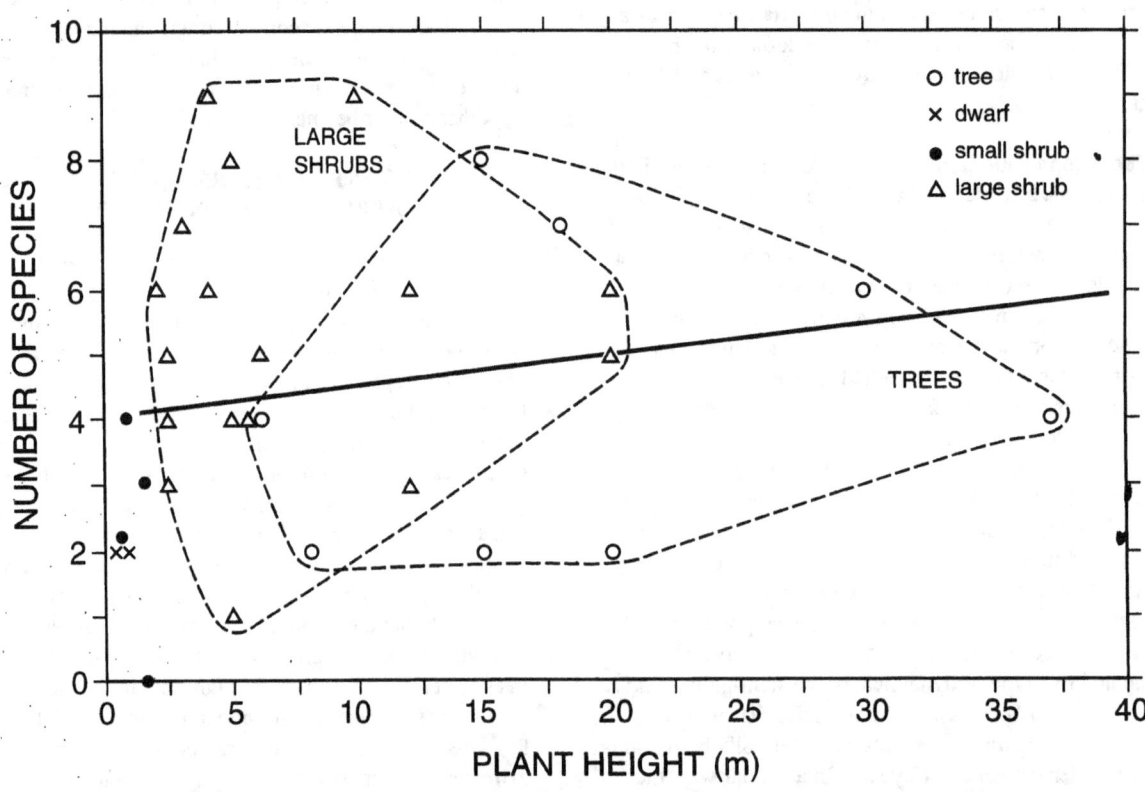

Figure 1.—*The relationship between plant species height, including growth form, with the number of gall-inducing sawfly species per host plant species with a large geographic range in the Palearctic flora and fauna. The regression of number of species of sawflies on height of the host plant is not significant (n = 31, r^2 = 0.03). Open circles show data for trees and are enveloped in a dashed line. Triangles show data for large shrubs with a dashed-line envelope. Solid circles are for small shrubs, and an x is for data on dwarf shrubs.*

fact, large shrubs supported the highest richness of sawflies, at nine sawfly species per plant species, and a greater percentage of species were represented above the regression line than for trees (63 percent of large shrub species versus 24 percent for tree species). Small and dwarf shrubs were not well represented in the large geographical range category, but undoubtedly influenced the trend of the regression. Removing the relevant six species, the regression became slightly negative but remained nonsignificant.

For willows with a moderate geographic range the relationships are very similar, with almost no general trend in species richness, but some shrubs with higher richness than trees (fig. 2). Among large shrubs 71 percent of species were above the regression line while for trees 50 percent were above the line. Even some dwarf and small shrubs supported sawfly species equivalent to most of the trees in the range of one to four species per host plant species.

Willow species with a small geographic range were generally relatively low in stature and only four trees were in this category (fig. 3). However, the trend seen in willows with large ranges is reflected again, but with a mean number of gall-forming species at about 1.0 galler per plant species compared to about 2.8 species for plants with moderate range, and about 4.5 species for plants with large geographic ranges.

We see in these relationships no support for the two main hypotheses on plant architecture in relation to herbivore species richness. The size per se hypothesis is not supported and the resource diversity hypothesis is not supported. The reduced apparency hypothesis comes closer to a viable explanation of the patterns observed. Trees are no richer in sawfly galling species in general than shrubs that are much smaller. However, trees are colonized by all the kinds of gall-forming sawflies and they do not in general support fewer species as we predicted. Also, when we consider individual genera of sawflies, all three genera show no significant trend in species richness in relation to plant height, indicating that trees pose no special challenge for *Euura* species as we predicted (fig. 4). Similar trends are observed on willow species with large and medium geographic ranges. On willows with small geographic ranges, numbers of sawfly species per genus were so low as to make the equivalent analysis unuseful.

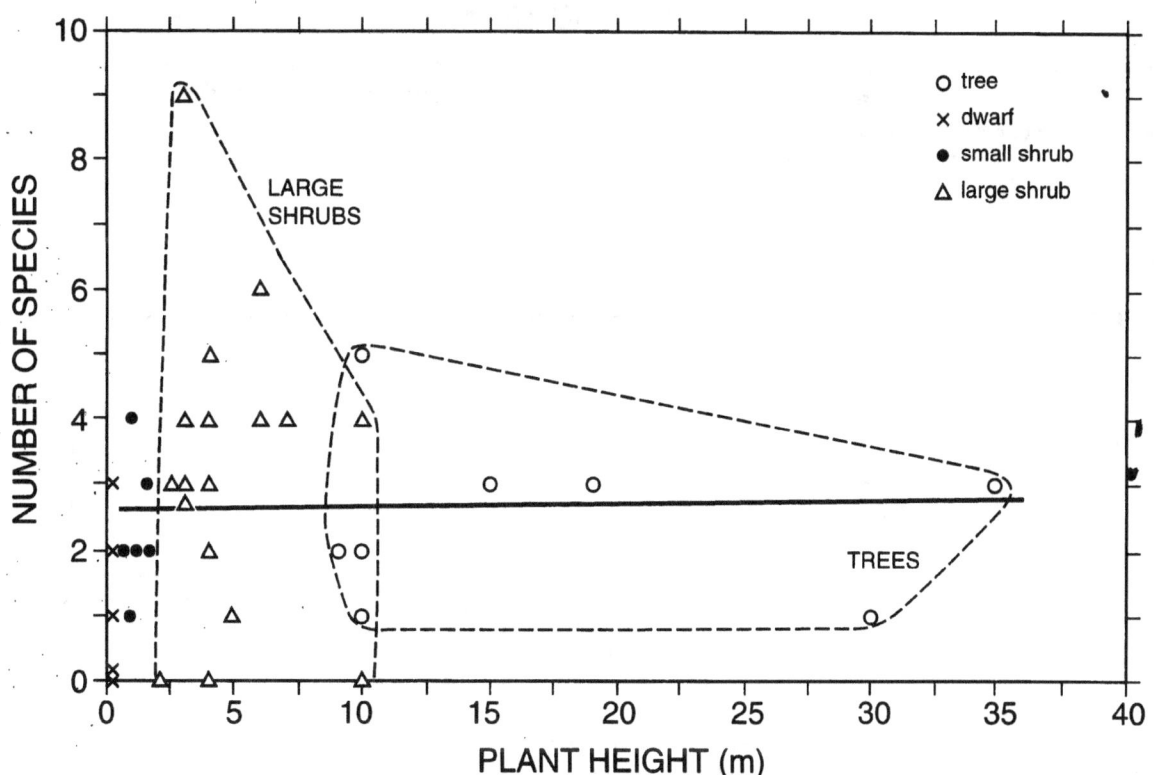

Figure 2.—*Relationships as depicted in Figure 1 for willow host plants with a moderate geographic range. Symbols are the same, and the regression line has no significant explanatory power, as in Figure 1 (n = 36, r^2 = 0.00042).*

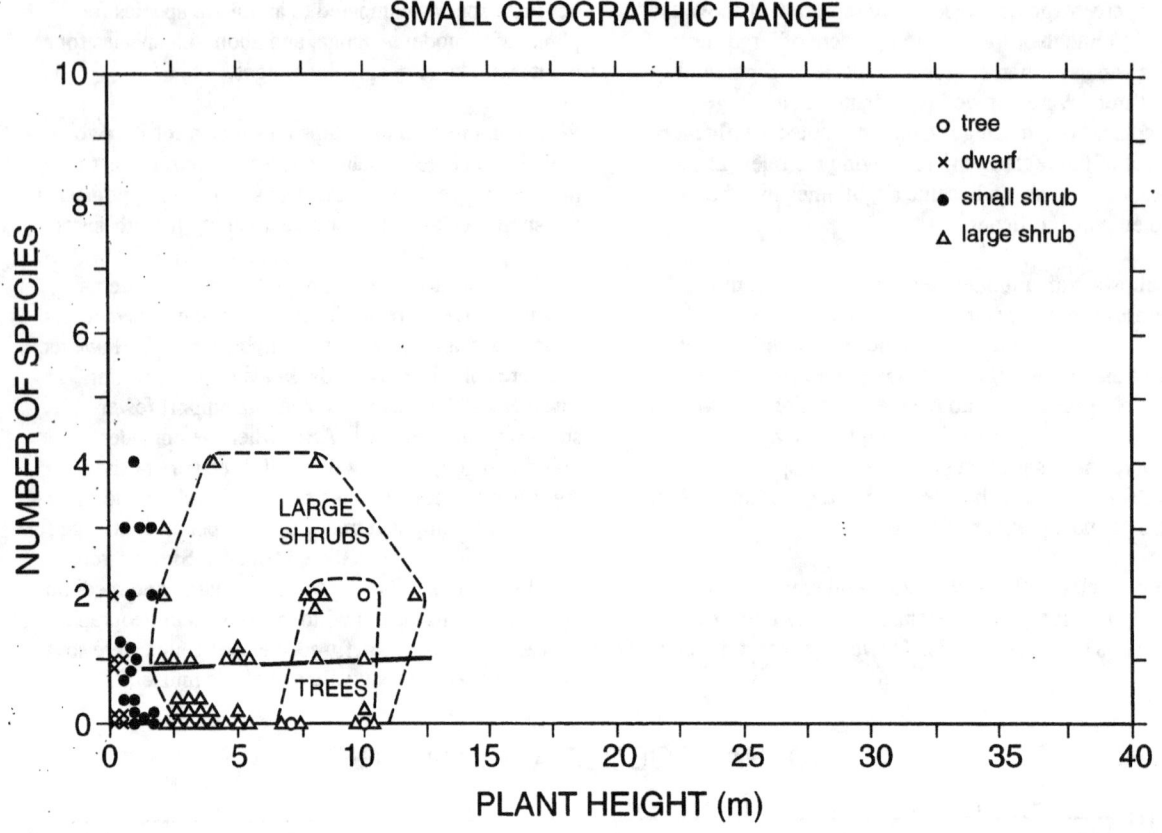

Figure 3.—*Relationships as shown in Figures 1 and 2 for host plants with a small geographic range. Symbols remain the same and the regression is again not significant (n = 68, r^2 = 0.0016) and the scales of the graph are the same to promote ready comparison.*

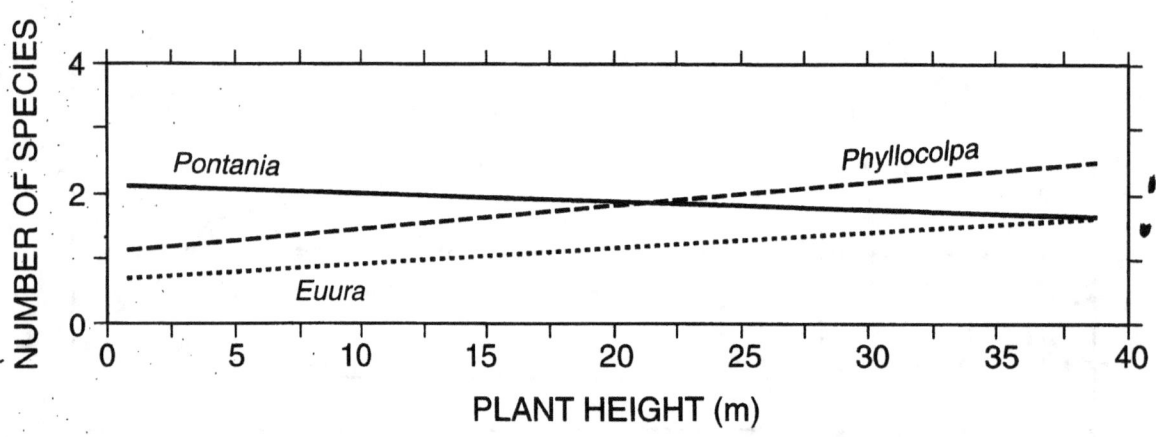

Figure 4.—*The number of species of sawflies per genus in relation to host plant height for the genera* Pontania, Phyllocolpa, *and* Euura. *None of the regressions is significant.*

We may conclude that the reduced apparency hypothesis is consistent with the data presented, to the extent that trees appear to have an apparency reduced to the general level of the much smaller shrubs. This apparency seems to be effective for the genera of gall-inducing sawflies, *Pontania*, *Phyllocolpa*, and *Euura*, and is not particularly influential on the genus *Euura*. So far as we can tell, this study represents the first case in which plant architecture has no measurable explanatory power in relation to the number of insect herbivores each plant species supports.

When general trends of increasing species richness with increasing architectural complexity are considered, from shrubs to trees, various sets of data indicate an increase in species number of about 2x or more (Lawton 1983). For microlepidoptera on British plants a mean of about six species per plant species is recorded on shrubs and about 11 on trees, and for leaf-mining Lepidoptera on European plants, about seven species occur on shrubs and 17 on trees (Price 1980). For insect herbivores on British plant genera, maximum differences between richness on shrubs and trees occurs on the plant species with the largest geographic range, with shrubs supporting about 40 insect herbivore species and trees about 120 species (Strong and Levin 1979). This difference declines to zero at low host plant geographic ranges.

These data indicate that reduced apparency of willow trees to gall-inducing sawflies has a strong effect. Shrubs of all kinds of willows with a large geographic range support a mean of 4.5 gall species per plant species, so we should expect about 9 gall species per tree species if the general trends prevail. However, willow trees with an equivalent range support an average of only 4.4 gall species per tree species. The reduced apparency of trees resulting evidently from apical dominance in growth form and ontogenetic and physiological aging seem to have negated any usual effects of plant architecture on the richness of the insect fauna per host plant species.

WHAT DOES ACCOUNT FOR VARIATION IN SAWFLY SPECIES NUMBER PER WILLOW HOST SPECIES?

Geographic range of host species is clearly an important factor accounting for sawfly species richness. Mean richness declined from 4.5 species per plant species with large geographic range, to 2.8 species on medium-range hosts and 1.0 species on small-range hosts (cf. figs. 1–3). In a stepwise logistic regression the size of the host species geographic distribution (SD) accounted for most of the explained variance in sawfly species richness (table 1). Next most important was the geographic locality (GL) such as Mediterranean and South Atlantic, European Arctic and alpine, and Siberian and EuroSiberian Arctic and alpine as classified in Skvortsov (1968). Small additional variances were accounted for by patchiness (P) of the willow species over their geographic ranges, and growth form (GF) of the willow species.

The full regression model accounted for 56 percent of the variance in the number of gall-inducing herbivores per willow host species: Number of sawfly species = constant + SD + GL + P + GF (table 1). Providing an account of more than 50 percent of the herbivore richness per plant species, and adding 17 percent of the explanation beyond the strong influence of plant geographic distribution is moderately better than some equivalent attempts. For example, geographic range of host plant species accounted for 32 percent of the number of agromyzid leafminers per host plant species in the family Apiaceae (= Umbelliferae) in Britain (Lawton and Price 1979), and at best only about 50 percent of the variance was accounted for using additional factors of plant size, leaf form, life history, and taxonomic isolation. Size of host plant added about 2 percent to the explanation of agromyzid species richness on these herbaceous plants varying in height from the small at 5 cm to the large at

Table 1.—*Variables in willow species geography, systematics and architecture used in stepwise logistic regression to account for variation in the number of gall-inducing sawfly species per host willow species.*

Variable	Categories	Increase in r^2	P-value
Size of geographic distribution (SD)	Large, medium, small	39%*	0.001
Geographic locality (GL)	11 geographic regions from Skvortsov (1968)	10%*	0.001
Patchiness (P)	Continuous, medium, patchy	4%*	0.005
Growth form (GF)	Dwarf, small, and large shrubs, trees	3%*	0.05
Dominance in community	4 categories: high to low	1%	N.S.
Willow species height	Continuous variable	$\simeq 0$	N.S.

*Cumulative variances accounted for which entered into the final model.

3.5 m in the family Apiaceae, a 70x span in heights with all species herbaceous. On the other hand, plant height accounted for an additional 7 percent of the variance of richness of macrolepidopteran species on Finnish trees and shrubs, after the removal of plant frequency or abundance as the main factor (Neuvonen and Niemalä 1981). With host plant abundance and plant height, 64 percent of the variance in macrolepidopteran species per host plant species could be accounted for, and a total 71 percent of the variance was accounted for when the number of relatives of the host plant was added as a third factor. Certainly, gall-forming Lepidoptera were not a significant factor in defining these patterns.

Most notable with reference to plant architecture of willow species is its very low explanatory power for herbivore species richness. Growth form adds a minor 3 percent to the final model and willow species height does not contribute significantly, as we saw in figures 1-4, even though the range in height is large, from 0.10 m to 37 m in the genus *Salix*. That a 370x span from the shortest to the tallest species of willow has no explanatory power on the insect herbivores colonizing these species is remarkable. This raises the question on what other kinds of herbivores do not respond to plant architectural traits.

FUTURE STUDIES

This study raises a series of questions worth study in the future. Can we find any pattern of colonization of willows by gall-inducing sawflies, or is the best simulation of radiation provided by a random colonization model? What other kinds of herbivores are likely to show patterns consistent with the reduced apparency hypothesis? We predict that those species requiring rapid plant growth, such as gall formers and shoot borers, are likely to show patterns similar to the sawflies in this study. Will the reduced apparency hypothesis be supported by broader studies on gall formers on shrubs and trees in a complete and well-studied flora and fauna, such as in Britain or any of the Fennoscandian countries? And will the pattern on the genus *Salix* found for sawflies be the same for cecidomyiids?

There can be little doubt that plant architecture is a major factor in the adaptive radiation of insect herbivores, for module size and other variables are critical for many species (Lawton 1983, Price 1997). We have noted many cases in which galling sawflies utilize rapidly growing modules that result in long shoots (Price and Roininen 1993; Price *et al.* 1995, 1998b). Perhaps to find clear patterns in the adaptive radiation of gall-forming sawflies on willows in relation to plant architecture, the relevant gradient is one of module growth rate and shoot length. Module size is highly variable within species and among species as plants age, over landscapes, and over altitudinal and latitudinal gradients. Therefore, the challenge is considerable if we wish to find patterns of sawfly colonization and adaptive radiation in relation to the details of plant architecture.

ACKNOWLEDGMENTS

We thank the following agencies for research support during this study: The U.S. National Science Foundation (Grant DEB-9318188), the Finnish Academy and the Russian Academy of Sciences.

LITERATURE CITED

Feeny, P. 1975. Biochemical coevolution between plants and their insect herbivores. In: Gilbert, L.E.; Raven, P.H., eds. Coevolution of animals and plants. Austin, TX: University of Texas Press: 3-19.

Fernandes, G.W.; Price, P.W. 1991. Comparison of tropical and temperate galling species richness: the roles of environmental harshness and plant nutrient status. In: Price, P.W.; Lewinsohn, T.M.; Fernandes, G.W.; Benson, W.W., eds. Plant-animal interactions: evolutionary ecology in tropical and temperate regions. New York: Wiley: 91-115.

Lawton, J.H. 1983. Plant architecture and the diversity of phytophagous insects. Annual Review of Entomology. 28: 23-39.

Lawton, J.H.; MacGarvin, M. 1986. The organization of herbivore communities. In: Kikkawa, J.; Anderson, D.J., eds. Community ecology: pattern and process. Oxford: Blackwell Scientific Publications: 163-186.

Lawton, J.H.; Price, P.W. 1979. Species richness of parasites on hosts: Agromyzid flies on the British Umbelliferae. Journal of Animal Ecology. 48: 619-637.

Lawton, J.H.; Schroder, D. 1977. Effects of plant type, size of geographical range and taxonomic isolation on number of insect species associated with British plants. Nature. 265: 137-140.

Neuvonen, S.; Niemalä, P. 1981. Species richness of Macrolepidoptera on Finnish deciduous trees and shrubs. Oecologia. 51: 364-370.

Nyman, T.; Roininen, H.; Vuorinen, J.A. 1998. Evolution of different gall types in willow-feeding sawflies (Hymenoptera: Tenthredinidae). Evolution. In press.

Price, P.W. 1980. Evolutionary biology of parasites. Princeton: Princeton University Press.

Price, P.W. 1997. Insect ecology. 3d ed. New York: John Wiley and Sons.

Price, P.W.; Roininen, H. 1993. Adaptive radiation in gall induction. In: Wagner, M.R.; Raffa, K.F., eds. Sawfly life history adaptations to woody plants. San Diego, CA: Academic Press: 229-257.

Price, P.W.; Craig, T.P.; Roininen, H. 1995. Working toward theory on galling sawfly population dynamics. In: Cappuccino, N.; Price, P.W., eds. Population dynamics: new approaches and synthesis. San Diego, CA: Academic Press: 321-338.

Price, P.W.; Roininen, H.; Tahvanainen, J. 1997. Willow tree shoot module length and the attack and survival pattern of a shoot-galling sawfly, *Euura atra* L. (Hymenoptera: Tenthredinidae). Entomol. Fennica. 8: 113-119.

Price, P.W.; Fernandes, G.W.; Lara, A.C.F.; Brawn, J.; Barrios, H.; Wright, M.G.; Ribeiro, S.P.; Rothcliff, N. 1998a. Global patterns in local number of insect galling species. Journal of Biogeography. In press.

Price, P.W.; Craig, T.P.; Hunter, M.D. 1998b. Population ecology of a gall-inducing sawfly, *Euura lasiolepis*, and relatives. In: Dempster, J.P.; McLean, I.F.G., eds. Insect populations: In theory and practice. London: Chapman and Hall. In press.

Roininen, H.; Price, P.W.; Tahvanainen, J. 1993. Colonization and extinction in a population of the shoot-galling sawfly, *Euura amerinae*. Oikos. 68: 448-454.

Skvortsov, A.K. 1968. Willows of the USSR: a taxonomical and geographical revision. Moskovsoe obschchestvo ispytatelei prirody, Moscow. (In Russian).

Smith, D.R. 1979. Suborder Symphyta. In: Krombein, K.V.; Hurd, P.D.; Smith, D.R.; Burks, B.D., eds. Catalog of Hymenoptera in America north of Mexico. Vol. 1. Washington, DC: Smithsonian Institution Press: 3-137.

Strong, D.R.; Levin, D.A. 1979. Species richness of plant parasites and growth form of their hosts. American Naturalist. 114: 1-22.

Zinovjev, A.G. 1982. A contribution to the knowledge of the gall-making sawflies in the genus *Pontania* Costa (Hymenoptera, Tenthredinidae). In: Lev, P.A., ed. Hymenoptera of the Far East. Vladivostok: 18-25. (In Russian).

Zinovjev, A.G. 1985. On the systematics of the sawflies in the genus *Pontania* O. Costa (Hymenoptera, Tenthredinidae). Subgenus *Eupontania* subg. n. Trudy Zoological Akademie Nanka SSSR. 132: 3-16. (In Russian).

Zinovjev, A.G. 1993a. Subgenera and Palaearctic species-groups of the genus *Pontania*, with notes on the taxonomy of some European species of the *viminalis*-group (Hymenoptera: Tenthredinidae). Zoosystematica Rossica. 2(1): 145-154.

Zinovjev, A.G. 1993b. Host-plant specificity of the gall-making sawflies of the genus *Pontania* O. Costa (Hymenoptera: Tenthredinidae). In: Reznik, S.Y., ed. Food specificity of insects. Proceedings of the Zoological Institute, Vol. 193. Gidrometeoizdat, St. Petersburg: 108-139. (In Russian).

Zinovjev, A.G. 1998. Palearctic sawflies of the genus Pontania Costa (Hymenoptera: Tenthredinidae) and their host-plant specificity. In: Csóka, Gyuri; Mattson, William J.; Stone, Graham N.; Price, Peter W., eds. The biology of gall-inducing arthropods: proceedings of the international symposium; 1997 August 14-19; Mátraküred, Hungary. Gen. Tech. Rep. NC-199. St. Paul, MN: U.S. Department of Agriculture, Forest Service, North Central Research Station: 204-225.

PALEARCTIC SAWFLIES OF THE GENUS *PONTANIA* COSTA (HYMENOPTERA: TENTHREDINIDAE) AND THEIR HOST-PLANT SPECIFICITY[1]

A. G. Zinovjev

Zoological Institute, Academy of Sciences, St.Petersburg 199034, Russia

Abstract.—Host-plant relationships and specificity in *Pontania* (Hymenoptera, Tenthredinidae) are discussed. The list of their host-plant species including more than 200 combinations is given for species from the subgenera *Eupontania*, *Pontania* s.str., and *Phyllocolpa* (in the Palearctic region). I assume that the extent of host specificity in *Pontania* corresponds to the type of gall formation, which is characteristic for species groups and subgenera. Leaf-rolling species of the subgenus *Phyllocolpa* are less specialized than gall-making species of the subgenera *Pontania* and *Eupontania*. Species of the latter subgenus are usually monophagous, but some of them rarely occur on subsidiary host plants and some of local populations might be associated with different host plant species. Each species group of the subgenera *Eupontania* and *Pontania* s.str. is characterized by a spectrum of host-plants. Related species of gall-making *Pontania* are usually associated with related willow species. In a particular area each species of *Salix* may be a main host plant for not more than one sawfly species from each species group of gall-making *Pontania*. That niche-exclusion among related species of gall-making *Pontania* might be attributed to speciation connected with a shift from one willow species to another. I admit that a change of a host-plant has been observed in J.W. Heslop-Harrison's experiments (the shift of *P. aestiva* Thomson from *S. myrsinifolia* to *S. purpurea* * *S. viminalis*).

INTRODUCTION

The larvae of sawflies of the family Tenthredinidae are either free-feeding on different plants or living in galls, mines, fruits, etc. The largest number of gall-makers are known from the subfamily Nematinae (Hymenoptera, Tenthredinidae). Closed galls or leaf rolls ("open galls") on willows and other Salicaceae are induced by *Pontania* Costa (including the subgenus *Phyllocolpa* Benson) and *Euura* Newman, which comprise the subtribe Euurina. The species of *Decanematus* Malaise (= *Amauronematus viduatus* group) from the subtribe Nematina (tribe Nematini) inhabit growing willow buds at early larval stages, although mature larvae of these sawflies are occasionally free-feeding. Probably, *Micronematus monogyniae* Hartig (subtribe Pristiphorina of Nematini) is also a gall-maker: its larvae live under folded leaf margins of *Prunus spinosa*. One can also consider the sawflies of the tribe Pristolini to be gall-makers in a broader sense. A nearctic species, *Pristola macnabi* Ross lives inside berries of *Vaccinium membranaceum*. Larvae of the European representative of this tribe, *Bacconematus pumilio* (Konow) feed inside enlarged berries of *Ribes nigrum*. Beyond the subfamily Nematinae, gall-makers are also known in the tribe Lycaotini (an European species *Hoplocampoides xylostei* Giraud and Nearctic one *Blennogeneris spissipes* Cresson). Both induce galls on Caprifoliaceae: *Lonicera* and *Symphoricarpos*. Gall-makers are also encountered in the family Xyelidae. Besides, a number of free-living sawflies from Tenthredinidae as well as from Argidae and Cimbicidae are known to induce underdeveloped galls, that is, abnormally grown plant tissues at the place

---

[1] *This paper is primarily an English translation of another publication (Zinovjev 1993b). Some minor changes have been made, mainly concerning the taxonomy. The lists of host plants have also been updated. For the completeness, I add the list of host plants for the species of Phyllocolpa, but exclude Nearctic species (they will be treated elsewhere).*

of egg insertion. The galls of *Euura* from the subgenus *Gemmura* E.L. Smith, as well as those caused by *Decanematus* and probably also by *Blennogeneris*, constitute entire plant organs (buds, shoots, or fruits) modified by the sawfly. According to Slepyan (1973), they should be called "parazitoidal teratomorphs".

Here, I will only discuss the host-plant specificity of the genus *Pontania* Costa, which is widely distributed throughout the Holarctic Region. This genus may be divided into three subgenera (or genera) and a number of species groups. There are more than 30 species of *Pontania* known in Europe, Asia, and North America, have not been adequately explored. So far, it is hard to evaluate the total number of species and even species groups. In Eurasia, eleven species groups are recognized (Benson 1941, 1960a; Kopelke 1986, 1989b, 1991; Zinovjev 1982, 1985, 1993a; Zinovjev in Zhelochovtsev 1988): three in the subgenus *Phyllocolpa* Benson (*piliserra*-, *leucapsis*-, and *leucosticta*-groups); three more in the subgenus *Pontania* s. str. (*crassispina*-, *dolichura*-, and *proxima*-); and some five in the subgenus *Eupontania* Zinovjev (*relictana*-, *vesicator*-, *herbaceae*-, *polaris*-, and *viminalis*-groups).

The larvae of *Pontania* s.l. may either inhabit closed leaf galls of variable shape or live inside rolled leaves or under folded leaf margins. The subgenus *Phyllocolpa* consists entirely of leaf-rolling species. Species of the *crassispina*-group from the subgenus *Pontania* s. str. are also leaf-rollers. The two other groups of *Pontania* s. str. as well as all of *Eupontania* are typical gall-makers. Traditionally, the sawflies living under rolled leaf margins have been called leaf-rollers, whereas those producing closed galls, gall-makers. I am going to follow this convenient terminology. Usually females of sawflies induce gall development during egg-laying. However, this is absolutely true just for *proxima*- and *dolichura*-groups (that is, for the subgenus *Pontania* s. str.). In *Eupontania*, larvae also play an important role in gall development, while females only initiate it.

Species of *Pontania* s.l. are associated exclusively with Salicaceae, mainly willows. According to A.K. Skvortsov (1968), Salicaceae are divided in three genera: *Populus*, *Chosenia*, and *Salix*. It is quite obvious that the poplars are more primitive, standing somewhat apart of other Salicaceae. Close filiation of *Chosenia* and *Salix* is also beyond question. It is as well acceptable to treat *Chosenia* as a subgenus of *Salix*. The genus *Salix* includes ca. 350 species united in 3 subgenera and about 30 sections (the classification of the willows here follows that by A.K. Skvortsov). The subgenus *Salix* is the most primitive of all the willows and the closest to the poplars. Most willows of the subgenus *Salix* are rather large trees (or tall shrubs). They are distributed primarily in warm temperate regions, occasionally in the tropics. Species of the more advanced subgenus *Vetrix* are widespread from Southeast Asia to Arctic regions and represented by shrubs and trees of moderate sizes. Distinctiveness of the third subgenus, *Chamaetia*, is still being a subject of discussion, and sometimes *Chamaetia* are included in *Vetrix*. The representatives of *Chamaetia* are mostly low or even prostrate shrubs and undershrubs restricted primarily to arctic and alpine regions.

MATERIALS AND METHODS

This review is based on the literature data, study of galls in herbaria of a number of institutions, as well as my own experiments and collections from different parts of European Russia, the Urals, Russian Far East, Northeastern Siberia, the Baltic States, Ukraine, and Finland. Cultivated (potted) willow plants were used for cross-experiments and studies of *Pontania* host-specificity. Occasionally, I succeeded in obtaining quite normal, mature galls (although somewhat undersized) on willow cuttings growing merely in water with some soil added. I was able to identify the majority of host plants excluding most difficult cases when I sought help from Prof. A.K. Skvortsov (the Main Botanical Garden, Moscow).

Since species taxonomy of both *Pontania* and *Salix* is rather intricate, their identification is not always reliable. Therefore, old or doubtful data were either not included in the tables or assigned question marks. One can find those records in the well-known reference books by C. Houard (1908-1913) and H. Buhr (1964-1965). Accidental host-plants were also distinguished by question marks, since a possibility of complete sawfly development in these galls remains unproved. The literature data from the following works were included in the tables without references: Benes (1967, 1968), Benson (1940, 1941, 1958, 1960a, b, 1962), Heslop-Harrison (1937), Kopelke (1985, 1986, 1987, 1989a, b, 1990a, b, 1991), Lacourt (1973), Vikberg (1965, 1970), Yukawa and Masuda (1996). The species of *Pontania*, which are so far known to me only as galls or leaf rolls, were marked "galls"; those known only from herbarium materials were marked "herb". All data based on doubtful identification of herbarium specimens were excluded. The data on *Pontania* found on willow hybrids were excluded as well[2].

---

[2] *Interspecific hybrids are rather common in willows. However, the opinion on their high frequency is largely exaggerated. According to M. Wichura's estimations, the relative frequency of willow hybrids in nature usually does not exceed 0.5 percent and is mostly far less than that. Many species do not produce hybrids at all, neither in nature nor in experiments. Besides, many distinct willow species have been mistakenly treated as hybrids (Wichura 1865, Buser 1940, Skvortsov 1968, and others).*

Sometimes, it was hard to distinguish between groups of *Pontania* in herbarium material, especially between the *viminalis-* and *polaris-*group, because of the existence of intermediate morphological forms. Therefore, I do not separate the *polaris-*group from *viminalis-*group in the table 9.

HOST PLANT RANGES OF THE *PONTANIA* SPECIES-GROUPS

Tables 1-9 show the range of *Pontania* s.l. host plants. Of all *Pontania*, the subgenus *Phyllocolpa* has the widest host plant range. Its species occur on willows from the subgenera *Salix*, *Vetrix*, and occasionally *Chamaetia* as well as on *Chosenia* and *Populus* (table 1). There is one species of the leaf-rolling *Pontania* which is found on poplars (*P. populi* Marlatt) in North America. The gall-making *Pontania* s.str. and *Eupontania* never occur on poplars.

Willows from the subgenus *Vetrix* are most susceptible to gall-making *Pontania* s.l. Those from the subgenus *Salix* are less susceptible: the sawflies from the *viminalis-*and *herbaceae-*groups occur on these willows only as very rare exceptions; those from the *relictana-*, *vesicator-*, and *dolichura-*groups, only on particular willow species (see tables 4, 6, 7). As for the willows from the subgenus *Chamaetia*, these may be hosts just for a few groups of *Pontania* s.l. (table 1). For example, such taxa as *Phyllocolpa* and the *proxima-*group are very unusual on willows from *Chamaetia*. However, rather then avoid the willows, the sawflies might merely escape the harsh climatic conditions of the alpine zone and northern latitudes, where the majority of *Chamaetia* grow. The prostrate habit in most of *Chamaetia* might be another cause of their low susceptibility to leaf-rolling *Pontania*.

The sawflies of the *viminalis-*group, the most advanced in *Eupontania*, are associated with *Vetrix* and some *Chamaetia* (table 9). It is as well possible to trace correlations between sawflies and their host plants within the *viminalis-*group, the largest one of the subgenus *Eupontania*. A complex of species closely related to *P. arcticornis* Konow is restricted to a group of willows from closely related sections *Arbuscella*, *Vimen*, *Subviminales*, and *Villosae*.

The *polaris-*group is mainly confined to willows from *Chamaetia*, undershrubs or very low, prostrate shrubs. This species-group is distinguished by galls prominent on both sides of leaf blades. Galls of this kind are to be considered a plesiomorphic state, as compared to those of the *viminalis-*group. In fact, the host plants of these *Pontania*, willows from *Chamaetia*, also do exhibit some primitive features in comparison with *Vetrix* (Skvortsov 1968).

The host plant range of the *dolichura-* and *crassispina-*groups of *Pontania* s.str. is close to that of the *viminalis-*group. However, both the *dolichura-* and *crassispina-*groups have not been found on willows from some sections of *Vetrix* and *Chamaetia* (tables 1, 3, 4). On the contrary, they occur on *Chosenia* as well as some willows of the subgenus *Salix* (*S. cardiophylla* and *S. pierotii*). The *herbaceae-*group has a much narrower host plant range (tables 1, 8). They are mostly confined to willows from closely related sections: *Vimen*, *Arbuscella*, and *Villosae* (the subgenus *Vetrix*). There is only one species in this group which occurs on *S. herbacea* (the subgenus *Chamaetia*). One can also notice an association between Palearctic representatives of the *vesicator-*group and willows of the section *Helix* (table 7). However, this might be merely a coincidence. The whole picture might change considerably when treating *Pontania* on the scale of the world fauna.

As for the *relictana-* and *proxima-*groups, perhaps, the less advanced ones of *Eupontania* and *Pontania* s.str., these are also confined to certain host-plant taxa, though this relation is not that obvious (tables 1, 5, 6). Sawflies from the *proxima-*group are common on willows from some sections of the subgenus *Salix* as well as on those from *Vetrix* (tables 1, 5). Of three species formerly included in the *relictana-*group (Zinovjev 1982, 1985, 1993b), the undescribed one from Siberia on *Salix viminalis* (section *Vimen*) appears not to belong there. The other two by all means are closely related. They occur on plants from different genera (one of them on *Chosenia*, another on *Salix cardiophylla*). However, of all willows, *S. cardiophylla* is the closest to *Chosenia*, they are even able to hybridize.

HOST PLANT RELATIONSHIPS IN *PONTANIA*

The confinement patterns discussed above may be attributed to a number of reasons. Theoretically, there exists a possibility of parallel evolution in the willows and sawflies. However, in most instances, it is more likely that sawflies choose willows closely related to original host plants merely because of a biochemical affinity in willow species of close filiation. This affinity makes it possible for sawflies to shift on from one host to another. On the other hand, the striking confinement of the gall-making *Pontania* (at least the most advanced of them) to the most advanced Salicaceae may hardly be attributed to a mere coincidence. It may be rather associated with historical factors, such as different evolutionary ages of *Pontania* groups.

Each of the species groups of *Pontania* s.l. has its own characteristic range of host plants. Although ranges of groups may considerably overlap (tables 1-9), there is not any overlapping of host ranges between closely related species within each group. Close species within the *viminalis-*, *vesicator-*, *relictana-*, and *dolichura-*groups (presumably, also from the *proxima-* and *herbaceae-*groups) are found each only on a particular willow species. A few different species of gall-making *Pontania* may simultaneously occur on the same host willow species in a particular geographical region, yet all of them would belong to different groups. For example, four species of sawflies which belong to four different groups (*proxima-*, *dolichura-*, *herbaceae-*, and *viminalis-*) have been found in the Far East on *S. schwerinii*; three species (belonging to *dolichura-*, *viminalis-*, and *vesicator-*groups) are encountered on the European *S. purpurea*.

This notion about the impossibility of coexistence of closely related gall-makers from the same group on one host species might be a simplification; yet nearly no exceptions are known to me (tables 4-9). The literature data on findings of closely related *Pontania* on same willow species might be attributed to misidentification (either of willow or sawfly species). In my own experience, it was only once that I reared two gall-makers of the same group on one willow species. These were *P. pedunculi* Hartig and *P.* cf. *arcticornis* from the *viminalis-*group on *S. aurita* from the flood plain of the Volkhov River. However, due to striking differences in their morphology, these species appear to be most distant within the *viminalis-*group. Besides, *S. aurita* was obviously just a subsidiary host plant for *P.* cf. *arcticornis*, which is normally associated with *S. cinerea* (at that particular site both *S. aurita* and *S. cinerea* were abundant).

The absence of closely related sawfly species on same host plants could be treated as an extreme case of the competitive elimination. However, it is hardly reasonable to explain the complete differentiation of ecological niches in these *Pontania* by the direct competition. Some other highly specialized and closely related *Nematinae* are known to exist on the same host. For instance, in a particular habitat, on a single plant specimen, one may find larvae of some closely related *Decanematus* species and leaf-rolling *Pontania* (e.g., *P. nudipectus* Vikberg and *P. tuberculata* Benson from the *crassispina-*group on *S. phylicifolia*). There is no reason to assume that the competition between species in these groups is less pronounced than in those of the true gall-making *Pontania*.

Another way to explain the principle of the niche-exclusion in the gall-making *Pontania* is to consider the possibility for closely related sawfly species to mate when they breed on the same plant. Then the selection might lead to an ecological divergence of species. However, this hybridization might as well take place if species of sawflies live on different willows, as long as these willow species grow together in the same habitat. Any long-lasting coexistence of closely related species on a single host plant might be only possible if at least one of them is parthenogenetic.

The very possibility to attribute the discussed phenomenon to this reason is not improbable. The reproductive isolation process in the gall-making *Pontania* as well as in other *Nematinae* should be a subject of a special study. However, the niche isolation in gall-making *Pontania* is most likely to be associated with their intimate adaptations to plant species. Among all the exo- and endophytic *Nematinae*, the true gall-makers exhibit most intricate relations with their host plants (namely, these are *Eupontania* as well as *proxima-* and *dolichura-*groups from *Pontania* s.str., and also the genus *Euura*). One may assume that a species or at least any particular population of the gall-making *Pontania* may become adapted only to a single host plant species or to a group of very closely related species. Then one may treat the absence of any other closely related gall-makers on any particular host plant as merely a result of their speciation process associated with the divergence to plant species.

THE EXPERIMENTAL STUDIES OF PLANT SPECIFICITY IN *PONTANIA*

The study of plant specificity in *Pontania* has been considerably complicated by the intricate systematics of this group, as well as extremely obscure diagnostic characters which are not always reliable. For a long time, an opinion has dominated that in Europe there were just a few species of *Pontania*, each with a very wide host plant range. In his latest works, Benson (1960a, 1960b, 1962) came to the conclusion that *P. viminalis* L., as well as *P. crassipes* Thomson, and *P. dolichura* Thomson were oligophagous species. However, all of the available experimental data prove that the gall-making *Pontania* are highly specialized.

The study by M. Carleton (1939), which R. Benson referred to as "classical", clearly demonstrated the existence of two "biological races" in *P. proxima* Lepeletier, one of them living on *S. fragilis*, another on *S. triandra*. Both forms exhibited definite preferences during egg-laying. Females originating from *S. fragilis* could hardly lay eggs on *S. triandra* when *S. fragilis* was unavailable for them. More than two-thirds of them merely refrained from egg-laying. Of those 130 galls which nevertheless appeared on *S. triandra*, 52 were underdeveloped, many of them containing no eggs. The other 78 produced just a few undersized weak females incapable of egg-laying. All galls that developed on *S. triandra* retained shape and color typical for galls on *S. fragilis*, although they were undersized. Genetic determination of these characters was beyond question. The reciprocal experiment, that is, trying to make *S. fragilis*-originating females lay eggs on *S. triandra*, had a similar result, although it was less impressive because of a smaller sample size. In the latter case, imagoes on *S. fragilis* were not obtained by M. Carleton at all. Later on, R. Benson (1941) recognized some small, inconspicuous morphological differences between those "races" and segregated them in distinct species within the *proxima*-group. These species were: *P. proxima* Lepeletier feeding on *S. alba*, *S. fragilis*, and their hybrids; *P. triandra* Benson, absolutely restricted to *S. triandra*; and *P. bridgmannii* Cameron, confined to *S. caprea* and other closely related willow species (the biology of the latter sawfly species was never studied in an experiment). All three species are sympatric, parthenogenetic, and confined to willows from different sections: *Salix*, *Amygdalinae*, and *Vetrix*, accordingly. There are also galls found on willows from other sections of the subgenus *Vetrix*. All of them presumably belong to *P. bridgmanni* (see table 5).

J.-P. Kopelke (1986) obtained similar results in his experimental study of three species from Central Europe, formerly treated as *P. dolichura*. All three of them proved to be monophagous. Either they rejected alien willow species or the galls were underdeveloped, although they retained the color typical for these species on their normal host plants. These underdeveloped galls even failed to produce mature larvae. The morphological differences found by J.-P. Kopelke were demonstrated on just a few specimens. Still, these differences are sufficient for making a decision about distinctiveness of the studied species: *P. dolichura* sensu Kopelke on *S. purpurea* (Sect. *Helix*), *P. nigricantis* Kopelke on *S. nigricans* (*S. myrsinifolia*, Sect. *Nigricantes*), and *P. helveticae* Kopelke on *S. helvetica* (Sect. *Villosae*). All of them are bisexual, which makes them different from the *proxima*-group.

These experiments were done using comparatively small number of specimens. Although they demonstrated distinctiveness of the species studied by M. Carleton and J.-P. Kopelke, they still did not exclude a possibility for sawflies to change their hosts. The experiments proved only that the probability of these changes was extremely low. In this connection, the old experiments of J.W. Heslop-Harrison (1927) deserve special attention, since he used thousands of galls and obtained a very large number of sawflies.

J.W. Heslop-Harrison conducted his research in England, studying "*Pontania salicis* Christ" (*P. viminalis* auct.). The first series of trials was held on a large willow plantation where more than 20 willow species and hybrids were growing. Cocoons from galls grown on *S. purpurea* were dispersed there during the autumn. In the following year, galls appeared only on *S. purpurea*. According to J.W. Heslop-Harrison, there were no galls even on hybrids *S. purpurea* * *S. viminalis*.

Transferring cocoons from *S. phylicifolia* had somewhat different results. This time, galls were successfully developing not only on *S. phylicifolia*, but also *S. andersoniana* Smith (which to my mind might actually be *S. myrsinifolia*). Moreover, some galls (1 to 8 percent) were found on hybrids of *S. purpurea* and *S. viminalis*. They were underdeveloped during the first and second season. However, on the third year mature larvae were found in seven galls that developed on the hybrid willows (on *S. phylicifolia*, there were 60 mature galls recorded; on *S. andersoniana*, 572; there were no galls found on other willow species).

The second series of experiments on host change are still of more interest. They were conducted on a small, territorially isolated plantation of *S. rubra* (*S. purpurea* * *S. viminalis*). Twice, cocoons were taken from a natural population of *S. andersoniana* and removed to the plantation of hybrid willows. During the early years, nearly all of the appearing galls stayed underdeveloped. However, according to J.W. Heslop-Harrison, a few years later, the population got adapted to the new host plant and succeeded to survive. More than that, when eventually some specimens of *S. andersoniana* were planted on the experimental lot, galls never appeared on the former host willow.

These experiments were conducted in conditions close to natural, so that an absolute isolation could not be provided. Therefore, an accidental invasion of sawflies belonging to other species was possible during the trials. Some time later, R. Benson (1940) described a species from northern England and Scotland (*P. harrisoni* Benson), which appeared to be close to *P. viminalis*. It produced galls of an irregular shape on *S. purpurea* as well as its natural hybrids (*S. purpurea* * *S. viminalis*). It was reasonable to attribute the results of the second set of J.W. Heslop-Harrison's experiments to the invasion of *P. harrisoni* to the plantation of willow hybrids, *S. purpurea* * *S. viminalis* (cf. Benson 1950). However, I would accept that J.W. Heslop-Harrison really succeeded in modelling of a host plant change. Obviously, the experiment was successful due to a large number (hundreds or maybe even thousands) of involved specimens. As stated by J.W. Heslop-Harrison, the shift resulted in an extremely high mortality which was close to 100 percent during the first few years. At the same time, one must keep in mind that he could misidentify or misunderstand some of the species of *Pontania* and *Salix*.

According to our present knowledge, *P. viminalis* L. represents a complex of closely related species. Three species appear to have been involved in his experiments: *P. viminalis* L. living on *S. purpurea*, *P. aestiva* Thomson (*P. viminalis* sensu Vikberg) on *S. myrsinifolia*, and probably also *P. arcticornis* Konow associated with *S. phylicifolia*. The populations of *P. aestiva* from the British Islands should be treated as a separate subspecies, *P. aestiva harrisoni* Benson (Zinovjev 1994). The irregularity of the gall shape in specimens examined by R. Benson might be attributed to their development on a subsidiary host plant (in my own experiments with *P. viminalis*, the three galls developed on the unusual host plant, *Salix babylonica*, were also of quite unusual shape; as for J.W. Heslop-Harrison, unfortunately, he never mentioned the shape of galls in the description of his experiments).

It is as well probable that J.W. Heslop-Harrison had trouble distinguishing *S. myrsinifolia* (*S. andersoniana*) from *S. phylicifolia*, which appears to be a common problem with many of researchers. According to V. Vikberg, in Finland, each of these two willows is a host plant of its own specific sawfly, both of them belonging to the *viminalis*-group: *P. arcticornis* Konow (*P. phylicifoliae* Forsius) and *P. aestiva* Thomson (*P. viminalis* sensu Vikberg). The former species was encountered only on *S. phylicifolia*; the latter, exclusively on *S. myrsinifolia*. The same relationship was encountered in the vicinity of St. Petersburg: *P. arcticornis* was never found there on *S. myrsinifolia*, whereas *P. aestiva* was recorded on *S. phylicifolia* not more than two times (the identity of these two specimens of willows with *S. phylicifolia* is not confirmed due to missing floriferous buds). Since both sawfly species are common in that region and easy to distinguish even by their galls, one can treat these results with confidence. Yet, according to other authors (Heslop-Harrison 1937; Benson 1958, 1960a, 1962; Kangas 1985; and others), each of the two sawfly species may occur either on *S. phylicifolia* or on *S. myrsinifolia* (*S. nigricans*). One can imagine two possible explanations of this contradiction (they may as well work for J.W. Heslop-Harrison experimental results). There may exist some geographical variability of *Pontania* species regarding their host specificity as well as variability of willow species regarding resistance to sawflies. However, much more probable is the assumption that in these cases we merely deal with wrong identifications of the host plants and occasionally also the sawflies.

In fact, delimiting of *S. phylicifolia* and *S. myrsinifolia* (*S. andersoniana*, *S. nigricans*, *S. borealis*) constitutes some real difficulties. Both are extremely polymorphic, look much alike each other, and are able to hybridize. Starting from C. Linnaeus, a number of authors either treated these species as a single one or admitted the introgressive hybridization on a large scale between them. Indeed, natural hybrids between the two species are not infrequent. Besides, the occurrence of their hybrids might be higher in disturbed habitats. However, as it was emphasized by A.K. Skvortsov (1968), the notion about the overall hybridization is attributed to a misunderstanding of the species. In any event, it is necessary to conduct a special study concerning the occurrence of *Pontania* species on both willow species and their hybrids, particularly, in Britain.

To my mind, J.W. Heslop-Harrison succeeded in modelling of the host plant change when shifting *P. aestiva* Thomson (*P. harrisoni* Benson) from *S. myrsinifolia* (*S. andersoniana*) to *S. rubra* (*S. purpurea* * *S. viminalis*). J.W. Heslop-Harrison treated the results of his own experiments from a purely Lamarkian point of view. However, it is quite clear that the high extent of mortality indicated a harsh natural selection, which had to be associated with significant changes in the population structure. However, there remain questions if the shift was really irreversible and if a real change in behavioral reactions or instincts of the sawfly took place, as it was stated by J.W. Heslop-Harrison. This part of his conclusions appears to be doubtful. Also, it remains unclear if the control specimens of *S. andersoniana* were taken from the same population that was first used for gall collecting.

A number of authors have supported the idea about an extreme species-specificity in gall-making *Pontania* (Caltagirone 1964; Benes 1967, 1968; Smith 1970). Observing willow thickets in California, E.L. Smith (1970) found some differences in susceptibility of particular clones of one willow species to certain *Pontania* and *Euura*. These differences might be attributed to an adaptation (preference) of a sawfly to a particular willow clone (or different resistance in willow clones). An alternative explanation is that susceptibility of a particular plant specimen may increase in the course of repeated infestation (Smith 1970). This possibility also should be taken into consideration when analyzing experiments like those conducted by J.W. Heslop-Harrison.

Unfortunately, an intricate mechanism of strict specialization in gall-making sawflies has never been a subject of a study. The strict host specificity of *Pontania* may be attributed either to their preferences while egg-laying or to variability in plant resistance. The latter may result in an abnormal development of galls and large mortality of larvae.

Resistance mechanisms of host plants may be quite different. According to my own observations, in some few days after the egg-laying, there appears a slight swelling 1 to 3 mm in diameter. Whenever galls grow on an inappropriate plant, usually they fail to develop any further. That happened, for instance, when I reared females of *P. viminalis* on *S. purpurea* and then put them on *S. babylonica*, which is a completely alien species from a different subgenus. The females laid eggs quite eagerly, however, in most cases, the only result was a small swelling not larger than 3 mm. Though I did not dissect those underdeveloped galls, I suppose that the termination of gall development was caused by the death of sawflies at the stage of either an egg or larva of the first instar. Of the total of more than a hundred, only three distorted galls developed to maturity. Those produced just two dwarf infertile females (Zinovjev 1994: figs. 10-12).

During my trials with other sawfly species, I could not even observe the initiation of gall development, and eggs died during the very first day. Moreover, when *P. aestiva* females that were reared on *S. myrsinifolia* occasionally laid eggs on *S. phylicifolia*, the next day the eggs got completely necrosed together with adjacent leaf tissues.

DISCUSSION

The gall-making *Pontania* appear to belong to either very strict oligophages or monophages, although their monophagy is not always absolute. In the majority of experiments, females revealed their preferences during the egg-laying, however, so far it is unknown if that character is of a genotypic nature. In case a sawfly succeeds to infect an unusual, alien willow species, galls mostly are either underdeveloped or not initiated at all. Yet, even for the most specialized sawfly species, one cannot fully exclude a possibility of gall developing, even a complete one, on foreign willows.

Also, it is not impossible that in some *Pontania* species, there may exist diverse populations restricted to different host plants. Apparently, *P. aestiva harrisoni* on the southern limit of its distributional area, is reprented by such local populations adapted to unusual hosts (*S. purpurea* and *S. purpurea* * *S. viminalis*). Variability of populations is well-known in other highly specialized insect phytophages (see, for example, Fox and Morrow 1981), however, regarding *Pontania*, it still needs to be proved.

The restriction of particular *Pontania* species to their host plants may be manifested to a variable extent. One may suppose that it depends upon the mode of gall formation. It is a known fact that *Phyllocolpa*, which are leaf-rollers, are less specialized than true gall-makers (Smith 1970). Although *Phyllocolpa* species were never studied experimentally, there is no doubt that some of them are oligophagous. To take an example, *P. (Ph.) acutiserra* Lindqvist has been encountered on two willows belonging to different sections (*Villosae* and *Glaucae*) and even to different subgenera (*Vetrix* and *Chamaetia*). Within the European fauna, it is *P. (Ph.) leucapsis* Tischbein that has the vastest range of host species. It may occur on a number of species belonging to different sections of the subgenus *Vetrix* and even on willows from the subgenus *Salix*. Some of *Phyllocolpa* species are restricted to willows belonging to particular sections. For instance, *P. (Ph.) piliserra* Thomson and *P. (Ph.) anglica* Cameron are found on willows from the section *Vimen*; *P. (Ph.) coriacea* Benson and *P. (Ph.) leucosticta* Hartig, on some willows from the section *Vetrix*. Among the Palearctic *Phyllocolpa*, to my knowledge, there are only two undescribed species that can be confidently named monophagous. Obviously, their monophagy is attributed to an isolated taxonomical position of their host plants, *Chosenia arbutifolia* and *Salix cardiophylla*. The latter belongs to the monotypic section *Urbanianae* (occasionally, it is even treated as a separate genus *Toisusu*). Generally, the species belonging to *Phyllocolpa* are not very much different in their host specificity from those belonging to exophytic Nematini, among which one may also find some strictly specialized species.

Gall-making sawflies of the subgenera *Eupontania* and *Pontania* s. str. (*proxima*- and *dolichura*-group) are confined either to particular willow species or to some closely related species from the same section. Very few of them, if any, may develop on willows belonging to closely related sections. There are only a few known records of developing on willows of distant relation. However, it appears much more probable that we deal here either with sibling species or one species living on subsidiary host plants (in other words, distantly related willows can be host plants for the same sawfly species only as rare exceptions). To take an example, J.-P. Kopelke (1989b) showed that *P. polaris* Malaise (*P. crassipes* Thomson sensu Benson) had to be divided in a number of species. Also, some distinct species are still hiding under the name of *P. arcticornis* Konow. I have found the species belonging to this complex on willows from related sections (*Arbuscella*, *Vimen*, *Subviminales*, and *Villosae*) and also on *S. cinerea* from the section *Vetrix* (Zinovjev 1995). Host specificity of *P. pedunculi* Hartig is not yet clear. This species is associated with willows from the section *Vetrix*, however, not all of them. In European Russia, it was never found on *S. cinerea*, occurring only on *S. aurita*, *S. starkeana*, and *S. caprea*. Any literature data recording this species on *S. cinerea* appears to be erroneous. This may be probably attributed to misidentifications of either the sawfly, or willow, or both of them. On the other hand, this sawfly species has been reared on *S. myrtilloides*, which belongs to another section and even a different subgenus. It has been described as *P. myrtilloidica* Kopelke, 1991. However, we still need to exclude a possibility that *S. myrtilloides* is merely a subsidiary host for *P. pedunculi*. It is worth saying, that, according to A.K. Skvortsov (1968), the taxonomical position of the entire section *Myrtilloides* is rather obscure: it might be merely a comparatively recent derivative of the subgenus *Vetrix*, particularly the section *Vetrix* (or *Incubaceae*).

As mentioned above, the subgenera *Eupontania* and *Pontania* differ in the mode of gall formation. In *Pontania* s. str. (*proxima-* and *dolichura*-group), the galls are produced exclusively by females, while in *Eupontania* (at least in the majority of species) the gall development is controlled by females and, later on, by larvae. As a result, a more tight connection between the sawfly and its host must develop in *Eupontania*. Therefore, one may further assume that the species of *Eupontania* are generally more host-specific than that of *Pontania* s. str.

Some *Eupontania* might develop on several willow species, yet the majority of them are monophagous. Sometimes, one may even find a few distinct sawfly species from the same group developing on closely related willows, all in the same habitat (see tables 6, 9). Closely related sawflies developing on related willow species are unknown in *Pontania* s. str., which makes them different from *Eupontania*. The species from *proxima-* and *dolichura*-groups might be associated with entire willow sections rather than particular species. Such species as *P*. (s.str.) *triandrae* and those from *dolichura*-group studied by J.-P. Kopelke (1986) may just appear to be monophagous: in particular regions, host plants of these sawflies are the only representatives of their sections.

Differences in specificity between the species of *Pontania* s.l. may result in their different capability to shift from one host to another, i.e., their different occurrence on subsidiary host species. In the subgenus *Eupontania*, findings of sawfly species on subsidiary hosts, appear to be extremely rare. The records of *P. viminalis* on willows of the subgenus *Salix* may be treated as a relevant example. Apparently, in the subgenus *Pontania* s. str. (*proxima-* and *dolichura*-groups) this phenomenon is more common. In these two groups, populations of a particular species are more likely to get adapted to a different host plant species.

The following observations may serve as an illustration. In the Upper Kolyma Basin ('Aborigen' Research Station), I found once a few galls of a sawfly belonging to *dolichura*-group on *S. pulchra*. Those galls encountered on a single plant looked underdeveloped, although each of them had an exit hole. At the same habitat, galls from *dolichura*-group were frequent on *S. myrtilloides* (which belongs to a different subgenus). In an experiment, a female reared from those galls on *S. myrtilloides* produced galls on *S. phylicifolia* (an European species, which is very close to *S. pulchra*). One of the galls even developed to maturity.

Analyzing these results, one may suggest that *S. pulchra* served as a subsidiary host to the sawfly living regularly on *S. myrtilloides*. Yet it is quite possible that in other parts of its distributional area *S. pulchra* might be the primary host either for the same or another *Pontania* species (see table 4).

One more example of a presumable geographical host-plant shift in the *dolichura*-group may be demonstrated on an undescribed species associated with some willows from the section *Vetrix*. Galls on willows of this section (except *S. iliensis*) are quite peculiar: short and roundish. No doubt that they are produced by a distinct species (or closely related species). In the northern Sikhote-Alin Mountains, I found them on *S. taraikensis*; in Siberia and northern Finland, on *S. bebbiana*. In southern Finland (Vikberg 1970) and also probably in Armenia (according to gall description by Mirzoyan 1970), they were encountered on *S. caprea* (see table 4). It is worth saying that these three willow species belong to two different subsections (*Laeves* and *Substriatae*); they are by no means vicarious and sometimes occur together in same habitats.

The existence of geographical populations of the same species associated with different host species is as well possible in *Eupontania*. Particularly, it appears to be true for *P. arcticornis* Konow. However, for *Eupontania*, the geographical change of host plants might be an exception rather than a rule. On rare occasions, when an area of a sawfly embraces geographical ranges of two vicarious host species (or subspecies), one can notice considerable morphological differences between populations, like those in the subspecies of *P. acutifoliae* Zinovjev (Zinovjev 1993c).

CONCLUSIONS

The sawflies of the genus *Pontania* s.l. are encountered exclusively on willows and other Salicaceae. The genus comprises a number of distinct species groups that appear to be monophyletic. Many species are either monophagous or strictly oligophagous. The extent of host specificity in *Pontania* species is presumably contingent with the gall type and mode of gall formation, which is characteristic for each species group. Among *Pontania* s.l., those less specialized are *Phyllocolpa;* those most specialized are *Eupontania*. *Pontania* s.str. presumably occupy an intermediate position regarding this character.

According to my own experimental data and field observations, the monophagy in *Pontania* may be conditional. Occasionally, the gall initiation and development may take place on unusual, subsidiary hosts. Also, local populations may get adapted to different host plants. However, findings of galls on alien host plants appear to be extremely rare, at least in *Eupontania*. One may assume that local populations adapted to different host plants are as well rare in the latter subgenus.

Generally, for species of *Pontania* s.l., there exists a possibility to enlarge (or narrow down) the host range in particular parts of the distributional areas. For the gall-making *Pontania*, the geographical variability of this kind appears to be less probable and in every particular case has to be confirmed experimentally. The niche-exclusion principle is valid for the gall-making *Pontania*. That means that one can never find two representatives of the same species group of the gall-making *Pontania* on the same host plant in a particular area. Also, closely related species are nearly always encountered on different host plant species. This phenomenon is probably attributed to an extremely narrow specialization of gall-making *Pontania* and, consequently, their speciation, which is associated with the divergence over different host plant species. On the other hand, this might be partially connected with the competition as well as mechanisms of reproductive isolation in related species of *Pontania*. Special investigations are necessary to prove the latter hypothesis.

One can also notice some interesting peculiarities when analyzing host-plant ranges in complexes of closely related species within the largest *viminalis*-group. In a number of cases, closely related *Pontania* are associated with host plants of close filiation, i.e., confined to willows belonging to either one section or related sections.

Among the willow species, these are representatives of the subgenus *Vetrix*, which are most susceptible to *Pontania*. More primitive Salicaceae are less susceptible to the gall-making *Pontania*. These willows are host species for less advanced *Eupontania* (from the *relictana*-, and *vesicator*-groups), as well as some *Pontania* s.str. and *Phyllocolpa*. Some more advanced groups of sawflies occur on more advanced willows, at least in the subgenus *Eupontania*.

Relationships of this kind are often attributed to the parallel evolution of parasites (in this particular case, the gall-makers) and their hosts. By all means, one cannot exclude a possibility of occasional parallel evolution of sawflies and their host plants. However, the speciation in sawflies might as well take place independently from the evolution of host plants, that is, by means of shift from one willow species to another. Although the shift may more likely involve closely related willows, it might as well be accomplished on distantly related host plants. According to our current knowledge, *Pontania* s.l. are phytogenetically much younger than *Salix*. The restriction of advanced gall-making *Pontania* to advanced willow subgenera (*Vetrix* and *Chamaetia*) could then be attributed to filiation of *Pontania* during a later time period when these younger willow subgenera started to dominate in a genus *Salix*.

ACKNOWLEDGEMENTS

I am greatly indebted for contributions of collected materials to A. Antropov (Moscow), E.L. Kaimuk (Yakutsk), M. Viitasaari and V. Vikberg (Finland), A.D. Liston (Germany and Scotland), I. Gauld and N. Springate (England), K. Benes (Czechia); J.-P. Kopelke, H.H. Weiffenbach, and A. Taeger (Germany); J. Lacourt (France); R. Danielsson (Sweden); D.R. Smith and W.J. Pulawski (U.S.A.); H. Goulet (Canada). I am grateful to curators of botanical collections in St.Petersburg, Moscow, Vladivostok, and Yuzhno-Sakhalinsk. I particularly appreciate the help of A.K. Skvortsov (the Main Botanical Garden R.A.S., Moscow).

LITERATURE CITED

Benes, K. 1967. Czechoslovak species of *Pontania crassipes*-group (Hymenoptera, Tenthredinidae). Acta entomologica bohemoslovaca. 64(5): 371-382.

Benes, K. 1968. Galls and larvae of the European species of genera *Phyllocolpa* and *Pontania* (Hymenoptera, Tenthredinidae). Acta entomologica bohemoslovaca. 65(2): 112-137.

Benson, R.B. 1940. Further sawflies of the genus *Pontania* Costa (Hym. Symphyta) in Britain. The Entomologist's Monthly Magazine. 76(911): 88-94.

Benson, R.B. 1941. On some *Pontania* species, with a revision of the *proxima* and *herbaceae* groups (Hymenoptera, Symphyta). The Proceedings of the Royal entomological Society of London. Series B. Taxonomy. 10(8): 131-136.

Benson, R.B. 1950. An introduction to the natural history of British sawflies (Hymenoptera, Symphyta). Transactions of the Society for British Entomology. 10(2): 45-142.

Benson, R.B. 1958. Hymenoptera. 2. Symphyta. Section (c). London. 139-252. [Handbooks for the identification of British Insects; 6, pt 2(c)].

Benson, R.B. 1960a. Studies in *Pontania* (Hym., Tenthredinidae). Bulletin of the British Museum (Natural History). Entomology. 8(9): 369-384.

Benson, R.B. 1960b. Some more high-alpine sawflies (Hymenoptera, Tenthredinidae). Mitteilungen der Schweizerischen entomologischen Gesellschaft. 33(3): 173-182.

Benson, R.B. 1962. Holarctic sawflies (Hymenoptera: Symphyta). Bulletin of the British Museum (Natural History). Entomology. 12(9): 379-409.

Buhr, H. 1964-1965. Bestimmungstabellen der Gallen (Zoo-und Phytocecidien) an Pflanzen Mittel-und Nordeuropas. 2. G. Fischer, Jena, 1572 p.

Buser, R. 1940. Kritische Beitrage zur Kenntnis der schweizerichen Weiden. Berichte. Schweizerische Botanische Gesellschaft. 50: 567-788.

Caltagirone, E.L. 1964. Notes on the biology, parasites and inquilines of *Pontania pacifica*, a leaf-gall incitant on *Salix lasiolepis*. Annals. Entomological Society of America. 57(3): 279-291.

Carleton, M. 1939. The biology of *Pontania proxima* Lep., the bean gall sawfly of willows. Journal Linnean Society of London. Zoology. 40: 575-624.

Fox, L.R.; Morrow, P.A. 1981. Specialization: species property or local phenomenon?. Science. 211(4485): 887-893.

Heslop-Harrison, J.W. 1927. Experiments in the egg-laying instincts of the sawfly, *Pontania salicis* Christ., and their bearing on the inheritance of acquired characters; with some remarks on a new principle in evolution. The Proceedings of the Royal entomological Society of London. Series B. Taxonomy. 101(707): 115-126.

Heslop-Harrison, J.M. 1937. The gallmaking sawflies of the genus *Pontania* in Durham and Northumberland. Entomologist. 70(887): 73-75.

Houard, C. 1908-1913. Les Zoocécidies des plantes d'Europe et du Bassin de la Méditeranée. 1-3. Librairie Scientifique A. Hermann et Fils, Paris, 1550 p.

Kangas, J.K. 1985. Pälkäneen sahapistiäisfauna 1953-1983. Pälkäne, Pälkäne-Seuran julkaisuja nro 5. 113 p.

Kopelke, J.-P. 1985. Über die Biologie und Parasiten der gallenbildenden Blattwespenarten *Pontania dolichura* (Thoms. 1871), *P. vesicator* (Bremi 1849) und *P. viminalis* (L. 1758) (Hymenoptera: Tenthredinidae). Faunistish-Ökologische Mitteilungen. 5(11/12): 331-344.

Kopelke, J.-P. 1986. Zur Taxonomie und Biologie neuer Pontania Arten der *dolichura*-Gruppe (Insecta: Hymenoptera: Tenthredinidae). Senckenbergiana biologica. 67(1/3): 51-71.

Kopelke, J.-P. 1987. Die Rolle der Brutparasiten in Feindkomplex der gallenbildenden *Pontania*-Arten (Hymenoptera: Tenthredinidae: Nematinae). Zoologische Jahrbucher. Abteilung für Systematik, Ökologie und Geographie der Tierre. 114(4): 487-508.

Kopelke, J.-P. 1989a. Der taxonomische Status von *Pontania crassipes* (Thomson, 1871) (Insecta: Hymenoptera: Tenthredinidae). Senckenbergiana biologica. 69(1/3): 29-39.

Kopelke, J.-P. 1989b. Mittel-und nordeuropaischen Arten der Gattung *Pontania* Costa 1859 aus der *herbacea*- und *polaris*-Gruppe (Insecta: Hymenoptera: Tenthredinidae). Senckenbergiana biologica. 69(1/3): 41-72.

Kopelke, J.-P. 1990a (1989). Der taxonomische Status von *Pontania dolichura* (Thomson, 1871) (Insecta: Hymenoptera: Tenthredinidae). Senckenbergiana biologica. 70(4/6): 271-279.

Kopelke, J.-P. 1990b. Wirtspezifität als Differenzierungskriterium bei gallenbildenden Blattwespenarten der Gattung *Pontania* O. Costa (Hymenoptera: Tenthredinidae: Nematinae). Mitteilungen der deutschen Gesellschaft für allgemeine und angewandte Entomologie, Giessen. 7: 527-534.

Kopelke, J.-P. 1991 (1990). Die Arten der *viminalis*-Gruppe, Gattung *Pontania* O.Costa 1859, Mittel-und Nordeuropas (Insecta: Hymenoptera: Tenthredinidae). Senckenbergiana biologica. 71(1/3): 65-128.

Lacourt, J. 1973. Deux nouvelles espèces de *Nematinae* du Maroc (Hymenoptera, Tenthredinidae). Du Bulletin de la Société des Sciences Naturelles et Physiques du Maroc. 53: 189-192.

Mirzoyan, S.A. 1970. Sawflies and horntails the pests of the trees and shrubs in Armenia. Proceeding of the Institute of Plant Protection, Erevan. 1: 95-127 (In Russian).

Skvortsov, A.K. 1968. Willows of the USSR: a taxonomical and geographical revision. Moscow, Nauka. 262 p. [Materialy k poznaniyu fauny i flory SSSR. N.S. Otd. Bot. 15 (23).] (In Russian with English summary)

Slepyan, E.I. 1973. [Neoplasms and their agents in plants. Gallogenesis and parasitic teratogenesis.] Leningrad, Nauka. 512 p. (In Russian)

Smith, E.L. 1970. Biosystematics and morphology of Symphyta. II. Biology of gall-making nematine sawflies in the California Region. Annals. Entomological Society of America. 63(1): 36-51.

Togashi, I. & Usuba S. 1980. On the species of the genus *Euura* Newman (Hymenoptera, Tenthredinidae) from Japan. Kontyu. 48(4): 521-525.

Vikberg, V. 1965. *Pontania nudipectus* sp.n. (Hym., Tenthredinidae), a new leaf-roller from Eastern Fennoscandia. Annales Entomologici Fennici. 31(1): 53-60.

Vikberg, V. 1970. The genus *Pontania* O. Costa (Hymenoptera, Tenthredinidae) in the Kilpisjärvi district, Finnish Lapland. Annales Entomologici Fennici. 36(1): 10-24.

Weiffenbach, H. 1975. Tenthredinidenstudien III (Hymenoptera). Entomologische Zeitschrift. 85(6): 57-59.

Wichura, M. 1865. Die Bastardbefruchtung in Pflanzenreich erläutern an den Bastarden der Weiden. Breslau, 96 S.

Yukawa, J.; Masuda H., eds. 1996. Insect and mite galls of Japan in colors. Zenkoku-Nouson-Kyouiku-Kyoukai, Tokyo. 829 p. (In Japanese)

Zinovjev, A.G. 1982 (1981). A contribution to the knowledge of the gall-making sawflies of the genus *Pontania* Costa (Hymenoptera, Tenthredinidae). In: Ler, P.A.; ed. Pereponchatokrylye Dal'nego Vostoka [Hymenoptera of the Far East]. Vladivostok. Institute of Biology and Pedology, Academy of Sciences of the USSR: 18-25. (In Russian)

Zinovjev, A.G. 1985. On the systematics of the sawflies of the genus *Pontania* O. Costa (Hymenoptera, Tenthredinidae). Subgenus *Eupontania* subg. n. Proceedings of the Zoological Institute, Academy of Sciences of the USSR. 132: 3-16. (In Russian).

Zinovjev, A.G. 1993a. Subgenera and Palaearctic species groups of the genus *Pontania*, with notes on the taxonomy of some European species of the *viminalis*-group (Hymenoptera: Tenthredinidae). Zoosystematica Rossica. 2(1): 145-154.

Zinovjev, A.G. 1993b. Host-plant specificity of the gall-making sawflies of the genus *Pontania* O. Costa (Hymenoptera, Tenthredinidae). In: Reznik, S.Ya., ed. [Food specialization of insects (ecological, physiological, evolutionary aspects). Proceedings of the Zoological Institute, Academy of Sciences of the USSR. 193. St.Petersburg, Gidrometeoizdat: 108-139. (In Russian)

Zinovjev, A.G. 1993c. Geographic variation of *Pontania acutifolia* (Hymenoptera, Tenthredinidae) and possibility of parallel evolution of the gall-maker and its host-plant. Zoologicheskiy Zhurnal. 72(8): 36-50. (in Russian with English summary). Translated in Entomological Review. 1994. 73(1): 142-155.

Zinovjev, A.G. 1994. Taxonomy and biology of two related species of gall-making sawflies from the *Pontania viminalis*-group (Hymenoptera: Tenthredinidae). Entomologica Scandinavica. 25(2): 231-240.

Zinovjev, A.G. 1995. The gall-making species of *Pontania* subgenus *Eupontania* (Hymenoptera, Tenthredinidae) of eastern Fennoscandia and their host plant specificity. Acta Zoologica Fennica. 199: 49-53.

Zhelochovtsev, A.N. 1988. Suborder Symphyta (Chalastogastra). In: Zhelochovtsev, A.N.; Tobias, V.I.; Kozlov, M.A., eds. [Keys to the insects of the European part of the USSR], vol. 3 (Hymenoptera), pt. 6. Leningrad, Nauka: 7-234. (In Russian)

Table 1.—*The Ranges of host plants for Palearctic species groups of the genus* Pontania

Host-plant taxa	1	2 4	2 5	2 6	3 7	3 8	3 9	3 10	3 11
Populus	+	-	-	-	-	-	-	-	-
Chosenia	+	+	+	-	+	-	-	-	-
Salix									
Subg. *Salix*									
Sect. *Urbanianae*	+	-	+	-	+	-	-	-	-
Sect. *Humboldtianae*	+	-	-	-	-	-	-	-	-
Sect. *Amygdalinae*	+	-	-	+	-	-	-	-	-
Sect. *Pentandrae*	+	-	-	+	-	-	-	-	?
Sect. *Salix*	+	-	-	+	-	-	-	-	?
Sect. *Subalbae*	+	+	-	?	-	+	-	-	?
Subg. *Chamaetia*									
Sect. *Chamaetia*	-	-	-	-	-	-	-	+	+
Sect. *Retusae*	-	-	+	-	-	-	+	+	-
Sect. *Myrtosalix*	+	+	+	-	-	-	-	+	+
Sect. *Glaucae*	+	+	+	?	-	-	-	+	+
Sect. *Myrtilloides*	-	+	+	-	-	-	-	+	+
Subg. *Vetrix*									
Sect. *Hastatae*	+	+	+	+	-	-	-	-	+
Sect. *Glabrella*	-	-	+	-	-	-	-	-	+
Sect. *Nigricantes*	+	-	+	?	-	?	-	-	+
Sect. *Vetrix*	+	+	+	+	-	-	-	-	+
Sect. *Arbuscella*	+	+	+	+	-	+	+	+	+
Sect. *Vimen*	+	+	+	+	+	-	+	-	+
Sect. *Villosae*	+	+	+	-	-	-	+	-	+
Sect. *Subviminales*	+	-	+	-	-	-	-	-	+
Sect. *Canae*	-	-	+	-	-	-	-	-	+
Sect. *Lanatae*	+	-	+	-	-	-	-	-	+
Sect. *Daphnella*	+	+	-	-	-	-	-	-	+
Sect. *Incubaceae*	-	+	+	-	-	-	-	-	+
Sect. *Helix*	+	+	+	-	-	+	-	-	+
Sect. *Cheilophilae*	-	-	-	-	-	-	-	-	?

1 - Subg. *Phyllocolpa*
2 - Subg. *Pontania*
3 - Subg. *Eupontania*
4 - *crassispina* species-group
5 - *dolichura* species-group
6 - *proxima* species-group
7 - *relictana* species-group
8 - *vesicator* species-group
9 - *herbaceae* species-group
10 - *polaris* species-group
11 - *viminalis* species-group

Table 2.—*List of host plants for the subgenus* Phyllocolpa

Host plant genera, subgenera, sections	Host plant species	*Pontania* species
Populus	*suaveolens*	*P.* cf. *excavata* Marlatt (**new record**)
Chosenia	*arbutifolia*	*P.* sp.
Salix		
Subgenus *Salix*		
Sect. *Humboldtianae*	*acmophylla*	**New record** (herb.)
Sect. *Amygdalinae*	*triandra* ssp. *triandra*	*P. puella* Thomson (needs confirmation)
Sect. *Urbanianae*	*cardiophylla*	*P.* sp.
Sect. *Pentandrae*	*pentandra*	*P. excavata* Marlatt
	pseudopentandra	**New record**
Sect. *Salix*	*alba*	*P. puella, P. leucapsis* Tischbein
	fragilis and hybrids with *alba*	*P. puella, P. leucapsis* Tischbein
Sect. *Subalbae*	*pierotii*	*P.* sp.
Subgenus *Chamaetia*		
Sect. *Glaucae*	*glauca*	*P. acutiserra* Lindqvist
	arctica	**New record**
Sect. *Myrsinites*	*myrsinites*	**New record**
Subgenus *Vetrix*		
Sect. *Hastatae*	*pyrolifolia*	**New record** (herb.)
	reinii	**New record** (herb.)
	jenisseensis	**New record** (herb.)
Sect. *Nigricantes*	*myrsinifolia* (= *nigricans*)	*P. leucapsis* Tischbein, perhaps *P. leucosticta* Hartig
Sect. *Vetrix*	*silesiaca*	*P. coriaceae* Benson ?
	caprea	*P. leucosticta, P. leucapsis* Tischbein
	cinerea	*P. leucosticta, P. coriacea* Benson, *P. leucapsis* Tischbein
	atrocinerea	*P. leucapsis* Tischbein, *P. leucosticta*
	aurita	*P. leucapsis* Tischbein, *P. leucosticta, P. coriacea* Benson
	starkeana	*P. leucapsis* Tischbein
	bebbiana	**New record**
	taraikensis	**New record**
	abscondita	**New record**
Sect. *Arbuscella*	*phylicifolia*	*P. leucapsis* Tischbein, as an exception also *P. excavata* Marlatt
	pulchra	**New record**
	kazbekensis	**New record** (herb.)
	rhamnifolia	**New record**
Sect. *Vimen*	*viminalis*	*P. anglica* Cameron, *P. piliserra* Thomson, *P. scotaspis* Förster
	schwerinii	*P. scotaspis* Förster and *P. anglica* Cameron
	dasyclados	*P. piliserra* Thomson, *P. anglica* Cameron
	udensis	*P.* spp.
Sect. *Subviminales*	*gracilistyla*	**New record**
Sect. *Villosae*	*lapponum*	*P. acutiserra* Lindqvist
	alaxensis	**New record**
Sect. *Lanatae*	*lanata*	**New record**
Sect. *Daphnella*	*daphnoides* and *acutifolia*	*P. leucapsis* Tischbein
	rorida	**New record**
	kangensis	**New record**

(table 2 continued on next page)

(*table 2 continued*)

Host plant genera, subgenera, sections	Host plant species	*Pontania* species
Sect. *Helix*	*kochiana*	**New record** (herb.)
	purpurea	P. *leucapsis* Tischbein
	miyabeana	**New record**
	integra	**New record**
	gilgiana	**New record** (herb.)
	tenuijulis	**New record** (herb.)
	pycnostachya	**new record** (herb.)
	niedzwieckii	**New record** (herb.)

Table 3.—*List of host plants for* Pontania crassispina-*group*

Host plant genera, subgenera, sections	Host plant species	*Pontania* species	Distribution
Chosenia	*arbutifolia*	*Pontania* sp.n.1	Russian Far East, Japan
Salix			
Subg. *Salix*			
Sect. *Urbanianae*	*cardiophylla*	*Pontania* sp.n.	South of the Russian Far East (Maritime Province). **New record**
Sect. *Subalbae*	*pierotii* (*eriocarpa*)	*Pontania* sp.n.2	South of the Russian Far East, Japan
Subg. *Chamaetia*			
Sect. *Myrtilloides*	*myrtilloides*	P. *tuberculata* Bens. ?	Magadan Reg.
Sect. *Glaucae*	*glauca*	P. *crassispina* Thoms.	Lappland, Northern Yakutia
	arctica (?)	P. *crassispina* Thoms. ?	Kamchatka (herb.)
	sphenophylla	P. *crassispina* Thoms. ?	Magadan Reg.
Sect. *Myrtosalix*	*saxatilis*	P. *tuberculata* Bens. ?	Magadan Reg.
Subg. *Vetrix*			
Sect. *Hastatae*	*hastata*	P. *tuberculata* Bens.	Northern Europe, Lower Lena River
Sect. *Vetrix*	*caprea*	P. *joergenseni* Ensl.	Finland (Kangas, 1985) and Leningrad Reg.
	starkeana	P. *cf. tuberculata*	Leningrad Reg.
	bebbiana	P. *tuberculata* Bens. ?	Magadan Reg.
Sect. *Arbuscella*	*phylicifolia*	P. *nudipectus* Vikberg	Fennoscandia, Leningrad Reg.
		P. *tuberculata* Benson	Fennoscandia, Leningrad Reg.
	pulchra	*Pontania* sp.	Magadan Reg.: Upper Kolyma River
	pulchra	P.? *tuberculata* Bens.	Anadyr Basin (galls)
	boganidensis (including *S. dshugdshurica*)	P.? *tuberculata* Bens.	Magadan Reg.
	rhamnifolia		Yakutsk Reg. **New record**
Sect. *Vimen*	*udensis*	*Pontania* sp.n.3	Russian Far East (Maritime Province)
Sect. *Villosae*	*alaxensis*		Northern Magadan Reg. (galls)
	krylovii		Northern Magadan Reg. (galls)
Sect. *Daphnella*	*acutifolia*	P.cf. *purpureae*	Moscow Reg.: Oka River (galls)
Sect. *Incubaceae*	*brachypoda*	P. sp.	Yakutia
Sect. *Helix*	*kochiana*		Transbaikalia (herb.); **new record**.
	purpurea	P. *purpureae* Cam.	Middle Europe, Latvia, S. of Pscov, Carpathians
	integra	P. cf. *tuberculata* Bens.	Maritime Province: Posjet Bay, Kunashir Island: eastern shore of Lake Goryachee (galls)

Table 4.—*List of host plants for* Pontania dolichura-*group*

Host plant genera, subgenera, sections	Host plant species	*Pontania* species	Distribution
Chosenia	arbutifolia	*Pontania* sp.n.4	Russian Far East
Sect. *Urbanianae*	*S. cardiophylla* (*urbaniana*)	*Pontania* sp.	Maritime Province, Sakhalin, Kunashir Island, Japan
Sect. *Retusae*	*S. herbacea*		Norway: Vardo, Veranger Peninsula, 22 VI 1989, leg. S.Schmidt, H.Goulet, and N.Springate (together with galls on *S. lapponum* and *S. lapponum* x *S. herbacea*, det. A.K. Skvortsov)
	S. polaris		Nortnern Europe; Nortnern Magadan Reg. (rare)
	S. retusa (*kitaibeliana*)		Mountains of Middle Europe
Sect. *Myrtilloides*	*S. myrtilloides*		Magadan Reg., Upper Kolyma River (common)
	S. fuscescens		Norhern Yakutia (Tiksi). **New Record**
Sect. *Glaucae*	*S. glauca* (*glaucosericea*)	*P. glaucae* Kopelke	North and Middle Europe
Sect. *Myrtosalix*	*S. myrsinites*		Nortnern Europe
	S. alpina		Mountains of Middle Europe
	S. saxatilis		Magadan Reg., Upper Kolyma River (common)
Sect. *Hastatae*	*S. hastata*		Nortnern Europe
	S. pyrolifolia		Yakutia: Nyurba (herb.), around Yakutsk
Sect. *Glabrella*	*S. glabra*		Mountains of Middle Europe
Sect. *Nigricantes*	*S. myrsinifolia* (*nigricans*)	*P. nigricantis* Kopelke	Northern and Middle Europe
Sect. *Vetrix*	*S. caprea*	*Pontania* sp.n. 5	Finland (North of Inari and Muddusjärvi); ? Armenia (Mirzoyan 1970)
	? *S. cinerea*		Europe (needs confirmation)
	? *S. aurita*		Europe (needs confirmation)
	S. bebbiana (*xerophila*)	*Pontania* sp.n. 5	Northern Finland; Yakutia, Magadan Reg., Kamchatka (Kluchi, 11.VIII.78 (single leaf with pair of galls)
	S. taraikensis	*Pontania* sp.n. 5?	Northern Sikhote-Alin: Tumnin (galls)
	S. iliensis	*P.* cf. *dolichura*	Caucasus (herb.)
Sect. *Arbuscella*	*S. phylicifolia*	*P. dolichura* Thomson (*P. femoralis* Cameron)	Nortnern Europe; Arckangelsk Reg. (herb.)
	S. pulchra		Magadan Reg.: Upper Kolyma River (rare galls), Anadyr River (galls)
	S. arbuscula		Nortnern Europe
	S. saposhnikovii		Altai Mts. (herb.). **New record**
	S. boganidensis (incl. *dshugdshurica*)	*Pontania* sp. n. 6.	Magadan Reg.
	S. rhamnifolia		Sayan Mts. (herb.). **New record**
Sect. *Vimen*	*S. viminalis*		Arkhangelsk (galls)
	S. schwerinii		Russian Far East
	S. udensis (*sachalinensis*)		Russian Far East

(*table 4 continued on next page*)

(table 4 continued)

Host plant genera, subgenera, sections	Host plant species	*Pontania* species	Distribution
Sect. *Subviminales*	S. gracilistyla		Maritime Province (common)
Sect. *Canae*	S. elaeagnos	P. elaeagnocola Kopelke	Middle Europe
Sect. *Villosae*	S. lapponum		Europe
	S. alaxensis		Nortnern Magadan Reg. (galls)
	S. krylovii		Nortnern Magadan Reg. (galls)
	S. helvetica	P. helveticae Kopelke	Middle Europe
Sect. *Lanatae*	S. lanata		Nortnern Europe
Sect. *Dapnella*	S. dapnoides		Europe ?
Sect. *Incubaceae*	S. rosmarinifolia		Leningrad Reg., Estonia
	S. brachypoda		Yakutia (herb.)
Sect. *Helix*	S. coesia		Altai (herb.)
	S. purpurea	P. virilis Zirngiebl (*dolichura* auct. in part, *rifana* Lacourt)	Middle Europe and North Africa, Baltic Republics, Carpathians
	S. vinogradovii?		Lower Dnepr River, Southern Volga Basin (herb.)
	S. elbursensis		Caucasus (herb.)
	S. miyabeana		Khabarovsk, Southern Maritime Province: Kraskino (galls)
	S. integra		Southern Maritime Province: Ryazanovka, Khasan (galls)
	S. tenuijulis		Middle Asia (herb.)
	S. linearifolia		Middle Asia (herb.)
	S. capusii		Middle Asia (herb.)
Incertae sedis	S. japonica	*Pontania shibayanagii* (Togashi et Usuba)	Japan (Togashi, Usuba 1980: as *Euura shibayanagii*)

Table 5.—*List of host plants for* Pontania proxima-*group*

Host plant genera, subgenera, sections	Host plant species	Distribution
Salix		
Subg. *Salix*		
Sect. *Amygdalinae*	*S. triandra* ssp. *triandra*	Middle Europe, Caucasus (herb.), Leningrad and Moscow Reg., Middle Urals
	S. triandra ssp. *nipponica*	Maritime Province, Amur Reg.
Sect. *Pentandrae*	*S. pentandra*	? West Europe (very rare or willow misidentified)
Sect. *Salix*	*S. alba*	Europe. North America (adventive)
	S. excelsa	Middle Asia (herb.)
	S. fragilis	Europe. North America (adventive)
Sect. *Subalbae*	*S. babylonica*	? Middle Europe (very rare or willow misidentified)
Subg. *Vetrix*		
Sect. *Hastatae*	*S. hastata*	West Europe
	S. pyrolifolia	Yakutia. **New record**.
Sect. *Nigricantes*	*S. myrsinifolia*	Europe, Middle Ural
Sect. *Vetrix*	*S. silesiaca*	Middle Europe
	S. appendiculata	Alps
	S. caprea	West Europe, Caucasus (herb.), Leningrad Reg., Middle Urals, Russian Far East
	S. cinerea	West Europe, Middle Urals
	S. atrocinerea	West Europe
	S. aurita	West Europe
	S. taraikensis	Russian Far East
Sect. *Arbuscella*	*S. phylicifolia*	West Europe, Leningrad Reg. (not very common)
	S. foetida	Alps
Sect. *Vimen*	*S. schwerinii*	Russian Far East (common)
	S. pantosericea	Caucasus (herb.)
	S. udensis	Russian Far East (common), Japan

Remarks: Old records of *Pontania proxima* on arcto-alpine willows of the subgenus *Chamaetia* are obviously erroneous and should belong to species of the *polaris*-group. However, on herbarium specimens identified as *S. arctica* (*pallasii*), *S. kurilensis*, and *S. nakamurana*, collected on the Kurils and Kamchatka I have seen galls that might belong to this group. These data need confirmation. Three species of this group are recorded in Europe: *P. bridgmanni* Cameron on willows of the subgenus *Vetrix*, *P. triandrae* Benson on *Salix triandra*, and *P. proxima* Lepeletier on *Salix alba*, *S. fragilis*, and, perhaps as a as rare exception, on willows from the related sections (*S. pentandra* and *S. babylonica*).

Table 6.—*List of host plants for* Pontania relictana-*group*

Host plant genera, subgenera, sections	Host plant species	*Pontania* species	Distribution
Chosenia	*Ch. arbutifolia*	*P. relictana* Zinovjev	Kamchatka, Magadan Reg., Northern Sikhote-Alin: Tumnin
Salix			
Subgenus *Salix*			
Sect. *Urbanianae*	*S. cardiophyla*	*P. mirabilis* Zinovjev	Southern Maritime Province and Southern Sakhalin

Remarks: Similar galls are known on *Salix viminalis* in Northeastern European Russia and Siberia. This species probably does not belong to the *relictana*-group. It might be more closely related to some Nearctic species.

Table 7.—*List of host plants for* Pontania vesicator-*group*

Host plant genera, subgenera, sections	Host plant species	*Pontania* species	Distribution
Sect. *Subalbae*	*S. pierotii*	*P. mandshurica* Zinovjev	Maritime Province and Amur Reg.
	S. babylonica	*P.* cf. *mandshurica*	Japan (Yukawa & Masuda 1996: *Pontania* sp. B)
Sect. *Arbuscella*	*S. phylicifolia*	*P. pustulator* Forsius[1]	Nortnern Europe; Russia: Arkhangelsk Reg., Karelia
Sect. *Helix*	*S. purpurea*	*P. vesicator* Bremi[2]	Middle Europe, Baltic Republics, Carpathians
	S. miyabeana	*P. amurensis* Zinovjev	Maritime Province, Amur Reg., Transbaikalia
	S. integra	*P. integra* Zinovjev	Southern Maritime Province, Japan
	S. tenuijulis	perhaps *P. nostra* Zhelochovtsev	Tian-Shan (herb.)
	S. pycnostachia	*P. nostra* Zhelochovtsev	Tian-Shan, Pamirs
Incertae sedis	*S. chaenomeloides*	*P.* sp.	Japan (Yukawa & Masuda 1996: *Pontania* sp. A)

Remarks: Galls, similar to those of the *vesicator*-group are known from northeastern Siberia and North America on *Salix alaxensis*. This sawfly appears mot to be closely related to any of the Palearctic species.

[1] *Salix myrsinifolia* (*nigricans*) was listed as another host for *P. pustulator* due to a misidentification of the willow.

[2] *P. vesicator* was also recorded on other, not closely related willows, however, these old records have to be attrbuted to species from other groups of *Pontania*.

Table 8.—*List of host plants for* Pontania herbaceae-*group*

Host plant genera, subgenera, sections	Host plant species	*Pontania* species	Distribution
Salix			
Subgenus *Chamaetia*			
Sect. *Retusae*	*S. herbacea*	*P. herbaceae* Cameron	West Europe
Subgenus *Vetrix*			
Sect. *Arbuscella*[1]	*S. arbuscula*	*P. arbusculae* Benson	Scotland, Kolguyev Island (herb.)
	S. foetida	*P. foetidae* Kopelke	Alps
	S. saposhnikovii		Altai (herb.)
	S. boganidensis (including *dshugdshurica*)	*Pontania* sp.	Magadan Reg.
	S. rhamnifolia	*Pontania* sp.	West of Laike Baikal
Sect. *Vimen*	*S. viminalis*	*Pontania* sp.	From Baltic Sea to Yakutia
	S. schwerinii	*Pontania* sp.	Russian Far East
	S. udensis	*Pontania* sp.	Russian Far East
Sect. *Villosae*	*S. lapponum*	*P. lapponica* Malaise	Scandinavia, Kola Peninsula (herb.), ? Scotland
	S. helvetica	*P. maculosa* Kopelke	Alps

[1] Sawflies of this group are characteristic for the subsection *Arbusculae*, but never occur on other willows from the section *Arbuscella*.

Table 9.—*List of host plants for* Pontania viminalis- *and* polaris-*groups*[1]

Host plant genera, subgenera, sections	Host plant species	*Pontania* species	Distribution
Salix			
Subgenus *Salix*			All records *Pontania viminalis* auct. on *Salix alba*, *S. fragilis*, *S. babylonica*, and *S. triandra* are either based on willow misidentification or represent incidental hosts.
Subgenus *Chamaetia*			
Sect. *Chamaetia*	*S. reticulata*	*P. reticulatae* Malaise (*viminalis*-group)	Northern Europe and Mts. of Middle Europe
	S. reticulata	*P.? arctica* MacGillivray (*polaris*-group)	Arctic Siberia
	S. vestita	species of (*polaris*-group)	Altai Mts. (herb.)
Sect. *Retusae*	*S. herbacea*	*P. polaris* Malaise (*polaris*-group)	North Europe and Mts. of Middle Europe, Kolguev Iskand (herb.: larva examined)
	S. turczninowii	*P.? polaris* Malaise (*polaris*-group)	Sayan Mts. (herb.: larva examined)
	S. polaris	*P. polaris* Malaise (*polaris*-group)	Northern Europe; Kolguyev Island (herb.), Novaya Zemlya (herb.), Chukotka and Northern Magadan Region
	S. nummularia	*P.? polaris* Malaise (*polaris*-group)	Lower Lena River, Polar Urals
	S. retusa (*kitaibeliana*)	*P. retusae* Benson (*polaris*-group)	Mts. of Middle Europe
Sect. *Myrtilloides*	*S. myrtilloides*	*P. myrtilloidica* Kopelke, 1991 (very similar to *pedunculi* Hartig) (*viminalis*-group)	Norway, Leningrad Reg. (rare), Amur Reg., Magadan Reg.
	S. fuscescens	species from the *polaris*-group	Magadan Reg., Kamchatka, and Chukotka
Sect. *Glaucae*	*S. glauca*	*P. nivalis* Vikberg (*viminalis*-group)	Scandinavia and Mts. of Middle Europe, Khibins, Altai (herb.)
		species from the *polaris*-group	Arctic Siberia (galls)
	S. arctica	species of the *polaris*- and perhaps also *viminalis*-groups	Polar Urals and Northern Siberia
	S. pyrenaica	species from the *polaris*-group	Pyrenees
	S. reptans	species from the *polaris*-group	Lower Lena River
Sect. *Myrtosalix*	*S. myrsinites*	*P. myrsiniticola* Kopelke, 1991[2]	West Europe
	S. alpina	*P. montivaga* Kopelke, 1991 (*viminalis*-group); probably also another species from the *polaris*-group	Mts. of the Middle Europe
	S. saxatilis	species from *viminalis*-group	Magadan Reg.
		species from *polaris*-group	Magadan Reg.
	S. chamissonis	species from the *polaris*-group	Chukotka (herb.)
	S. rectijulis	species from the *polaris*-group	Altai, Sayan Mts. (herb.)
	S. breviserrata	*breviserratae* Kopelke (*polaris*-group)	Mts. of Middle Europe

(*table 9 continued on next page*)

(table 9 continued)

Host plant genera, subgenera, sections	Host plant species	*Pontania* species	Distribution
Subg. *Vetrix*			
Sect. *Hastatae*	S. hastata	P. hastatae Vikberg	Mts. of Middle Europe; Khibins, Yakutia, Magadan Reg., Chukotka (herb., galls)
	S. pyrolifolia	*Pontania* sp.n.	Siberia, Yakutia (herb.)
Sect. *Glabrella*	S. glabra	P. cf. viminalis	Mts. of Middle Europe (Weiffenbach, 1975: as *P. viminalis*)
	S. reinii		Iturup Island (herb.)
Sect. *Nigricantes*	S. myrsinifolia (nigricans, borealis)	P. aestivus Thomson (*P. viminalis* sensu Vikberg; *P. norvegica* Kopelke, 1991; *P. varia* Kopelke, 1991)	Most of Europe reaching Middle Urals in the east
Sect. *Vetrix*	S. silesiaca	P. pedunculi Hartig	Middle Europe
	S. caprea	P. pedunculi Hartig	Most of Europe, similar galls are known also from Russian Far East
	S. cinerea	*Pontania* cf. *arcticornis*	Germany, Finland, Middle Urals, Nizhnij Novgorod, Pscov, Volkhov River
	S. atrocinerea	P. pedunculi Hartig?	West Europe
	S. aurita	P. pedunculi Hartig; also as an exception P. cf. arcticornis	Most of Europe
	S. starkeana	P. pedunculi Hartig	Leningrad Reg.
	S. bebbiana (xerophula)	P. ? pedunculi Hartig	Yakutia, Russian Far East
	S. taraikensis	P. cf. pedunculi	Yakutia, Russian Far East
	S. abscondita	*Pontania* sp.n.	Yakutia (herb.)
	S. iliensis		Caucasus (herb.)
Sect. *Arbuscella*	S. phylicifolia	P. arcticornis Konow	Most of Northern Europe; reaching Northern England and Latvia in the north
	S. pulchra	P. cf. arcticornis	Siberia
		Pontania of the *polaris*-group	Northern Siberia
	S. arbuscula		Nortnern Europe
	S. foetida		Mts. of Middle Europe
	S. kazbekensis		Caucasus (herb.)
	S. saposhnikovii		Altai Mts. (herb.)
	S. boganidensis	P.? arcticornis Konow	Magadan Reg.
	S. rhamnifolia		East Siberia
Sect. *Vimen*	S. viminalis	P. cf. arcticornis Konow	Yakutia (probably belong to a species living on *S. schwerinii* farther east)
	S. schwerinii	P. cf. arcticornis	Yakutia, Russian Far East
	S. dasyclados	P. cf. arcticornis	from Leningrad Reg. to Yakutia
	S. sajanensis		Altai, Sayan Mts. (herb.)
	S. udensis	P. cf. arcticornis	Russian Far East, Japan
Sect. *Subviminales*	S. gracilistyla	P. cf. arcticornis	Maritime Province and Amur Reg, Japan
Sect. *Canae*	S. elaeagnos	P. kriechbaumeri Konow	Middle Europe

(table 9 continued on next page)

(*table 9 continued*)

Host plant genera, subgenera, sections	Host plant species	*Pontania* species	Distribution
Sect. *Villosae*	*S. lapponum*	*P. samolad* Malaise	Nortnern Europe; Kola Peninsula and Volchov River
	S. alaxensis	*P.* cf. *arcticornis*	Northern Magadan Reg. (galls)
	S. krylovii	*P.* cf. *arcticornis*	Magadan Reg. (herb., galls)
	S. helvetica		Middle Europe ?
Sect. *Lanatae*	*S. lanata*	*P. glabrifrons* Benson	Nortnern Europe, northeastern Siberia
Sect. *Daphnella*	*S. daphnoides*	*P. acutifoliae daphnoides* Zinovjev	Middle Europe, sandy dunes on Baltic Sea shore (Zinovjev 1993c)
	S. acutifolia	*P. acutifoliae acutifoliae* Zinovjev	On sandy sites in European Russia and Ukraine
		P. acutifoliae daphnoides Zinovjev	Sandy dunes in Baltic Republics, Leningrad Reg., sandy shores of Lake Ladoga in Karelia (Zinovjev 1993c)
	S. rorida	*P.* cf. *viminalis*	Yakutia, Amur Reg., Maritime Province, Sakhalin Island
	S. kangensis	*P.* cf. *viminalis*	Southern Maritime Province
Sect. *Incubaceae*	*S. repens*	*P. collactanea* Förster.	West Europe
	S. rosmarinifolia	*P. collactanea* Förster.	West Europe, Leningrad Reg.
	S. brachypoda	*Pontania* sp.	Yakutia, Maritime Province, Amur Reg.
Sect. *Helix*	*S. coesia*		? Alps; Altai, Sayan Mts. (herb.)
	S. kochiana		Altai (herb.)
	S. purpurea	*P. viminalis* L.	Middle Europe including England in the north, Pskov Reg., Latvia, Lithuania, Carpathians, Crimea Peninsula. Kiev and Borisovka (adventive)
	S. vinigradovii		Volga Basin (herb.)
	S. elbursensis		Caucasus (herb.)
	S. miyabeana	*P.* cf. *viminalis*	Maritime Province, Amur Reg.
	S. tenuijulis		Middle Asia (herb.)
	S. kirilowiana		Middle Asia (herb.)
	S. niedzwieckii		Middle Asia (herb.)
	S. capusii		Middle Asia (herb.)
	S. caspica ? (or *S. vinogradovii*)		Belgorod Reg.: Borisovka (adventive)
	S. ledebourana		Altai Mts., Tuva (herb.)
Sect. *Cheilophilae*	*S. wilhelmsiana*		Middle Asia (herb.: very small galls)
Incertae sedis	*S. japonica*		Japan

[1] The species of the *polaris*-group are known mainly on dwarf willows of the subgenus *Chamaetia* and northern dwarf form of *Salix pulchra*.

[2] Probably belongs to the *polaris*-group. The name might be a synonym of *Pontania algida* Benson described from British Islands.

HOST-PLANT ASSOCIATIONS AND SPECIFICITY AMONG CYNIPID GALL-INDUCING WASPS OF EASTERN USA

Warren G. Abrahamson[1,2], George Melika[1,3], Robert Scrafford[1] and György Csóka[4]

[1] Department of Biology, Bucknell University, Lewisburg, PA 17837 USA
[2] Archbold Biological Station, P.O. Box 2057, Lake Placid, FL 33852 USA
[3] Present address: Natural History Department, Savaria Museum, Kisfaludy S. u. 9, 9700 Szombathely, Hungary
[4] Hungarian Forest Research Institute, Department of Forest Protection, 3232 Mátrafüred, P.O. Box 2, Hungary

Abstract.—Two separate field surveys of cynipid gall-inducer occurrences were conducted in Florida, North Carolina, and Pennsylvania, USA. All cynipids demonstrated strong host species and organ fidelity and several gallers showed preferences based on habitat or host-plant size. One result of this specialization is effective niche partitioning among cynipids. The host-association patterns of these specialist herbivores should reflect similarities among oaks, thus we clustered oak species according to their cynipid distributions. Clearly, cynipids distinguish small differences among their hosts. A dendrogram of oak species based on cynipid distributions was reasonably congruent with botanical arrangements. Cynipid occurrences can offer information helpful to resolving plant systematic problems.

INTRODUCTION

Many authors (e.g., Felt 1917, 1940; Mani 1964) have commented on the high level of specificity of gall inducers to their host plants and to specific plant organs. Although gall-inducing animals are broadly scattered across taxonomic lines (i.e., four kingdoms, two phyla, two arthropod classes, and six insect orders), relatively few plant groups account for most galls. Over 90 percent occur on dicotyledonous plants and most of these infest only three families, Rosaceae, Compositae, and Fagaceae (Mani 1964). This specificity is in part a consequence of the gall-inducer's intimate relationship with its host plant (Shorthouse 1982, Raman 1993). Gall insects must be able to induce the appropriate host-plant reactions to produce the cell proliferations that become specific galls (Abrahamson and Weis 1987). Thus, these insects are sensitive to small differences in host-plant physiology, chemistry, development, and phenology. As a result, gall inducers may discriminate among closely related species of host plants (Floate et al. 1996). Indeed, biologists have used gall identity as a clue to plant identity and vice versa (Felt 1917, 1940; Raman 1996). For example, studies of hybrids in *Quercus* (Aguilar and Boecklen 1992, *Salix* (Fritz et al. 1994), and *Populus* (Floate and Whitham 1995) have demonstrated that gall inducers can partially or almost completely distinguish hybrid host plants from parental hosts in hybrid zones. Furthermore, cecidomyiid and tephritid gall inducers were shown recently to discern subspecies of the composite, *Chrysothamnus nauseosus* (Floate et al. 1996). It is likely that the gall inducers respond to subsets of their host-plant's genome that may not necessarily be expressed in traditionally utilized plant taxonomic traits. Therefore, the distributions of gall inducers on plants may provide independent tests for plant phylogenies generated from morphological, chemical, or genetic information. Similarly, gall occurrences may enable discrimination of morphologically similar plant species or hybrids from their parental species (Floate et al. 1996).

The fidelity of cynipids (tribus Cynipini, Cynipidae: Hymenoptera) to oaks demonstrates the degree of monophagy among gall inducers. In North America, for instance, 87 percent of all cynipids attack species of *Quercus* (Fagaceae) and the host range of each wasp species is generally restricted to one or a few related oak species (Abrahamson and Weis 1987). Burks (1979) listed 485 species of oak gall-inducing cynipids for North

America north of Mexico and additional species have been described subsequently. Nearly 300 of these species occur in the eastern USA making this region particularly suitable for cynipid/oak studies. Although the final species total is likely to be somewhat less than these numbers (because of currently unpaired alternate generations), these numbers do suggest the richness of the Cynipini in North America.

Cynipid/oak interactions represent an excellent model on which to base examinations of herbivore preference, factors affecting host association, host shifts and host-race formation, and the processes of speciation. If we can understand the development of the exceptional species richness of cynipid oak gallers, we would gain crucial insights into the processes of speciation. Furthermore, the precision of cynipid host associations makes this taxon a good candidate to examine the processes by which herbivores radiate to novel host plants. Cynipids appear to be very good taxonomists perhaps because of the fitness advantage of placing offspring on a host plant that enables high performance. Consequently, cynipid distributions should reflect similarities and relatedness in host plants and thus may mirror the phylogenetic relationships of oak species. Unfortunately, the taxonomy and phylogenetic relationships of these insects are incomplete and numerous species are yet to be described. However, our understanding of cynipid taxonomy and phylogeny is rapidly progressing (Abrahamson *et al.* 1998, Cook *et al.* 1998, Drown and Brown 1998, Shorthouse 1998). Furthermore, the complex taxonomy of oaks has inhibited markedly our limited comprehension of cynipid host associations (Trelease 1924, West 1948, Muller 1951). Efforts to clarify the phylogenetic relationships of species of *Quercus* will make the rewards of cynipid/oak studies considerably greater.

This study represents an initial attempt to clarify the host specificity and host-association patterns of cynipid gall inducers by examining niche partitioning among this speciose taxon and similarities among the cynipid communities of oak hosts. Examining the similarities among cynipid distributions across oak species allowed us to generate clusters of 'cynipid-similar' oak species. These patterns offer insights into oak relatedness and phylogeny. Finally, our dendrogram of oak species is contrasted with relationships suggested by botanical traits alone.

METHODS

Tree Architecture and Gall Occurrence

We measured ramet height (to the nearest 0.1 meter), height of the lowest canopy, crown diameter (two crown diameters were measured and averaged), and stem basal diameter (nearest mm) of 1,021 ramets of the evergreen red oaks *Q. myrtifolia* (n = 306) and *Q. inopina* (n = 300), and the evergreen white oak *Q. geminata* (n = 415) from three vegetative associations (sandhill, sand pine scrub, and scrubby flatwoods) at the Archbold Biological Station (ABS), Lake Placid, FL (27°11'N, 81°21'W). All sampling was conducted during January and February 1995 and January 1996. Consequently, our censuses represent the richness and abundance of galls persistent during those months. Sampled ramets were selected systematically and equal numbers were sampled in each height category (0.1-1, 1.1-2 meters, etc.).

Two observers searched each oak ramet for cynipid galls for 1, 2, or 5 minutes depending on stem height (Trees ≤1 meter = 1 minute, 1.1-2 meters = 2 minutes, >2 meters = 5 minutes). Small ramets were well sampled (species richness was accurately censused) while larger ramets may have been undersampled. However, Cornell (1986) and our experience found that times of 5 minutes were sufficient to accurately determine species richness even on large trees. Trees greater than 4 meters in height are uncommon in the Lake Wales Ridge dwarf oak communities at ABS.

Scrubby flatwoods consist of a low shrubby association of predominantly evergreen, xeromorphic oaks including *Quercus inopina*, *Q. chapmanii*, *Q. geminata*, as well as abundant understory palms *Serenoa repens* and *Sabal etonia*. Tree presence is variable, typically comprised of widely scattered *Pinus elliottii* var. *densa* and *P. clausa*. The mean height of vegetation of the shrubby vegetation varies with local site conditions and time since fire but is typically 1-2 meters (Abrahamson *et al.* 1984, Abrahamson and Hartnett 1990, Abrahamson and Abrahamson 1996a). Sand pine scrub designates a three-tiered community characterized by a canopy of nearly even-aged *P. clausa*, an intermediate canopy of *Carya floridana*, *Lyonia ferruginea*, *Q. myrtifolia*, *Q. geminata*, *Q. chapmanii*, *Persea humilis*, and *Ilex opaca* var. *arenicola*, and an understory of *S. repens*, *S. etonia*, grasses and herbs (Abrahamson *et al.* 1984, Myers 1990, Abrahamson and Abrahamson 1996b). Southern ridge sandhills, also a three-layered community, has an overstory of *P. elliottii* var. *densa*. A lower deciduous canopy contains *Q. laevis* and *C. floridana*, while the shrub and understory layers are composed of *Q. myrtifolia*, *Q. geminata*, and *Q. chapmanii*, *L. ferruginea*, *S. repens*, *S. etonia*, and others (Abrahamson *et al.* 1984, Myers 1990). The climate of the study area is characterized by hot, wet summers and mild, dry winters. The highest monthly mean (27.5°C) occurs in August and the lowest (16°C) in January. The rainy season normally extends from June through September, with an average of 61 percent of the annual precipitation (1374 mm) falling during this 4-month period (Abrahamson *et al.* 1984).

Host-Association Patterns

To ascertain the host associations of eastern USA cynipid gall inducers, we determined the presence of cynipid gall inducers on all oak species encountered in Florida (from November 1994-May 1995 and during January 1996), North Carolina (mid-October 1995), and Pennsylvania (October 1995-August 1996). We surveyed oaks in a wide variety of vegetative associations to encounter as many oak species as possible. We sampled the following locations on the Florida peninsula: Deering Estate Pineland in Dade County; Jonathan Dickinson State Park in Martin County; and Merritt Island National Wildlife Refuge, Coconut Point Sanctuary, Malabar Scrub Sanctuary, and Enchanted Forest Sanctuary in Brevard County; Archbold Biological Station and Avon Park Air Force Base in Highlands County; Myakka River State Park in Sarasota County; Lake Manatee Recreation Area in Manatee County; and Cedar Key in Levy County. During late April and early May of 1995, we sampled oaks at localities in the Florida panhandle including: Eglin Air Force Base in Okaloosa County, DeFuniak Springs area in Walton County, Torreya State Park in Liberty County, Florida Caverns State Park in Jackson County, and Wakulla Springs State Park in Wakulla County. Surveys were conducted in North Carolina in Carteret, Craven, Jones, and Pamlico Counties. We intensively sampled oaks in many habitats over time in central Pennsylvania including: Lycoming, Montour, Northumberland, Snyder, Sullivan, and Union Counties. These host-association data were supplemented with information available in the literature (e.g., Weld 1951, 1952, 1959; Burks 1979).

RESULTS

Tree Architecture and Gall Occurrence

We surveyed the three most strongly evergreen oak hosts (*Q. myrtifolia, Q. inopina, Q. geminata*) that occur at ABS across habitats (sandhill, sand pine scrub, and scrubby flatwoods. Two of these oaks, *Q. myrtifolia* and *Q. geminata*, are found in varying densities in each of these three associations, while *Q. inopina* is restricted to scrubby flatwoods (Abrahamson et al. 1984; Abrahamson and Hartnett 1990; Myers 1990; Abrahamson and Abrahamson 1996a, b). There were sufficient data on 24 species of cynipid gall inducers to examine their preferences for host-plant species, habitat, or host size.

All 24 species showed strong preference for their host plant. No cynipid that attacked the white oak *Q. geminata* was ever found on the red oaks, *Q. myrtifolia* or *Q. inopina*, or vice versa. Of the cynipids infesting both *Q. myrtifolia* and *Q. inopina*, three species showed significant preferences for one host (table 1). For example, *Amphibolips murata* and *Callirhytis quercusmedullae* showed preferences for *Q. myrtifolia* over *Q. inopina*. However, although *Callirhytis difficilis* attacked both of these oak hosts, it preferred the narrow endemic *Q. inopina* when both hosts were available. Three cynipids exhibited significant preferences for habitat including *Andricus quercuslanigera* and *Callirhytis quercusventricosa* for sand pine scrub while *Amphibolips murata* favored open-canopied scrubby flatwoods (table 1). These habitat preferences were not simply a result of host-plant abundance differences. For example, even though *Amphibolips murata*'s preferred host, *Q. myrtifolia*, was markedly more abundant in sand pine scrub than in scrubby flatwoods (Abrahamson et al. 1984), this cynipid reached its highest abundances in the open-canopied scrubby flatwoods (fig. 1).

The abundances of eight cynipids were related positively to oak size (table 2). We used tree heights, crown diameters, and lowest crown heights to approximate the crown surface areas and volumes of each sampled tree. Each of these measures plus tree stem diameter were regressed on natural-log transformed cynipid abundance. Although most of the significant r^2 values reported are low, they do show that some cynipid species develop larger populations per tree as trees become larger and easier to colonize. Alternatively, cynipids may develop larger populations on larger trees that have become more architecturally complex. Given that many other factors (e.g., natural enemy levels, surrounding vegetation, overstory canopy, isolation of the stand from other similar vegetation, host genotype, edaphic factors) likely also influence cynipid gall-inducer abundance, it is noteworthy that these simple measures of host architecture explain any variation. For instance, two stem-galling cynipids, *Callirhytis difficilis* and *C. quercusmedullae*, developed markedly greater populations on larger oak ramets (table 2). This relationship was particularly clear for *C. quercusmedullae*, due to its abundance on *Q. myrtifolia* (fig. 2). Equally informative, however, is that other cynipids do not show such size-related preferences. Such differences in host-association patterns suggest that each cynipid gall inducer may have a unique relationship with its host plant and that members of this taxon are specialized to particular host species and their resources.

Niche-Space Partitioning Among Cynipids

A principal component analysis of Florida cynipid gall inducers suggested that these insects partition niche space by associating with specific oaks, attacking different host organs, having various lifecycles, and emerging and ovipositing in distinct seasons (fig. 3). For this analysis, we used the Florida subset of data because

Table 1.—*Host species and habitat preferences of cynipid gall-inducing species attacking the red oaks* Quercus myrtifolia *and* Q. inopina, *and the white oak* Q. geminata *sampled in long-unburned stands of sandhill, sand pine scrub, and scrubby flatwoods during January and February 1995 and January 1996 at the Archbold Biological Station, Highlands County, Florida. Results shown are from one-way anovas with Tukey post-hoc tests.*

Host-organ attacked/ Cynipid species	Host species	Habitat
Bud:		
Amphibolips murata	*Q. myrtifolia* $F_{1,161} = 7.3$ $p < 0.008$	scrubby flatwoods $F_{1,161} = 5.4$ $p = 0.021$
Leaf:		
Andricus quercuslanigera	—	sand pine scrub $F_{1,181} = 19.0$ $p < 0.001$
Stem:		
Callirhytis difficilis	*Q. inopina* $F_{1,118} = 16.1$ $p < 0.001$	—
Callirhytis quercusmedullae	*Q. myrtifolia* $F_{1,269} = 10.8$ $p < 0.001$	—
Callirhytis quercusventricosa	—	sand pine scrub $F_{1,78} = 5.3$ $p = 0.023$

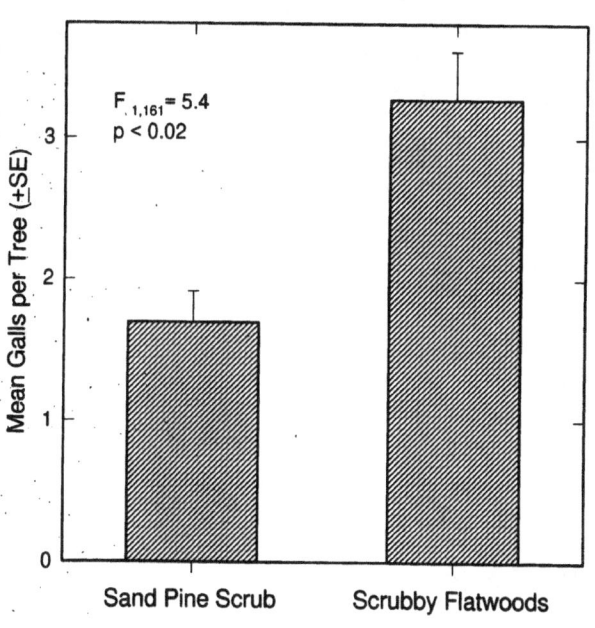

Figure 1.—*Mean ($\pm$ standard error) numbers of* Amphibolips murata *galls per tree in sand pine scrub and scrubby flatwoods vegetative associations at the Archbold Biological Station in south-central Florida, USA. The density of* A. murata's *red-oak host,* Quercus myrtifolia, *is considerably higher in sand pine scrub than in scrubby flatwoods (Abrahamson et al. 1984). Statistical results are from a one-way anova.*

Table 2.—*Proportion (r^2) of natural-log transformed gall occurrence on the red oaks* Quercus myrtifolia *and* Q. inopina, *and on the white oak* Q. geminata *explained by measures of host architecture including tree stem basal diameter, height, crown surface area, and crown volume. Oaks were sampled in long-unburned stands of sandhill, sand pine scrub, and scrubby flatwoods during January and February 1995 and January 1996 at the Archbold Biological Station, Highlands County, FL. — = not significant at p = 0.05. * = proposed new species. Sample sizes of censused trees:* Q. myrtifolia $n = 306$, Q. inopina $n = 300$, Q. geminata $n = 415$.

Host-organ attacked/ Cynipid species	Host species	Basal diameter	Tree height	Surface area	Crown volume
Bud:					
Amphibolips murata	*Q. myrtifolia*	0.04	0.03	0.08	0.07
Amphibolips murata	*Q. inopina* & *Q. myrtifolia*	0.07	0.07	0.12	0.11
Andricus quercusfoliatus	*Q. geminata*	—	0.02	0.02	0.02
Stem:					
*Andricus abrahamsoni**	*Q. inopina*	—	—	0.05	—
Callirhytis difficilis	*Q. inopina*	0.30	0.18	0.22	0.28
Callirhytis quercusmedullae	*Q. myrtifolia*	0.18	0.28	0.25	0.24
Callirhytis quercusmedullae	*Q. inopina*	—	0.30	—	—
Callirhytis quercusmedullae	*Q. inopina* & *Q. myrtifolia*	0.17	0.25	0.24	0.23
Callirhytis quercusbatatoides	*Q. geminata*	—	0.04	0.07	0.07
Disholcaspis quercusvirens	*Q. geminata*	0.04	0.05	0.03	0.03
Leaf:					
Neuroterus quercusminutissimus	*Q. geminata*	—	0.08	—	—

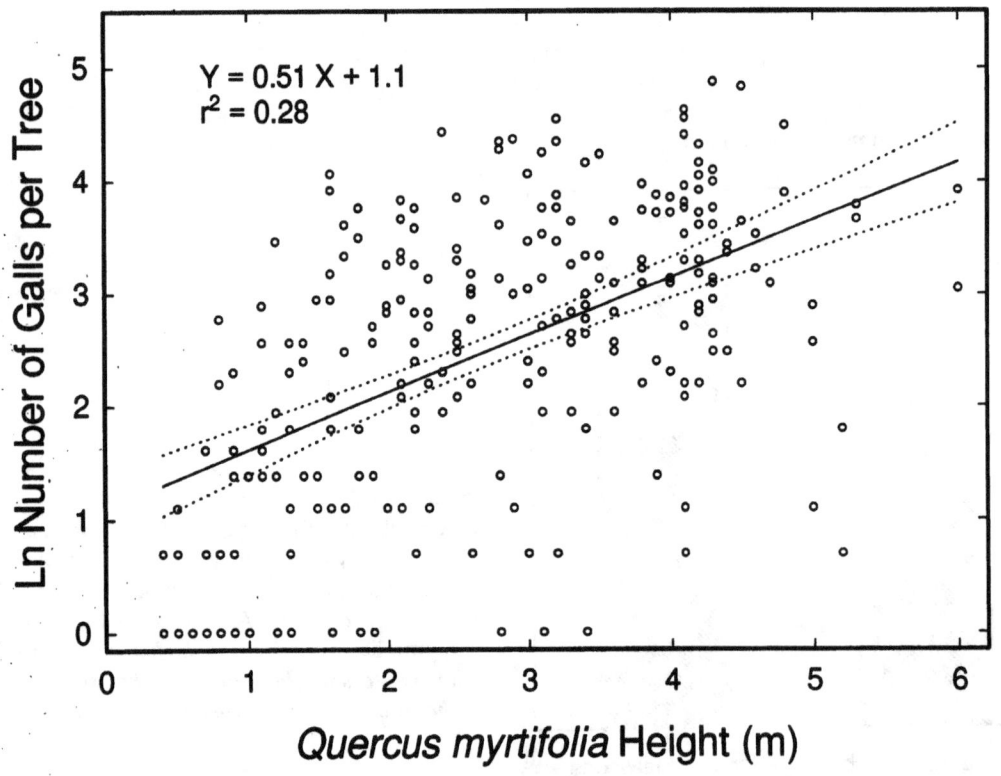

Figure 2.—*Linear regression ± 95 percent confidence intervals (dotted lines) of the natural log number of* Callirhytis quercusmedullae *galls on trees of various heights (m) of the red oak* Q. myrtifolia.

this sampling and rearing of insects was our most intensive. Furthermore, this subset includes data for a total of 127 species of cynipid gall inducers; a more manageable number than the 247 cynipids included in our entire eastern USA data set. For each of the 127 cynipids, the principal component analysis compared (with equal weighting) (1) their presence/absence on each of the 34 possible eastern USA host oaks, (2) presence/absence on each of the following host organs (catkin, acorn, bud, stem-woody swellings, stem-hidden under bark, stem-detachable, leaf-detachable, and leaf-integral); (3) season of emergence (spring, fall), and (4) lifecycle (unisexual, bisexual). The purpose of this analysis was to find combinations of the host oak, organ, season, and lifecycle variables that are correlated so that a reduced number of uncorrelated component axes are generated. This lack of correlation among component axes means that the axes measure different "dimensions" in the data (Manly 1986). We then examined how the individual cynipids loaded on the various components.

Cynipids were widely distributed across the volume defined by the first three component axes (fig. 3). A relatively high number of axes, 27, had eigenvalues over 1.0 which indicates that many of the cynipid life-history patterns were unique and uncorrelated to patterns of other cynipids. A principal component analysis can reduce a large number of original variables to a small number of transformed component variables only when many of the original variables are correlated with one another. However, in our analysis only the first five axes individually explained more than 5 percent of the variance and consequently we will focus only on these component axes. The first three axes explained 15.7,

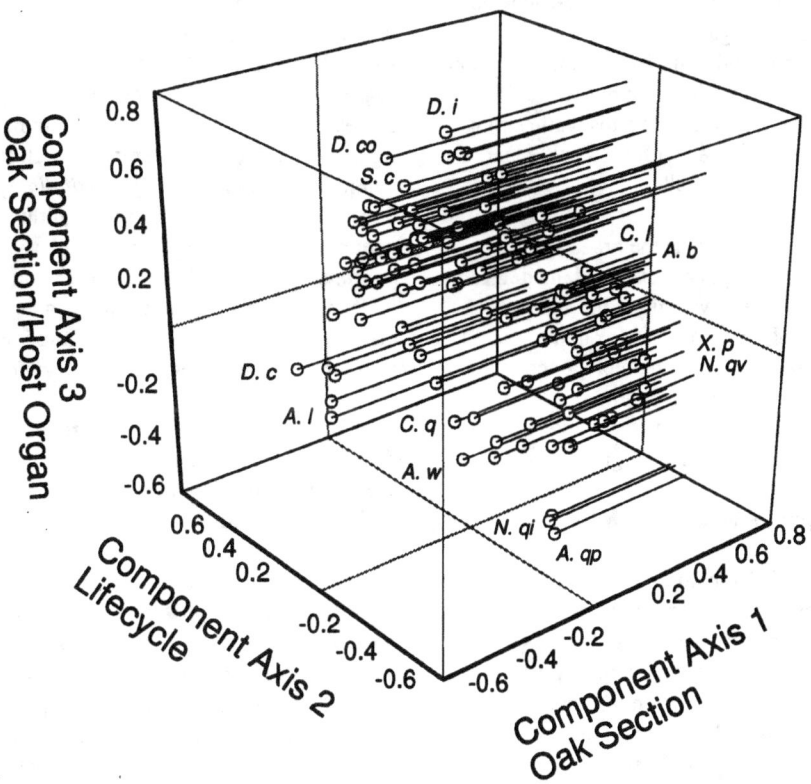

Figure 3.—*Component plot in Varimax-rotated space (Kaiser normalized) for axes 1, 2, and 3 of a principal component analysis based on the host-oak species, host-plant organ attacked, season of attack, and lifecycle for 127 cynipid gall-inducing insects of Florida. Acronyms for selected species shown: Ab = Andricus brooksvillei*, Al = A. lyoni*, Aqp = A. quercuspetiolicola, Aw = A. winegarneri*, Cl = Callirhytis lantana, Cq = C. quercusfoliae, Dc = Dryocosmus cinerae, Dco = D. cornigera, Di = D. imbricariae, Nqi = Neuroterus quercusirregularis, Nqu = N. quercusverrucarum, Sc = Sphaeroteras carolina. * = proposed new species.*

231

13.4, and 6.7 percent, respectively of the variance while axes 4 and 5 explained 6.0, and 5.4 percent of variance, respectively, (total 47.2 percent). Eigenvalues for axis 1 = 20.0, axis 2 = 17.1, axis 3 = 8.5, axis 4 = 7.7, and axis 5 = 6.8.

Component axis 1 was primarily associated with the section of oak (*Lobatae*-red vs. *Quercus*-white) to which the cynipid's host oak(s) belonged. In no case, did a cynipid that associated with a red oak attack a white oak and vice versa. Cynipids that had strong, positive loadings on axis 1 were hosted by white oaks while those with strong negative loadings attacked red oaks. Axis 2 appeared related to the lifecycle of the cynipid. Cynipids with unisexual generations tended to load positively on axis 2 while those with bisexual generations tended to load negatively. Axis 3 was related to both the section of host oak and to the host-organ attacked. Strong positive-loading cynipids tended to form detachable stem galls on red oaks while those with negative loadings tended to develop detachable leaf galls on white oaks. Axes 4 and 5 both were related to the host-organ attacked. Cynipids loading positively on axis 4 typically formed detachable leaf, detachable stem, or bud galls while negative loadings came from cynipids that initiated woody stem or integral leaf galls. Cynipids with strong positive loadings on axis 5 formed integral leaf galls while strong negative loadings were associated with detachable leaf, detachable stem, or woody stem galls.

The striking result of this analysis is the broad distribution of cynipids on these component axes and the degree to which individual cynipids have unique life-history patterns (fig. 3). No cynipid species, for example, shared the niche space defined by the principal component axes. This result suggests evolutionary divergence and specialization of these gall inducers as they have radiated to attack oak species and their various organs.

Patterns of Cynipid Association with Host Oaks

The narrow host specificity of cynipid gall inducers means that oak species or at least groups of oaks are infested by specific communities of cynipids. By examining the similarities among the cynipid communities that attack oak species, we can gain insights into which oak species are most alike and most unlike (at least from the perspective of cynipid gall inducers). While it is yet unknown whether cynipid host-association patterns reflect oak phylogeny, it is very likely that cynipid species using multiple hosts tend to associate with oak species that are most similar in chemistry, physiology, anatomy, morphology, development, and phenology.

We analyzed the presence and absence of 247 cynipid species that occur in the eastern USA on 34 oak species using a principal component analysis. The purpose of this analysis was to find combinations of the 247 cynipid host-association patterns that are similar so as to produce a reduced number of uncorrelated component axes. Next, we examined how the oak host species loaded on these axes. The data set used included the above Florida data plus data from the site surveys conducted in North Carolina and Pennsylvania. Although our North Carolina surveys indicate only those cynipids with galls visible in autumn, our Pennsylvania samples were conducted throughout the year. Unlike the above analysis of the Florida data subset, this analysis utilized only cynipid presence/absence on host oaks. Data on host-organ attacked, season of emergence, and lifecycle were not used.

Cynipid communities clearly separated according to whether they infested members of the red oak or white oak sections within the three-dimensional volume defined by first three principal component axes (fig. 4). No cynipid attacked a member of both sections. Cynipid communities of red oaks formed a large cluster that varied considerably along axis 1 while communities on white oaks formed two distinct clusters that showed substantial variation along axis 3. Axis 2 rather strongly segregated the communities on red oaks with positive scores from those on white oaks with negative scores. In figure 4, the left side of the red oak cluster contained the cynipid communities on *Q. myrtifolia*, *Q. inopina*, and *Q. laurifolia* while the right portion of this cluster included communities attacking *Q. rubra*, *Q. velutina*, and *Q. coccinea*. Among white oaks, the community on *Q. alba* is represented by the lowest white-oak symbol while the distinct cluster of white-oak communities nearest the top of figure 4 includes those on *Q. virginiana*, *Q. geminata*, and *Q. minima*. The cynipid communities on these latter three oaks are very similar and are unique relative to other white oaks. We never found any of the cynipid species that attacked these three species on any other oak during our 7 months of field work in Florida.

In this analysis, 10 component axes had eigenvalues over 1.0 which indicates that many of the cynipid communities on various host oaks are unique and thus not correlated with one another. However only the first five axes individually explained more than 5 percent of the variance. The first three axes explained 17.0, 8.3, and 7.9 percent of the variance, respectively, while axes 4 and 5 explained 6.9 and 5.4 percent, respectively (total 45.5 percent). Eigenvalues for axis 1 = 5.8, axis 2 = 2.8, axis 3 = 2.7, axis 4 = 2.4, and axis 5 = 1.8. Cynipid communities of red oaks had strongly negative (*Q. inopina*, *Q. myrtifolia*) to strongly positive loadings (*Q. velutina*, *Q. rubra*, *Q. ilicifolia*, *Q. coccinea*) on component axis 1. In contrast, all white oak cynipid communities had negative loadings with the loadings of white oaks varying from slightly negative (*Q. michauxii*, *Q.*

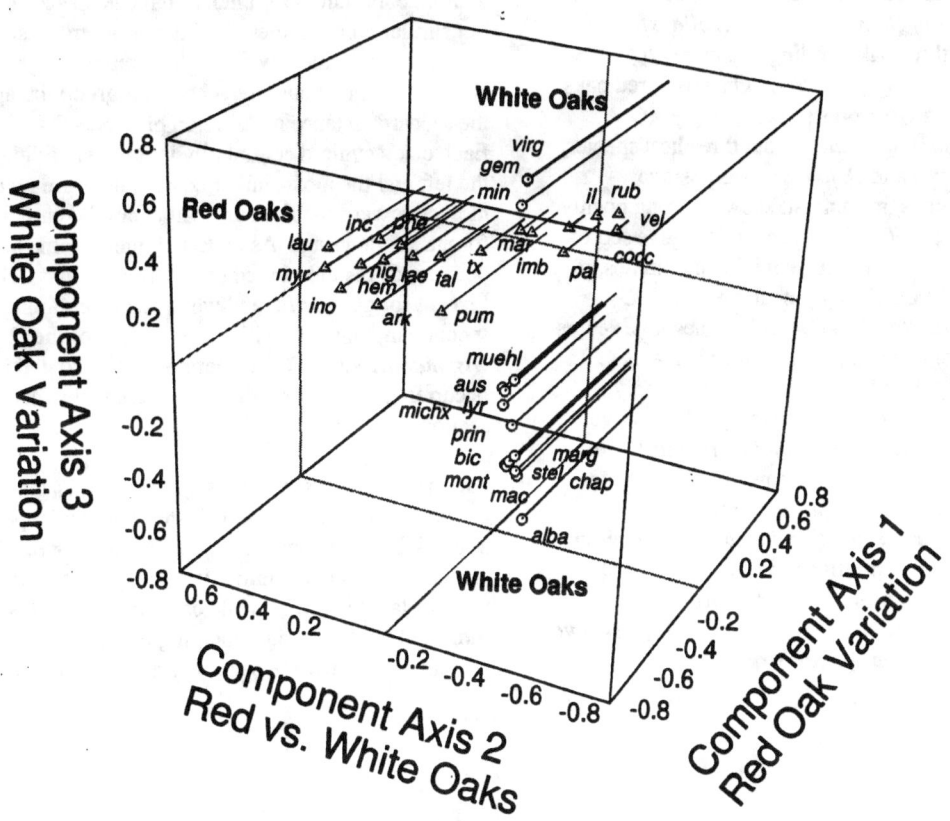

Figure 4.—*Component plot of eastern USA oaks in Varimax-rotated space (Kaiser normalized) for axes 1, 2, and 3 of a principal component analysis based on the communities of cynipid gall-inducing insects that attack each oak species. Red oak positions are shown with open triangles and white oaks with open circles. Acronyms for cynipid communities on red oaks: ark = Q. arkansana, cocc = Q. coccinea, fal = Q. falcata, hem = Q. hemisphaerica, il = Q. ilicifolia, imb = Q. imbricaria, inc = Q. incana, ino = Q. inopina, lae = Q. laevis, lau = Q. laurifolia, mar = Q. marilandica, myr = Q. myrtifolia, nig = Q. nigra, pal = Q. palustris, pum = Q. pumila, phe = Q. phellos, rub = Q. rubra, tx = Q. texana, and vel = Q. velutina. Acronyms of communities on white oaks: alba = Q. alba, aus = Q. austrina, bic = Q. bicolor, chap = Q. chapmanii, gem = Q. geminata, lyr = Q. lyrata, mac = Q. macrocarpa, marg = Q. margaretta, michx = Q. michauxii, min = Q. minima, mont = Q. montana, muehl = Q. muehlenbergii, prin = Q. prinoides, stel = Q. stellata, and virg = Q. virginiana.*

lyrata) to strongly negative *(Q. alba, Q. chapmanii)*. White oaks also had negative loadings on axis 2 with scores similar to those on axis 1. In contrast, gall communities of red oaks generally loaded positively on axis 2 *(Q. laurifolia, Q. incana, Q. myrtifolia)* with cynipids of only three oaks loading negatively *(Q. coccinea, Q. velutina, Q. palustris)*. On axis 3, red oaks in general had negative scores (most strongly—*Q. inopina, Q. palustris, Q. velutina*) but three host species, *Q. texana, Q. laevis,* and *Q. falcata,* had positive loadings. White oaks, in contrast, showed strong positive *(Q. macrocarpa, Q. alba, Q. bicolor)* to negative scores *(Q. geminata, Q. virginiana)*. Cynipid communities of white oaks varied from strongly positive scores *(Q. stellata, Q. margaretta, Q. chapmanii)* to negative scores *(Q. prinoides, Q. geminata, Q. virginiana)* on axis 4. In contrast, red oak gall inducers varied from strongly negative *(Q. coccinea, Q. inopina)* to weakly positive loadings *(Q. texana, Q. laevis, Q. falcata)*. Finally, the gall communities of white oaks showed strong segregation on axis 5 with strong positive loadings by *Q. virginiana, Q. geminata,* and *Q. minima.* The remaining white oaks had weakly positive to negative scores. In contrast, gall communities of red oaks varied from weakly positive *(Q. texana, Q. laevis)* to weakly negative *(Q. inopina, Q. coccinea, Q. palustris).*

A hierarchical cluster analysis was used to define 'objective' groups of cynipid communities according to their host associations and to reduce the many unique cynipid communities on individual oak species to fewer, but similar communities. The dendrogram illustrated in figure 5 was formed by an agglomeration of cynipid communities that employed between-group linkage and the Jaccard distance measure on presence/absence data. Each oak's cynipid community is shown individually on the left and the most similar communities are gradually merged toward the right until all cynipid communities form a single group. As expected, the patterns of cynipid occurrence resulted in red oaks forming a large cluster (except for *Q. pumila*) and white oaks combining in a second large cluster (with the exception of the *Q. virginiana* complex). Furthermore, cynipid gall inducer occurrence patterns produced clusters within each oak section.

Among white oaks, cynipids attacking *Q. virginiana* and *Q. geminata* formed a tight cluster that overlapped appreciably with cynipids infesting *Q. minima*. This cluster of cynipid communities is very unique and does not overlap with other white oaks or red oaks (as indicated by the far right merging of this cluster). A second group of white oaks sharing cynipids was

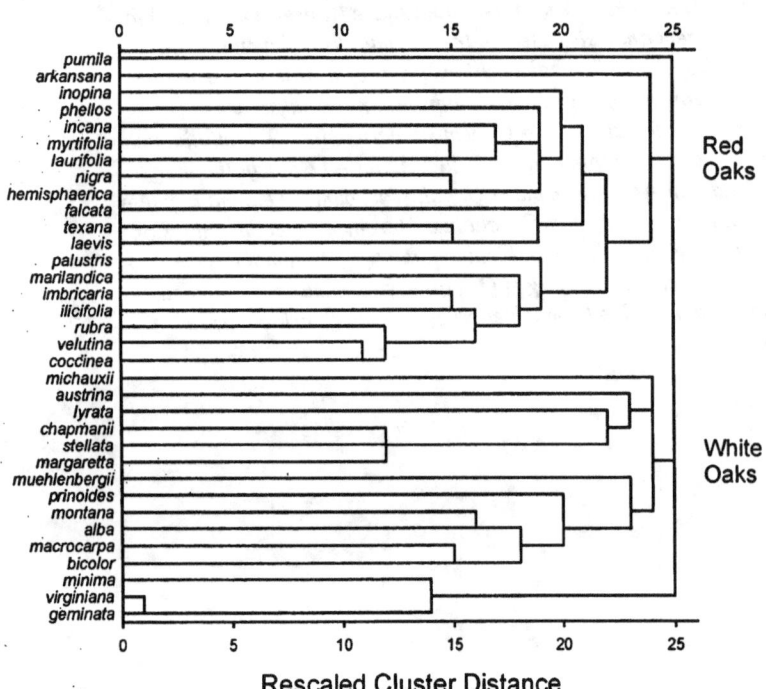

Figure 5.—*Dendrogram of eastern USA host oaks clustered according to their cynipid gall-inducer communities. The dendrogram illustrates the results of a hierarchical cluster analysis using between-group linkage and the Jaccard distance measure on binary (presence/absence) data.*

composed of *Q. bicolor* and *Q. macrocarpa* which was joined by *Q. alba* and *Q. montana*. At greater distance to this four host subunit *Q. prinoides* and *Q. muehlenbergii* merge. A third cluster of cynipid communities included those on *Q. stellata*, *Q. margaretta*, and *Q. chapmanii*. *Q. lyrata* merged at a greater distance and at still greater distances with *Q. austrina* and *Q. michauxii*.

The variation among cynipids attacking red oaks produced numerous clusters. The tightest cluster included *Q. coccinea*, *Q. velutina*, and *Q. rubra*. This subunit merged with a second subunit composed of *Q. ilicifolia* and *Q. imbricaria* and these merged subunits were joined with *Q. marilandica* and *Q. palustris* at greater distances. A second cluster was formed by cynipid species infesting *Q. laevis*, *Q. texana*, and *Q. falcata*, with the greatest overlap of cynipids on the first two species. A third large cluster was formed from cynipids attacking *Q. hemisphaerica* and *Q. nigra* on one hand and those infesting *Q. laurifolia*, *Q. myrtifolia*, *Q. incana*, and *Q. phellos* on the other. *Q. inopina*, a Florida endemic believed to be closely related to *Q. myrtifolia* (Johnson and Abrahamson 1982), has a somewhat unique community of cynipids so it joins this third cluster at a still greater distance. *Q. arkansana* and the shrub *Q. pumila* have rather unique cynipid gall communities and consequently merge with other red oaks at larger distances.

DISCUSSION

Specialization and Diversification

Our results illustrate the high degree of specialization and diversification within the Cynipinae attacking species of *Quercus*. Our ABS survey showed that several cynipids possess preferences for specific habitats (e.g., *Amphibolips murata*) or for host plants of particular sizes (e.g., *Callirhytis quercusmedullae*). Habitat preferences are likely the consequence of many correlated variables including those related to microhabitat conditions (e.g., site openness, light availability) that may affect flight, oviposition, and other behaviors. Preferences for larger trees may be an outgrowth of the availability of appropriate host organs since seven out of eight cynipids that exhibited positive host-size relationships were bud or stem gall inducers. Consequently, the abundances of these habitat-specific and size-dependent cynipids should vary considerably from locality to locality depending on site factors. One site factor, time since last fire, is an especially important parameter for the uplands we sampled at ABS since these habitats have a long evolutionary history of repeated fire (Abrahamson and Hartnett 1990; Myers 1990; Abrahamson and Abrahamson 1996a, b). Consequently, the architecture of oaks is modified markedly during fire-recovery periods compared to the architecture of oaks during fire-free periods (Abrahamson 1984a, b). Cynipid species that prefer larger oaks are uncommon in recently burned sites (WGA & GM, pers. observ.). However, a few species (e.g., *Andricus quercusfoliatus* that attacks *Q. geminata*) were more frequent on resprouting shoots following fire. This cynipid was also particularly abundant on trees growing along fire lanes that pass through long-unburned scrub and sandhill (WGA & GM, pers. observ.) suggesting a preference for open sites and edges. However, many of the cynipids sampled at ABS did not show preferences for either habitat or for host size.

The strongest preferences we identified were for host organ and host species. For example, cynipids that initiated galls on the section *Quercus* (white oaks) were never found infesting oaks in the section *Lobatae* (red oaks) and vice versa. Furthermore, cynipids were sufficiently constant in their attack of specific host organs that these gall inducers can be classified according to the plant organ attacked (e.g., root, catkin, acorn, bud, woody stem, stem under bark, detachable stem, integral leaf, detachable leaf).

The degree of specialization among these gall inducers was particularly evident from the principal component analysis performed on 127 Florida cynipids. This analysis showed that cynipids distributed widely across the component axes with limited overlap occurring even on only the first three component axes. Furthermore, the high number of component axes with eigenvalues above 1 demonstrated the high degree of uncorrelated cynipid host associations, lifecycles, and organ galled. Cynipids appear to have radiated such that individual species utilize only specific host organs and oak species. In addition, cynipids attacking the same species and organ often partition resources by attacking in different seasons. For example, *Amphibolips murata* and *A. bucknellensis* (proposed new species, Melika and Abrahamson, in prep.) both attack buds of *Q. incana*, *Q. laurifolia*, and *Q. myrtifolia* as well as other red oaks. However, *A. murata* galls mature in autumn with adult emergence beginning in November and continuing through Florida's winter. In contrast, *A. bucknellensis* galls develop from early May to June with adult emergence occurring in late May and June. Such partitioning of resources appears common among cynipids and likely has been driven by selection acting to reduce competition among these gall inducers.

The remarkable degree of differentiation among these herbivores may in part result from host shifts that have enabled host-race formation and subsequent speciation (Bush 1975, 1994; Abrahamson and Weis 1997). Recent systematic efforts on these gall inducers and efforts to clarify their host-association relationships (Melika and

Abrahamson 1997, in press, submitted a, b, c, d) will make this taxon especially useful for detailed studies exploring the fundamentals of speciation. Several examples illustrate that specialist herbivores similar to cynipids can be composed of host-associated populations or host races, each of which exploit different host-plant species (Bush 1975, 1994; Wood 1980; Berlocher 1995; Craig et al. 1993; Brown et al. 1995, 1996; Abrahamson and Weis 1997). In spite of the controversy that these studies have generated on how host-race formation may influence the processes of speciation, we have yet to resolve the fundamental requisite processes of speciation. Gall-inducing cynipids present an excellent model system for exploration of speciation processes because of their remarkable diversification and specialization.

Distinguishing Oak Relationships via Cynipid Communities

Our analyses of the distributions of 247 eastern USA cynipids on 34 oak species enabled us to generate similarity relationships among the cynipid communities on these oak species. Our principal component analysis of cynipid distributions clearly distinguished between members of the red oak and white oak sections. In contrast, many botanical traits (e.g., wood anatomy, phenolics) are unable to reliably segregate oak species to their appropriate sections (Tillson and Muller 1942, Li and Hsiao 1973). Furthermore, cynipid distributions segregated oaks within sections. The relatively high number of component axes with eigenvalues over 1 indicates the distinctiveness of cynipid communities attacking each oak species. These data indicate that cynipids can distinguish small differences among oak species and can offer information helpful to resolving plant systematic problems.

Similarly, our hierarchical cluster analysis provided groups of oaks based on the similarities of their cynipid gallers. The grouping patterns evident in the cluster analysis were congruent with those generated by the principal component analysis suggesting that many aspects of these groupings are distinct and represent biological reality. For example, both analyses tightly grouped the cynipid communities on *Q. virginiana*, *Q. geminata*, and *Q. minima* while differentiating these oaks from all other oaks. This clustering of species parallels botanical classification (Trelease 1924; Li and Hsiao 1973, 1976a). For example, a cluster analysis of white oaks based on pollen exine traits tightly grouped *Q. geminata* and *Q. minima* and placed *Q. virginiana* relatively close to these species (Solomon 1983a, b). Furthermore, *Q. geminata* (Wunderlin 1982, Clewell 1985) and *Q. minima* (Solomon 1983a, b) once were considered only varieties of the more widespread *Q. virginiana*.

Unfortunately no modern comprehensive phylogeny of oak species exists, however, several studies using limited numbers of oak taxa are available to compare to our cynipid-based clusters of oaks. Guttman and Weigt (1989) evaluated the systematic relationships of 18 *Quercus* species including 10 red oaks and 8 white oaks using allozyme variation. Among white oaks, their analysis clustered *Q. alba* and *Q. macrocarpa* with a second cluster of *Q. muehlenbergii* and *Q. prinus* (=*montana*). Similarly, the grouping of *Q. lyrata* and *Q. stellata* was clustered with *Q. virginiana* and *Q. bicolor*. Our arrangement has many similarities to Guttman and Weigt's (1989) dendrogram in that cynipid distributions suggest grouping *Q. alba*, *Q. montana*, *Q. muehlenbergii*, and *Q. macrocarpa* but differed by also including *Q. bicolor*. Furthermore, our clustering suggested that *Q. alba* was most similar to *Q. montana*, not *Q. macrocarpa* which cynipids suggest is most similar to *Q. bicolor*. This grouping of *Q. macrocarpa* and *Q. bicolor* agrees with Trelease's (1924) original Macrocarpae series but differs from more recent suggestions that these species be separated (Muller 1951; Li and Hsiao 1973, 1976a, b). Cynipid distributions indicate that *Q. muehlenbergii* is more distantly related to other species in this group than the allozymes examined by Guttman and Weigt (1989). Our clustering of *Q. lyrata* and *Q. stellata* agrees with Guttman and Weigt's (1989) dendrogram as it does with the proposed more distant relationship of *Q. virginiana*.

Comparison of our cynipid-based grouping also can be made with Hardin's (1979a, b) classification of trichome variation among 42 species of eastern USA oaks. Overall, his scheme for variation among white oaks is largely congruent with our grouping of white oaks. Species within several of our clusters shared trichome features including (1) *Q. stellata*, *Q. margaretta*, and *Q. chapmanii*, (2) *Q. macrocarpa* and *Q. bicolor*, and (3) *Quercus montana*, *Q. muehlenbergii*, and *Q. prinoides*. However, our arrangement of *Q. michauxii* did not predict its shared trichome features with *Q. montana*, *Q. muehlenbergii*, and *Q. prinoides*. It is possible that such disagreements stem from the conservation of trichome character states in spite of alterations to numerous other characters. Cynipids probably evaluate a suite of host-plant traits and react to the composite of character states rather than to individual characteristics. It is likely that the dendrogram generated from cynipid distributions may more accurately assess oak relationships than classifications based on only one or two traits (e.g., pollen exine, trichomes).

Trelease's (1924) and Li and Hsiao's (1973, 1975, 1976a, b) arrangements of red oaks share some features with our cynipid-based arrangement. For example, all arrangements recognize the similarities of *Q. laurifolia*, *Q. myrtifolia*, and *Q. incana* in addition to the affinities of

Q. rubra and *Q. coccinea*. Furthermore, cynipid distributions support the suggestion that *Q. ilicifolia* (Li and Hsiao 1976b) and *Q. velutina* (Li and Hsiao 1976a, b) are not as distinct as Trelease (1924) proposed. Indeed, community similarities of oak-galling cynipids support Muller's (1951) and Jensen's (1977a, b) proposals that *Q. velutina* has affinities with *Q. coccinea* and *Q. rubra*. Our arrangement also conforms to Muller's (1951) placement (see also Li and Hsiao 1973, 1975, 1976a) of *Q. nigra* with the cluster of *Q. laurifolia*, *Q. myrtifolia*, and *Q. incana* but at a distance similar to that of *Q. phellos*. Cynipid distributions provide support for Li and Hsiao's (1975) suggestion of separating *Q. imbricaria* from this cluster. Cynipids perceive *Q. imbricaria* as more similar to *Q. ilicifolia*. Our findings also agree with the results of an allozyme study (Manos and Fairbrothers 1987) of red oaks which suggested that *Q. palustris* differs considerably from the more closely related *Q. coccinea*, *Q. rubra*, *Q. velutina*, and *Q. ilicifolia*. However, the distinctiveness of *Q. palustris* might well disappear if studies proposing the uniqueness of *Q. palustris* (Manos and Fairbrothers 1987, Guttman and Weigt 1989) had included additional oaks since other work (Jensen 1977a) offers that *Q. palustris* is allied with two unexamined oaks, *Q. shumardii* and *Q. nuttallii*.

Hardin's (1979a) trichome types also organize a number of red oaks similar to our cynipid-based ordering. Both studies suggested that *Q. laurifolia*, *Q. myrtifolia*, and *Q. phellos* are related but cynipid distributions suggest that *Q. pumila* and *Q. imbricaria* should be separated from this group. Similar to Hardin (1979a), cynipids perceive *Q. nigra* and *Q. hemisphaerica* as well as *Q. rubra* and *Q. coccinea* as quite alike. Likewise, our arrangement of red oaks using cynipid distributions is congruent with some features of Solomon's (1983b) clustering based on pollen exine traits. The similarity of (1) *Q. falcata* and *Q. laevis*, (2) *Q. laurifolia* and *Q. phellos*, and (3) *Q. ilicifolia* and *Q. velutina* is evident in both orderings.

However, our cynipid-based arrangement of red oaks disagrees with botanical character-based arrangements in several ways. For example, cynipid distributions do not support the clustering of *Q. arkansana* with *Q. marilandica* and *Q. laevis* (Trelease 1924; Li and Hiao 1973, 1976a; Hardin 1979a). These gallers view *Q. marilandica* as more similar to *Q. imbricaria* and *Q. ilicifolia* and *Q. arkansana* is seen as a rather distinctive species. Guttman and Weigt's (1989) arrangement differs from ours by grouping *Q. falcata* with *Q. rubra* as well as *Q. velutina* with *Q. nigra*.

It is likely that some of the differences in arrangements result from the inclusion and exclusion of various species. If the evolutionarily closest species to an included species have been excluded with the analysis, more distantly related species will be clustered. For example, cynipid distributions suggested similarities among *Q. falcata*, *Q. laevis*, and *Q. texana*, however, these latter two species were not included in studies such as Guttman and Weigt's (1989).

Geographic factors also confound any arrangement based on herbivores such as gall inducers. Because oak and cynipid species vary in their geographic distributions, not all combinations of host association are possible. This problem limits our ability to use insect herbivores to provide plant systematic insights across geographic areas the size of North America or Europe and Asia.

We must add the cautionary note that the arrangement of oak species based on cynipid distributions is phenetic and thus cannot be interpreted as the definitive indication of evolutionary relationships. Nevertheless, because the cynipids are responding to a suite of biological traits that vary across their host plants, the patterns of phenetic variation in host associations must reflect in part phylogenetic relationships. We look forward to the day that we will be able to contrast arrangements of oaks based on cynipid distributions with DNA-sequence derived oak phylogenies.

ACKNOWLEDGMENTS

We express our deep appreciation to C. Abrahamson, J. Abrahamson, K. Ball, L. Blazure, R. Bowman, M. Chipaloski, D. Cronin, M. Deyrup, M. Frey, J. Fitzpatrick, R. Hammer, A. Johnson, I. Kralick, J. MacBeth, R. Melika, E. Menges, A. Menke, R. Peet, R. Roberts, P. Schmaltzer, A. Schotz, C. Shantz, C. Winegarner, and M. Winegarner for field assistance, and/or technical advice. The initial ideas for this study developed from WGA's conversations and field work with T. Craig and J. Itami. Support was provided by NSF grant and supplements BRS-9107150 to WGA, the David Burpee endowment of Bucknell University, the Archbold Biological Station, and a Smithsonian Institution National Museum of Natural History grant to GM.

LITERATURE CITED

Abrahamson, W.G. 1984a. Post-fire recovery of Florida Lake Wales Ridge vegetation. American Journal of Botany. 71: 9-21.

Abrahamson, W.G. 1984b. Species responses to fire on the Florida Lake Wales Ridge. American Journal of Botany. 71: 35-42.

Abrahamson, W.G.; Abrahamson, C.R. 1996a. Effects of fire on long-unburned Florida uplands. Journal of Vegetation Science. 7: 565-574.

Abrahamson, W.G.; Abrahamson, J.R. 1996b. Effects of a low-intensity, winter burn on long-unburned sand pine scrub. Natural Areas Journal. 16: 171-183.

Abrahamson, W.G.; Hartnett, D.C. 1990. Pine flatwoods and dry prairies. In: Myers, R.L.; Ewel, J.J., eds. Ecosystems of Florida. Orlando, FL: University of Central Florida Press: 103-149.

Abrahamson, W.G.; Weis, A.E. 1987. Nutritional ecology of arthropod gall makers. In: Slansky, F., Jr.; Rodriquez, J.G., eds. Nutritional ecology of insects, mites, spiders, and related invertebrates. New York, NY: John Wiley & Sons, Inc.: 235-258.

Abrahamson, W.G.; Weis, A.E. 1997. Evolutionary ecology across three trophic levels: goldenrods, gallmakers, and natural enemies. Monogr. in Pop. Biol. 29. Princeton University Press. 456 p.

Abrahamson, W.G.; Johnson, A.F.; Layne, J.N.; Peroni, P. 1984. Vegetation of the Archbold Biological Station, Florida: an example of the southern Lake Wales Ridge. Florida Scientific. 47: 209-250.

Abrahamson, W.G.; Melika, G.; Scrafford, R.; Csóka. 1998. Host-plant associations and specificity among cynipid gall-inducing wasps of eastern USA. In: Csóka, Gyuri; Mattson, William J.; Stone, Graham N.; Price, Peter W., eds. The biology of gall-inducing arthropods: proceedings of the international symposium; 1997 August 14-19; Mátraküred, Hungary. Gen. Tech. Rep. NC-199. St. Paul, MN: U.S. Department of Agriculture, Forest Service, North Central Research Station: 226-240.

Aguilar, J.M.; Boecklen, W.J. 1992. Patterns of herbivory in the *Quercus grisea* X *Quercus gambelii* species complex. Oikos. 64: 498-504.

Berlocher, S.H. 1995. Population structure of *Rhagoletis mendax*, the blueberry maggot. Heredity. 74: 542-555.

Brown, J.M.; Abrahamson, W.G.; Packer, R.A.; Way, P.A. 1995. The role of natural-enemy escape in a gallmaker host-plant shift. Oecologia. 104: 52-60.

Brown, J.M.; Abrahamson, W.G.; Way, P.A. 1996. Mitochondrial DNA phylogeography of host races of the goldenrod ball gallmaker *Eurosta solidaginis* (Diptera: Tephritidae). Evolution. 50: 777-786.

Burks, B.D. 1979. Superfamily Cynipoidea. In: Krombein, K.V.; Hurd, P.D., Jr.; Smith, D.R.; Burks, B.D., eds. Catalog of Hymenoptera in America north of Mexico. Vol. 1. Symphyta and Apocrita. Washington, DC: Smithsonian Institution Press: 1045-1107.

Bush, G.L. 1975. Modes of animal speciation. Annual Review of Ecology and Systematics. 6: 339-357.

Bush, G.L. 1994. Sympatric speciation in animals: new wine in old bottles. Trends in Ecology and Evolution. 9: 285-288.

Clewell, A.F. 1985. Guide to the vascular plants of the Florida panhandle. Tallahassee, FL: University Presses of Florida.

Cook, J.M.; Stone, G.N.; Rowe, A. 1998. Patterns in the evolution of gall structure and life cycles in oak gall wasps (Hymenoptera: Cynipidae). In: Csóka, Gyuri; Mattson, William J.; Stone, Graham N.; Price, Peter W., eds. The biology of gall-inducing arthropods: proceedings of the international symposium; 1997 August 14-19; Mátraküred, Hungary. Gen. Tech. Rep. NC-199. St. Paul, MN: U.S. Department of Agriculture, Forest Service, North Central Research Station: 261-279.

Cornell, H.V. 1986. Oak species attributes and host size influence cynipine wasp species richness. Ecology. 67: 1582-1592.

Craig, T.P.; Itami, J.K.; Abrahamson, W.G.; Horner, J.D. 1993. Behavioral evidence for host-race formation in *Eurosta solidaginis*. Evolution. 47: 1696-1710.

Drown, D.M.; Brown, J.M. 1998. Molecular phylogeny of North American oak-galling Cynipini (Hymenoptera: Cynipidae) supports need for generic revision. In: Csóka, Gyuri; Mattson, William J.; Stone, Graham N.; Price, Peter W., eds. The biology of gall-inducing arthropods: proceedings of the international symposium; 1997 August 14-19; Mátraküred, Hungary. Gen. Tech. Rep. NC-199. St. Paul, MN: U.S. Department of Agriculture, Forest Service, North Central Research Station: 241-246.

Felt, E.P. 1917. Key to American insect galls. New York State Museum Bull. 200. Albany, NY.

Felt, E.P. 1940. Plant galls and gall makers. Hafner, NY.

Floate, K.D.; Whitham, T.G. 1995. Insects as traits in plant systematics: their use in discriminating between hybrid cottonwoods. Canadian Journal of Botany. 73: 1-13.

Floate, K.D.; Fernandes, G.W.; Nilsson, J.A. 1996. Distinguishing intrapopulational categories of plants by their insect faunas: galls on rabbitbush. Oecologia. 105: 221-229.

Fritz, R.S.; Nichols-Orians, C.M.; Brunsfeld, S.J. 1994. Interspecific hybridization of plants and resistance to herbivores: hypotheses, genetics, and variable responses in a diverse herbivore community. Oecologia. 97: 106-117.

Guttman, S.I.; Weigt, L.A. 1989. Electrophoretic evidence of relationships among *Quercus* (oaks) of eastern North America. Candian Journal of Botany. 67: 339-351.

Hardin, J.W. 1979a. Patterns of variation in foliar trichomes of eastern North American *Quercus*. American Journal of Botany. 66: 576-585.

Hardin, J.W. 1979b. Atlas of foliar surface features in woody plants, I. Vestiture and trichome types of eastern North American *Quercus*. Bulletin of the Torrey Botanical Club. 106: 313-325.

Jensen, R.J. 1977a. Numerical analysis of the scarlet oak complex (*Quercus* subgen. *Erythrobalanus*) in the eastern United States: relationships above the species level. Systematic Botany. 2: 122-133.

Jensen, R.J. 1977b. A preliminary numerical analysis of the red oak complex in Michigan and Wisconsin. Taxon. 26: 399-407.

Johnson, A.F.; Abrahamson, W.G. 1982. *Quercus inopina*: a species to be recognized from south-central Florida. Bulletin of the Torrey Botanical Club. 109: 392-395.

Li, H.L.; Hsiao, J.Y. 1973. A preliminary study of the chemosystematics of American oaks: phenolic characters of leaves. Bartonia. 42: 5-13.

Li, H.L.; Hsiao, J.Y. 1975. A chemosystematic study of the series Laurifoliae of the red oaks: phenolics of leaves. Bartonia. 43: 25-28.

Li, H.L.; Hsiao, J.Y. 1976a. A preliminary study of the chemosystematics of American oaks: phenolic characters of young twigs. Bartonia. 44: 14-21.

Li, H.L.; Hsiao, J.Y. 1976b. A preliminary study of the chemosystematics of American oaks: phenolic characters of staminate catkins. Bartonia. 44: 8-13.

Mani, M.S. 1964. Ecology of plant galls. The Hague: Dr. W. Junk Publisher.

Manly, B.F.J. 1986. Multivariate statistical methods: a primer. New York, NY: Chapman and Hall. 159 p.

Manos, P.S.; Fairbrothers, D.E. 1987. Allozyme variation in populations of six northeastern American red oaks (Fagaceae: *Quercus* subg. *Erythrobalanus*). Systematic Botany. 12: 365-373.

Melika, G.; Abrahamson, W.G. 1997. Descriptions of four new species of cynipid gall wasps of the genus *Neuroterus* (Hymenoptera: Cynipidae) with redescriptions of some known species from the eastern United States. Proceedings of the Entomological Society of Washington. 99: 560-573.

Melika, G.; Abrahamson, W.G. In Press. Synonymy of two genera (*Eumayria* and *Trisoleniella*) of cynipid gall wasps (Hymenoptera: Cynipidae) and description of a new genus, *Eumayriella*. Proceedings of the Entomological Society of Washington.

Melika, G.; Abrahamson, W.G. Submitted a. Revision of the world genera of the tribus Cynipini (Hymenoptera, Cynipidae). Journal of Hymenoptera Research.

Melika, G.; Abrahamson, W.G. Submitted b. Revision of cynipid gall wasps of the genus *Bassettia* Ashmead (Hymenoptera: Cynipidae) with descriptions of new species. Proceedings of the Entomological Society of Washington.

Melika, G.; Abrahamson, W.G. Submitted c. Revision of cynipid gall wasps of the genus *Loxaulus* (Hymenoptera: Cynipidae) with descriptions of new species. Proceedings of the Entomological Society of Washington.

Melika, G.; Abrahamson, W.G. Submitted d. Revision of cynipid oak gallmaker wasps of the genus *Zopheroteras* with description of a new species (Hymenoptera: Cynipidae). Proceedings of the Entomological Society of Washington.

Muller, C.H. 1951. The oaks of Texas. Contributions Texas Research Foundation. 1: 21-323.

Myers, R. 1990. Scrub and high pine. In: Myers, R.L.; Ewel, J.J., eds. Ecosystems of Florida. Orlando, FL: University of Central Florida Press: 150-193.

Raman, A. 1993. Chemical ecology of gall insect—host plant interactions: substances that influence the nutrition and resistance of insects and the growth of galls. In: Ananthakrishnan, T.N.; Raman, A., eds. Chemical ecology of phytophagous insects. New Delhi: Oxford and IBM Publishing Co. Pvt. Ltd.: 227-250.

Raman, A. 1996. Nutritional diversity in gall-inducing insects and their evolutionary relationships with flowering plants. International Journal of Ecology and Environmental Sciences. 22: 133-143.

Shorthouse, J.D. 1982. Resource exploitation by gall wasps of the genus *Diplolepis*. In: Visser, J.H.; Minks, A.K., eds. Proceedings of the 5th International symposium of insect-plant relationships. Wageningen: Pudoc: 193-198.

Shorthouse, J.D. 1998. Role of *Periclistis* (Hymenoptera: Cynipidae) inquilines in leaf galls of Diplolepis (Hymenoptera: Cynipidae) on wild roses in Canada. In: Csóka, Gyuri; Mattson, William J.; Stone, Graham N.; Price, Peter W., eds. The biology of gall-inducing arthropods: proceedings of the international symposium; 1997 August 14-19; Mátraküred, Hungary. Gen. Tech. Rep. NC-199. St. Paul, MN: U.S. Department of Agriculture, Forest Service, North Central Research Station: 61-81.

Solomon, A.M. 1983a. Pollen morphology and plant taxonomy of white oaks in eastern North America. American Journal of Botany. 70: 481-494.

Solomon, A.M. 1983b. Pollen morphology and plant taxonomy of red oaks in eastern North America. American Journal of Botany. 70: 495-507.

Tillson, A.H.; Muller, C.H. 1942. Anatomical and taxonomic approaches to subgeneric segregation in American *Quercus*. American Journal of Botany. 29: 523-529.

Trelease, W. 1924. The American oaks. Memoirs of the National Academy of Science. 20: 1-255.

Weld, L.H. 1951. Superfamily Cynipoidea. In: Muesebeck, C.F.W.; Krombein, K.V.; Townes, H.K., eds. Hymenoptera of America north of Mexico. Agric. Monogr. 2. Washington, DC: United States Government Printing Office: 594-654.

Weld, L.H. 1952. New American cynipid wasps from galls. Proceedings of the United States National Museum. 102: 315-342.

Weld, L.H. 1959. Cynipid galls of the Eastern United States. Ann Arbor, MI: Privately printed: 1-124.

West, E. 1948. The oaks of Florida. Journal of the New York Botanical Garden. 49: 273-283.

Wood, T.K. 1980. Divergence in the *Enchenopa binotata* Say complex. In: Wilson, M.R.; Nault, L.R., eds. Proceedings of the 2d international workshop on leafhoppers and planthoppers of economic importance. London: CIE: 361-368.

Wunderlin, R.P. 1982. Guide to the vascular plants of central Florida. Tampa, FL: University Presses of Florida.

MOLECULAR PHYLOGENY OF NORTH AMERICAN OAK-GALLING CYNIPINI (HYMENOPTERA: CYNIPIDAE) SUPPORTS NEED FOR GENERIC REVISION

Devin M. Drown[1] and Jonathan M. Brown[2]

Department of Biology, Grinnell College. P.O. Box 805, Grinnell, IA 50112-0806 USA
[1]Present address: U of Utah, Dept of Biology, 204 S. Biology Bldg, Salt Lake City, UT 84112-0840 USA
[2]To whom correspondence should be addressed.

Abstract.—Generic classification of the members of the tribe Cynipini (Hymenoptera: Cynipidae) is made difficult by the presence of adult morphological variation among members of alternating unisexual and bisexual generations. Here we apply a cladistic approach to taxonomic definition, using a DNA sequence phylogeny to test the hypothesis that previously defined genera are monophyletic. Our analysis of a preliminary data set comprised of 300 bp of sequence from the cytochrome oxidase I mitochondrial gene indicates that members of some genera (e.g., *Callirhytis* and *Andricus*) do not form monophyletic groups, supporting the view of Abrahamson et al. (1998) that previously used diagnostic criteria should be reconsidered. Sequence divergence was at appropriate levels for phylogenetic inference (2-13 percent within the Cynipini); however, extreme AT-bias (71 percent), particularly at 3rd positions in codons (95 percent) suggests the need for intensive taxon sampling, as well as more DNA characters, to avoid misleading assumptions of homology from convergent changes at these positions.

INTRODUCTION

The recent surge of interest in phylogenetic tests of biological hypotheses has highlighted the important need for the modern taxonomists to produce classifications that accurately reflect evolutionary relationships. Here, we take a cladistic approach to the definition of genera in the tribe Cynipini, the most speciose tribe in the hymenopteran family Cynipidae, testing the hypothesis that genera are monophyletic. Abrahamson et al. (1998) describe the history of systematic study of this group and suggest that the definition of genera has been the primary difficulty in classification. They attribute this to the use of diagnostic characters that are variable between the alternate asexual and sexual generations found in many species. These authors also criticize the use of host plant association and gall characteristics in defining genera, as these characters sometimes vary among populations of the same species. The search for morphological characteristics that accurately indicate common ancestry can be aided by combining morphological and molecular sources of information. Aspects of cynipid morphology (or gall morphology) may be subject to adaptive changes during species radiation, and thus prone to convergent evolution when species experience the same selective pressures. Molecular sequence characters from mitochondrial genes are unlikely to be subject to the same selection pressures as gall or adult morphology, and thus provide a good independent source of information to test for homoplasy of such characters (e.g., Brown et al. 1994a, b; Cook et al. 1998).

Here, we report on our test of the appropriateness of the cytochrome oxidase I (COI) coding region in the mitochondrial DNA to test hypotheses of monophyly of genera in the Cynipini. A 300 bp region was sequenced for ten genera, which included 19 different species. We also obtained an adjacent 360 bp region upstream (within COI) for 10 of these species. The cytochrome oxidase genes have shown appropriate rates of sequence evolution to be useful for phylogenetic inference at the generic and species level in many different insect orders (e.g., Lepidoptera: Brower 1994; Brown et al. 1994a, b; Diptera: Gleason et al. 1997, Spicer 1995; Coleoptera: Funk et al. 1995, Juan et al. 1995, Normark 1996).

METHODS

G. Melika and W. Abrahamson collected adult specimens between January and May 1996; adult specimens were stored at -70°C and larval specimens in absolute ethanol at 3°C until DNA extraction. We report here on a total of 19 species, which came from 10 different genera (table 1). All species were from North America.

Table 1.—*Species of Cynipini in current study ("*" have 660 bp total sequence).*

Species
* *Adleria quercusstrobilanus*
* *Acraspis echini*
Amphibolips globus
* *Amphibolips murata*
Amphibolips quercusinanis
Andricus quercusfoliatus
* *Andricus quercuslanigera*
Callirhytis abrahamsoni
Callirhytis quercusbatatoides
Callirhytis quercusventricosa
Callirhytis seminator
* *Disholcaspis quercusvirens*
* *Dryocosmus imbricariae*
Dryocosmus quercusnotha
* *Dryocosmus quercuspalustris*
* *Neuroterus distortus*
* *Neuroterus pallidus*
Philonix nigra
* *Sphaeroteras carolina*

Laboratory Methods

We extracted total DNA from specimens using two different methods. First, we used the IsoQuick Nucleic Acid Extraction kit method (MicroProbe Corporation) which utilizes a non-corrosive extraction reagent containing a nuclease-binding matrix. Early on in the study, we switched to the more traditional chloroform extraction method (Brown 1994b) because PCR amplification was not consistent with our original preps. The majority of samples were then prepared using this technique. The total DNA preps were visualized on an agarose gel and photographed to check for relative concentration and possible contamination.

We used the polymerase chain reaction (PCR) (Saiki *et al.* 1988) to amplify a specific region of the mitochondrial DNA (mtDNA) within the cytochrome oxidase I (COI) gene. A MJ Research thermo cycler was used for PCR with the following program: 2 min. at 94°C once, 30 s at 93°C, 1 min. at 47°C, and 2 min. at 72°C for 30 cycles, and 5 min. at 72°C. We used a 2.5 mM $MgCl_2$ concentration. The PCR product was visualized on a 1 percent agarose gel. We used two primers to generate the double stranded PCR product: S2183, 5' CAACATTTATTTTGATTTTTTGG 3'; and A2566, 5' GCAACTACATAATA(A/T)GT(A/G)TGATGTA 3'. The S2183 was designed by members of the Richard Harrison lab at Cornell University using their own sequences and those of the published sequences of *Drosophila yakuba* and *Apis mellifera* (Simon *et al.* 1994). We designed A2566 by comparison of sequences of various Hymenoptera and *Drosophila melanogaster* (Lewis *et al.* 1995) from Genebank (L26569-L26579, U16710-U16718).

In order to obtain more sequence data for this region, we designed an internal primer based on the Hymenoptera sequence and the newly found cynipid sequences. This primer, TATAATAAGCTCG(G/A)GTATCAATATC, is designated A2344, and we used the same PCR method, but increased the annealing temperature to 49°C. Another PCR fragment upstream of this region was made via S1751 (GGATCACCTGATATAGCATTCCC) and A2344. The annealing temperature was again modified to 52°C.

We obtained sequence data by standard manual Sanger sequencing using ^{35}S and the Sequenase version 2.0 DNA sequencing kit for PCR products (U.S. Biochemical). The first PCR products were sequenced using either the A2566 or A2344 primer in the sequencing reaction. The fragments were made continuous by matching their overlapping regions. This resulted in a sequence of 300 bp spanning nucleotide positions 2195 to 2495 in COI when aligned to *Drosophila yakuba* (Clary and Wolstenholme 1985). The second PCR product was sequenced using S1751 resulting in a sequence of 360 bp.

Data Analysis

We used SeqSpeak (Conover 1991) and SeqPup (Gilbert 1995) to enter and align the 300bp sequences from all 19 taxa. These were then aligned with four Hymenoptera outgroups [*Bombus terrestris* (Pedersen unpubl. data), *Orthogonalys pulchella*, *Schlettererius cinctipes*, and *Xiphydria mellipes* (Dowton and Austin unpubl. data)]. We chose these outgroups based amount of sequence overlap that they had with our new sequence and their relatively good distribution within Hymenoptera. PAUP 3.1.1 (Swofford 1993) was used to search for the most parsimonious trees. A Heuristic search was done using the random addition method with 25 replicates.

RESULTS

Sequence divergence based on pairwise comparisons of the 19 Cynipini taxa (table 2) was less than 12 percent, values typical of within-genus comparisons in other insects (e.g., Brown et al. 1994b). [Analysis of the 10 taxa for which 660bp was available was showed only a slightly higher maximum divergence (table 2).] As expected, highest divergence was found at 3rd positions in codons, where substitutions are often synonymous. A+T base composition was 76 percent (table 3), a level typical for mitochondrial genes in insects (e.g., Liu and Beckenbach 1992, Brown et al. 1994b, Spicer 1995); base composition was particularly A+T-biased at third positions (95 percent; table 3), as has been found for protein-coding mitochondrial genes in other Hymenoptera [e.g., *Apis mellifera* (Crozier and Crozier 1993)]. This strong A+T bias has interesting effects on frequency of substitutions at these lower levels of divergence. We reconstructed the classes of substitutions over one of the most parsimonious trees (see below), excluding the relatively distant outgroups, using MacClade 3.05 (Maddison and Maddison 1995). Contrary to the common discovery of a transition bias at low levels of sequence divergence (e.g., Brown et al. 1994b, Liu and Beckenbach 1992), extreme A+T bias at third positions resulted in A↔T transversions being equal in number to all transitions combined (70 each).

The heuristic search using equal weighting of the 300bp data set returned 2 most parsimonious trees (369 steps; C.I. (excl. uninf.) = 0.461; fig. 1). All species of Cynipini formed a monophyletic group; however, this was usually not the case when multiple species from a genus were present (e.g., with *Callirhytis*, *Andricus* and *Neuroterus*).

DISCUSSION

The results of this preliminary analysis indicate that while the cytochrome oxidase I gene exhibits appropriate levels of variation for use in phylogenetic inference, two aspects of sequence evolution in these genes, A+T bias and strong constraint at the amino acid level, raise concerns over inference methods. As 3rd codon position substitutions are often synonymous, they are typically the first to differentiate after lineages split. However, with effectively only 2 character states at 3rd positions (A and T), instead of 4, reversals are likely with increasing divergence time. Thus differences at 3rd positions are unlikely to be informative with highly divergent taxa. By weighting 3rd codon positions changes less than changes at other positions, one can favor conservative changes without discarding information from the less conservative 3rd positions. However, it is possible that selection at the amino acid level may constrain evolution at 1st and 2nd positions such that differences accumulate

Table 2.—*Percent sequence divergence among 19 cynipini taxa. Values in parentheses are for 660 bp data set (with only 10 taxa).*

Sequence divergence	All positions	1st position	2nd position	3rd position
Minimum	1.9 (7.4)	0 (1.9)	0 (0.5)	5.9 (13.8)
Maximum	11.3 (13.1)	5.2 (8.3)	5.1 (7.1)	32.0 (31.2)

Table 3.—*Percent base composition in the 300 bp region of COI in Cynipidae*

Base	Codon position			
	All	1st	2nd	3rd
C	10.0	6.3	21.4	2.3
G	14.5	23.3	17.1	2.9
A	34.9	42.0	17.9	45.0
T	40.6	28.5	43.6	49.8
A+T	75.5	70.5	61.5	94.8

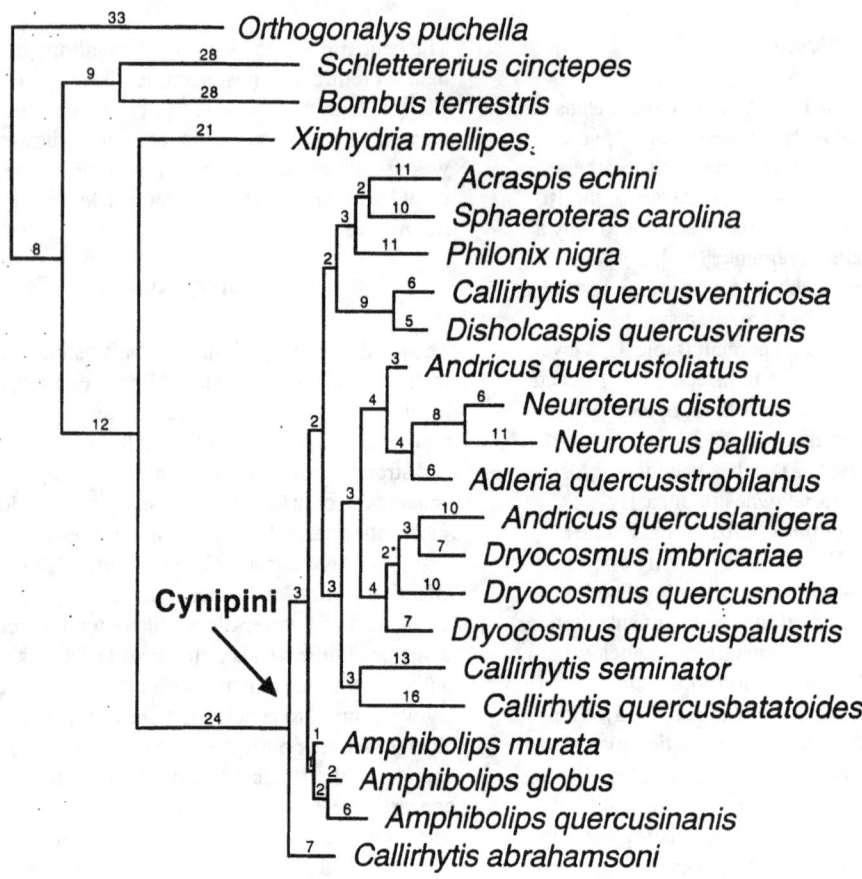

Figure 1.— *One of the two most parsimonious trees [tree length = 369, C.I (excluding uniform.) = 0.461]. Branch lengths are proportional to the number of inferred substitutions (using PAUP 3.0, assuming accelarated transformation), which are also shown above each branch. The node marked with an "*" does not appear in the other MP tree.*

at these sites only after very long periods of divergence. [For example, our data indicate only four sites at 1st and 2nd positions that are phylogenetically informative within the sampled Cynipini.] Thus such genes may only be useful at very low and very high levels of divergence.

Strong A+T bias at 3rd codon positions also means that the common assumptions about transition:transversion ratios at low levels of divergence will be violated, as A↔T transversions will be more common than transitions. This has implications for the use of "transversion parsimony" (Swofford and Olson 1990) which ignores information from transitions in favor of the presumably more conservative transversions. With such high A+T bias at 3rd positions, transversions as a class are not conservative, and thus the use of transversion parsimony should result in errors in phylogenetic reconstruction. At best, transversion parsimony discards useful information from transitions. Weighted parsimony is only necessary when homoplasies cannot be easily identified, for example when multiple substitutions occur on long branches in the phylogeny (Felsenstein 1981). This suggests that parsimony analyses using genes subject to homoplasy (all genes?) should have intense sampling of taxa, which breaks up the long branches and allows the identification of homoplasic changes. We believe this is a better approach because characters that exhibit homoplasy across the entire range of group under investigation, and thus do not provide support for deep nodes, can nonetheless prove informative at shallow nodes of a phylogeny. Along with good taxon sampling, homoplasies may be identified by combination with independent sources of data, such as morphological traits or a more slowly evolving nuclear gene. It is this strategy which we are currently undertaking in inferring the phylogeny of the Cynipini.

We do not attempt to present a conclusive analysis of the monophyly of genera here because of the small number of DNA characters and the current lack of adequate taxon sampling (in particular we lack Eurasian species for holarctic taxa). However, we do have enough taxa from a few genera to consider evidence of monophyly, and to consider whether the gene is providing information consistent with the morphological reanalysis of Abrahamson et al. (1998).

Amphibolips is an example of a genus showing strict monophyly of three species sampled on the most parsimonious tree (fig. 1). This genus is diagnosed by a morphological synapomorphy, a coarsely rugose and dull scutum. Although only two species have been sampled, *Neuroterus* also is monophyletic; Abrahamson et al. (1998) note that this genus is also diagnosed by a morphological synapomorphy, the absence of the scutoscutellar suture. It is significant that even the small molecular data set is congruent with strong morphological evidence for monophyly of some genera.

The three other genera multiply sampled in our molecular data set do not show such strong evidence of monophyly. Three *Dryocosmus* species form a paraphyletic group, as *Andricus quercuslanigera* emerges from within the cluster of *Dryocosmus* species (although the latter genus is monophyletic in trees only one step longer than the most parsimonious trees). Abrahamson et al. (1998) synonymize *Dryocosmus* with *Andricus* on the basis of its lack of diagnostic traits not also found in *Andricus* species; the paraphyletic relationship with *Andricus* in the molecular phylogeny is consistent with this decision. The two *Andricus* species sampled do not cluster as a monophyletic group, and would require a tree five steps longer to do so. The four species of *Callirhytis* are dispersed widely across the most parsimonious tree, and require a tree of eight additional steps to form a monophyletic group. While Melika and Abrahamson recognize *Callirhytis*, they diagnose it on the basis of a transversely rugose scutum, as opposed to previous treatments which used the presence of simple tarsal claw [see Abrahamson et al. (1998) for arguments against the use of tarsal claw characters for generic diagnosis]. As a consequence, they synonymize many North American *Callirhytis* species with *Andricus*, including the four species in the molecular phylogeny. Increased taxon sampling, particularly of Eurasian species of *Callirhytis* and *Andricus*, will be required to test rigorously any new morphological criteria for these historically problematic genera.

CONCLUSIONS

These preliminary data indicate that sequences from the mitochondrial COI gene, at best in combination with other sources of information, will prove useful for reconstructing the history of radiations in the Cynipini. The congruence of the small number of molecular characters with the morphological studies of Melika and Abrahamson lead us to expect that analyses including intensive sampling within genera and a larger number of molecular characteristics will support the need for a generic revision of the Cynipini. The development of robust hypotheses of phylogeny will provide the framework for examining the evolution of ecological associations and adaptations during the extensive radiation of this group of gallmakers.

ACKNOWLEDGMENTS

We wish to thank G. Melika and W. Abrahamson for providing specimens and expert advise for this study. This research was supported by NSF DEB-9107150, amendment No.3, and Grinnell College.

LITERATURE CITATIONS

Abrahamson, W.G.; Melika, G.; Scrafford, R.; Csóka, G. 1998. Host-plant associations and specificity among cynipid gall-inducing wasps of eastern USA. In: Csóka, Gyuri; Mattson, William J.; Stone, Graham N.; Price, Peter W., eds. The biology of gall-inducing arthropods: proceedings of the international symposium; 1997 August 14-19; Mátraküred, Hungary. Gen. Tech. Rep. NC-199. St. Paul, MN: U.S. Department of Agriculture, Forest Service, North Central Research Station: 226-240.

Brower, A.V.Z. 1994. Phylogeny of *Heliconius* butterflies inferred from mitochondrial DNA sequences (Lepidoptera: Nymphalidae). Molecular Phylogenetics and Evolution. 3: 159-174.

Brown, J.M.; Pellmyr, O.; Thompson, J.N.; Harrison, R.G. 1994a. Mitochondrial DNA phylogeny of the Prodoxidae (Lepidoptera: Incurvarioidea) indicates rapid ecological diversification of the yucca moths. Annals of the Entomological Society of America. 87: 795-802.

Brown, J.M.; Pellmyr, O.; Thompson, J.N.; Harrison, R.G. 1994b. Phylogeny of *Greya* (Lepidoptera: Prodoxidae), based on nucleotide sequence variation in mitochondrial cytochrome oxidase I and II: congruence with morphological data. Molecular Biology and Evolution. 11: 128-141.

Clary, D.O.; Wolstenholme, D.R. 1985. The mitochondrial DNA molecule of *Drosophila yakuba*: nucleotide sequence, gene organization, and genetic code. Journal of Molecular Evolution. 22: 252-271.

Conover, Keith. 1991. SeqSpeak: DNA Sequence Editor. version 1.0 for Mac. Nova Scotia, Canada: Dalhousie University, Halifax.

Cook, J.M.; Stone, G.N.; Rowe, A. 1998. Patterns in the evolution of gall structure and life cycles in oak gall wasps (Hymenoptera: Cynipidae). In: Csóka, Gyuri; Mattson, William J.; Stone, Graham N.; Price, Peter W., eds. The biology of gall-inducing arthropods: proceedings of the international symposium; 1997 August 14-19; Mátraküred, Hungary. Gen. Tech. Rep. NC-199. St. Paul, MN: U.S. Department of Agriculture, Forest Service, North Central Research Station: 261-279.

Crozier, R.H.; Crozier, Y.C. 1993. The mitochondrial genome of the honeybee *Apis mellifera*: complete sequence and genome organization. Genetics. 133: 97-117.

Felsenstein, J. 1981. A likelihood approach to character weighting and what it tells us about parsimony and compatibility. Biological Journal of the Linnean Society. 16: 183-196.

Funk, D.J.; Futuyma, D.J.; Orti, G.; Meyer, A. 1995. Mitochondrial DNA sequences and multiple data sets: a phylogenetic study of phytophagous beetles (Chrysomelidae: *Ophraella*). Molecular Biology and Evolution. 12: 627-640.

Gilbert, D.G. 1995. SeqPup version 0.6f for Mac. Bloomington, IN: Department of Biology, Indiana University.

Gleason, J.M.; Caccone, A.; Moriyama, E.N.; White, K.P.; Powell, J.R. 1997. Mitochondrial DNA phylogenies for the *Drosophila obscura* group. Evolution. 51: 433-440.

Juan, C.; Oromi, P.; Hewitt, G.M. 1995. Mitochondrial DNA phylogeny and sequential colonization of Canary Islands by darkling beetles of the genus *Pimelia* (Tenebrionidae). Proceedings of the Royal Society of London (Series B). 261: 173-180.

Lewis, D.L.; Farr, C.L.; Kaguni, L.S. 1995. *Drosophila melanogaster* mitochondrial DNA: completion of the nucleotide sequence and evolutionary comparisons. Insect Molecular Biology. 4: 263-278.

Liu, H.; Beckenbach, A.T. 1992. Evolution of the mitochondrial cytochrome oxidase II gene among 10 orders of insects. Molecular Phylogenetics and Evolution. 1: 41-52.

Maddison, W.P.; Maddison, D.R. 1995. MacClade: analysis of phylogeny and character evolution. Sunderland, MA: Sinauer Associates, Inc.

Normark, B.B. 1996. Phylogeny and evolution of parthenogenetic weevils of the *Aramigus tessellatus* species complex (Coleoptera: Curculionidae: Naupactini): evidence from mitochondrial DNA sequences. Evolution. 50: 734-745.

Saiki, R.K.; Gelfand, D.H.; Stoffel, S.; Scharf, S.J.; Higuchi, R.; Horn, G.T.; Mullis, K.B.; Erlich, H.A. 1988. Primer-direct enzymatic amplification of DNA with a thermostable DNA polymerase. Science. 239: 4487-4491.

Simon, C.; Frati, F.; Beckenbach, A.; Crespi, B.; Liu, H.; Flook, P. 1994. Evolution, weighting and phylogenetic utility of mitochondrial gene sequences and a compilation of conserved polymerase chain reaction primers. Annals of the Entomological Society of America. 87: 651-701.

Spicer, G.S. 1995. Phylogenetic utility of mitochonrial cytochrome cxidase gene: molecular evolution of the *Drosophila buzzatii* species complex. Journal of Molecular Evolution. 41: 749-759.

Swofford, D.L. 1993. PAUP: phylogenetic analysis using parsimony, version 3.1.1. Champaign, IL: Illinois Natural History Survey.

Swofford, D.L.; Olson, G.J. 1990. Phylogeny reconstruction. In: Hillis, D.M.; Moritz, G., eds. Molecular systematics. Sunderland, MA: Sinauer Associations, Inc.: 411-501.

MOLECULAR PHYLOGENY OF THE GENUS *DIPLOLEPIS* (HYMENOPTERA: CYNIPIDAE)

Olivier Plantard[1], Joseph D. Shorthouse[2] and Jean-Yves Rasplus[1]

[1] Equipe INRA-CNRS Biologie évolutive des hyménoptères parasites, Laboratoire Populations, Génétique et Evolution, 91198 Gif-sur-Yvette Cedex, FRANCE
[2] Department of Biology, Laurentian University, Sudbury, Ontario, P3E 2C6 CANADA
(present address of OP: Laboratoire de Zoologie, INRA, Domaine de la Motte, B.P. 29, 35650 Le Rheu, France; present address of JYR : Laboratoire de Modélisation et Biologie Evolutive, INRA, 488 rue de la Croix de Lavit, 34090 Montpellier Cedex, France)

Abstract.—Phylogenetic relationships among 18 species of gall-inducing cynipid wasps on *Rosa* (Hymenoptera/Cynipidae/Rhoditini/*Diplolepis* and *Liebelia* genera) were investigated using partial sequences of two mitochondrial genes. Six species groups were found with high bootstrap value whatever the phylogenetic method or data set used. Most of the groups were composed of species sharing common adult and gall characters as well as a common geographic range. The strong association observed between the number of chambers per gall and the organ bearing the gall is discussed. Our results suggest a nearctic origin for the genus *Diplolepis* with a subsequent colonization of Asia and Europe.

INTRODUCTION

Because gall-inducing organisms are intimately associated with their host-plants and are at the center of species rich component communities based upon them (Askew 1961; Brooks and Shorthouse 1997; Shorthouse 1973, 1993; Hawkins and Goeden 1984), they provide interesting biological models to investigate various topics in ecology such as plant-insect relationships and community ecology. These subjects, as well as numerous others tackled in cecidology, would greatly benefit from a better knowledge of the evolutionary history of the gall inducers, because of the long common history of the other inhabitants associated with the systems (host plants, gall inducers, inquilines and parasitoids). However, the evolution of gall-inducing arthropods has rarely been investigated to date (but see Stern 1995 for aphid gallers, and Cook et al. 1998). We present here the phylogenetic relationships among various species of the genus *Diplolepis* using DNA sequences of two mitochondrial genes to pursue several aims.

First, we intend to clarify the debated taxonomic status of several problematic taxa (including *Liebelia*, *Lytorhodites*, *Nipporhodites*, *Diplolepis fructuum*, *D. nervosa*). We also attempt to gain new insight into the evolution of gall morphology by mapping gall characteristics on the *Diplolepis* phylogeny (are gall characteristics well conserved in a given lineage? How many times has a given character state evolved in the genus?). Furthermore, the relationship of the phylogeny of the genus with geographic distribution patterns can help us to reconstruct phylogeography (Where did the genus originate? How were the different mainlands colonized?). The evolution of parthenogenetic *versus* sexual species (Do they form two monophyletic lineages? How often has a change in reproductive strategy occurred in the genus?) is investigated in another paper in this volume (Plantard and Solignac 1998).

Presentation of the Taxonomy and Biology of the Group Studied

The tribe Rhoditini forms a monophyletic group, well characterized by a crenulate furrow across the mesopleuron, an autapomorphic character of the tribe according to Liljeblad and Ronquist (in press). Furthermore, all species in this tribe are restricted to inducing galls on species of *Rosa*. Galls of various sizes and shapes are produced on different organs of the bushes (leaves, bud, stem, fruit, root...) but most species of *Diplolepis* are host, organ and tissue specific (Shorthouse 1982, 1993). The relative position of the tribe Rhoditini in the Family Cynipidae was recently described by Liljeblad and Ronquist (in press) (fig. 1). The sister group of the Rhoditini is the tribe Eschatocerini, a monobasic tribe with three described species inducing galls on *Acacia* and *Prosopis* (Leguminosae) in South America (Diaz 1981, Weld 1952). [Rhoditini+ Eschatocerini] forms a separate clade derived from some genera of "Aylacini". [Rhoditini+Eschatocerini] is the sister group of [Pediaspini+Cynipini]. In the tribe Rhoditini, two genera are currently recognized, namely *Liebelia* Kieffer and *Diplolepis* Geoffroy (Vyrzhikovskaja 1963; Ronquist 1994, 1995; Liljeblad and Ronquist in press; but see Weld (1952) who considered *Liebelia* (*Nipporhodites*) *fukudae* (Shinjii) as a species of *Diplolepis*).

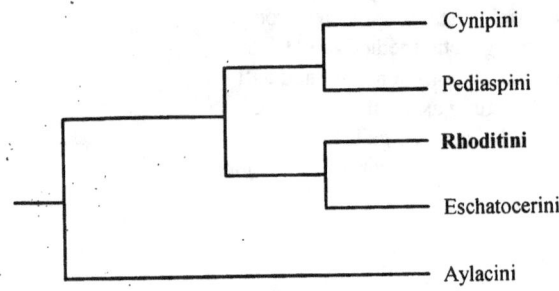

Figure 1.—*Relative position of the tribe Rhoditini in the family Cynipidae (Liljeblad and Ronquist, In press)*

In *Liebelia*, Vyrzhikovskaja (1963) distinguished two subgenera: *Liebelia* subgenus *Liebelia* (with one species in Europe—*L.(L.) cavarai* Kieffer—and three other species in Central Asia), and the subgenus *Liebelia Nipporhodites* Sakagami (with four species from Central Asia and one species from Japan).

The genus *Diplolepis* is more speciose; about 42 species having been described, including 30 from North America (Burks 1979, Dailey and Campbell 1973, Shorthouse and Ritchie 1984, Shorthouse 1993), six from Europe (Eady and Quinlan 1963, Pujade 1993), five additional species from Central Asia (Dalla Torre and Kieffer 1910, Belizin 1957, Vyrzhikovskaja 1963) and one additional species from Japan (Dalla Torre and Kieffer 1910, Yasumatsu and Taketani 1967). Pujade (1993) has made a revision of the European species and found six valid species. However, the taxonomy of the genus *Diplolepis*, especially the Nearctic species, remains poorly known, as numerous species were described on the basis only of gall characteristics (Shorthouse 1993). It is also likely that additional species will be found in central and eastern Asia (Vyrzhikovskaja 1963).

MATERIALS AND METHODS

Sampling of Taxa

We attempted to obtain as many species as possible, from the widest geographical range and representing the largest biological diversity (i.e., characteristics of galls, reproduction systems...) exhibited by the genus. A total of 22 taxa of Rhoditini were included in our analysis (table 1). Species of *Diplolepis* from three continents were used. All six European species are represented; although there are some species from the USA, where the genus is most speciose, that we were unable to obtain. Two taxa were obtained from Japan, along with *Liebelia fukudae*. Other endemic species from Asia are missing from our analyses. We were unable to obtain other species of *Liebelia* (nor the central Asiatic or the unique European species of that genus, *L. cavarai*, from galls collected in Sardinia, but to our knowledge, not sampled since). *Liposthenes glechomae* was chosen as an outgroup, because of its position in the tribe Aylacini.

Mitochondrial DNA Sequences

DNA was extracted from adults (most Canadian species) or larvae (*D. californica, D. ignota, D. variabilis,* and the Japanese species), stored in 96 percent alcohol, and sent to the laboratory at Gif-sur-Yvette, or on adults frozen at -80°C, immediately after they exited from galls (European species). Total DNA was prepared from individual adults by SDS/proteinase K digestion and phenol-chloroform extraction according to Kocher et al. (1989). DNA was resuspended in 50 µl of ultra pure water. PCRs were performed with 4 µl of 1/10 dilution DNA in 50 µl. Two mitochondrial genes, a coding one (Cytochrome B gene) and a non-coding one (small ribosomal subunit - 12S - gene) were partly sequenced.

For the Cytochrome B gene, the CP1-CP2 primers designed by Harry et al. (submitted) were used for most of the species. PCR conditions were as follows: $MgCl_2$ = 2.5 mM; 30 cycles of 92°C denaturation (1 min), 50°C annealing (1 min), and 72°C extension (1 min) with an initial denaturation of 5 min at 92°C and a final extension of 5 min at 72°C. CB1-CB2 primers (alias CB-J-10933 and CB-N-11367 in Simon et al. 1994) designed by Y.C. Crozier (Crozier and Crozier 1992) were used for the few species for which amplification was unsuccessful using CP1-CP2 (same PCR conditions but with annealing at 48°C). For the 12 S gene, the 12Sbi and 12Sai primers

Table 1.—*Author names of the taxa used in the study and the locality in which they were collected.*

Species	Populations (department)
Diplolepis centifoliae (Hartig)	Fontainebleau (France)
Diplolepis eglanteriae (Hartig)	Ouzouère sur Loire (France)
Diplolepis fructuum (Rübsaamen)	Taleghan (Iran)
Diplolepis mayri (Schlechtendal)	Gap (France)
Diplolepis nervosa (Curtis)	Causse de la Selle (France)
Diplolepis rosae (L.)	Chinon (France)
Diplolepis spinosissimae (Giraud)	Fontainebleau (France)
Diplolepis bicolor (Harris)	Manitoulin I. (Ontario)
Diplolepis californica (Beutenmueller)	Cosumnes river preserve (California)
Diplolepis fusiformans (Ashmead)	Chelmsford (Ontario)
Diplolepis ignota (Osten Sacken)	Lethbridge (Alberta)
Diplolepis nebulosa (Bassett)	Cranbrook (British Columbia)
Diplolepis nodulosa (Beutenmueller)	Manitoulin I. (Ontario)
Diplolepis polita (Ashmead)	Chelmsford (Ontario)
Diplolepis radicum (Osten Sacken)	Manitoulin I. (Ontario)
Diplolepis rosaefolii (Cockerell)	Manitoulin I. (Ontario)
Diplolepis spinosa (Ashmead)	Chelmsford (Ontario)
Diplolepis triforma Shorthouse and Ritchie	Chelmsford (Ontario)
Diplolepis variabilis (Bassett)	Penticton (British Colombia)
Diplolepis sp. near *spinosissimae*	Aomori Apple Exp. Stn (Japan)
Diplolepis sp. near *centifoliae*	Aomori Apple Exp. Stn (Japan)
Liebelia fukudae (Shinjii)	Ajigasawa (Japan)
Liposthenes glechomae (Kieffer)	Lancieux (France)

(alias SR-J-14233 and SR-N-14588 in Simon et al. 1994) designed by Kocher et al. (1989) were used (same PCR conditions except $MgCl_2$ = 3.5 mM). Each strand of PCR products were directly sequenced by manual sequencing at the beginning of the study, followed by sequencing of PCR products purified using a Qiaquick purification kit and an ABI 373 automated sequencer using the dye terminator cycle sequencing kit (Perkin Elmer).

There was no insertion or deletion in the fragment of the Cytochrome B gene from all species studied and alignment was therefore straight forward. For the 12 S rRNA gene fragment, sequences were aligned manually using the secondary structure of the small ribosomal subunit (Kjer 1995). One region was excluded because of its excessively high AT content, leading to uncertain alignments.

Phylogenetic Analyses

Cytochrome B and 12 S characters were analyzed separately and in a combined data set, providing three different data sets. Aligned sequences were analyzed using three different methods for phylogenetic reconstruction. Neighbor-joining analyses (Saitou and Nei 1987) were performed with MEGA 1.0 software (Kumar et al. 1993) with 500 replications, using Kimura-2-parameter distance. Maximum Parsimony analyses (Farris 1970) were performed using PAUP 3.1.1 software (Swofford 1993) with the default options. A heuristic search with 100 replicates was performed for each data set. The trees were rooted by a basal polytomy with the outgroup defined as being *Liposthenes glechomae*. Maximum Likelihood analyses was performed with PUZZLE 3.1 software (Strimmer and von Haeseler 1996) with the default options.

Definition of Gall Characteristics

Three different sets of specific unordered characters with three to six states were coded for the galls of each species: (1) "number of larval chambers per gall" with three possible states: unilocular (single chamber per gall), plurilocular (several chambers per gall), or galls induced by the same species with varying numbers of chambers. (2) "organ bearing the gall" with six possible states: leaf *or* bud (the entire bud is incorporated into the gall) *or* stem (stem grows beyond the gall) *or* fruit, *or* collar *or* variable inside a given species. (3) "surface of the gall" with five possible states: smooth, *or* spiny *or* with aborted leaves *or* with branched hairs *or* variable inside a given species. These different character states were mapped onto the "working phylogeny". Because of the relatively significant phylogenetic distance between the outgroup used and the ingroup, character states of *Liposthenes glechomae* were not taken into account to infer ancestral character states in Rhoditini. MacClade was used to reconstruct the most parsimoniously evolution of gall characteristics on the maximum likelihood tree.

RESULTS

The final lengths of the 12S fragment (after removal of an AT rich region with uncertain alignment) and the Cytochrome B fragment analyzed were 359 and 386 bp respectively. The variability was 27.6 and 50.6 percent and the percentage of informative sites was 15.5 and 36.5 percent, respectively. A strong bias in nucleotide composition was observed with respectively 84.21 and 77.3 percent of A+T, a common feature of insect mitochondrial genes (Simon et al. 1994). Sequences are available from the authors on request.

Pairwise Comparisons of Cytochrome B Sequences

While most of the closely related taxa have at least 25 substitutions in pairwise comparisons, three groups of taxa exhibit far smaller number of substitutions (fig. 2, table 2). Only one transition was found between *D. centifoliae* and a very similar species from Japan (the former species has not been recorded previously in that country). Those two taxa are obviously conspecific. *D. centifoliae* and *D. nervosa* differ by only seven substitutions, suggesting that *D. nervosa* is a synonym of *D. centifoliae* (the validity of *D. nervosa* was already subject to debate as only very weak morphological differences such as leg color are recognized between the two). *D. nebulosa* and *D. ignota* differ from each other by only one substitution and *D. ignota* and *D. variabilis* by three. Such a small number of differences is more typical of intraspecific variation than of interspecific differences; although, these results need to be confirmed by further morphological investigations. *D. sp.* near *centifoliae* from Japan, *D. sp.* near *spinosissimae* from Japan, *D. nervosa* and *D. ignota* have not been included in subsequent analysis.

Phylogenetic Reconstruction

The trees recovered have very similar topologies, whatever the data set and whatever the phylogenetic reconstruction method used (figs. 3, 4, and 5). Two features are common to all of these trees. First, in all analyses *Liebelia fukudae* falls outside of the *Diplolepis* clade, confirming that *Liebelia* is a distinct genus from *Diplolepis*. But elsewhere, the genus *Lytorhodites*, represented in our sample by *D. spinosa* (= *Rhodites multispinosus* Gillette) and *D. nebulosa*, appears to be polyphyletic and non-valid as these two species belong to different species groups. *Lytorhodites*, defined by Kieffer on the base of a few morphological characters such as radial cell (opened) or the absence of a fovea at the base of scutellum, was previously considered non-valid by Weld (1952).

The second characteristic of the trees is that six groups of species can be recognized in most of the analyses, forming clades with strong bootstrap values (usually better than 80 percent). These species groups are as follows:

The nearctic species can be partitioned in four groups: the *D. nebulosa* group (also including *D. ignota*), the *D. polita* group (including *D. bicolor*), the *D. rosaefolii* group (including *D. fusiformans*) and the "flanged-femur" group (including *D. radicum, D. spinosa, D. californica*, and *D. triforma*). The Palaearctic species form two species groups: the *D. eglanteriae* group (including *D. centifoliae*) and the *D. rosae* group (including *D. fructuum, D. mayri* and *D. spinosissimae*, the latter species forming a separate lineage outside the *D. rosae* group in the analyses based on the Cytochrome B data set (figs. 3b, 4b, and 5b).

Although the relationships among all six species groups are poorly resolved (cf. the basal polytomy of the trees), the flanged-femur group plus the two European species groups form a distinct clade in three out of the four analyses with the best resolution (figs. 3c, 5b, and 5c). In particular, all three analyses based on the combined data set exhibit such a clade (only the Maximum Parsimony analysis has a bootstrap value of less than 50 percent for this clade; fig. 4c). Inside this clade, the two European species groups are associated in two analyses (Cytochrome B data set using Neighbor Joining or Maximum Likelihood with bootstrap values of 69 and 50 percent, respectively; figs. 3b and 5b).

Comparison of Trees Obtained Using Data of the Cytochrome B or 12 S Gene

The trees recovered using only the 12 S data set or only the Cytochrome B data set show only slight variation in tree topology; in particular, the species groups remain the same. This congruence justifies the combination of the two data sets. There is some evidence that the 12S data sets gives slightly less resolved trees than those obtained with Cytochrome B, as expected from their overall variability. For example, *D. polita* and *D. bicolor* are associated with bootstrap values of less than 50 percent in the three analyses based on the 12S data set (figs. 3a, 4a, and 5a) whereas they form a robust clade with bootstrap values of 75 and 91 percent with the Cytochrome B data set (fig. 3b and 5b).

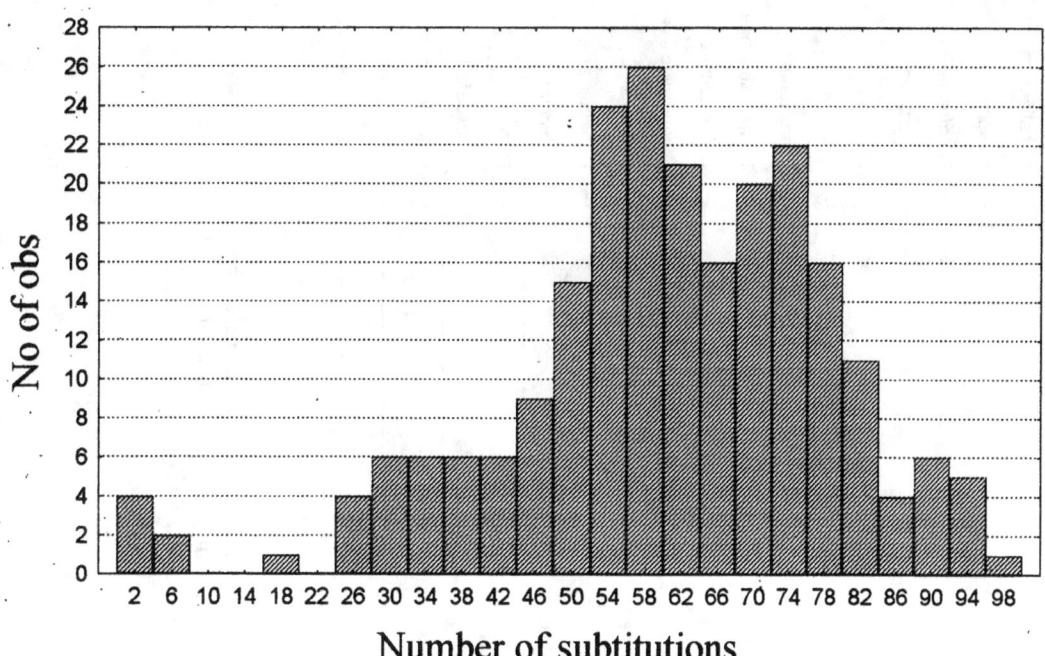

Figure 2.—*Distribution histogram of the number of substitutions in pairwise comparisons of the Cytochrome B gene sequences.*

Table 2.—*Taxa that are the most closely related (Cytochrome B).*

Pairwise comparison	Number of substitutions
D. centifoliae / *D. sp* near *centifoliae* (spherical leaf gall from Japan)	1
D. centifoliae / *D. nervosa*	7
D. nervosa / *D. sp* near *centifoliae* (spherical leaf gall from Japan)	8
D. nebulosa / *D. variabilis*	1
D. nebulosa / *D. variabilis*	3
D. ignota / *D. variabilis*	3
D. spinosissimae / *D. sp* near *spinosissimae* (irregular leaf gall from Japan)	18
D. rosaefolii / *D. fusiformans*	25
D. mayri / *D. fructuum*	32

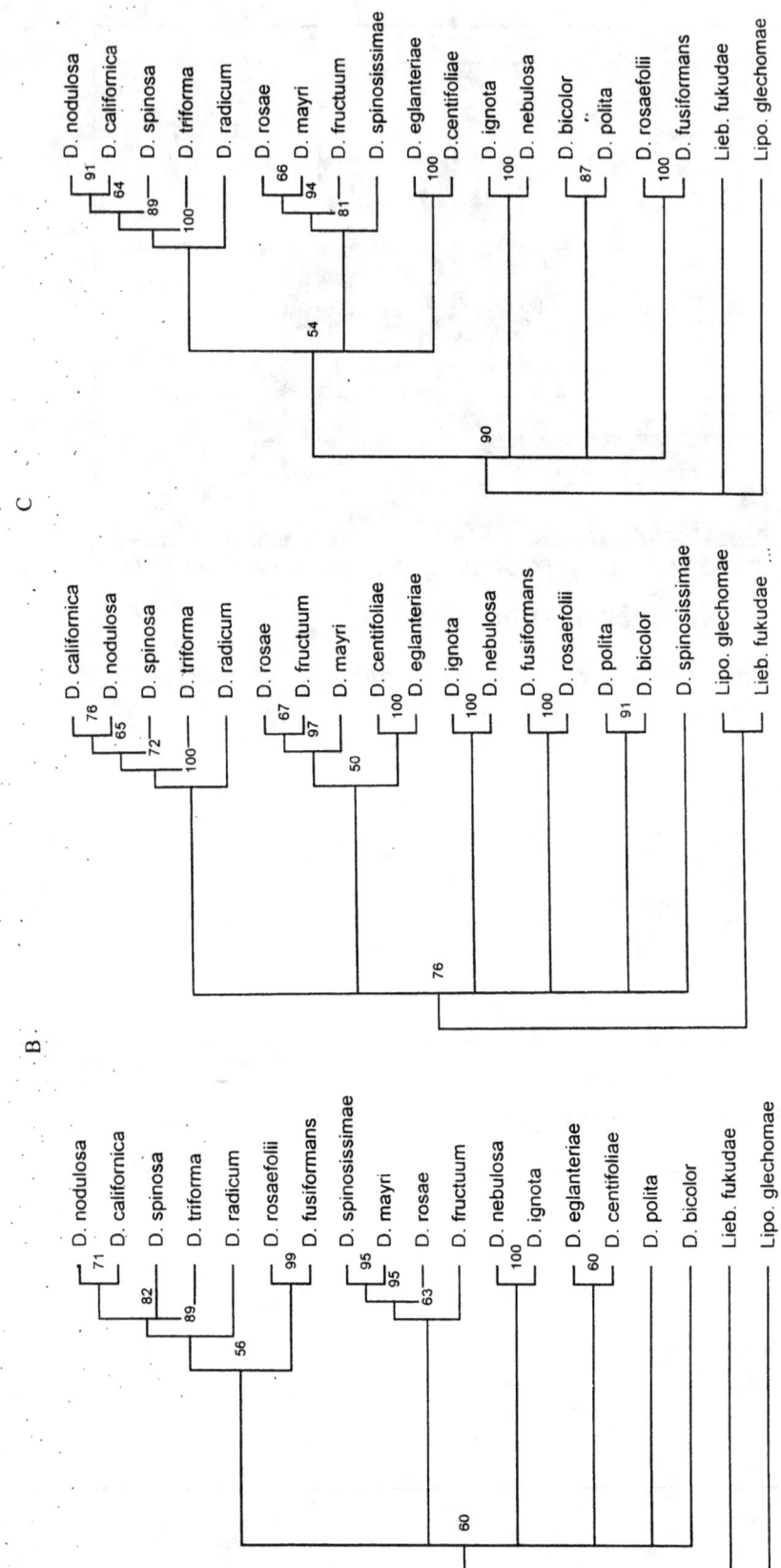

Figure 3.—Neighbor-Joining trees obtained with the 3 data sets, using MEGA with Kimura 2-parameter distance. Bootstrap values based on 200 replicates are shown at nodes. The cut-off command has been used to collapse branches with a bootstrap value of less than 50 percent a) tree based on the 12S data set; b) tree based on the Cytochrome B data set; c) tree recovered using the combined data set (12S + Cytochrome B).

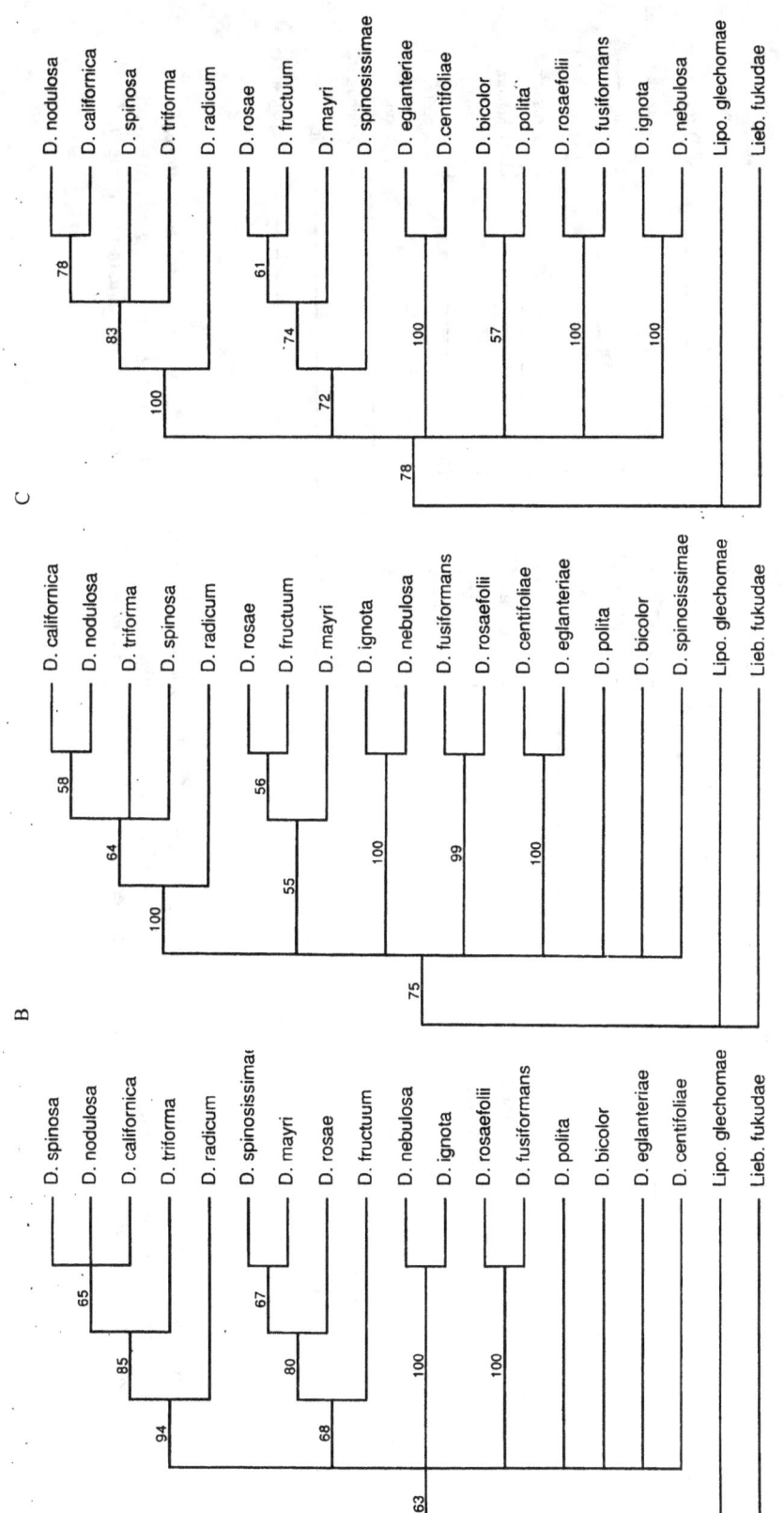

Figure 4.—Maximum Parsimony trees obtained with the 3 data sets, using PAUP. Bootstrap values based on 100 replicates are shown at nodes. Only groups with a frequency of more than 50 percent are retained. a) strict consensus tree based on 78 most parsimonious trees (CI = 0.651, RI = 0.654) recovered using the 12S data set; b) most parsimonious tree (CI = 0.589, RI = 0.654) recovered using the Cytochrome B data set; c) most parsimonious tree (CI = 0.460, RI = 0.560) tree recovered using the combined data set (12S + Cytochrome B).

253

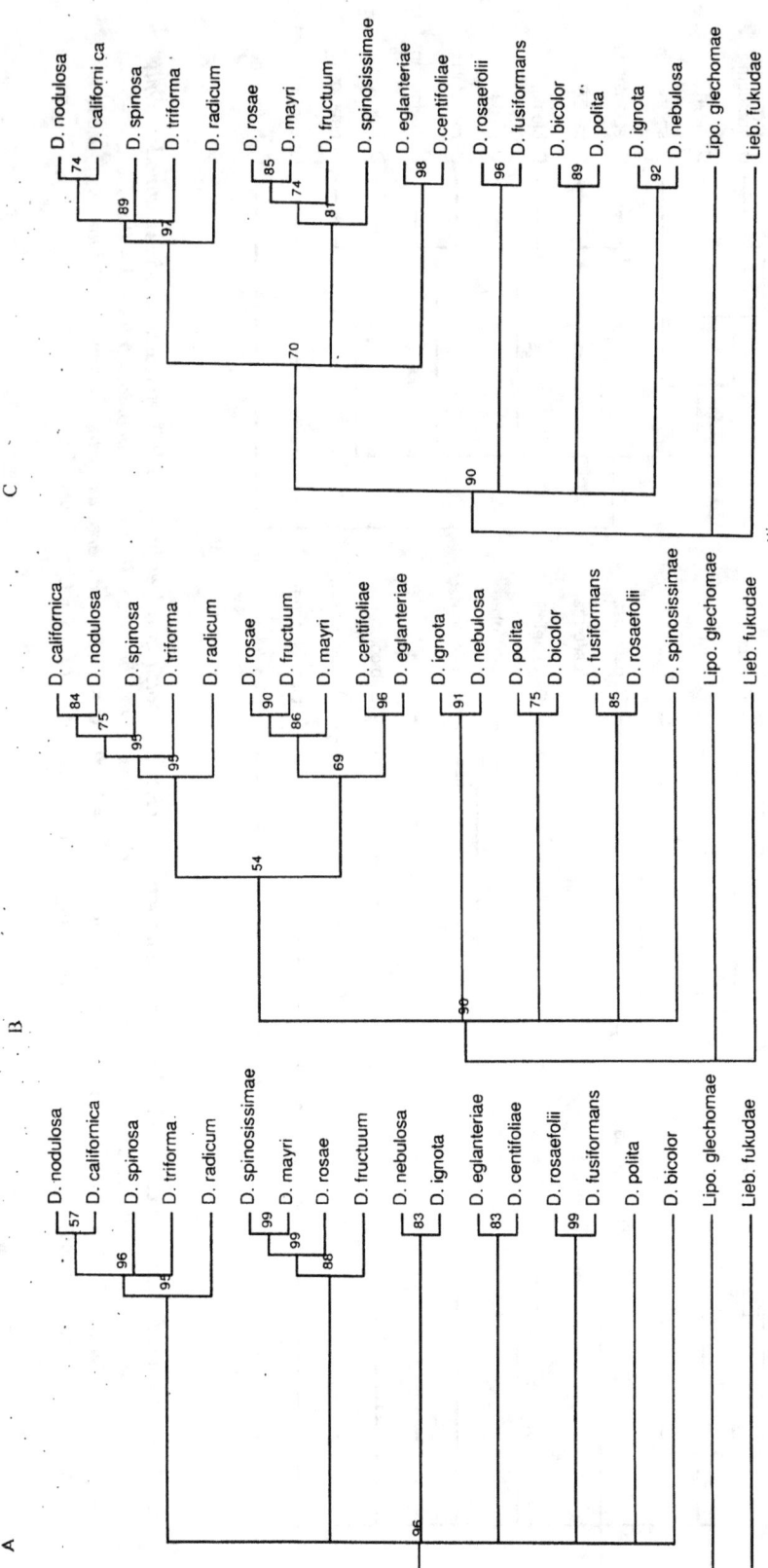

Figure 5.—Maximum Likelihood trees obtained with the 3 data sets, using PUZZLE with the default assumptions. Support values for each internal branch (same practical meaning as bootstrap values) based on 1000 puzzling steps (meaning comparable to the number of bootstrap replicates) are shown at nodes. a) tree recovered using the 12S data set; b) tree recovered using the Cytochrome B data set; c) tree recovered using the combined data set (12S + Cytochrome B).

Among species groups, the topology remains the same for the flanged-femur species group whatever the data set used (namely *D. radicum* being the most basal species, followed by *D. triforma*, then by *D. spinosa*, being the sister group of *D. californica* + *D. triforma*). In the *D. rosae* group, *D. fructuum* appears to be the most basal species with *D. spinosissimae* and *D. mayri* the most derived ones in all analyses based on 12S data set (figs. 3a, 4a, and 5a). In contrast, *D. fructuum* and *D. rosae* are the most derived taxa and *D. spinosissimae* is not associated with this group when using the Cytochrome B data set (figs. 3b, 4b, and 5b).

Comparison of the Trees Obtained Using the Different Phylogenetic Reconstruction Methods

For a given data set, very similar results are found whatever the phylogenetic reconstruction method used. The tree based on the combined data set with the Maximum Likelihood method (fig. 5c) has been chosen as the "working phylogeny" because of its slightly better resolution.

DISCUSSION

Our results clearly identify *Liebelia fukudae* as belonging to a distinct genus from *Diplolepis*, confirming the interpretation of Vyrzhikovskaja (1963) but contrasting with the opinion of Weld (1952). However, additional species of the genus *Liebelia* (in particular those considered to belong to the other subgenus *L. Liebelia*) are needed for a complete revision of the Rhoditini tribe.

Although there is no morphology-based phylogeny of the genus *Diplolepis* to date, the species showing closest morphological features (i.e., *D. centifoliae* and *D. eglanteriae*; *D. fructuum*, *D. mayri* and *D. rosae*; *D. polita* and *D. bicolor*) are always associated in our analyses. The association of all species of *Diplolepis* with a flanged-femur (this conspicuous character—first reported by Dailey and Campbell 1973 in *D. inconspicuis*, and subsequently observed in several other Nearctic species by Shorthouse and Ritchie 1984—constitutes a unique synapomorphy in the tribe) in the same clade, confirms the relevance of our phylogeny.

Phylogeography of the *Rhoditini* Tribe

In all the resolved trees recovered with the combined data set and whatever the type of analysis performed, the three Nearctic species groups (namely *nebulosa*, *polita*, *rosaefolii*) are ancestral (fig. 6). This suggests that *Diplolepis* had a Nearctic rather than Palaearctic origin. After evolving in North America, we suggest that the genus dispersed across Beringia to eastern Asia and from there to western Europe. The trans-Pacific Bering bridge has intermittently connected the western Nearctic and the eastern Palaearctic during much of the Tertiary until 3-5 mya ago and has played an important role in plant and animal dispersal (Cox 1974).

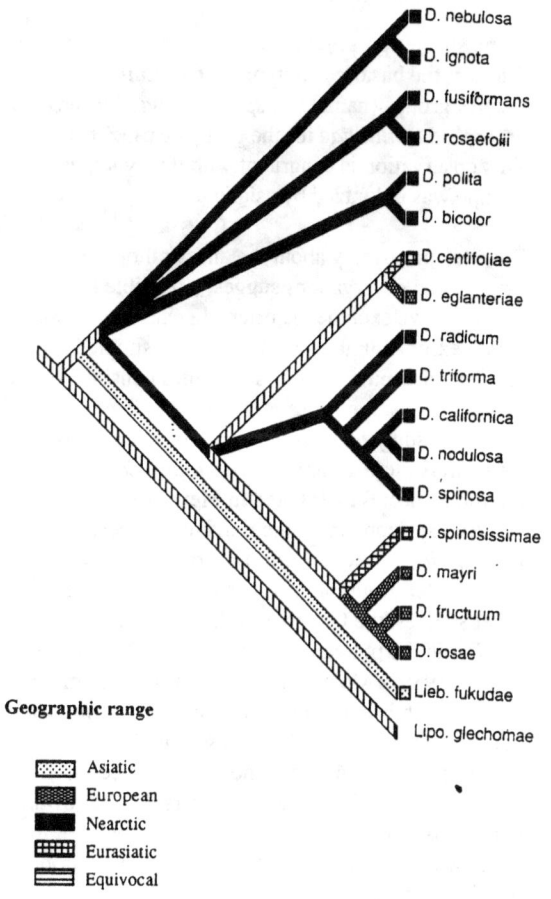

Figure 6.—*Estimated phylogeny of the Rhoditini (tree recovered using Maximum Likelihood method analysis of the combined data set) on which the range of each species has been mapped.*

This hypothesis is in general agreement with the following arguments:

- The dispersal starting point in Holarctic-distributed animals is more frequently the western Nearctic region than East Nearctic or Palaearctic (Enghoff 1995). However, *Diplolepis* may constitute an interesting case, as genera dispersal between continents has rarely been reported even though such distributions are more frequent in family-clades.

- The highest species richness in *Diplolepis* occurs in North America, where about three-quarters of the known species are found. However, this finding may be due to a lack of collecting in central Asia.

- The Eschatocerini, sister group of the Rhoditini, is a strictly neotropical group associated with *Acacia*. This suggests that the common ancestor of Eschatocerini and Rhoditini could have been a neotropical or nearctic taxon.

- Some species present in both Asia and in Europe are found at the basal position of the two European species groups (namely *D. spinosissimae* for the *rosae* group and *centifoliae* for the group bearing its name). This observation is congruent with the hypothesis that Europe was colonized by asiatic taxa.

- Despite controversy about the ancestral origin of *Rosa*, some observations suggest a possible Nearctic origin. If we exclude the oriental genus *Hulthemia*, which exhibits unique characters (i.e., simple leave, absence of stipules) and is sometimes considered as a distinct genus, the genus *Rosa sensu stricto* is divided into three subgenera *Platyrhodon* (one species only found in Eastern-central Asia), *Hesperodos* (containing three species distributed in Mexico and California) and *Eurosa* (containing the remaining species). *Hesperodos* shows ancestral characteristics and could be the sister group of the other *Rosa*. Unlike all the other species, the flowers have no disks (Boulenger 1937) and the pollen exine has no ridges (Ueda and Tomita 1989). Moreover the chromosome number is $n = 7$, which is considered the ancestral chromosome number, while this number is also found in *Platyrhodon* and in some other *Eurosa* species. Asia is generally considered as the dispersal center of the genus *Rosa*, the highest diversity being found in the Chinese and Himalayan regions. The exact number of Nearctic species of *Rosa* varies from 20 to 170 depending on the authors.

- Oldest fossils of *Rosa* are known from Oligocene in Colorado and Oregon in the USA. This indicates that the genus has been present in the Nearctic region for at least 32 million years.

In conclusion, a detailed phylogenetic and phylogeographic investigation of the genus *Rosa* is needed and would be very helpful for understanding the history of the tribe Rhoditini. However, to explain the actual range of the genus *Liebelia* (absent from North America), with respect to this hypothesis we have to assume that either all Nearctic species of *Liebelia* have disappeared, they have yet to be described or they are not recognized as belonging to the genus. A taxonomic revision of the Rhoditini is urgently needed and it would also be of great interest to discover some Rhoditini associated with species of *Hesperodos*.

Mapping of Gall Characteristics Onto the *Diplolepis* Phylogeny

Evolution of the Number of Larval Chambers Per Gall

The unilocular (single-chambered) vs. plurilocular (multichambered) state of cynipid galls is determined by the egg laying behavior of the female (see Cook *et al.* 1998). If she lays one egg per bud or leaflet, or even several eggs on the same leaflet but sufficiently distant from each other (species inducing unilocular galls often induce several galls on the same leaflet), plurilocular galls will not be produced (except *D. ignota* and *D. variabilis* for which some galls appear plurilocular because adjoining chambers coalesce). In the case of plurilocular galls such as those of *D. spinosa*, numerous eggs are laid in clusters at the base of unforced buds. Such behavior appears to have arisen only once (in the ancestor of the flanged-femur species group plus the two European species groups, but this assumption implies a reversion for the European pea galls induced by *D. centifoliae* ? that are unilocular) or twice (in the flanged-femur species group and in the *D. rosae* species group) as it is found in only two lineages (fig. 7).

However some exceptions are found in those two groups: *D. nodulosa* is the only species inducing unilocular galls in the flanged-femur group. This species appears to be the most derived of this species group in all analyses, suggesting that this character is a recent reversion. When six additional Nearctic species of *Diplolepis* not included in our molecular analyses but known to have a flanged-femur are considered (*D. ashmeadi*, *D. terrigena*, *D. tuberculatrix*, *D. tumida*, *D. inconspicuis*, and *D. verna*) to belong to this species group, again all those species induce plurilocular galls, except *D. inconspicuis* and *D. verna* (the latter producing both unilocular and plurilocular galls and being closely allied morphologically to *D. nodulosa*) (Shorthouse and Ritchie 1984). In the *D. rosae* group, most galls of *D. spinosissimae* are unilocular (80 percent, O.P. unpubl. data). However, this species is often (i.e., in all analyses using the combined data set) basal in the *D. rosae* group and could thus constitute an intermediate behavior between the two other types.

Evolution of the "Organ Bearing the Gall" Character

As with the previous character, the organ bearing the gall is also dependent on behavior of the female. Collar galls such as those of *D. radicum* are induced on the tips or sides of sucker shoots either at the surface or beneath the surface of the soil. Fruit galls, such as those of *D.*

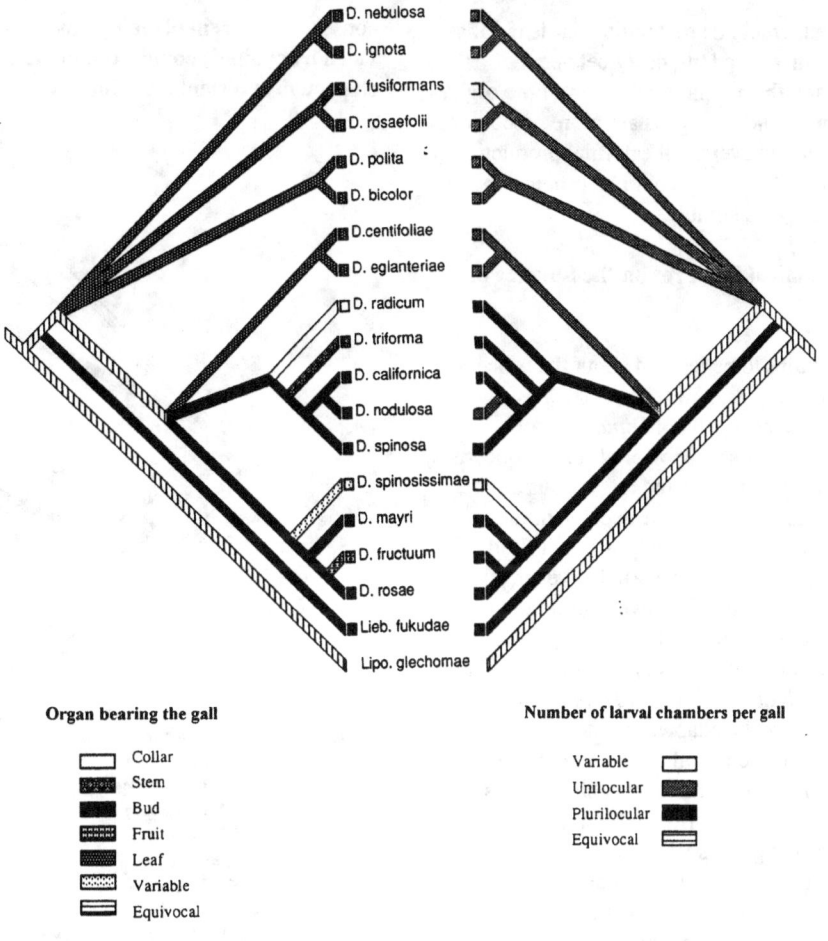

Figure 7.—*Estimated phylogeny of the Rhoditini (tree recovered using Maximum Likelihood analysis of the combined data set) on which the number of larval chambers per gall and the organ bearing the gall has been mapped for each species.*

fructuum, result from oviposition in flower buds. Among the unilocular leaf galls, the surface of the leaflet on which the egg is laid is responsible for the position of the gall on the leaf. For example, galls of *D. polita* appear on the adaxial surface and galls of *D. centifoliae* appear on the abaxial surface. Again, the three most basal species groups are less variable (intra- and inter-group comparisons) than the three derived ones, as most species are leaf gallers (except *D. fusiformans* which is a true stem galler, see fig. 7). It must be noted that as all galls result from oviposition in buds (except some stem galls such as those of *D. spinosissimae*), the differences between the various types we have identified are artificial. An interaction between the number of larvae per gall and the organ bearing the gall clearly appears when comparing the distribution of both characters (fig. 7). Most unilocular galls are found on leaves (except *D. nodulosa*, some unilocular *D. spinosissimae* galls and some *D. fusiformans* galls), whereas plurilocular galls are always found on other organs (except some plurilocular *D. spinosissimae* galls). A similar pattern is found in other European Cynipini associated to oaks; here all leaf galls are unilocular, except for those of *Plagiotrochus quercusilicis* and some coalescent galls of *Andricus curvator*. However, unlike the galls of Rhoditini, numerous bud galls on oaks are unilocular. Note also that the smallest galls of *D. rosae* are unilocular and are found on leaflets and are not considered bud galls! These two characters could consequently be linked. A larger number of larvae per bud could be responsible for enhanced levels of the cecidogenic stimulus overrunning the normal development of the bud and could be responsible for the absence of leaves. Of interest, larvae and adults of *D. spinosissimae* are the smallest of all species of *Diplolepis* (Dalla Torre and Kieffer 1910) and this smaller size could be responsible for a smaller cecidogenic stimulus, explaining the plurilocular leaf galls of this species. If this hypothesis is correct, the character "organ bearing the gall", (especially in the comparison of bud/leaf galls) would not have evolutionary significance, but rather would be a consequence of the egg laying behavior. Another hypothesis, though not exclusive of the first, would be that unilocular leaf galls are produced when there is a long delay between induction of galls and the normal production of the organ by the bud and that plurilocular galls are more precocious in their development and overrun the normal production of organs by the bud. This delay could be proportional to the amount of cecidogenic stimulus.

Evolution of Anatomical Features on the Surface of Galls

Although much remains to be learned about the process of cecidogenesis in cynipids, gall features such as shape (not examined here as most galls of *Diplolepis* are spherical or globular—but see Cook *et al.* 1998 concerning gall shape in oak gall wasps) and the various structures on the surface are thought to be dependent on activities of the larvae rather than ovipositional fluids (Dreger-Jauffret and Shorthouse 1992). However, like the other characters, some species have conserved character states, whereas others exhibit intraspecific polymorphism. For example, galls of *D. rosae* invariably have thin, branching hairs and galls of *D. polita* always have straight spines (see the chapter by Shorthouse 1998). In contrast, only some galls of *D. nervosa* have "horns" and only some galls of *D. mayri* have spines. Some of these structures are normally produced by organs that produce the galls, such as spines on stems also being found on galls of *D. triforma* and some *D. spinosissimae*, or aborted leaves on the surface of young *D. californica* galls. In this case, presence or absence of structures on the surface of galls can vary depending on whether or not tissue morphogenesis is stopped. Inquilines of the genus *Periclistus* inhabiting the galls of some species may be more densely spined than galls without inquilines (Shorthouse and Ritchie 1984).

Again, the three more basal species groups exhibit no intra-group variation, whereas the three derived groups show high variability (fig. 8). Though we have failed to include it in our molecular study, the Nearctic species *D. bassetti* also produces long, green filaments forming a moss-like mass similar to the galls of *D. rosae* (Beutenmueller 1918). As all Nearctic and European species are found in distinct clades, it's very unlikely that *D. bassetti* and *D. rosae* share a direct common ancestor. These complex structures are convergent in galls of *D. bassetti* and *D. rosae* (note that similar hairs are found on the surface of oak galls induced by *Andricus caputmedusae*); and seem to have evolved independently in separate lineages. It is possible that such hairs have evolved to reduce attack by parasitoids (Berdegue *et al.* 1996, Gross and Price 1988, Jeffries and Lawton 1984) and would explain its emergence in distant lineages (see also Cook *et al.* 1998). The reactivation of ancestral but conserved genes in plant genomes due to stimuli provided by cynipids could explain the convergence of these characters in distant plant lineages.

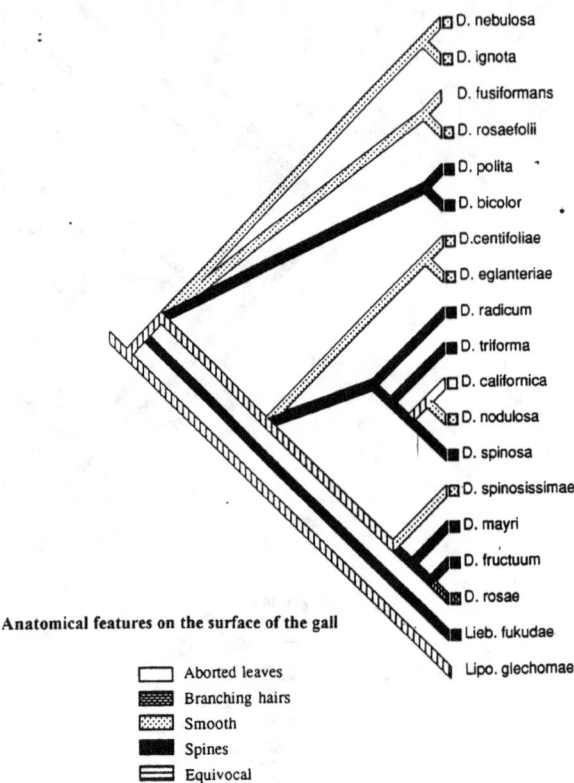

Figure 8.—*Estimated phylogeny of the Rhoditini (tree recovered using Maximum Likelihood analysis of the combined data set) on which the anatomical features on the surface of the gall of each species has been mapped.*

ACKNOWLEDGMENTS

We are grateful to Norio Sekita (Japan), Kathy Schick (California, USA), Fredrik Ronquist (Sweden), Hossein Goldansaze (Iran), and the Laboratoire de modélisation et biologie évolutive (Montpellier, France) for providing samples. We also thank Scott Brooks, formerly of Laurentian University in Sudbury, who helped collect Canadian galls and rear inhabitants. We also have greatly benefited from the technical help of Gwenaelle Mondor, Isabelle Le Clinche and Dominique Vautrin, all of Gif-sur-Yvette, in the molecular biology work. Olivier Langella and Jean-François Silvain of Gif-sur-Yvette and Silvain Piry of Montpellier generously helped in the computer analyses. This work was supported in part by grants from the Ministere de l'Environnement awarded to OP and JYR and the Natural Sciences and Engineering Research Council of Canada and the Laurentian University Research Fund awarded to JDS.

LITERATURE CITED

Askew, R.R. 1961. On the biology of the inhabitants of oak galls of Cynipidae (Hymenoptera) in Britain. Transactions of the Society for British Entomology. 14: 238-267.

Belizin, V.I. 1957. Gall wasps (Hymenoptera: Cynipidae) in the USSR fauna developed on roses. Entomologicheskoe Obozrenie. 36: 925-934.

Berdegue, M.; Trumble, J.T.; Hare, J.D.; Redak, R.A. 1996. Is it enemy-free space? The evidence for terrestrial insects and freshwater arthropods. Ecological Entomology. 21: 203-217.

Beutenmueller, W. 1918. New species from Oregon. The Canadian Entomologist. 50: 305-309.

Boulenger, G.A. 1937. Monographie du *Rhodites* genre *Hesperodos* Cocker. Bulletin du Jardin Botanique de l'Etat Bruxelles. 14: 227-239.

Brooks, S.E.; Shorthouse, J.D. 1997. Biology of the rose galler *Diplolepis nodulosa* (Hymenoptera: Cynipidae) and its associated component community in Central Ontario. The Canadian Entomologist. In press.

Burks, B.D. 1979. Cynipoidea. In: Krombein, K.V.; Hurd, P.D., Jr.; Smith, D.R.; Burks, B.D., eds. Catalog of Hymenoptera of North America north of Mexico. Vol. 1. Washington, DC: Smithsonian Institution Press: 1045-1107.

Cook, J.M.; Stone, G.N.; Rowe, A. 1998. Patterns in the evolution of gall structure and life cycles in oak gall wasps (Hymenoptera: Cynipidae). In: Csóka, Gyuri; Mattson, William J.; Stone, Graham N.; Price, Peter W., eds. The biology of gall-inducing arthropods: proceedings of the international symposium; 1997 August 14-19; Mátraküred, Hungary. Gen. Tech. Rep. NC-199. St. Paul, MN: U.S. Department of Agriculture, Forest Service, North Central Research Station: 261-279.

Cox, C.B. 1974. Vertebrate paleodistributional patterns and continental drift. Journal of Biogeography. 1: 75-94.

Crozier, R.H.; Crozier, Y.C. 1992. The Cytochrome b and ATPase genes of honeybee mitochondrial DNA. Molecular Biology and Evolution. 9: 474-482.

Dailey, D.C.; Campbell, L. 1973. A new species of *Diplolepis* from California. The Pan-Pacific Entomologist. 49: 174-176.

Dalla Torre, K.W.; Kieffer, J.J. 1910. Hymenoptera: Cynipidae. - Friedlander und Sohn, Berlin.

Diaz, N.B. 1981. Cinipoideos galigenos e inquilinos de la Republica Argentina. Revista de la Sociedad Entomologica Argentina. 39: 221-226.

Dreger-Jauffret, F.; Shorthouse, J.D. 1992. Diversity of gall-inducing insects and their galls. In: Shorthouse, J.D.; Rohfritsch, O., eds. Biology of insect-induced galls. Oxford: Oxford University Press: 8-33.

Eady, R.D.; Quinlan, J. 1963. Hymenoptera: Cynipoidea. Key to families and subfamilies and Cynipinae (including galls). London: Royal Entomological Society of London.

Enghoff, E. 1995. Historical biogeography of the Holarctic: area relationships, ancestral areas, and dispersal of non-marine animals. Cladistics. 11: 223-263.

Farris, J.S. 1970. Methods for computing Wagner trees. Systematic Zoology. 18: 374-385.

Gross, P.; Price, P.W. 1988. Plant influences on parasitism of two leafminers: a test of enemy-free space. Ecology. 69: 1506-1516.

Harry, M.; Solignac, M.; Lachaise D. (submitted) Molecular evidence for multiple homoplasies of complex adaptive syndromes in fig-breeding *Lissocephala* (Drosophilidae). Submitted to Molecular Phylogenetics and Evolution.

Hawkins, B.A.; Goeden, R.D. 1984. Organization of a parasitoid community associated with a complex of galls on *Atriplex* spp. in southern California. Ecological Entomology. 9: 271-292.

Jeffries, M.J.; Lawton, J.H. 1984. Enemy-free space and the structure of ecological communities. Biological Journal of the Linnean Society. 23: 269-286.

Kjer, K.M. 1995. Use of rRNA secondary structure in phylogenetic studies to identify homologous positions: an example of alignment and data presentation from the frogs. Molecular Phylogenetics and Evolution. 4: 314-330.

Kocher, T.D.; Thomas, W.K.; Meyer, A.; Edwards, S.V.; Pääbo, S.; Villablanca, F.X.; Wilson, A.C. 1989. Dynamics of mitochondrial DNA evolution in animals: amplification and sequencing with conserved primers. Proceedings of the National Academy of Sciences U.S.A. 86: 6196-6200.

Kumar, S.; Tamura, K.; Nei, M. 1993. MEGA: Molecular Evolutionary Genetics Analysis, version 1.01. University Park, PA: The Pennsylvania State University.

Liljeblad, J.; Ronquist, F. (In press) A phylogenetic analysis of higher-level gall wasp relationships (Hymenoptera: Cynipidae). Systematic Entomology.

Plantard, O.; Solignac, M. 1998. *Wolbachia*-induced thelytoky in cynipids. In: Csóka, Gyuri; Mattson, William J.; Stone, Graham N.; Price, Peter W., eds. The biology of gall-inducing arthropods: proceedings of the international symposium; 1997 August 14-19; Mátrakured, Hungary. Gen. Tech. Rep. NC-199. St. Paul, MN: U.S. Department of Agriculture, Forest Service, North Central Research Station: 111-121.

Pujade i Villar, J. 1993. Revisio de les especies del genere *Diplolepis* de l'Europa centro-occidental (Hym. Cynipidae) amb una especial atencio a la peninsula Iberica. Historia Animalium. 2: 57-76.

Ronquist, F. 1994. Evolution of parasitism among closely related species: phylogenetic relationships and the origin of inquilinism in gall wasps (Hymenoptera, Cynipidae). Evolution. 48: 241-266.

Ronquist, F. 1995. Phylogeny and early evolution of the Cynipoidea (Hymenoptera). Systematic Entomology. 20: 309-335.

Saitou, N.; Nei, M. 1987. The neighbor-joining method: a new method for reconstructing phylogenetic trees. Molecular Biology and Evolution. 4: 1406-1425.

Shorthouse, J.D. 1973. The insect community associated with rose galls of *Diplolepis polita* (Cynipoidea, Hymenoptera). Quaestiones Entomologicae. 9: 55-98.

Shorthouse, J.D. 1982. Resource exploitation by gall wasps of the genus *Diplolepis*. In: Proceedings of the 5th International symposium on insect-plant relationships. Pudoc. Wageningen: Pudoc: 193-198.

Shorthouse, J.D. 1993. Adaptations of gall wasps of the genus *Diplolepis* (Hymenoptera: Cynipidae) and the role of gall anatomy in Cynipid systematics. Memoirs of the Entomological Society of Canada. 165E: 139-163.

Shorthouse, J.D. 1998. Role of *Periclistis* (Hymenoptera: Cynipidae) inquilines in leaf galls of *Diplolepis* (Hymenoptera: Cynipidae) on wild roses in Canada. In: Csóka, Gyuri; Mattson, William J.; Stone, Graham N.; Price, Peter W., eds. The biology of gall-inducing arthropods: proceedings of the international symposium; 1997 August 14-19; Mátrakured, Hungary. Gen. Tech. Rep. NC-199. St. Paul, MN: U.S. Department of Agriculture, Forest Service, North Central Research Station: 61-81.

Shorthouse, J.D.; Ritchie, A.J. 1984. Description and biology of a new species of *Diplolepis* Fourcroy (Hymenoptera: Cynipidae) inducing galls on the stems of *Rosa acicularis*. The Canadian Entomologist. 116: 1623-1636.

Simon, C.; Frati, F.; Beckenbach, A.; Crespi, B.; Liu, H.; Flook, P. 1994. Evolution, weighting, and phylogenetic utility of mitochondrial gene sequences and a compilation of conserved polymerase chain reaction primers. Annals of the Entomological Society of America. 87: 651-701.

Stern, D.L. 1995. Phylogenetic evidence that aphids, rather than plants, determine gall morphology. Proceedings of the Royal Society of London B. 260: 85-89.

Strimmer, K.; von Haeseler, A. 1996. Quartet puzzling: a quartet maximum likelihood method for reconstructing tree topologies. Molecular Biology and Evolution. 13: 964-969.

Swofford, D.L. 1993. PAUP: Phylogenetic Analysis Using Parsimony, version 3.1.1. Champain, IL: Illinois Natural History Survey.

Ueda Y., Tomita H. 1989. Morphometric analysis of pollen exine patterns in roses. Journal of the Japanese Society of Horticultural Sciences. 58: 211-220.

Vyrzhikovskaja, A.V. 1963. New gall wasps (Hymenoptera, Cynipidae) from dog-rose in Central Asia and Kazakhstan. Entomologicheskoe Obozrenie. 42: 651-659.

Weld, L.H. 1952. Cynipoidea (Hymenoptera). Ann Arbor, MI: [Privately published].

Yasumatsu, K.; Taketani, A. 1967. Some remarks on the commonly known species of the genus *Diplolepis* Geoffroy in Japan. Esakia. 6: 77-86.

PATTERNS IN THE EVOLUTION OF GALL STRUCTURE AND LIFE CYCLES IN OAK GALL WASPS (HYMENOPTERA: CYNIPIDAE)

James M. Cook[1], Graham N. Stone,[2] and Alex Rowe[3]

[1] Department of Biology, Imperial College, Silwood Park, Ascot, Berkshire SL5 7PY UK.
[2] Institute of Cell, Animal and Population Biology, University of Edinburgh, Ashworth Laboratories, King's Buildings, West Mains Road, Edinburgh EH9 3JT.
[3] Institute of Molecular Medicine, John Radcliffe Hospital, Oxford, UK.

Abstract.—We present an analysis of the phylogenetic distribution of gall and life cycle characters within cynipid gallwasps of the genus *Andricus* (Hymenoperta: Cynipidae). Our intention is to identify which of these traits have proved evolutionarily labile during the radiation of oak cynipids, and so may have been associated with speciation events, and which (if any) are conserved within lineages. We show that:

1. While it is generally true that closely related species induce galls of similar structure, some gall traits have evolved repeatedly. One trait (an air space separating the larval chamber from the outer wall) has evolved independently in two entirely different groups of species in sexual and asexual generation galls.
2. Multiple transitions between alternative galling sites (acorn, bud, catkin, etc.) have occurred in both sexual and asexual generation galls. The most frequent shifts are from bud-galling to catkin-galling in the sexual generation. For species with alternation of generations, a shift from the ancestral galling sites in one generation is not correlated with a shift in the other.
3. There have been repeated evolutionary departures from an ancestral life cycle involving alternate sexual and asexual generations.
4. The transition from a life cycle involving two generations on oaks in the same section, *Quercus*, to a life cycle in which sexual generation galls develop on section *Cerris* oaks may have occurred only once in *Andricus*.

INTRODUCTION

Oak cynipids (Hymenoptera: Cynipidae, tribe Cynipini) are notable for the structural diversity of the galls they induce (Shorthouse and Rohfritsch 1992; Williams 1994; Stone and Cook, submitted). They also display a range of galling sites on the plant, host plant associations and life cycles (Askew 1961, 1984; Ambrus 1974). In this chapter we use a molecular phylogeny of a selected group of european oak cynipid species to examine patterns of evolution of these traits. Our intention is to identify which of these traits have proved evolutionarily labile during the radiation of oak cynipids, and which (if any) are conserved within lineages. We then discuss observed patterns in the light of existing theories on the evolution of life cycles and gall structures.

The study taxa

In this study we have concentrated on the genus *Andricus*, which is the largest genus of oak cynipids, containing up to 96 species in the western Palaearctic (Askew 1984; George Melika, pers. comm.). Species within the genus induce a wide diversity of gall structures on almost all parts of an oak tree, and also show the full range of oak host associations and life cycle structures found in the Cynipini as a whole. *Andricus* is thus a suitable taxon within which to examine issues relevant to the evolution of oak cynipids in general. Here we provide a brief introduction to the traits of interest in *Andricus*, and the alternative character states for these traits. We then summarize the questions arising from observed groupings of character states in *Andricus* species.

1. Gall structures

Adult oak gall wasps are all morphologically similar, small black or brown insects ranging from less than 1mg to ca. 15 mg in weight (Eady and Quinlan 1963). In contrast, their galls are spectacularly diverse (fig. 1: Askew 1984, Dreger-Jauffret and Shorthouse 1992). Cynipid gall structures are characteristic of the gall-inducer, rather than of the host plant (Ambrus 1974, Rohfritsch 1992), and result from cynipid traits expressed at two stages in the wasp's life cycle (Stone and Cook, submitted). Larval secretions, as yet uncharacterized (Schönrogge 1998), are thought to control the type and structure of plant tissues forming the gall (Rohfritsh 1992). Maternal egg-laying behavior determines how many larval cells develop within a single gall structure, and so whether the gall is unilocular (containing a single larval chamber) or multilocular (containing many larval chambers). Gall structures, though constructed of plant tissues, thus represent the extended phenotypes of gall wasp genes (Dawkins 1983, Stern 1995, Crespi et al. 1997).

Structurally, cynipid galls can be divided into two parts—the larval chamber and the outer gall. The larval chamber, similar in all cynipid galls (Bronner 1992), is lined with nutritive plant tissues on which the larva feeds, and surrounded by a thin wall of schlerenchyma. The cynipid larva completes its entire development within this chamber. The diversity of cynipid galls is the result of variation in gall tissues which develop outside the larval chamber, including surrounding layers of woody or spongy tissue, complex air spaces within the gall, and surface coats of sticky resins, hairs or spines (fig. 1). Particular outer gall structures are often common to sets of sympatric gall wasp species developing on the same part of the same oak species in the same place and time (Weld 1957, 1959, 1960; Ambrus 1974; Askew 1984).

An additional feature of oak cynipid gall structure is that a single species may form two very differet galls. Many species of oak cynipid gall wasp have two generations each year (see below). The galls of the two generations are often radically different in structure (fig. 1; compare parts 1 and 13 for *Andricus kollari,* parts 2 and 9 for *A. quercuscalicis*), and produce adults of very different size; where both generations are present in the life cycle, sexual generation galls and adults are usually smaller than their asexual counterparts (Ambrus 1974; Askew 1984; Pujade 1994, 1997). In *Andricus*, sexual generation galls usually develop far more rapidly than associated asexual generations, and are often tiny, inconspicuous structures (fig. 1, parts 1 and 2). Perhaps because of their more rapid development, *Andricus* sexual generation galls are structurally far more uniform than asexual generation galls.

2. Gall location

Members of the genus *Andricus* induce galls on a wide diversity of oak plant organs, including acorns, buds, catkins, leaves, stems and roots. As with gall structure, the majority of species induce gall formation on a characteristic plant organ, and very few Western Palaearctic species gall more than one type of plant tissue. In *Andricus* species which have both sexual and asexual generations, the two galls are often induced on different plant organs (table 1). The majority of known sexual generation *Andricus* galls are induced on buds or catkins (fig. 1, parts 1 and 2), with fewer species galling stems (Ambrus 1974, Pujade 1994). Asexual generation galls are also most abundant on buds, with lower numbers on acorns, and very few species attacking catkins or roots. Leaves are very rarely galled by *Andricus* species (although members of other oak cynipid genera commonly gall leaves) and only one european *Andricus* species (*A. gallaeurnaeformis*; fig. 1, part 20) galls leaves in both sexual and asexual generations.

3. Life cycle structure

Many (but not all) oak cynipids have two generations each year, one sexual and the other parthenogenetic (asexual). Where both generations exist, the sexual generation usually develops in the spring, and the asexual generation through the summer and autumn (Askew 1984). In Europe outside the Iberian peninsular, 20 *Andricus* species have both generations in their life cycle, 35 species have only a known asexual generation, and 10 species have only a known sexual generation. There are two alternative explanations for this diversity of life cycles. First, such diversity may be a genuine result of loss of either generation from species' life cycles over evolutionary time. Second, absence of a generation may reflect our continuing ignorance of cynipid life cycles; (a) known sexual and asexual generations which currently have different species names may in fact be alternating generations in the life cycle of a single species. (b) The tiny and inconspicuous sexual generation galls and insects may yet to be discovered for species currently known only from the larger, more conspicuous asexual generation adults and galls. The latter has been demonstrated recently for two European species—*A. gallaeurnaeformis* and *A. viscosus* (Folliot 1964; Pujade 1994, 1997). Notwithstanding such potential errors, enough is known of the majority of common European *Andricus* species to establish that species posessing only a sexual or asexual generation do exist.

4. Host oak associations

European oaks are members of two different oak species groups within the genus *Quercus*—the 'white' oaks of

Figure 1.—*Gall structures induced by representative* Andricus *species discussed in this study. Unless otherwise stated, drawings are approximately life-size. 1. Sexual gall of A. kollari. 2. Sexual gall of A. quercuscalicis on axis of a catkin. Actual size similar to previous drawing. 3. A. coriarius (asexual), cross-section. 4. A. caputmedusae (asexual), two thirds life size. 5. A. mayri (asexual), cross-section. 6. A. seckendorffi (asexual). 7. A. grossulariae (sexual) on catkin axis, cross-section, twice life size. 8. A. coronatus (asexual), cross-section. 9. A. quercuscalicis (asexual), cross-section. 10. A. viscosus (asexual), cross-section. 11. A. quercustozae (asexual), cross-section. 12. A. hungaricus (asexual), cross-section. 13. A. kollari (asexual), cross-section. 14. A. tinctoriusnostrus (asexual), cross-section. 15. A. caliciformis (asexual), cross-section. 16. A. polycerus (asexual), cross-section. 17. A. fecundator (asexual), cross-section. 18. A. hartigi (asexual). 19. A. conificus (asexual). 20. A. gallaeurnaeformis (asexual). 21. A. solitarius (asexual), twice life size. Drawings are reproduced from Ambrus (1974).*

Table 1.—*Gall and life cycle character states for Cynipid species in this study. Structure codes are explained in the Methods section of the text.*

Species	Location	Uni-/Multi locular	Structure type	Spines	Resinous coating	Host oak section
A. Sexual generation galls						
Genus *Andricus*						
A. burgundus Gir.	catkin	U	S1	no	no	Cerri
A. corruptrix Schldl.	bud	U	S1	no	no	Cerris
A. curvator Htg.	bud/leaf	U	S4	no	no	Quercus
A. fecundator Htg.	catkin	U	S1	no	no	Quercus
A. gallaeurnaeformis Fonsc.	leaf	U	S1	no	no	Quercus
A. gemmea Gir.	bud	U	S1	no	no	Cerris
A. grossulariae Gir.	catkin	U	S4	no	no	Cerris
A. inflator Htg.	bud	U	S4	no	no	Quercus
A. kollari Htg.	bud	U	S1	no	no	Cerris
A. lignicola Htg.	bud	U	S1	no	no	Cerris
A. quercuscalicis Burgsdorf	catkin	U	S1	no	no	Cerris
A. solitarius Fonsc.	catkin	U	S1	no	no	Quercus
A. tinctoriusnostrus Stef.	catkin	U	S1	no	no	Cerris
A. viscosus	see Note 1.					
Outgroups						
Biorhiza pallida Oliv.	bud	M	See Note 2	no	no	Quercus
Cynips divisa Htg.	leaf/catkin	U	S1	no	no	Quercus
Cynips quercus	bud	U	S1	no	no	Quercus
B. Asexual generation galls						
Genus *Andricus*						
A. caliciformis Gir.	shoot bud	U	S3	no	no	Quercus
A. caputmedusae Htg.	acorn	M	S5	yes	yes	Quercus
A. conglomeratus Gir.	bud	U	S3	no	no	Quercus
A. conificus Htg.	bud	U	S5	no	no	Quercus
A. coriarius Htg.	bud	M	S3	yes	no	Quercus
A. coronatus Gir.	bud	U	S4	yes/no	yes	Quercus
A. corruptrix Schldl.	bud	U	S3	no	no	Quercus
A. curvator Htg.	bud	U	S2	no	no	Quercus
A. viscosus Nieves-Aldrey	acorn	U	S4	no	yes	Quercus
A. fecundator Htg.	bud	U	S2	no	no	Quercus
A. gallaeurnaeformis Fonsc.	leaf	U	S5	no	no	Quercus
A. gemmea Gir.	bud	U	S5	no	no	Quercus
A. hartigi Marschal	bud	U	S5	yes	no	Quercus
A. hungaricus Htg.	bud	U	S4	no	no	Quercus
A. hystrix Trotter	bud	U	S1	yes	no	Quercus
A. inflator Htg.	bud	U	S2	no	no	Quercus
A. kollari Htg.	bud	U	S3	no	no	Quercus
A. lignicola Htg.	bud	U	S3	no	no	Quercus
A. lucidus Htg.	bud	M	S3	yes	yes	Quercus
A. mayri Wachtl.	catkin	M	S3	yes	yes	Quercus
A. polycerus Gir.	bud	U	S3	yes/no	no	Quercus
A. quercuscalicis Burgsdorf	acorn	U	S4	no	yes	Quercus
A. quercustozae Bosc.	bud	U	S4	no	yes	Quercus
A. seckendorffi Wachtl.	acorn	M	S3	yes	yes	Quercus
A. solitarius Fonsc.	bud	U	See Note 2	no	no	Quercus
A. tinctoriusnostrus Stef.	bud	U	S3	no	no	Quercus
Outgroups						
Aphelonyx cerricola Gir.	bud	U	S4	no	no	Cerris
Biorhiza pallida Oliv.	root	M	S3	no	no	Quercus
Cynips cornifex Htg.	leaf	U	See Note 2	no	no	Quercus
Cynips divisa Htg.	leaf	U	S3	no	no	Quercus
Cynips quercus Fourcr.	leaf	U	S3	no	no	Quercus
Diplolepis rosae Htg.	bud/leaf/fruit	M	S3	yes	no	Roses

Note 1. Character states for the sexual gall of *A. viscosus* are unknown
Note 2. The sexual gall of *Biorhiza pallida* is large, soft and spongy. The asexual galls of *A. solitarius* and *Cynips cornifex* are both spindle-shaped.

the section *Quercus*, and the 'black' oaks of the section *Cerris* (Nixon and Crepet 1985). Oaks characteristic of temperate altitudes and latitudes are typically members of the section *Quercus*, including *Q. faginea*, *Q. frainetto*, *Q. petraea*, *Q. pubescens*, *Q. pyrenaica*, and *Q. robur*. Oaks characteristic of more mesic, mediterranean habitats are typically members of the section *Cerris*, and include *Q. cerris*, *Q. coccifera*, *Q. libani*, and *Q. suber*. These two oak sections are to some extent associated with different cynipid genera. For example, in central and eastern Europe the genera *Aphelonyx*, *Chilaspis*, and *Dryocosmus* only attack *Quercus cerris*, while most species in the genera *Cynips*, *Neuroterus*, and *Andricus* are found on oaks in the section *Quercus* (Ambrus 1974; Csóka 1997). An interesting feature of the genus *Andricus* is that the vast majority of galls found on section *Cerris* oaks belong to sexual generations, including 9 of the 10 sexual generation-only species, and 6 of the 20 species known to have a two generation lifecycle. The remainder of the two-generation species have both generations on section *Quercus* oaks, and all of the asexual generation-only *Andricus* species attack oaks in this group.

Questions Adressed in this Study

A preliminary examination of the *Andricus* species with particular gall structures or life cycle types shows that species sharing one trait need not share others (table 1). For example, *Andricus lucidus* and *A. mayri* are similar in that both are known only from asexual generation galls which are multilocular, and have a sticky and spiny surface. The two differ, however, in that *A. lucidus* usually develops on a bud, while *A. mayri* develops on a catkin. As an alternative example, the asexual generation galls of *A. caputmedusae* and *A. viscosus* are structurally very different, but are both formed on acorns. A number of the species listed in table 1 form sexual generation galls on *Q. cerris*, but on a range of plant parts forming a diversity of structures. Because groupings on the basis of similarity in any one of the four traits introduced above are different, some similarities must result from convergence. So little is known of the gall-inducing process that it is impossible to predict *a priori* which of gall structure, gall location or life cycle might have changed very rarely in evolution, and which might have diverged repeatedly within lineages.

To address this issue we examine the distribution of species traits over a molecular phylogeny for 28 European *Andricus* species for which gall structure traits, life cycle types and oak associations are well-known (table 1). Of these species, 12 have both sexual and asexual generations in their life cycle (six with a sexual generation on *Q. cerris*, and six with a sexual generation on oaks in the section *Quercus*). Fourteen species have only a known asexual generation, and two species have only a known sexual generation. Gall structures and oak hosts are known for all but the sexual generation of *A. viscosus*, for which neither host oak nor gall structure are known.

There is considerable debate over the monophyletic status of *Andricus*, based both on a worldwide study of adult morphology (Abrahamson *et al.* 1998) and on sequence data for North American taxa (Drown and Brown 1998). In order to assess the monophyly of *Andricus* in the Western Palaearctic, we have therefore included members of the oak cynipid genera *Aphelonyx*, *Biorhiza*, and *Cynipis*. To root our molecular trees, we also sequenced the rose galler *Diplolepis rosae* in the tribe Rhoditini (Plantard *et al.* 1998), the sister group to the Cynipini/Pediaspini clade (Ronquist 1995, Liljeblad and Ronquist 1997).

METHODS

Establishing Ancestral States for *Andricus* Traits

(a) Gall structure:
Within the Cynipini, phylogenetic relationships between genera are largely unknown (Ronquist 1995, Liljeblad and Ronquist 1997), and ancestral states for basal *Andricus* species cannot be identified with certainty. Most cynipine sexual generation galls, including the vast majority of *Andricus* species, are tiny and consist only of the larval chamber with its thin outer wall of schlerenchyma. The majority of cynipine asexual generation galls outside *Andricus* are single-chambered, and lack extensive lignification of the outer gall, surface spines or sticky resins (Weld 1957, 1959, 1960; Ambrus 1974). In the absence of any further phylogenetic information, we apply parsimony and assume these to be the ancestral states for *Andricus* galls. These inferred ancestral states are further supported by outgroup comparison with the Pediaspinae sister group of the Cynipinae (Liljeblad and Ronquist 1997).

(b) Gall location.
The outgroup oak gallwasps show a range of gall locations for each generation (table 1). Lack of resolution of intergeneric relationships prevents any *a priori* assumptions about the oak tissues galled by an ancestral *Andricus* species.

(c) Life cycle structure.
The pediaspine sister group to the Cynipini has alternation between sexual and asexual generations, as do the majority of genera of oak cynipids. Such alternation seems to have evolved only once in cynipid evolution, and is not found in the the Rhoditini sister group to the Pediaspini/Cynipini clade, or in the paraphyletic Aylacini (Liljeblad and

Ronquist 1997). Notwithstanding the possible errors discussed above, it is therefore parsimonious to assume that this was the ancestral state in *Andricus*.

(d) Host oak associations.

With the exception of *Andricus* (and the possiblly *Fiorella marianii*), all genera of oak cynipids which have alternating generations form sexual and asexual galls on the same host oak taxa (Folliot 1964, Ambrus 1974, Askew 1984, Nieves-Aldrey 1987). This is also true for the pediaspine sister group of the the Cynipini (Ambrus 1974). It is therefore parsimonious to assume that the ancestral *Andricus* species had alternation of generations on the same oak host(s), and to regard the few *Andricus* species possessing alternation between oaks in the sections *Quercus* and *Cerris* as representing a derived state.

Scoring Character States

Character states for each species are listed in table 1. Gall locations, oak associations and life-cycle structures were determined by literature survey, long personal field experience, and unpublished data collected by Dr. Gyuri Csóka, Dr. José-Luis Nieves-Aldrey, and Dr. Juli Pujarde.

Gall structures were allocated to the following categories, illustrated for representative examples in figure 1. (i) Gall single-chambered *or* multi-chambered (compare fig. 1, parts 13-16 with parts 3 and 5). (ii) Gall surface smooth *or* bearing spines (compare fig. 1, part 13 with parts 3-6). (iii) Gall surface not sticky *or* covered in sticky resin. Single-chambered galls were divided further between 5 easily recognisable structural types:

S1. Gall consists of larval chamber only, without exterior structures (fig. 1, parts 1 and 2).
S2. Larval chamber surrounded by modified bud scales, but not completely enclosed by gall tissue (fig. 1, part 17).
S3. Larval chamber completely surrounded by, and in direct contact with, woody outer gall tissue (fig. 1, parts 13-16).
S4. Larval chamber completely enclosed, but separated from the outer gall by an air space (fig. 1, parts 7-12).
S5. Gall structure unique within the genus (fig. 1, parts 18-20).

Molecular Methods

DNA was extracted from single wasps using either a proteinase-K/SDS digestion followed by "salting out", or a simple chelex procedure (Werren *et al.* 1995). A 433 base pair fragment of the mitochondrial cytochrome b gene was then amplified using PCR (35 cycles of denaturation at 92¡C for 60s, annealing at 45-55¡C for 60s and extension at 72¡C for 90s) using the primers CB1 and CB2 (Jermin and Crozier 1994) in a 50µl reaction. 10µl of each PCR product was then electrophoresed in a 1.5 percent agarose gel to check the amplicon. 6µl of the remaining 40µl was used in standard ligation and transformation reactions using the TA-cloning kit (Invitrogen). Plasmid DNA was purified using Wizard miniprep kits (Promega) and sequenced using Taq-FS (Perkin-Elmer) chemistry and an ABI 373 sequencer. Both DNA strands were sequenced for all species.

Phylogenetic Analyses

Sequences were aligned by eye. All sequences were 433 bases long (sequences available from the authors by request), and no insertions or deletions needed to be inferred in order to achieve alignment. Two analytical techniques were used to generate phylogenies from the sequence data: maximum parsimony (Farris 1970) and neighbor joining (Saitou and Nei 1987). Maximum parsimony trees were generated using PAUP 3.1.1 (Swofford 1993) with codons weighted equally, and using *Diplolepis rosae* as an outgroup. One hundred random additions were used in a heuristic search, and the strict consensus of the eight resulting shortest trees (termed the MP tree) used as the working phylogeny (fig. 2). While changes were more common at third positions (454) than first (148) or second (51) positions, downweighting or exclusion of third positions did not alter the deeper branches of the trees and severely reduced resolution within the main *Andricus* clade. To illustrate branch lengths joining species within clades we also generated a second tree using neighbor joining in PHYLIP (Felsenstein 1993) (termed the NJ tree: fig. 3). For both tree-building techniques, bootstrap values were generated using 500 replicates of 100 random additions.

RESULTS

1. Status of the Genus *Andricus*

Our sequence data show the *Andricus* species to consist of two groups (figs. 2 and 3). All bar two species (*Andricus gallaeurnaeformis* and *A. hystrix*) form a group which, though poorly resolved around the base, is compatible with the existence of a monophyletic *Andricus* genus (figs. 2 and 3). Enforcing monophyly on this group of species resulted in MP trees of 620 steps, only six steps longer than the shortest unconstrained MP trees. Among these species, the greatest sequence divergence (between *A. inflator* and *A. coriarius*) is 14.5 percent. *Andricus gallaeurnaeformis* and *A. hystrix*, however, show considerable sequence divergence both from the other *Andricus* species (21.5-25.5 and 22.9-26.3 percent, respectively) and from each other (16.4 percent), and are separated from the other *Andricus* species by the

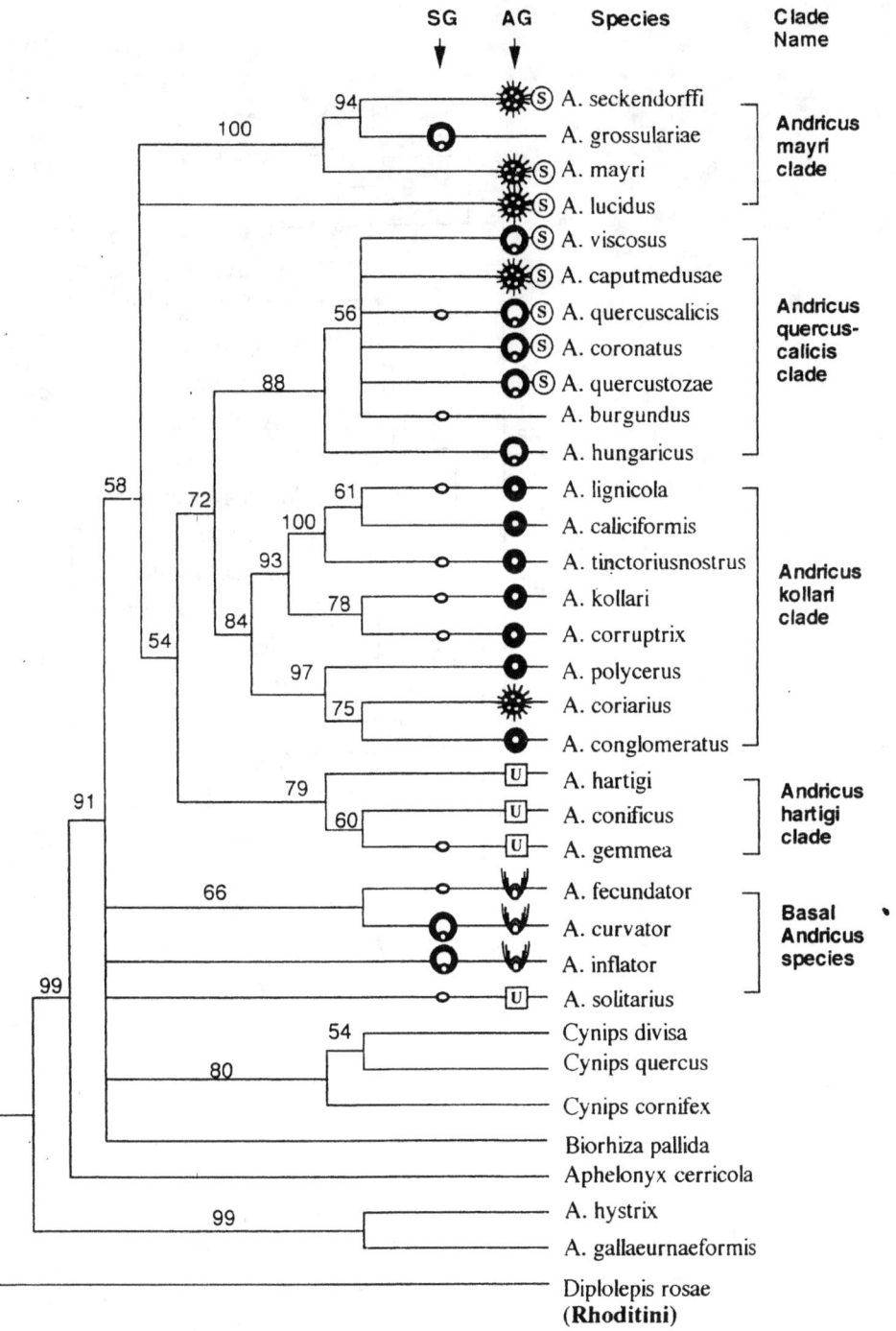

Figure 2.—Andricus *gall structures for sexual (SG) and asexual (AG) generations mapped over the strict concensus tree. All species with an abbreviated generic name are currently classified as members of* Andricus. *The minimum trees differed from each other only in the branching of the basal members of the main* Andricus *clade and the close outgroups (*Cynips, Biorhiza, *and* Aphelonyx*), and do not affect any of the groupings discussed in the text. Numbers shown at nodes are bootstrap percentages; branches without values were supported by less than 50 percent of bootstrap replicates.*

Key to gall structure symbols used on the phylogeny (see methods for explanation of S1-S5).

○ S1 W S2 ● S3 Q S4 U S5 ✻ multilocular, spiny Ⓢ sticky outer surface

267

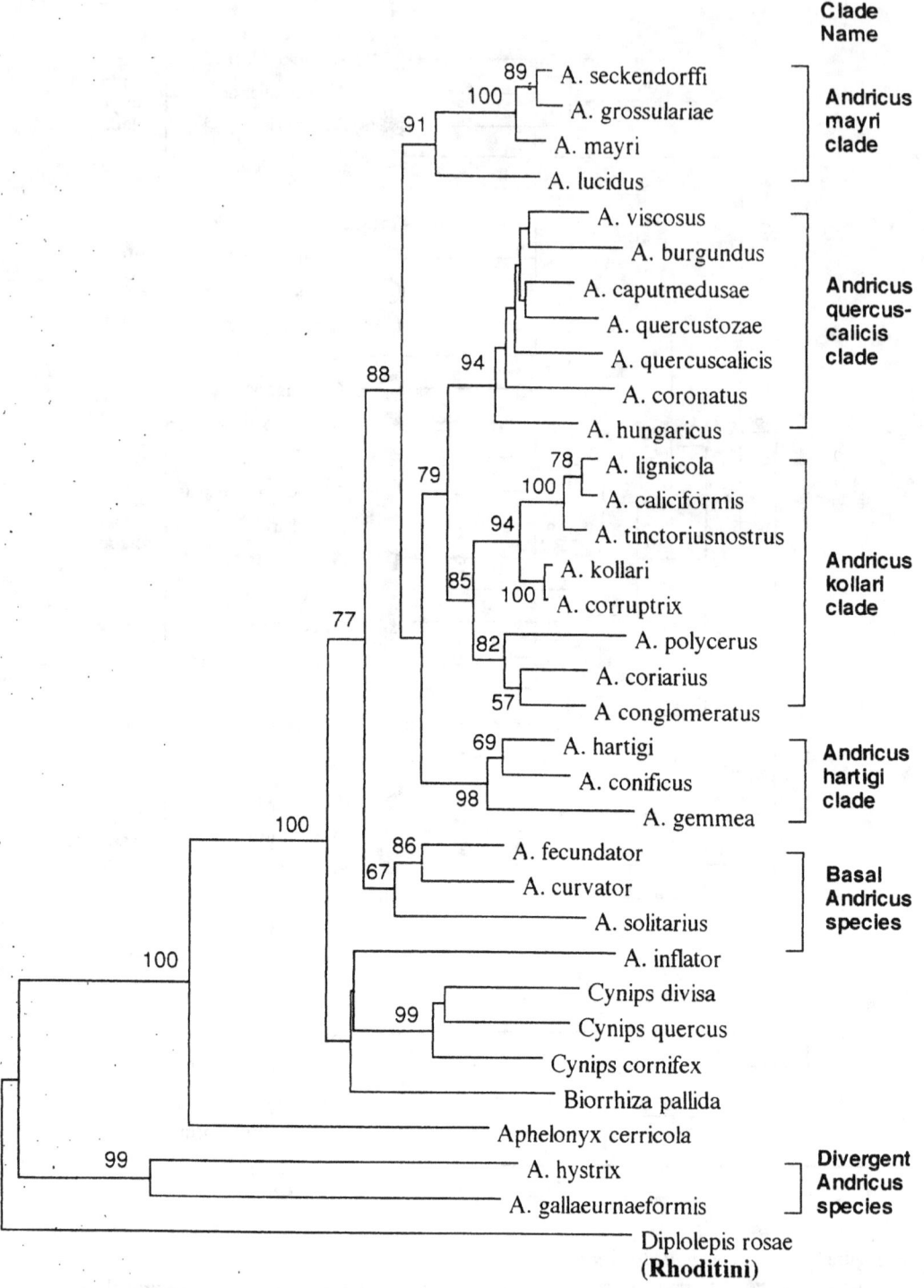

Figure 3.—*Neighbor joining tree for the selected gallwasp species generated using PHYLIP (Felsenstein 1993). All species with an abbreviated generic name are currently classified as members of Andricus. Numbers shown at nodes are bootstrap percentages; branches without values were supported by less than 50 percent of bootstrap replicates.*

non-*Andricus* cynipine species (fig. 3). Enforcing monophyly for all *Andricus* species results in an MP tree of 642 steps, 28 steps longer than the unconstrained MP tree. We suggest that the taxonomic status of these divergent species should be reassessed, and we exclude them from the rest of this analysis. For ease of reference, we divide *Andricus* (excluding the divergent *A. hystrix* and *A. gallaeurnaeformis*) into five groups. All except the first of these represent monophyletic clades with strong bootstrap support in both the MP and NJ trees (figs. 2 and 3).

1. The *A. mayri* clade.
2. The *A. quercuscalicis* clade.
3. The *A. kollari* clade.
4. The *A. hartigi* clade.
5. The basal species (*A. curvator, A. fecundator, A. inflator,* and *A. solitarius*).

Although *Andricus lucidus* emerges alone from a polytomy within the genus in the MP tree, it is a member of the *A. mayri* clade with high bootstrap support in the NJ tree (fig. 3). We therefore regard this species as a member of the *A. mayri* clade.

2. Phylogenetic Patterns in Gall Structure

The distribution of gall traits over the phylogeny is shown in figure 2.

(a) Galls of the sexual generation.
Only two structural types are induced by the 12 species in the main *Andricus* clade; nine species form the small, thin-walled structures corresponding to the presumed ancestral state (fig. 1, parts 1 and 2), and three species (*A. grossulariae, A. curvator,* and *A. inflator*) induce a more complex (presumed derived) structure (fig. 1, part 7). For each case of the derived state, sister groups possess the presumed ancestral state (fig. 2). The most parsimonious explanation of the observed distribution of this derived state is that it has evolved independently three times.

(b) Galls of the asexual generation.
The basal *Andricus* species include three in which the asexual gall has a single larval chamber surrounded by modified bud scale leaves (fig. 1, part 17; *A. inflator, A. fecundator,* and *A. curvator*). This character state has either evolved from a simpler state at least twice (once in *A. inflator* and once in the common ancestor of *A. fecundator* and *A. curvator*), or represents the ancestral state within the main *Andricus* clade. In the three remaining *Andricus* clades, the vast majority of the species in each clade share a common asexual gall structure (fig. 2).

(i) The *A. mayri* clade (see fig. 1, parts 5 and 6) all have multi-chambered asexual galls in which the larval chambers are entirely surrounded with extensive woody tissue. The gall surface is covered in spines, and coated in sticky resin.

(ii) The *A. quercuscalicis* clade contains six species with asexual generation galls, all but one of which contain a single larval chamber and have an airspace between this and the outer gall wall (fig. 1, parts 8-12). Five species have outer surfaces covered in sticky resin.

(iii) The *A. kollari* clade all have solid asexual galls with extensive development of a hard, woody outer gall entirely surrounding the larval chamber (fig. 1, parts 13-16). All but one species have a single larval chamber and lack a sticky surface coating or spines.

Although the clades shown in figure 2 do not consist of pure structural groups, closely related galls generally do induce galls of similar structure. The most parsimonious explanation of such a character distribution is that each clade has radiated from an ancestral species with the shared gall structure, and speciation has been accompanied by very few changes in gall structure.

There are two types of exception to the rule that closely related species induce similar galls. First, the *A. hartigi* clade (fig. 2) contains three species whose outer gall structures are unique within the genus. The only obvious shared structural character is the presence of a single larval chamber. In *A. hartigi* the larval chamber is surrounded by an airspace formed from a roof of furry, purple club-shaped spines (fig. 1, part 18). *A. conificus* has a well-developed outer gall which is spineless, fleshy and soft, pigmented bright white with longitudinal red markings, and there is no airspace around the larval chamber (fig. 1, part 19). *Andricus gemmea* is a small, fleshy gall whose surface is covered with red tubercles. While sequence divergence within each of the *A. kollari* and *A. quercuscalicis* clades between species inducing structurally similar asexual galls reaches 3.5-4.5 percent, structurally divergent *A. hartigi* and *A. conificus* differ by only 2.8 percent.

Second, the phylogenetic distribution of several traits suggests that they have evolved repeatedly during the radiation of *Andricus* (fig. 2):

(a) Multi-chambered galls (clustered laying of eggs by the mother) have evolved at least three times in asexual *Andricus* generations—once in the *A. mayri* clade (fig. 1, parts 5 and 6), once in *A. caputmedusae* (in the *A. quercuscalicis* clade; fig. 1 part 4), and once in *A. coriarius* (in the *A. kollari* clade; fig. 1, part 3). This trait has also evolved independently in the outgroup rose galler *Diplolepis rosae*.

(b) An outer coating of spines has evolved at least four times in asexual galls—once in the *A. mayri* clade (fig. 1, parts 5 and 6), once in *A. caputmedusae* (in the *A. quercuscalicis* clade; fig. 1, part 4), once in *A. coriarius* (in the *A. kollari* clade; fig. 1, part 3) and again in *A. hartigi* (fig. 1, part 18). This trait has evolved independently in the rose galler *Diplolepis rosae*, although here the gall is of a sexual generation.

(c) A surface covering of sticky resins for asexual galls has evolved at least twice within the genus—once in the *A. mayri* clade, and once in the *A. quercuscalicis* clade.

(d) An air space separating the inner cell from the outer gall wall has evolved at least five times in *Andricus*—once in the asexual generation galls of the *A. quercuscalicis* clade, once in *A. hartigi*, and three times in sexual generation galls. Outside *Andricus*, the same trait has evolved independently in *Aphelonyx cerricola*.

3. Phylogenetic Patterns in Gall Location

The phylogenetic distribution of gall locations is shown in figure 4.

(a) Sexual generation galls.
In the sampled species, sexual gall locations are fairly evenly split between catkins and buds. The diversity of locations in outgroups to *Andricus*, and the poor resolution of the phylogeny around the base of the genus, make it impossible to identify a single most likely ancestral state by outgroup comparison. If we assume that the ancestral *Andricus* species induced sexual galls on catkins, then a minimum of five independent changes to bud galling are required. The same minimum number of changes is required if we assume bud galling as the ancestral sexual generation gall location. Whichever of the ancestral assumptions is true, it is clear that gall location has shifted between buds and catkins relatively frequently in the genus.

(b) Asexual generation galls.
Again, the diversity of asexual gall sites in genera immediately outside *Andricus* in the phylogeny makes it is difficult to identify a single ancestral state by outgroup comparison. The basal *Andricus* species all induce bud galls, and if this trait is ancestral, it is retained in 19 of the 24 species in this study.

A shift in galling location from buds to acorns has occurred at least twice in *Andricus*—once in the *A. mayri* clade (for *A. seckendorffi*) and at least once in the *A. quercuscalicis* clade (for *A. caputmedusae*, *A. quercuscalicis*, and *A. viscosus*). *A. mayri* shows a shift from buds to catkins.

All of the members of the *A. hartigi* clade induce galls on lenticel buds on well-established stems and major branches, rather than terminal or lateral buds on young shoots. This suggests that the diversification of gall structures seen in this clade may have occurred after relocation to this novel galling site.

4. Phylogenetic Patterns in Life Cycle Structure

Basal *Andricus* species all show alternation of generations (fig. 5). The remaining clades include species showing a mixture of life cycle types. The *A. quercuscalicis* clade contains sexual-only, asexual-only and alternating species, the *A. mayri* and clade contains asexual-only and sexual-only species, and the *A. kollari* and *A. hartigi* clades contain both alternators and asexual-only species.

The distribution of life cycle types across the phylogeny suggests that evolutionary transitions in life cycle type must have occurred repeatedly in *Andricus*. The number of transition events inferred depends on assumptions of which sorts of life cycle transition are possible. We illustrate the implications of these assumptions by considering possible transitions within the *A. mayri* clade (fig. 5).

(a) The simplest assumptions are (i) that generations can only be lost from alternating life cycles, and not regained, and (ii) that sexual generations and asexual generations cannot evolve directly into each other. In this scenario, asexual-only species are derived from only either alternating or asexual-only ancestors, and sexual-only species only from either alternating or sexual-only ancestors. If we consider the *A. mayri* clade, the loss of a sexual generation in *A. seckendorffi* and of an asexual generation in *A. grossulariae* (both members of the *A. mayri* clade) must represent independent departures from a common ancestor possessing both generations. The loss of a sexual generation in *A. mayri*, sister taxon to this pair, would then represent a further independent event. Under these assumptions, a minimum of two losses of sexual and 11 losses of asexual generations, and thus 13 steps, are required to explain the observed distribution of life cycle types in the sampled *Andricus* species in the MP tree.

(b) An alternate set of assumptions is that lost generations can be restored to the life cycle, while maintaining the assumption that generations cannot evolve diresctly into each other. It would then be possible for the *A. mayri* clade, including *A. lucidus*, to have

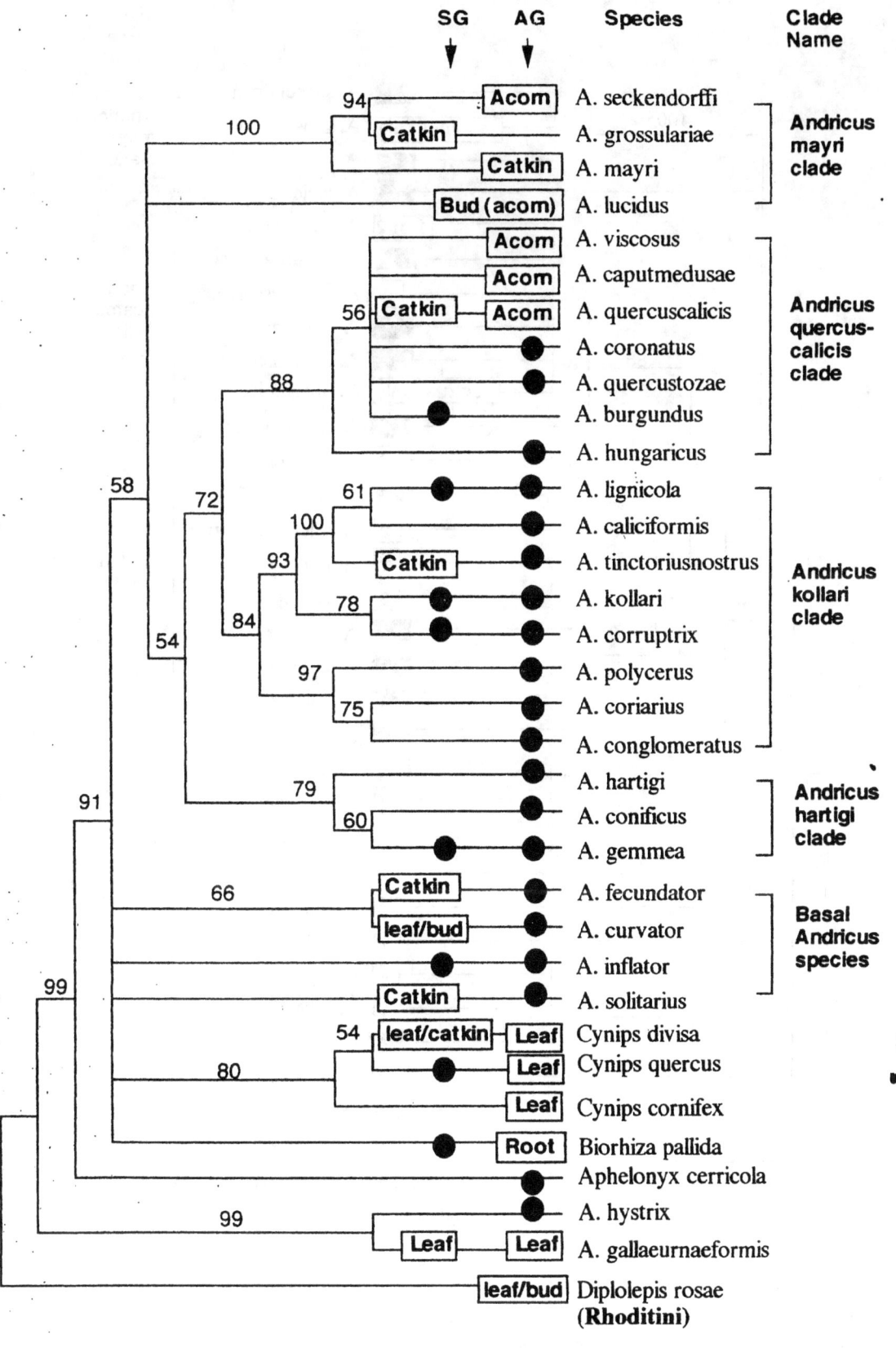

Figure 4.—*Distribution of gall locations for sexual (SG) and asexual (AG) generation galls over the MP tree. Shaded circles represent bud galls.*

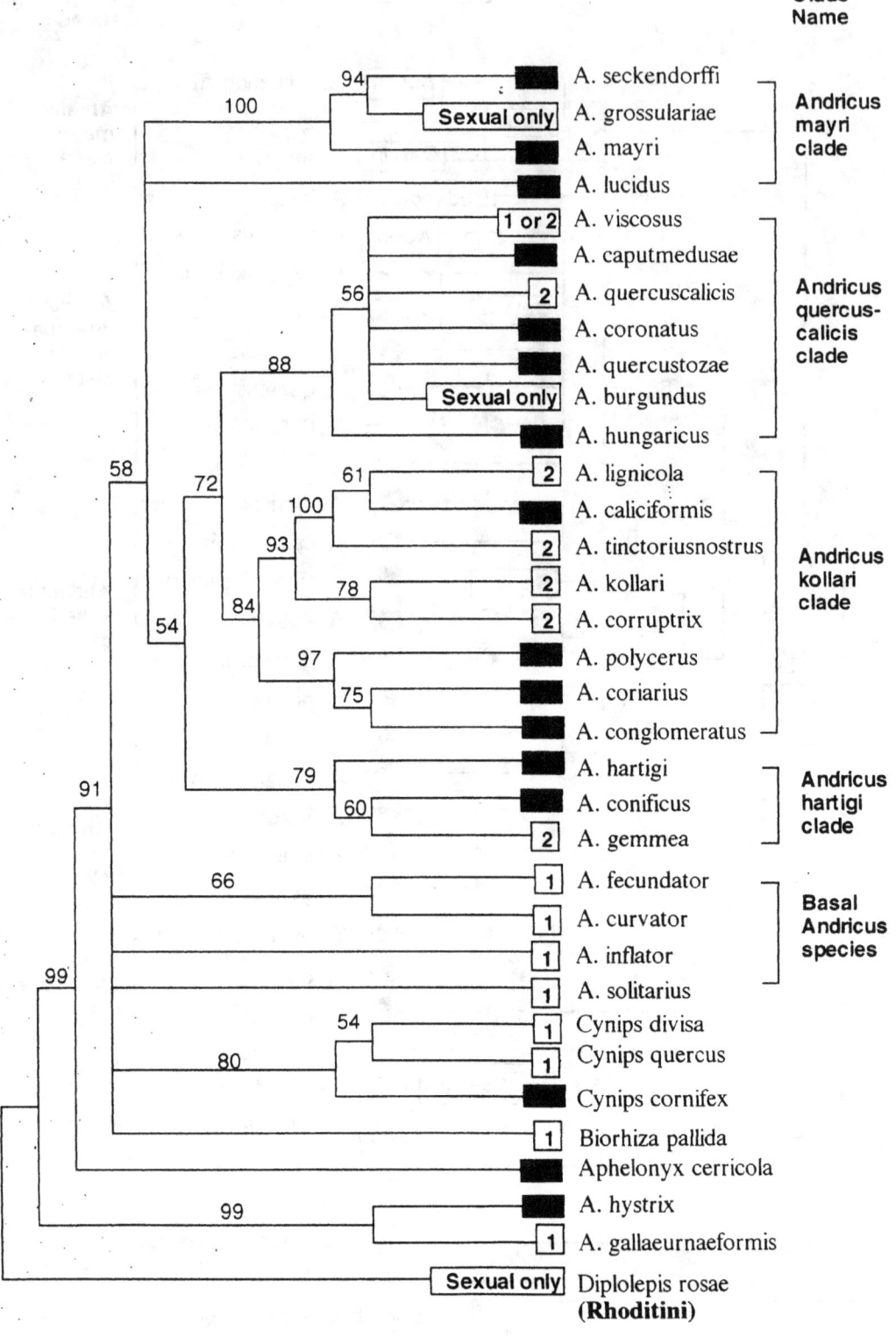

Figure 5.—*Distribution of life cycle types over the MP tree. The number 1 indicates a life cycle involving alternation of generations on oaks in the section* Quercus. *The number 2 indicates a life cycle involving alternation of generations between a sexual generation on section* Cerris *and an asexual generation on section* Quercus. *Black boxes indicate species for which only an asexual generation is known.* Andricus viscosus *is marked as '1 or 2' because the host of its sexual generation remains unknown.*

an asexual-only ancestor. The sexual only *A. grossulariae* would result from restoration of the sexual generation to an alternating ancestor, followed by loss of the asexual generation. Under these assumptions, sexual generations could have been lost in the common ancestor of the *A.mayri* clade and its sister group and regained to form the alternating life cycle on a minimum of six occasions (in the ancestor of *A. grossulariae*, in *A. viscosus*, *A. quercuscalicis*, the ancestor of *A. burgundus*, the ancestor of the clade containing *A. lignicola* and *A. corruptrix*, and in *A. gemmea*). This would be followed by further loss of asexual generations in *A. grossulariae* and *A. burgundus*, and by loss of the sexual generation in *A. caliciformis*. This yields a total of 10 steps.

(c) Finally, asexual and sexual generations could evolve into each other directly, without the need for an intermediate alternating life cycle. This would reduce the number of steps required to explain the evolution of sexual-only species within predominated asexual-only clades by 1 step for each of *A. grossulariae* and *A. burgundus*, yielding a minimum of eight steps.

The applicability of these different assumptions is discussed below. It is clear that two important conclusions are supported by our analysis. First, each life cycle type is distributed through the phylogeny, and not restricted to members of a specific clade, so transitions between life cycle types must have occurred repeatedly. Second, in at least one clade (the branch of the *A. kollari* containing *A. polycerus*, *A. coriarius*, and *A. gemmea*) application of parsimony would imply that asexual-only species have evolved from asexual-only ancestors.

Phylogenetic Patterns in Host Oak Association

All of the basal *Andricus* species have alternation of generations in which both asexual and asexual galls develop on oaks in the section *Quercus* (fig. 5). With the exception of *A. viscosus*, whose sexual generation host is unknown, all other members of *Andricus* sampled are either asexual-only species developing on *Quercus* oaks, or alternating species with an asexual generation on *Quercus* oaks and a sexual generation galls develop on *Cerris* oaks. The most parsimonious explanation of this pattern is that exploitation of *Cerris* oaks by *Andricus* evolved only once, in the common ancestor of these four clades, and that in the seven descendant species in which alternation of generations was retained, *Cerris/Quercus* alternation was also retained. This strongly predicts that the sexual generation gall of *A. viscosus*, when it is eventually identified in the field, will be found on *Cerris* oaks. In this interpretation, the phylogenetic spread of species with sexual generations on *Cerris* oaks does not represent multiple colonization events, but instead could be the result of frequent losses of sexual generations among the decendants of the *Cerris*-colonising ancestor.

A less parsimonious explanation is that colonization of *Cerris* section oaks by sexual generations has occurred repeatedly during the *Andricus* radiation. However, because asexual-only species provide no information with which to make out-group comparisons for sexual generation oak hosts, we cannot estimate how many colonization events (beyond the most parsimonious single colonization) might have taken place.

DISCUSSION

1. Patterns in the Evolution of Gall Structure

(a) Structural similarities within clades.

The general correlation between gall morphology and gall wasp phylogeny demonstrated here parallels findings in rose gall wasps (Plantard *et al.* 1998) and other taxa of gall-forming insects (Stern 1995, Crespi *et al.* 1997). Conservation of gall form within clades could result without inferring any adaptive significance for gall shape simply through low rates of generation of structural novelty. Two patterns in *Andricus*, however, suggest that novel gall structures have arisen relatively rapidly and repeatedly during the radiation of the genus. First, two of the clades whose other members induce structurally similar galls (the *A. kollari* clade and the *A. quercuscalicis* clade) contain species with atypical structures (*A. coriarius* and *A. caputmedusae*, respectively). Second, if a constant rate of sequence divergence over time is assumed within the genus, the three members of the *A. hartigi* clade have evolved radically divergent gall structures over a shorter timescale than was required for divergence of the *A. kollari* and *A. quercuscalicis* clades.

As an alternative to a non-adaptive hypothesis, shared characters may be maintained by strong stabilizing selection (Price *et al.* 1987). The most likely candidate for such selection is thought to be mortality inflicted by insect parasitoids. If the ancestor of a clade possessed a gall trait limiting mortality inflicted by an important generalist parasitoid, selection could act to retain that gall trait in all descendant species. Evidence for such an adaptive explanation is currently limited. While generalist parasitoids commonly inflict mortalities of 40-100 percent on oak gall wasps (Askew 1961, 1965; Washburn and Cornell 1981; Askew 1984; Schönrogge *et al.* 1995, 1996; Stone *et al.* 1995; Plantard *et al.* 1996), it remains unclear to what extent particular gall traits affect parasitoid attack rates. Protection of the gall former by ants recruited through nectar secretion remains perhaps the most elegant defensive trait whose efficacy has been demonstrated (Abe 1992, Washburn 1984). Of the characters shared by *Andricus* clades, high gall hardness and diameter have been shown to impede attack by certain parasitoid species in cynipid galls (Askew 1965, Washburn and Cornell 1979, Abe 1997) and other insect gall-former systems (Jones 1983, Weis 1982, Weis *et al.* 1985, Price and Clancy 1986, Craig *et al.* 1990, Craig 1994, Zwölfer and Arnold-Rinehart 1994). While defensive functions have been suggested for the other traits conserved within *Andricus* clades (sticky outer surfaces, an air space between the larval chamber and the outer wall, and surface spines; Askew 1984) their adaptive significance has yet to be demonstrated. Whatever the adaptive significance of these gall traits, conservation within clades shows that speciation in *Andricus* is rarely associated with large-scale changes in gall morphology. The patterns we describe suggest that at least one of the possible evolutionary scenarios proposed for diversification of gall structure—disruptive selection within clades (Price *et al.* 1987)—has been rare in *Andricus*.

(b) Convergent evolution of gall structures.

Four gall traits have evolved convergently in *Andricus*: (1) air spaces between the inner cell and the outer gall, (2) surface coatings of resins, (3) surface spines, and (4) production of a multi-chambered gall. Regardless of their adaptive significance, two interesting conclusions result from repeated evolution of traits. First, if we take the parsimonious view that ancestors of clades possessed the gall morphology now shared by most of the members of that clade, this pattern shows that similar galls can result from modification of quite different ancestral structures. Second, in the case of an air space between larval chamber and outer wall, the same character state has evolved in one set of species in sexual generation galls, and a non-overlapping set of species in asexual generation galls. This suggests that the evolution of gall form in these two generations is not tightly coupled.

A non-adaptive explanation for repeated trait evolution is that they represent a set of most probable morphologies resulting from the underlying mechanism of gall formation. To date, too little is known about the development of a diversity of cynipid galls for this possibility to be assessed. This type of explanation may well be important, however, in understanding the diversity of gall structures induced by eriophyid mites and pemphigine aphids (Price *et al.* 1987, Westphal 1992). Both groups of gall formers have no known enemies which attack them through the gall wall, and the selective hypotheses presented here for cynipid gall structures thus cannot currently apply to them. An alternative is that selective retention of advantageous gall traits has resulted in convergent evolution. The same traits conserved within *Andricus* clades show repeated evolution, and again mortality imposed by generalist parasitoids or predators is the most probable selective pressure. The challenge is now to assess which of the traits showing repeated evolution actually have any impact on natural enemy attack rates.

An interesting pattern in *Andricus* galls is the simultaneous evolution on three occasions (and again in *Diplolepis*, (Plantard *et al.* 1998) of the multi-chambered state and the presence of surface spines. One possibility is that spininess is an inevitable and non-adaptive consequence of the development of many larvae in the same structure. Not all multi-chambered *Andricus* galls are spiny, however (e.g., Ambrus 1974; see fig. 6 and discussion below), and the two traits are therefore not inevitably linked. Furthermore, because the number of chambers in a gall is maternally controlled, while spininess is controlled by the larva, it seems unlikely that these two gall traits are genetically linked. An alternative is that some many-chambered galls face particular selective pressures which have resulted in the evolution of additional defensive structures. Many-chambered galls are typically larger than single-chambered structures, and while increased size may confer partial protection from insect parasitoids, larger galls are attacked preferentially by opportunist vertebrate predators (Abrahamson *et al.* 1989, Weis *et al.* 1985, Weis 1993). Many oak cynipid galls are also attacked by vertebrate predators, including rodents and birds, particularly in hard winters (Csóka 1997). We suggest that the large spines present on multilocular galls such as *A. coriarius* are too large to be effective in defense against insect parasitoids, but may instead be an adaptation against vertebrate predators. Those multilocular galls which do not have spines fall into two groups (fig. 6). First, several cynipids in a range of genera induce spineless multilocular sexual generation galls in the spring (*Biorhiza pallida, Dryocosmus mayri*), covered in hairs (*Chiulaspis nitida*) or surrounded by distorted leaves (*A. cydoniae, A. multiplicatus*). Vertebrate predation may be less severe in the spring due to the abundance of alternative prey items, including small and thin-walled cynipid galls (Schönrogge, pers. comm.). Second, several multilocular asexual generation galls which lack spines are formed on roots (fig. 6, *A. quercusradicis* and *B. pallida*), and are therefore probably less exposed to opportunist predators.

2. Patterns in the Evolution of Gall Location

The patterns seen in *Andricus* suggest that gall location has changed relatively frequently in the radiation of the genus, particularly in the sexual generation. This suggests that the mechanisms associated with gall induction are probably not specific to particular oak tissues, but are effective in a variety of meristematic

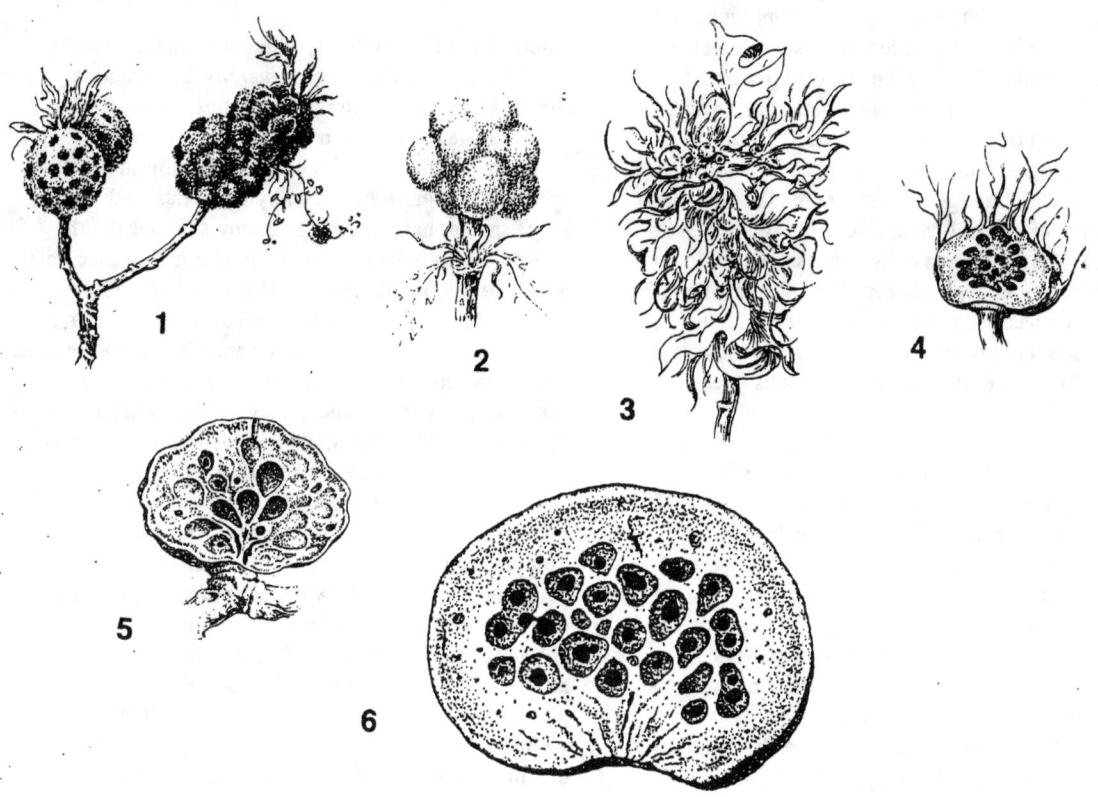

Figure 6.—*Cynipid oak galls which are multilocular and yet lack spines.* 1. The sexual generation of Chilaspis nitida *Gir.* on Q. cerris. 2. The sexual generation of Dryocosmus mayri *Mülln.* on Q. cerris. 3. The sexual generation (no known asexual generation) of Andricus multiplicatus *Gir.* 4. The sexual generation (no known asexual generation) of Andricus cydoniae *Gir.* on Q. cerris. 5. The sexual generation of Biorhiza pallida *Oliv.* on section Quercus oaks. 6. The asexual generation of Andricus quercusradicis *Fabr.* on section Quercus oaks.

tissues. Our observations are consistent with the ability of some *Andricus* species to induce structurally similar galls on different oak tissues; *A. curvator* is able to induce its sexual generation galls on leaves or buds, while *A. lucidus* is able to induce galls on buds or acorns. These results suggest in turn that gall location is probably more a function of female oviposition behavior than of specificity in the gall induction process.

The conservation of gall induction in lenticel buds in the *A. hartigi* clade is particularly interesting given the diversity of gall structures in this clade. Relatively few oak cynipids gall buds on mature branches and tree trunks, although this trait is found in other genera such as *Cynips, Dryocosmus,* and *Trigonaspis*. It is tempting to suggest that the structural diversity in the *A. hartigi* clade is related to this exploitation of a novel gall location within the genus.

A. gallaeurnaeformis is unique among western palaearctic *Andricus* species in galling leaves in both generations. The unusual gall structures induced in both generations of this species are both consistent with the distant relationship between *A. gallaeurnaeformis* and other *Andricus* species suggested by sequence divergence.

3. Patterns in the Evolution of Life Cycle Structure

The analysis of life cycle patterns in *Andricus* is relavent to the wider issue of the distribution of asexual taxa and the possible adapative significance of patterns of sex versus parthenogenesis. Two common and important assumptions are: (1) that asexual taxa do not revert to sex, and (2) that purely asexual taxa do not persist long in evolutionary time and do not produce diverse clades of asexual species (Hebert 1987, Judson and Normark 1996). We consider the impact of the *Andricus* data on these two paradigms.

Hebert (1987) pointed out that all major groups of cyclic parthenogens include obligately asexual members and *Andricus* is no exception with 35 asexual species in Europe. Gall wasps are the only cyclic parthenogens with regular alternation of the sexual and asexual generations, which occur at different times of year. Thus there is an alternation of seasons as well as reproductive mode. Because the seasonal pattern of asexual and sexual generations also applies to *Andricus* species with just one of the generations, it seems most likely that changes tend to involve *life cycle truncation* from species with both generations, rather than the direct conversion of sexual only to asexual only life cycles (or vice versa). This is seen in obligately asexual aphid clones derived from cyclic forms. The life cycle has a simple genetic basis such that a founder event could lead to asexual populations; in other cases environmental cues are required to trigger the sexual generations (Hebert 1987).

Asexuality in some chalcidoid wasps is induced by W*olbachia* bacteria, which are cytoplasmically transmitted intracellular parasites. At least in the laboratory, *Wolbachia* parthenogenesis can be reversed (Stouthamer et al. 1990) and a study of *Trichogramma* wasps also showed that there was considerable gene flow between conspecific sexual and parthenogenetic strains due to matings between normal males and 'parthenogenetic' females. *Wolbachia* infection has also recently been demonstrated as a cause of geographic parthenogenesis in gall wasps (Plantard and Solignac 1998). *Wolbachia* seem unlikely to be involved in cyclic parthenogenesis in *Andricus* but could possibly be a cause of obligate parthenogenesis. If so, reversals from *Wolbachia*-induced parthenogenesis might be possible. However, *Wolbachia* parthenogenesis would not explain the reduction from two generations to one and might also be expected to lead to some cases of asexual generations at the 'sexual' time of year.

Explaining the existence of clades containing only asexual species is always a challenge. It is always difficult to be fully confident that sexual generations are absent but it does seem likely that there exists at least one small clade of asexual species in our *Andricus* phylogeny (*A. polycerus*, *A. conglomeratus*, and *A. coriarius*). These species could have diverged from a common asexual ancestor—a process that is poorly understood and often considered heretical—or have been derived independently from a sexual ancestor that we have not sampled. Interpretation of *Andricus* life cycle changes depends critically on our assumptions and preconceptions about what is possible. Whatever assumptions we make, there have clearly been many changes. If we really believe that parthenogenesis is irreversible, and especially if we believe that asexual taxa are unlikely to speciate, then parsimony may not be an appropriate approach to the reconstruction of this type of evolutionary history.

4. Patterns in the Evolution of Oak Host Association

Alternation of generations, with both generations attacking oaks in the section *Quercus*, is strongly supported as the ancestral state in *Andricu*—both by character states present in the basal species and through outgroup comparison with other generation-alternating species. Within *Andricus*, a phylogenetically dispersed set of species have sexual generations on oaks in the section *Cerris*, either as part of an alternating lifecycle or in sexual-only life cycles. The most parsimonious interpretation is that the evolution of a life cycle involving alternation between *Quercus* and *Cerris* oaks evolved once in the common ancestor of these species, with subsequent loss of the the *Cerris*-galling generation in the asexual-only species. Gall induction on *Cerris* oaks has evolved independently a number of times in oak Cynipids: *Aphelonyx cerricola*, for example, is not only phylogenetically removed from the main *Andricus* grouping, but also differs in having only an asexual generation on *Q. cerris*, while all *Andricus* galls on *Q. cerris* are sexual generations. In addition to *Aphelonyx*, gall induction on *Cerris* oaks in european Cynipini is also found in *Callirhytis, Chilaspis, Dryocosmus, Neuroterus,* and *Plagiotrochus* (Ambrus 1974, Nieves-Aldrey 1987). Inference of the number of additional colonisation events of *Cerris* oaks represented by these genera must await more extensive phylogenetic analyses of the Cynipini as a whole. At least one completely independent colonization of *Cerris* oaks is represented by the genus *Synophrus*, which lies within the predominantly inquiline tribe Synergini, and is well separated phylogenetically from the Cynipini (Ronquist 1995, Liljeblad and Ronquist 1997).

ACKNOWLEDGEMENTS

We would like to thank Dr. Sue Kyes and Dr. Tim Anderson for their generous help and advice in this work, and Dr. Louis Miller and Dr. Karen Day for permission to work in their laboratories. We thank Dr. György Csóka and Dr. George Melika in Hungary, and Dr. Juli Pujade-i-Villar and Dr. José-Luis Nieves-Aldrey in Spain for their hospitality, help during field work, and samples of rare species. We thank Fredrik Ronquist, Olivier Plantard, Peter Price, and Karsten Schönrogge for their many helpful comments on earlier versions of this paper. This work was supported in part by the Royal Society.

LITERATURE CITED

Abe, Y. 1992. The advantage of attending ants and gall aggregation for the gall wasp *Andricus symbioticus* (Hymenoptera: Cynipidae). Oecologia. 89: 166-167.

Abe, Y. 1997. Well-developed gall tissues protecting the gall wasp *Andricus mukaigwae* (Mukaigwa) (Hymenoptera: Cynipidae) against the gall-inhabiting moth *Oedematopoda* sp. (Lepidoptera: Stathmopodidae). Applied Entomology and Zoology. 32: 135-141.

Abrahamson, W.G.; Melika, G.; Scrafford, R.; Csóka, G. 1998. Host-plant associations and specificity among cynipid gall-inducing wasps of eastern USA. In: Csóka, Gyuri; Mattson, William J.; Stone, Graham N.; Price, Peter W., eds. The biology of gall-inducing arthropods: proceedings of the international symposium; 1997 August 14-19; Mátrakūred, Hungary. Gen. Tech. Rep. NC-199. St. Paul, MN: U.S. Department of Agriculture, Forest Service, North Central Research Station: 226-240.

Abrahamson, W.G.; Sattler, J.F.; McCrea, K.D.; Weis A E. 1989. Variation in selection pressure on the goldenrod gall fly and the competitive interactions of its natural enemies. Oecologia. 79: 15-22.

Ambrus, B. 1974. Cynipid galls. In: Gusztáv, S., ed. Fauna Hungariae 12, part 1/a. Budapest, Hungary: Academic Press.

Askew, R.R. 1961. On the biology of the inhabitants of oak galls of Cynipidae (Hymenoptera) in Britain. Transactions of the Society for British Entomology. 14: 237-268.

Askew, R.R. 1965. The biology of the British species of the genus *Torymus* Dalman (Hymenoptera: Torymidae) associated with galls of Cynipidae (Hymenoptera) on oak, with special reference to alternation of forms. Transactions of the Society for British Entomology. 16: 217-232.

Askew, R.R. 1984. The biology of gallwasps. In: Ananthakrishnan, T.N., ed. The biology of galling insects. New Delhi: Oxford and IBH Publishing Co.: 223-271.

Bronner, R. 1992. The role of nutritive cells in the nutrition of cynipids and cecidomyiids. In: Shorthouse, J.D.; Rohfritsch, O., eds. Biology of insect-induced galls. Oxford: Oxford University Press: 118-140.

Cornell, H.V. 1983. The secondary chemistry and complex morphology of galls formed by the Cynipinae (Hymenoptera): why and how? American Midland Naturalist. 110: 220-234.

Craig, T.P. 1994. Effects of intraspecific plant variation on parasitoid communities. In: Hawkins, B.A.; Sheehan, W., eds. Parasitoid community ecology. Oxford: Oxford University Press: 205-227.

Craig, T.P.; Itami, J.K.; Price, P.W. 1990. The window of vulnerability of a shoot-galling sawfly to attack by a parasitoid. Ecology. 59: 297-308.

Crespi, B.J.; Carmean, D.A.; Chapman, T.W. 1997. Ecology and evolution of galling thrips and their allies. Annual Review of Entomology. 42: 51-71.

Csóka, G. 1997. Plant galls. Budapest, Hungary: Agroinform Publishing.

Dawkins, R. 1983. The extended phenotype: the long reach of the gene. Oxford: Oxford University Press.

Dreger-Jauffret, F.; Shorthouse, J.D. 1992. Diversity of gall-inducing insects and their galls. In: Shorthouse, J.D.; Rohfritsch, O., eds. Biology of insect-induced galls. Oxford: Oxford University Press: 8-34.

Drown, D.M.; Brown, J.M. 1998. Molecular phylogeny of North American oak-galling Cynipini (Hymenoptera: Cynipidae) supports need for generic revision. In: Csóka, Gyuri; Mattson, William J.; Stone, Graham N.; Price, Peter W., eds. The biology of gall-inducing arthropods: proceedings of the international symposium; 1997 August 14-19; Mátrakūred, Hungary. Gen. Tech. Rep. NC-199. St. Paul, MN: U.S. Department of Agriculture, Forest Service, North Central Research Station: 241-246.

Eady, R.D.; Quinlan, J. 1963. Hymenoptera, Cynipoidea. Handbooks for the identification of british insects 8, part 1(a). London, UK: Royal Entomological Society of London.

Felsenstein, J. 1993. PHYLIP (Phylogenetic inference package), version 3.5c. Seattle, WA: Department of Genetics, University of Washington.

Folliot, R. 1964. Contribution a l'tude de la biologie des Cynipides gallicoles (Hymenoptera: Cynipoidea). Annals des Sciences Naturels, Zoologie (Paris), Series 12, 5: 407-564.

Hebert, P. 1987. Genotypic characteristics of cyclic parthenogens and their obligately asexual derivaitives. In: Stearns, S.C., ed. The evolution of sex and its consequences. Birkhauser.

Jermin, L.S.; Crozier, R.H. 1994. The cytochrome-b region in the mitochondrial DNA of the ant *Tetraponera rufoniger* - sequence divergence in Hymenoptera may be associated with nucleotide content. Journal of Molecular Evolution. 38: 282-294.

Jones, D. 1983. The influence of host density and gall shape on the survivorship of *Diastrophus kinkaidii* Gill. (Hymenoptera: Cynipidae). Canadian Journal of Zoology. 61: 2138-2142.

Judson, O.P.; Normark, B.B. 1996. Ancient asexual scandals, Trends in Ecology and Evolution. 11: 41-45.

Liljeblad, J.; Ronquist, F. 1997. A phylogenetic analysis of higher-level gall wasp relationships. Systematic Entomology. (In press.)

Nieves-Aldrey, J.L. 1987. Estado actual del conocimiento de la subfamilia Cynipidae (Hymenoptera, Parasitica, Cynipidae) en la Península Ibérica. Eos. 63: 179-195.

Nixon, K.C.; Crepet, W.L. 1985. Preliminary phylogenetic reconstruction of *Quercus* at subgeneric and sectional levels. American Journal of Botany. 72: 934-935.

Plantard, O.; Rasplus, J.-Y.; Hochberg, M.E. 1996. Resource partitioning in the parasitoid assemblage of the oak galler *Neuroterus quercusbaccarum* L. (Hymenoptera: Cynipidae). Acta Oecologia. 17: 1-15.

Plantard, O.; Solignac, M. 1998. *Wolbachia*-induced thelytoky in cynipids. In: Csóka, Gyuri; Mattson, William J.; Stone, Graham N.; Price, Peter W., eds. The biology of gall-inducing arthropods: proceedings of the international symposium; 1997 August 14-19; Mátraküred, Hungary. Gen. Tech. Rep. NC-199. St. Paul, MN: U.S. Department of Agriculture, Forest Service, North Central Research Station: 111-121.

Plantard, O.; Shorthouse, J.D.; Rasplus, J.-Y. 1998. Molecular phylogeny of the genus *Diplolepis* (Hymenoptera: Cynipidae). In: Csóka, Gyuri; Mattson, William J.; Stone, Graham N.; Price, Peter W., eds. The biology of gall-inducing arthropods: proceedings of the international symposium; 1997 August 14-19; Mátraküred, Hungary. Gen. Tech. Rep. NC-199. St. Paul, MN: U.S. Department of Agriculture, Forest Service, North Central Research Station: 247-260.

Price, P.W.; Clancy, K.L. 1986. Interactions among three trophic levels: gall size and parasitoid attack. Ecology. 67: 1593-1600.

Price, P.W.; Fernandes, G.W.; Waring, G.L. 1987. Adaptive nature of insect galls. Environmental Entomology. 16: 15-24.

Pujade, J. 1994. Formes cinipo-cecidogenes detectades o que poden detectar-se, en les flors i els fruits de les fagacies a Andorra (Hymenoptera: Cynipidae: Cynipinae). Annals Inst. Est. Andorrans Centre de Barcelona, (1992).

Pujade, J. 1997. Malformacions produïdes per cynípids als borrons de fagacies detectades a Andorra. Annals Inst. Est. Andorrans Centre de Barcelona, (1995), 13-39.

Rohfritsch, O. 1992. Patterns in gall development. In: Shorthouse, J.D.; Rohfritsch, O., eds. Biology of insect-induced galls. Oxford: Oxford University Press: 60-87.

Ronquist, F. 1995. Phylogeny and early evolution of the Cynipoidea (Hymenoptera). Systematic Entomology. 20: 309-335.

Saitou, N.; Nei, M. 1987. The neighbor-joining method: a new method for reconstructing phylogenetic trees. Molecular Biology and Evolution. 4: 1406-1425.

Schönrogge, K.; Stone, G.N.; Crawley, M.J. 1995. Spatial and temporal variation in guild structure: parasitoids and inquilines of *Andricus quercuscalicis* (Hymenoptera: Cynipidae) in its native and alien ranges. Oikos. 72: 51-60.

Schönrogge, K.; Stone, G.N.; Crawley, M.J. 1996. Alien herbivores and native parasitoids: rapid development and structure of the parasitoid and inquiline complex in an invading gallwasp; *Andricus quercuscalicis* (Hymenoptera: Cynipidae). Ecological Entomology. 21: 71-80.

Schönrogge, K.; Harper, L.J.; Brooks, S.E.; Shorthouse, J.D.; Lichtenstein, C.P. 1998. Reprogramming plant development: two approaches to study the molecular mechanism of gall formation. In: Csóka, Gyuri; Mattson, William J.; Stone, Graham N.; Price, Peter W., eds. The biology of gall-inducing arthropods: proceedings of the international symposium; 1997 August 14-19; Mátraküred, Hungary. Gen. Tech. Rep. NC-199. St. Paul, MN: U.S. Department of Agriculture, Forest Service, North Central Research Station: 153-160.

Shorthouse, J.; Rohfritsch, O., eds. 1992. Biology of insect-induced galls. Oxford: Oxford University Press.

Stern, D.L. 1995. Phylogenetic evidence that aphids, rather than plants, determine gall morphology. Proceedings of the Royal Society of London, Series B 260: 85-89.

Stone, G.N.; Schönrogge, K.; Crawley, M.J.; Fraser, S. 1995. Geographic variation in the parasitoid community associated with an invading gallwasp, *Andricus quercuscalicis* (Hymenoptera: Cynipidae). Oecologia. 104: 207-217.

Stone, G.N.; Cook, J.M. (submitted). The structure of cynipid oak galls: patterns in the evolution of an extended phenotype. Proceedings of the Royal Society of London, Series B.

Stouthamer, R.; Luck, R.F.; Werren, J.H. 1990. Antibiotics cause parthenogenetic *Trichogramma* to revert to sex. Proceedings of the National Academy of Sciences of the USA. 87: 2424-2427.

Swofford, D.L. 1993. PAUP: Phylogenetic Analysis Using Parsimony, version 3.1. Formerly distributed by Illinois Natural History Survey, Champaign, Illinois.

Washburn, J.O. 1984. Mutualism between a cynipid wasp and ants. Ecology. 65: 654-656.

Washburn, J.O.; Cornell, H.V. 1979. Chalcid parasitoid attack on a gallwasp population (*Acraspis hirta* [Hymenoptera: Cynipidae]) on *Quercus prinus* (Fagaceae). Canadian Entomologist. 111: 391-400.

Washburn, J.O.; Cornell, H.V. 1981. Parasitoids, patches and phenology: their possible role in the local extinction of a cynipid wasp population. Ecology. 62: 1597-1607.

Weis, A.E. 1982. Resource utilization patterns in a community of gall-attacking parasitoids. Environmental Entomology. 11: 809-815.

Weis, A.E. 1993. Host gall size predicts host quality for the parasitoid *Eurytoma gigantea*, but can the parasitoid tell? Journal of Insect Behaviour. 5: 591-602.

Weis, A.E.; Abrahamson, W.G.; McCrea, K.D. 1985. Host gall size and oviposition success by the parasitoid *Eurytoma gigantea*. Ecological Entomology. 10: 341-348.

Weld, L.H. 1957. Cynipid galls of the Pacific Slope. Ann Arbor, Michigan. Privately printed. 64 p.

Weld, L.H. 1959. Cynipid galls of the eastern United States. Ann Arbor, Michigan. Privately printed. 124 p.

Weld, L.H. 1960. Cynipid galls of the Southwest. Ann Arbor, Michigan. Privately printed. 35 p.

Werren, J.H.; Windsor, D.; Guo, L.R. 1995. Distribution of *Wolbachia* among neotropical arthropods. Proceedings of the Royal Society of London, Series B 262: 197-204.

Westphal, E. 1992 Cecidogenesis and resistance phenomena in mite-induced galls. In: Shorthouse, J.D.; Rohfritsch, O., eds. Biology of insect-induced galls. Oxford: Oxford University Press: 141-156.

Williams, M.A., ed. 1994. Plant galls: organisms, interactions, populations. Special Volume of the British Systematics Association 49. Oxford: Oxford Science Publications.

Zwölfer, H.; Arnold-Rinehart, J. 1994. Parasitoids as a driving force in the evolution of the gall size of *Urophora* on Cardueae hosts. In: Shorthouse, J.D.; Rohfritsch, O., eds. Biology of insect-induced galls. Oxford: Oxford University Press: 245-258.

THE POPULATION GENETICS OF POSTGLACIAL INVASIONS OF NORTHERN EUROPE BY CYNIPID GALL WASPS (HYMENOPTERA: CYNIPIDAE)

Gyuri Csóka[1], Graham Stone[2], Rachel Atkinson[2], and Karsten Schönrogge[3].

[1] Forest Research Institute, Department of Forest Protection, 3232 Mátrafüred, P.O.Box 2, Hungary
[2] Oxford University Department of Zoology, South Parks Road, Oxford OX1 3PS, U.K.
[3] School of Biological Sciences, Queen Mary and Westfield College, Mile End Road, London E1 4NS, U.K.

Abstract.—We compare the genetic consequences of range expansion in three gall wasp species in the genus *Andricus*, comparing new data for *A. kollari* and *A. lignicola* with extensive published data for *A. quercuscalicis*. All three species have expanded their ranges into northern and western Europe from natural distributions in southern Europe following human introduction of Turkey oak, *Quercus cerris*. *Andricus lignicola* and *A. quercuscalicis* reached Britain through natural range expansion from Italy and the Balkans, while *Andricus kollari* was purposely introduced into Britain in large numbers in the first half of the 19th century from the eastern Mediterranean. Allozyme data show all three species to have lost genetic diversity during range expansion. British populations of *A. kollari* possess alleles absent from neighboring continental populations, and are genetically close to populations in the Balkans. These alleles may represent a 'genetic footprint' of the original introductions. In addition to a native distribution in southern-central Europe and the Balkans, *A. kollari* is also native to Iberia and north-western Africa. Eastern and western populations possess regionally distinctive alleles, suggesting that neither population is derived by sub-sampling from the other. The two parts of the range may represent genetically divergent glacial refuge populations dating from the last ice age.

INTRODUCTION

The geographic distributions of many organisms are probably inherently unstable, constrained not by physiological limitations but by historical accidents which impose barriers to their dispersal. When these barriers are removed or crossed, either naturally or through human intervention, a species may become invasive and expand its distribution until biological or physical factors once again limit its spread (Hewitt 1996, Williamson 1996, Eber and Brandl 1997). As they expand their range, invaders may experience changes in their ecological links with other organisms, generating new trophic linkages and affecting community structure (Cornell and Hawkins 1993; Godfray et al. 1995; Schönrogge et al. 1994a, 1996a, b). In addition, the genetic makeup of invading populations may change in response to selection and sampling effects, such as population bottlenecks (Hewitt 1996). Range expansion is a continuing feature of quaternary ecology, and understanding the biological processes accompanying invasion is important in applied theatres, including the control of many important pests and the planned release of genetically modified organisms (Williamson 1996).

Gall Wasps as Model Systems in the Study of Invasions

Some of the best-studied natural invasions are the colonization of northern and western Europe by a group of gall-inducing cynipid wasps (Hymenoptera: Cynipidae; Cynipini) native to southern Europe. The species are members of the genus *Andricus*, and form galls on oaks, *Quercus* spp. The gall wasp species in question all have a lifecycle which involves obligate alternation between two host oak taxa (Stone and Sunnucks 1993). A sexual spring generation develops on oaks in the section Cerris, particularly the Turkey oak,

Quercus cerris, while a parthenogenetic summer generation develops on oaks in the section Quercus, such as *Q. petraea*, *Q. pubescens*, and *Q. robur*. The asexual generation may be highly host-specific (as in *A. quercuscalicis*, which only attacks *Q. robur*), or attack a range of host species (as do *A. corruptrix*, *A. kollari*, and *A. lignicola*) in this oak section (Ambrus 1974, Nieves Aldrey 1987, Csóka 1997).

The geographic distribution of these host-alternating gall wasps is in part limited by the distribution of their oak hosts. While oaks in the section Quercus (such as *Q. petraea* and *Q. robur*) are widely distributed in Europe, members of the section Cerris are naturally limited to areas south of the Pyrenees, Alps and Tatra Mountains (Jalas and Suominen 1987). The two oak groups occur together naturally in southern Europe, the Mediterranean coast of north Africa, and Asia Minor, and these regions represent the natural, native distribution of gall wasps exhibiting the alternating life cycle.

The majority of host alternating *Andricus* species are found in the eastern part of this region (the extreme east of Austria, the Balkans, and Asia Minor), where the Turkey oak, *Q. cerris*, is the host of sexual generation galls (Ambrus 1974, Csóka 1997). Two of the host-alternating species which attack Turkey oak in eastern and central Europe (*A. kollari* and *A. gemmea*) are also found in the Iberian Peninsular and the foothills of the Atlas Mountains in north-west Africa (Houard 1912; Nieves-Aldrey 1987). However, there is no Turkey oak in the Iberian/African part of their distribution (Jalas and Suominen 1987), and the sexual generation galls of these species remain unknown in this region (Nieves-Aldrey 1987). *Andricus gemmea* and *A. kollari* populations in Iberia may have lost their sexual generations (in common with many other *Andricus* species in Europe), or more probably develop in as yet undiscovered galls on an alternative sexual generation host (Juli Pujade, pers. comm.). The cork oak, *Q. suber*, which is phylogenetically close to *Q. cerris*, is native to Iberia and north-western Africa (Jalas and Suominen 1987), and is the most likely potential host for the sexual generations of these species in the region.

Range expansion by host-alternating *Andricus* has been made possible by human dispersal of Turkey oak. For the last 400 years or so, Turkey oak has been planted extensively in parks and gardens, and is now found far to the north and west of its original range (Stone and Sunnucks 1993). Turkey oak is present in small quantities in planted forestry, particularly in France, and is able to self-seed with considerable success through much of its introduced range (Hails and Crawley 1991). This human dispersal has introduced a mosaic of Turkey oak, the sexual generation host, into an area naturally rich in hosts of the asexual generation. The gall wasp species with lifecycles which need Turkey oak were thus provided with the opportunity to spread westwards, and have proved highly successful in doing so. Four gall wasp species with this pattern of oak host alternation are now naturalized throughout northern Europe as far north and west as the United Kingdom and Ireland (Schönrogge et al. 1994b). One, the Marble gall wasp, *Andricus kollari*, was introduced intentionally in large numbers from the Levant in the first half of the 19th century, its galls once used as a raw material in the manufacture of ink (Askew 1984). As far as is known, the other three species reached Britain without human assistance. The Knopper gall wasp, *A. quercuscalicis*, reached Britain by the 1950's, while *A. corruptrix* and *A. lignicola* were first recorded in Britain in the 1970's (Collins et al. 1983, Askew 1984, Hails and Crawley 1991).

To date, most research effort has been concentrated on the knopper gall wasp, *Andricus quercuscalicis*, a species native to Italy and the Balkans. *A. quercuscalicis* has proved an excellent tool with which to examine many aspects of invasion across a patchy habitat. One question concerns the development of communities associated with the invader in its novel range (Collins et al. 1983; Hails 1989; Hails and Crawley 1991; Stone et al. 1995; Schönrogge et al. 1994a, b, 1996a, b). The second question, which we develop further here, concerns the genetic consequences of the invasion process (Stone and Sunnucks 1993, Sunnucks et al. 1994, Sunnucks and Stone 1996).

Population Genetic Patterns in *Andricus Quercuscalicis*

Westwards invasion by *A. quercuscalicis* has involved colonization of Turkey oak patches at ever greater distances from its native range. These patches could have been colonized independently by wasps arriving directly from the native range ('source-sink' model), or colonized sequentially, one after the other to the west ('stepping-stone' model). In 'source-sink' colonization, neighboring populations in the invaded range should show no particular genetic similarity to each other, while in 'stepping-stone' colonization populations in the invaded range should show genetic similarities to their neighbors. Colonization could also have involved many arrivals, or small numbers of founders. This is important because many colonists should bring a high proportion of the genetic variation present in the parent population, while colonization by small numbers of founders will only bring a small subset (Bryant et al. 1981, Barton and Charlesworth 1984, Berlocher 1984). For this reason, colonization by few founders is said to result in a 'genetic bottleneck'. As well as these genetic results of the mechanics of the invasion process, natural selection might also be at work (Florence et al. 1982; Gu and Danthanarayana 1992). Expansion into areas with climatic conditions which differ from those experienced in the native range might also have resulted in strong selective changes in the genetic composition of more northerly and westerly populations.

To distinguish between alternative models of the colonization process, we have looked at allozyme variation in over 1,100 individuals from 50 populations of *A. quercuscalicis* along a transect stretching from the Ukraine and Romania in the east to Ireland in the west (Stone and Sunnucks 1993, Sunnucks and Stone 1996). *A. quercuscalicis* has lost a great deal of genetic variability during the invasion. Populations in Ireland, at the limits of the distribution, have on average only slightly more than one allele per locus, while populations in south-eastern Europe have up to five alleles (table 1, fig. 1). Loss of alleles along the line of invasion happened in all of the enzyme systems studied. It is generally believed that natural selection would not result in loss of variability in all of the enzyme systems at once, and this pattern has probably been generated by the colonization process itself. The frequencies of alleles in each population can be used to generate a phenogram, or genetic similarity tree, linking populations. Populations in the same part of Europe group together on the same branch of the tree (fig. 2). Populations in the native range form the trunk of the tree, and more distant colonies (in the outer branches and twigs) have subsets of the variation found in their neighbors immediately to the south and east. Because the sites are related in this way, it is extremely unlikely that sites in the invaded range have been colonized independently from the native range (as in the 'source-sink' model above), and much more likely that the gall wasp migration has proceeded in a 'stepping-stone' manner, through foundation of new western populations from their more immediate eastern neighbors. The extent of the loss of variation points to frequent genetic bottlenecks during the invasion; as new populations were founded by small numbers of gall wasps, only a small subset of the variation was carried on by each new set of colonists. This 'stepping-stone' process, repeated many times during expansion across Europe, would result in the observed general loss of variability in all enzymes.

Predictions

The patterns observed in *A. quercuscalicis* provide a baseline for comparison with the genetic consequences of range expansion in the other host-alternators. *Andricus corruptrix* and *A. lignicola* have the same native range as *A. quercuscalicis*, and have spread naturally across the same set of Turkey oak patches. We therefore predict similar results of genetic subsampling in these species,

resulting in loss of genetic diversity with increasing distance from the native range. *Andricus kollari* represents a more complex situation. The genetic composition of populations in the invaded range is the result of both natural range expansion (as for the previous species) and human introduction in Britain. If the introduced genotypes have persisted since their introduction, we might therefore expect to find eastern mediterranean genotypes in Britain absent from western parts of mainland Europe. There is also the possibility that gene flow may have occurred between Britain and Iberian populations of *A. kollari*, via the Bay of Biscay. If Iberian and Eastern European populations of this species have diverged genetically to the extent that they possess different alleles, it may be possible to infer which of the alleles present in Britain have resulted through immigration from these two regions.

In this paper we summarize preliminary allozyme data for *A. lignicola* and *A. kollari*. We assess the extent to which patterns in *A. kollari* and *A. lignicola* parallel those seen in *A. quercuscalicis*, and ask whether direct introduction of *A. kollari* to Britain from the eastern Mediterranean has produced a different geographic pattern of genetic variation to those associated with natural range expansion. Finally, we assess whether genetic differentiation has occurred between the eastern (Balkan) and western (Iberian) populations of *A. kollari*, and address three questions: (a) is there evidence that one end of the range of this species is clearly derived from the other? (b) Is there evidence of barriers to gene flow between these two parts of the native range of this species?

METHODS

Sampling Sites

Sampling sites for *Andricus quercuscalicis* are described in detail in Stone and Sunnucks (1993). To allow comparison with the three other host alternators, samples of *A. quercuscalicis* have been pooled into regions in table 1. Sampling in *A. lignicola* is concentrated in Hungary (representing the native range) and north-western Europe (representing the margins of the invaded range). To achieve reasonable sample sizes for analysis, samples have been pooled into regions in table 2. Sampling for *A. kollari* is more extensive, and summary statistics are presented for each sample location by country/region in table 3.

Electrophoretic Methods

Adult gall wasps reared from galls were stored at -70°C until needed for electrophoresis. Cellulose acetate electrophoresis was chosen because of its low requirement for material: pilot studies showed that each locus

Table 1.—*Summary statistics for genetic diversity in populations of* Andricus quercuscalicis H_{DC} *is the mean heterozygosity per locus through direct count from the data, while H_{HW} is the mean heterozygosity calculated under the assumption of Hardy-Weinberg equilibrium. N is the sample size, and all errors are standard errors.*

Population	Mean n per locus (±SE)	Mean alleles per locus (±SE)	% loci poly-morphic	H_{DC} (%±SE)	H_{HW} (%±SE)
Native Range					
Rumania	19.7(0.2)	1.7(0.3)	46.2	22.3±7.4	22.5±7.6
Slovenia	50.7(0.2)	1.7(0.2)	46.2	19.8±6.8	20.9±7.3
Hungary	98.8(0.2)	2.2(0.4)	53.8	21.1±7.1	21.2±7.2
Ukraine	26.9(0.1)	1.8(0.3)	46.2	19.5±6.8	21.8±7.3
E Austria	86.2(0.5)	1.9(0.3)	53.8	22.1±7.2	21.6±7.2
N. Italy	53.9(0.1)	1.8(0.3)	53.8	17.8±6.8	17.3±6.5
Invaded Range					
S. Germany	31.9(0.1)	1.7(0.3)	46.2	18.4±6.5	21.0±7.2
N Germany	50.8(0.2)	1.6(0.2)	46.2	13.6±4.6	20.5±7.0
Belgium	45.0(0.0)	1.4(0.1)	38.5	10.1±4.9	10.2±4.7
Holland	23.5(0.2)	1.5(0.2)	38.5	12.3±5.0	13.4±5.9
France	87.6(0.2)	1.5(0.2)	38.5	12.7±5.0	13.5±5.3
UK					
Channel Isles	15.1(0.9)	1.3(0.1)	30.8	8.7±5.4	11.7±5.8
Southwest UK	126.5(0.5)	1.4(0.2)	30.8	8.4±4.7	9.1±5.1
Southeast UK	180.5(0.8)	1.4(0.2)	30.8	14.0±6.8	13.1±6.4
Central UK	24.0(0.0)	1.4(0.2)	30.8	10.3±5.0	12.3±5.9
East Anglia UK	23.9(0.1)	1.4(0.2)	30.8	9.4±4.7	12.9±6.2
North England UK	63.1(0.7)	1.4(0.2)	30.8	10.7±5.4	12.3±6.1
Ireland	97.7(0.2)	1.4(0.2)	30.8	4.6±2.7	6.2±3.5

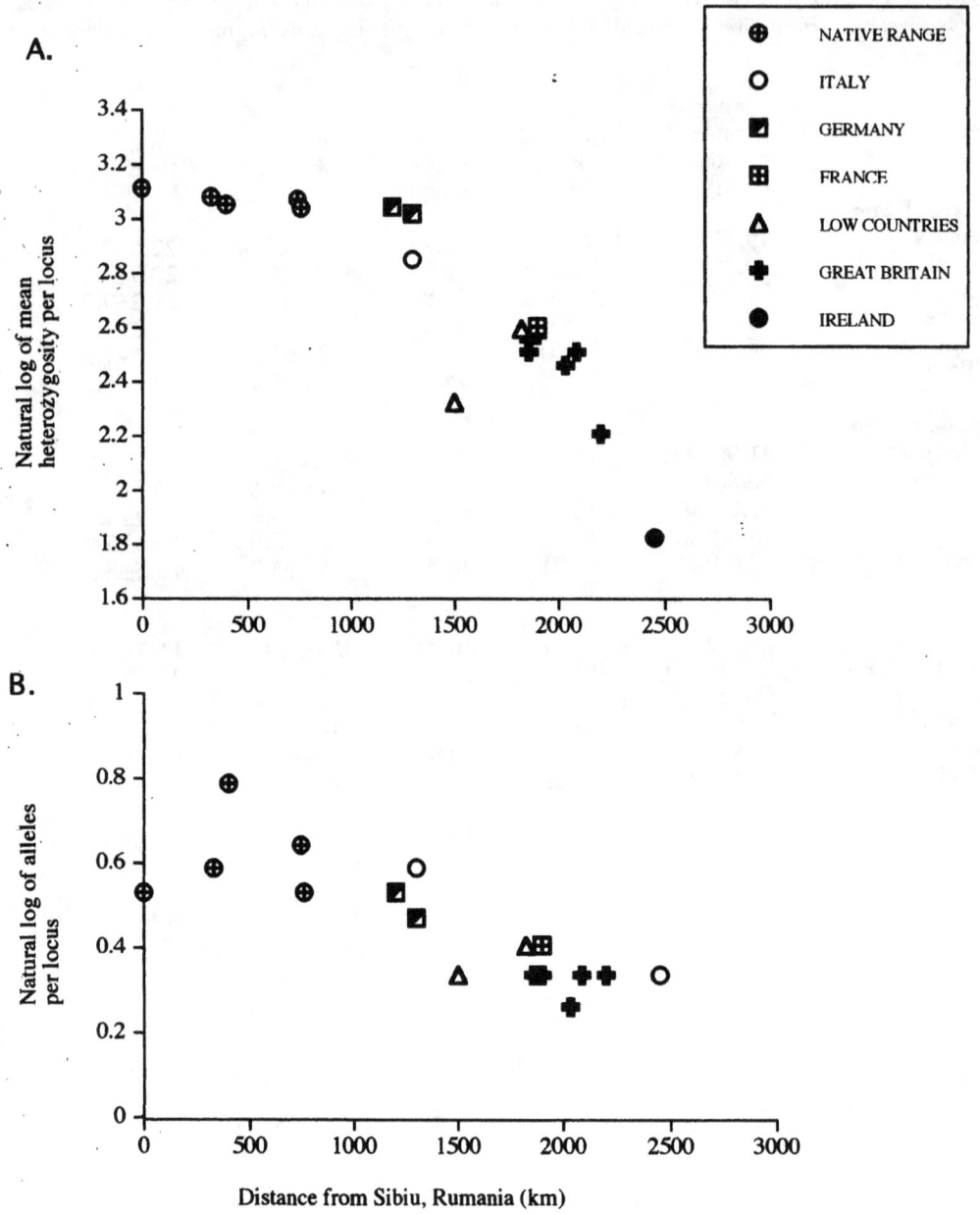

Figure 1.—*Changes in levels of population genetic variation with distance from the center of the native range (defined as Sibiu, Roumania, which is the site with the greatest number of alleles) in* Andricus quercuscalicis *(a) Mean heterozygosity per locus (H_{HW}), and (b) Mean number of alleles per locus.*

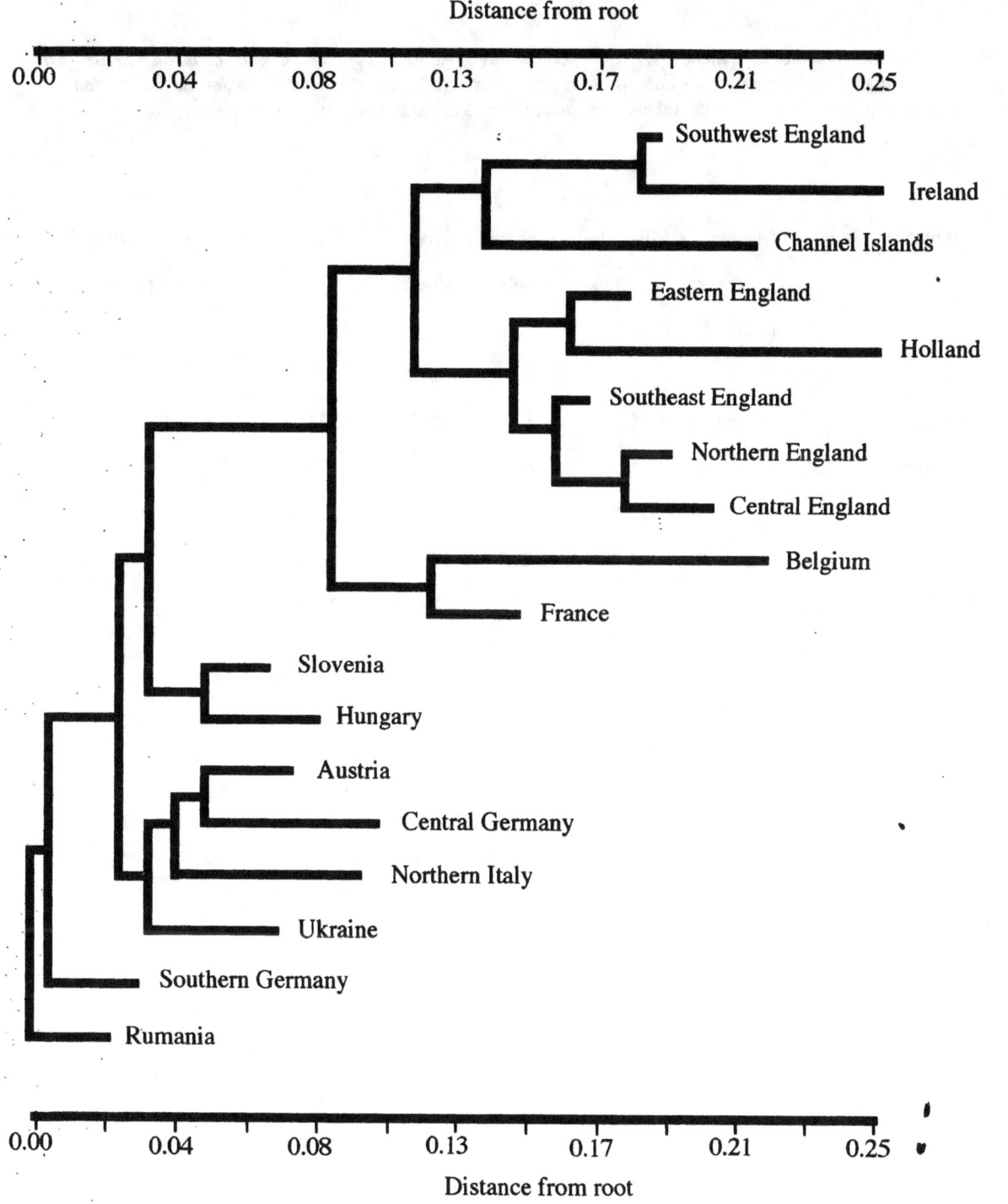

Figure 2.—*Phylogeographic relationships between populations of* Andircus quercuscalicis. *The phylogram is a Wagner tree built using Cavalli-Sforza and Edward's Arc distance.*

Table 2.—*Summary statistics for genetic diversity in populations of* Andricus lignicola. H_{DC} *is the mean heterozygosity per locus through direct count from the data, while H_{HW} is mean heterozygosity calculated under the assumption of Hardy-Weinberg equilibrium. N is the sample size, and all errors are standard errors. N = northern, and S = southern.*

Population	n	Mean n per locus (±SE)	Mean alleles per locus (±SE)	% loci polymorphic	H_{DC} (%±SE)	H_{HW} (%±SE)	Total alleles	Regionally private alleles
Hungary	108	107.6 (.3)	3.9 (.3)	100	25±7	33±8	35	6
N. France	23	23.0 (.0)	2.2 (.3)	66.7	27±8	30±8	20	0
S. UK	41	40.9 (.1)	2.0 (.3)	66.7	20±7	22±7	18	0
N. UK	147	146.4 (.2)	3.0 (.4)	88.9	26±8	30±7	27	2
Ireland	163	162.9 (.1)	2.8 (.4)	88.9	28±9	27±8	25	0
TOTAL	482							

Table 3.—*Summary statistics for genetic diversity in populations of* Andricus kollari. H_{DC} *is the mean heterozygosity per locus through direct count from the data, while H_{HW} is mean heterozygosity calculated under the assumption of Hardy-Weinberg equilibrium. N is the sample size, and all errors are standard errors.*

Population	Mean n per locus (±SE)	Mean alleles per locus (±SE)	% loci poly-morphic	H_{DC} (%±SE)	H_{HW} (%±SE)
Spain					
Madrid SP	20.0 (0)	1.5 (0.2)	45.5	14.5±7	14.2±6
Orense & Potes S	19.0 (0)	2.0 (0.3)	54.5	17.7±7	19.2±7
Salamanca SP	35.0 (0)	2.4 (0.3)	72.7	12.5±5	17.2±6.5
Quintanilla SP	23.0 (0)	1.8 (0.4)	45.5	14.2±6	15.4±6
Gudina SP	66.7 (0.2)	2.5 (0.2)	72.7	16.7±6	17.5±6
Eastern Native Range					
Rest of Hungary	29.8 (0.2)	2.6 (0.5)	72.7	23.4±9	25.9±10
Matrafured HU	34.2 (0.3)	2.6 (0.5)	72.7	25.3±9	26.6±9
Italy					
Appenines IT	59.4 (2.7)	2.7 (0.5)	63.6	22.7±9	24.9±9
Florence IT	15.8 (0.2)	2.4 (0.4)	63.6	25.510	25.4±10
Germany					
North Germany	37.0 (0.0)	2.0 (0.4)	54.5	15.2±6	17.0±7
South Germany	7.6 (0.2)	1.9 (0.4)	45.5	19.7±8	23.0±9.5
France					
Saumur N.FR	31.9 (0.1)	2.1 (0.4)	54.5	23.4±9	20.3±7
Cherbourg N.FR	16.6 (0.2)	2.0 (0.4)	54.5	15.5±8	21.8±9
Bordeaux W.FR	8.9 (0.7)	1.8 (0.3)	54.5	18.5±8	20.9±8
Agen S. FR	29.1 (0.8)	2.2 (0.4)	63.6	22.9±9	24.8±9
Nouvion N. FR	10.9 (1.0)	1.6 (0.3)	36.4	18.2±10	18.0±9
Auch S. FR	20.0 (1.0)	2.2 (0.4)	54.5	23.7±9	23.2±9
Low Countries					
Utrecht HOL	10.9 (0.1)	1.5 (0.2)	36.4	14.5±7	15.1±7.5
Belgium	13.5 (0.5)	1.7 (0.3)	36.4	19.3±9	18.0±9
UK					
Oxford ENG	33.0 (0.0)	1.9 (0.4)	45.5	18.2±8	18.1±8
Dunrobin SCOT	42.7 (1.5)	2.3 (0.4)	54.5	20.1±9	20.5±8.5
Puttenham ENG	22.0 (0.0)	2.0 (0.4)	45.5	15.3±6	19.6±8.5
Ireland					
Muckross IR	29.0 (0.0)	1.9 (0.4)	36.4	18.8±8	19.1±8
Rest of Ireland	7.0 (0.0)	1.5 (0.2)	27.3	9.1±5	12.6±7

could be scored from about one hundredth of the protein suspension obtained from an agamic female. Agamic females were used rather than sexual males or females for three reasons: (1) they weigh about 10 mg, 10 times as much as sexual females, (2) the agamic galls are easier to collect, and (3) the agamic generation consists only of diploid females rather than both haploid males and diploid females. Gels were run and stained following standard methods (Richardson et al. 1986) with some modifications to improve gel running and staining. Individual insects were homogenized in 50μ of homogenization buffer (Richardson et al. 1986), 15μ\%l of the resulting solution was diluted with another 15μ\%l of the buffer, and 0.5-1.0μ\ of this run on each gel. Initial screenings for allozyme variability in each species were based on experience with *A. quercuscalicis* (Stone and Sunnucks 1993). We scored 11 systems in *A. kollari*, with the addition of AK for *A. lignicola*, using three standard buffers and enzyme-specific substrate stains given by Richardson et al. (1986).

(a) 0.04M phosphate pH 6.1 for GOTm and GOTs, alpha GPD1 and 2, and PEPb.
(b) Tris malate EDTA pH8.6 for ME, MDHs, MDHm, and HK.
(c) Tris glycine pH 8.6 for GPI and PGM.

AK was scored in *A. lignicola*, but has yet to be scored for *A. kollari*. Gels were run at 200 volts in standard apparatus (Helena zip-zone system) for running times ranging from 20 minutes (HK) to 95 minutes (PEPb). Where possible, at least 12 individuals per population were examined electrophoretically.

Allele Frequency Analysis

We used the program BIOSYS version 1.0 (Swofford and Selander 1981) for population genetic analyses of allozyme frequency data. We use three estimators to describe the genetic diversity of populations or regions: (1) the total number of alleles present at the 11 loci screened, (2) the mean number of alleles per locus, and (3) the mean heterozygosity per locus expected from observed phenotypic frequencies under Hardy Weinberg equilibrium, H_{HW} (Nei and Roychoudhury 1974, Stone and Sunnucks 1993).

When analyzing the association between genetic measures and geographic location for *A. kollari* we have used the site in the native range where the maximum number of alleles was detected (Mátrafüred, Hungary) as a hypothetical origin. It must be remembered that *A. kollari* is native to both south-east Europe and the Iberian peninsular, and so the origin in this representation does not necessarily represent the center of the native range of this species. We present the data in this way solely to allow comparison with patterns seen in *A. quercuscalicis*, whose origin in the Balkans is known with certainty (Stone and Sunnucks 1993). In fitting a curve describing the relationship between measures of genetic variability and distance we have used natural log transformed values of heterozygosity and the mean number of alleles per locus because this model, unlike a simple linear model, allows values to approach an asymptotic value (corresponding to monomorphism at all loci) without becoming negative.

Genetic relationships between populations of *A. kollari* are illustrated using Wagner phenograms constructed using the Cavalli-Sforza and Edwards Arc measure of genetic distance, calculated in BIOSYS (see Stone and Sunnucks 1993 for justification of this approach). Phenetic trees illustrate the genetic differences between sites and demonstrate where any major genetic discontinuities exist. Patterns of clustering of sites within the tree can be used to distinguish between source-sink and stepping stone patterns of range expansion: strong correlations between the spatial arrangement of sites and genetic distance measures support a stepping-stone model, while lack of any such correlation is compatible with a source-sink model. More complex analyses of spatial patterns of genetic variation, including analyses of pairwise differences in Wright's F_{st} and spatial autocorrelation (Stone and Sunnucks 1993), are in progress but will not be presented here.

RESULTS

Andricus Lignicola

Andricus lignicola shows higher genetic diversity in its Balkan native range populations than any other host alternator, both in terms of mean numbers of alleles per locus (reaching 3.9 in Hungarian samples) and mean heterozygosity (reaching 33 percent in Hungarian samples) (fig. 3). Populations at the western limit of the invaded range lack six alleles found in Hungarian populations at low frequency, and although *A. lignicola* possesses greater genetic diversity than *A. quercuscalicis*, it shows a more marked reduction in alleles per locus between Hungary and southern England (to 51 percent of Hungarian levels in *A. lignicola* and to 64 percent in *A. quercuscalicis*). Loss of these six alleles has a less marked effect on mean heterozygosity, and populations in southern Britain show 67 percent of the values observed in Hungary, almost exactly the proportional drop observed in *A. quercuscalicis* (fig. 3).

Andricus Kollari

Centers of Genetic Diversity

The greatest genetic diversity in *Andricus kollari* is found in Spain and Hungary/Italy, the two regions in

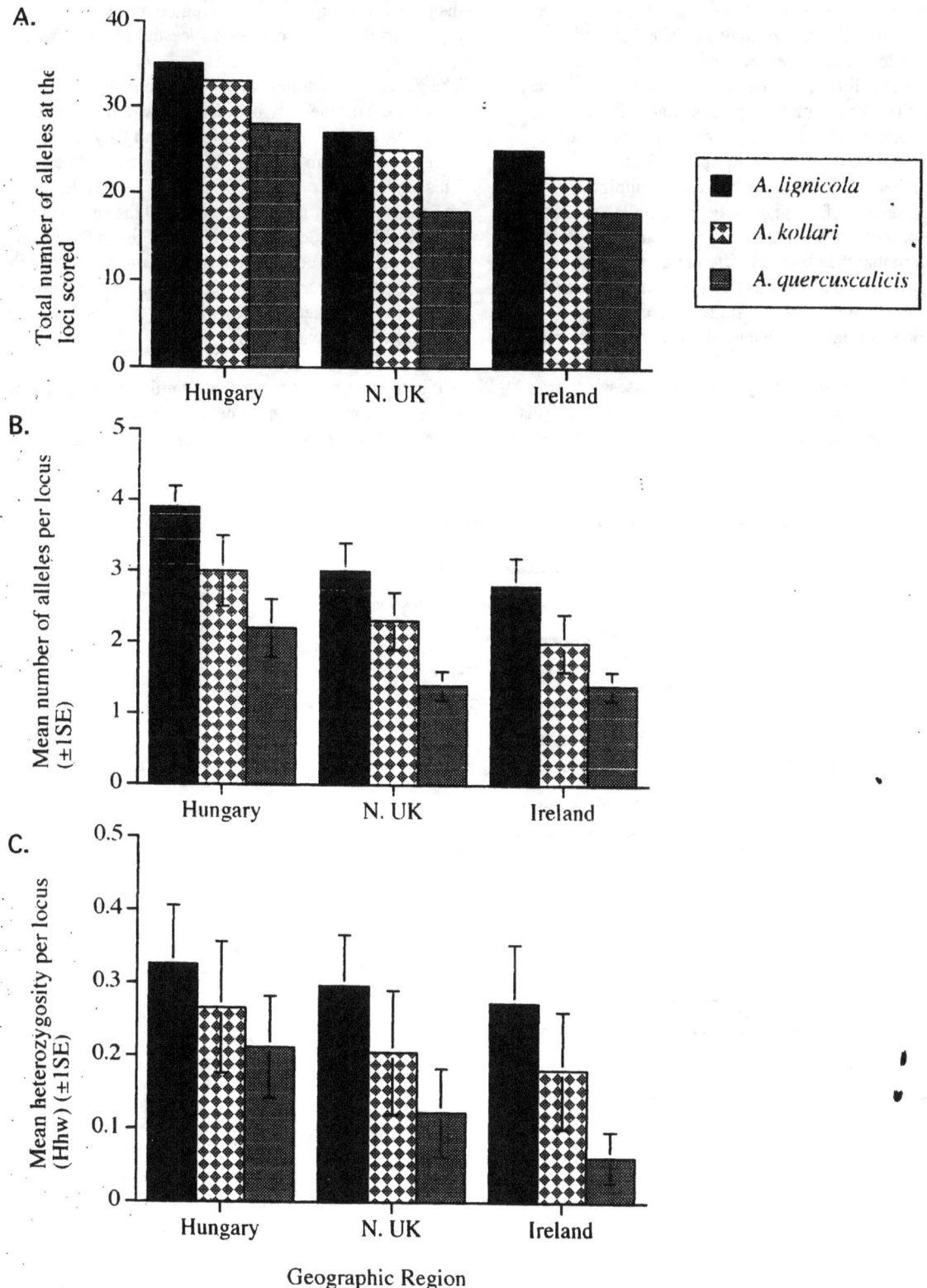

Figure 3.—Measures of genetic diversity in three geographic regions for A. lignicola, A. kollari and A. quercuscalicis: (a) Total numbers of alleles at the loci scored, (b) Mean numbers of alleles per locus, and (c) Mean heterozygosity per locus.

which the species is native. Spain has 37 alleles at the 11 loci sampled, and Hungary/Italy 34. There is evidence of genetic differentiation between these two endemic regions, in the form of regionally restricted (or 'private') alleles: Spanish populations possess four alleles absent from Hungary/Italy, and Hungarian/Italian populations possess two alleles absent from Spain. While it is possible that such differences represent sampling artifacts, several of the regionally private alleles reach frequencies of 3-8 percent in relatively large samples, suggesting that the observed differentiation is real. A genetic distance phenogram (fig. 4) shows the Spanish populations to lie in a cluster quite distinct from the Hungarian and Italian populations used to root the tree.

Because both Spain and Italy/Hungary possess private alleles, neither of the native range regions of *A. kollari* is obviously a genetic subset of the other. With the current data we thus cannot say which region more closely represents the shared common ancestral population.

The geographic limitations of the Spanish private alleles are very clear: the only population north of the Pyrenees to possess any of them is Bordeaux, and they are absent from sub-Pyrenean sites in France (Auch and Agen). This explains the position of Bordeaux within the cluster of Spanish populations in figure 4, and the separation between Spanish populations and the French Pyrenean sites (Auch and Agen), which lie closer to the root of the tree. These differences between populations north and south of the Pyrenees suggests that an effective barrier to gene flow lies between them.

While the two native regions are similar in these simple terms, other measures of genetic diversity reveal differences between them. Spanish populations often differ in

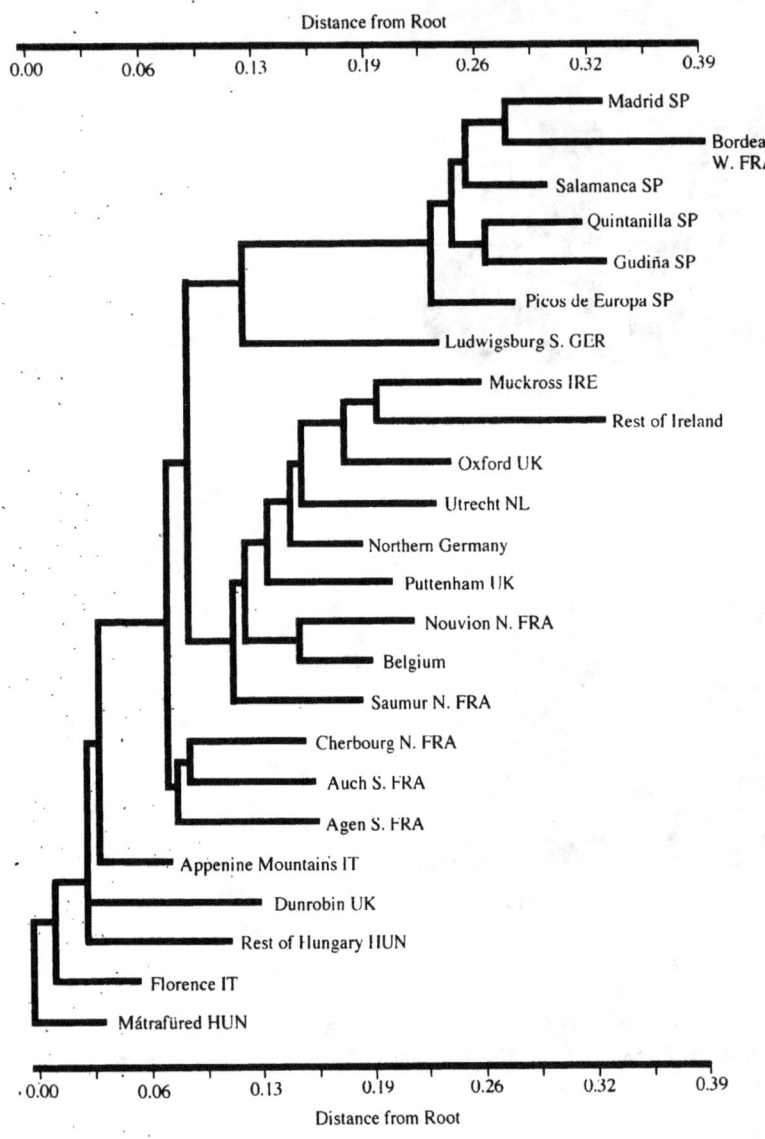

Figure 4.—*Phylogeographic relationships between populations of* Andricus kollari. *The phylogram is a Wagner tree built using Cavalli-Sforza and Edward's Arc distance.*

the alleles they possess at specific loci, while populations in Hungary and Italy are more similar in allele frequency. This results in higher mean levels of genetic diversity per locus in south-eastern Europe than in Iberia (table 3). Hungary and Italy have a mean of 2.6-2.7 alleles per locus over the 11 loci screened, and mean heterozygosity per locus of 25-27 percent (table 3), while values in Spain (particularly for heterozygosity) are rather lower.

Patterns in the Invaded Range

A. kollari populations in the invaded range show lower genetic diversity than endemic areas (table 3, fig. 3). However, the drop in diversity with distance from the site showing greatest richness (Mátrafüred, Hungary) is neither as marked nor as linear as the trend seen in *A. quercuscalicis* (compare fig. 1 with fig. 5). At least two factors contribute to this difference. First, unlike *A. quercuscalicis*, *A. kollari* has two centers of genetic diversity, and genetically diverse Spanish sites are some of those most distant from centers of diversity in the east. Second, even if Spanish sites are excluded from consideration, sites in Britain and France have a relatively higher proportion of the native range diversity than is seen in *A. quercuscalicis*. In *A. quercuscalicis*, no British population retains more than 63 percent of the mean heterozygosity and 64 percent of the alleles per locus recorded in Hungary. In *A. kollari*, however, populations near the north-western limit of the invaded range (such as Dunrobin, in northern Scotland) retain up to 77 percent of the mean heterozygosity per locus and 88 percent of the alleles per locus seen at the richest sites in Hungary and Italy.

Heterozygosity and numbers of alleles per locus are generally similar across much of northern Europe when individual sites are compared (table 3, fig. 5). Considered at a regional level, however, Britain has a greater diversity of alleles (29) than sites in Ireland (22), the Low Countries (23), and northern France (24). There are also regional differences in which alleles are present at a locus. British populations possess six alleles absent from the Low Countries and five absent from northern France, while both of these continental regions possess a single allele absent from Britain. These differences between populations either side of the channel could be artifacts associated with relatively small sample sizes. Alternatively, the higher allelic diversity and numbers of private alleles found in Britain may be the persistent genetic consequences of large scale introductions from the eastern Mediterranean. Were this to be the case, we would expect British populations to be more genetically similar to eastern native populations than to those in Spain. This is indeed the case; all 29 of the alleles found in Britain are also found in Hungary and Italy, while two of the alleles found in Britain are absent from Spain.

The high richness of alleles in Britain has a significant impact on the position of British sites in the genetic distance phenogram for *A. kollari* (fig. 4). Dunrobin, despite its position near the limit of the invaded range, has a high number of alleles and clusters close to the root of the tree, while British sites with lower diversity (such as Oxford) cluster with sites in north-western continental Europe and are further from the root. This pattern is in marked contrast to *A. quercuscalicis* (fig. 2), in which the British populations all show similarly low genetic diversity, and form a clear cluster distant from the root (fig. 2).

DISCUSSION

Genetic Diversity of Invading Species in Their Novel Range

All three of the gall wasp species discussed here (*A. kollari*, *A. lignicola*, and *A. quercuscalicis*) show lower genetic diversity in their invaded range than in areas where each species is native. This is a common feature of colonizing species in a wide range of taxa, and has been demonstrated in insects both following natural range expansion (Seitz and Komma 1984, Eber and Brandl 1997) and human introduction (Parker et al. 1977, Bryant et al. 1981, Berlocher 1984).

It is notable that despite lower sample sizes, *A. kollari* and *A. lignicola* both have higher genetic diversity in Britain than has been demonstrated for *A. quercuscalicis*. Further sampling in the invaded range will probably enlarge this difference, as rare alleles in the two less-sampled species are detected.

High diversity in British *A. lignicola* may be a direct result of higher native range diversity in this than in any other of the host alternators (fig. 3). Random subsampling during each colonization step, as proposed for *A. quercuscalicis* (Stone and Sunnucks 1993, Sunnucks and Stone 1996), would then yield higher genetic diversity at the margin of the invaded range. Further sampling of *A. lignicola* through its invaded range is necessary before this argument can reasonably be applied to this species.

Despite large introductions of *A. kollari* to Britain in the 1840's, levels of genetic diversity in British populations are comparable to those in the naturally invading species (fig. 3). British populations of *A. kollari* possess, however, a number of alleles shared with Balkan populations which are absent from populations in northern France, Belgium, and Holland. One interpretation of this pattern is that alleles introduced into Britain in the 1840's have persisted there, without crossing the Channel to continental populations. The ability of this

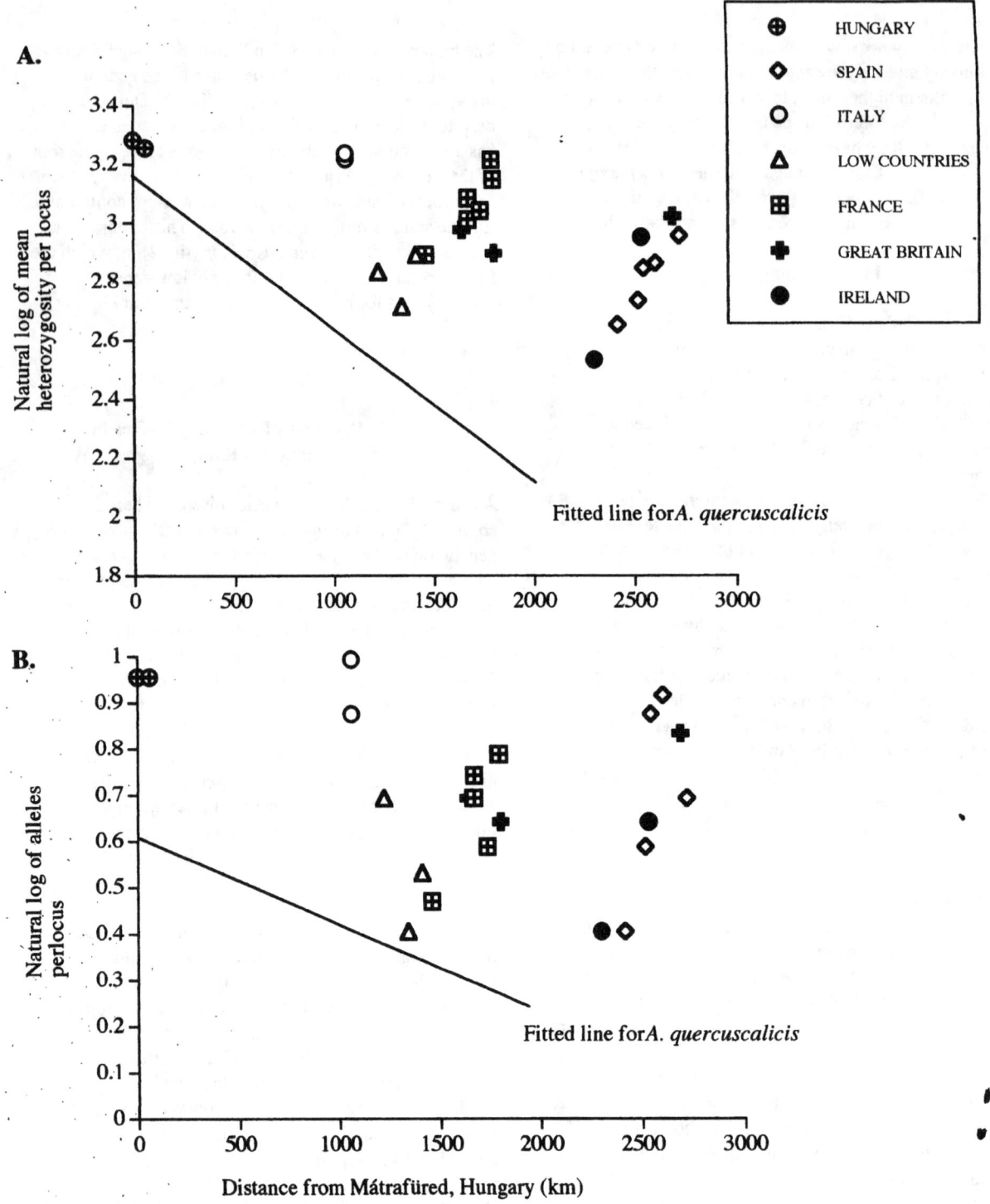

Figure 5.—*Changes in levels of population genetic variation with distance from Mátrafüred, Hungary (the site with the greatest number of alleles) in* Andricus kollari *(a) Mean heterozygosity per locus (H_{HW}), and (b) Mean number of alleles per locus. It should be remembered that this species has both eastern and western native populations, and the data are plotted in this way only to allow comparison with patterns in* A. quercuscalicis.

relatively narrow water barrier to limit migration is supported by the presence of alleles in populations of *A. kollari* on the continental side of the Channel which are absent from southern England. Further sampling on both sides of the Channel is required to remove the possibility that the observed patterns are artifacts resulting from relatively small sample sizes.

Genetic Differentiation Between Eastern and Western Native Range Populations of *A. Kollari*

The presence of alleles limited to either Spanish or Hungarian/Italian populations suggests (a) that there is little gene flow between these two limits of the European distribution of *A. kollari*. and (b) that neither region is obviously derived genetically from the other. One explanation for this pattern is that the two populations are derived from a shared common ancestor, but have been separated for long enough for regionally characteristic alleles to evolve. The eastern and western limits of the distribution of *A. kollari* in Europe correspond to two glacial refugia for a wide diversity of taxa during the last ice ages (Hewitt 1996). One possible cause of the observed genetic differences is that southward advance of the ice left two separated populations of *A. kollari* which diverged genetically while gene flow through intermediate areas was prevented. Such a separation has also been associated with regional differentiation in host oak associations for both sexual and asexual generations of *A. kollari* (Sternlicht 1968, Ambrus 1974, Nieves-Aldrey 1987, Csóka 1997). If utilization of a particular set of host oaks has become obligate for particular gene pools of *A. kollari*, then the observed genetic discontinuity at the Pyrenees may be due to limits on oak distributions rather than any inability of *A. kollari* to cross these mountains. Experimental work on the host-specificity of regional populations of *A. kollari* is required to assess this possibility.

ACKNOWLEDGEMENTS

We would like to thank all those who have helped us to collect samples for this work, including José-Luis Nieves-Aldrey, Juli Pujade-i-Villar, George Bableves Melika, Csaba Thúroczy, many members of the Hungarian Department of Forestry, Pat Walker, and especially Csóka Anikó, Csóka Ági, and Csóka Bence. Allozyme work by GNS was supported by the Royal Society, the British Ecological Society, and the EP Abraham Trust. RA is supported by a BBSRC studentship award.

LITERATURE CITED

Ambrus, B. 1974. Cynipid galls. Fauna Hungariae. S. Gusztáv, ed. 12, part 1/a. Budapest, Hungary: Academic Press.

Askew, R.R. 1984. The biology of gallwasps. In: Ananthakrishnan, T.N., ed. The biology of galling insects. New Delhi: Oxford and IBH Publishing Co.: 223-271.

Barton, N.H.; Charlesworth, B. 1984. Genetic revolutions, founder effects and speciation. Annual Review of Ecology and Systematics. 5: 133-164.

Berlocher, S.H. 1984. Genetic changes coinciding with the colonization of California by the walnut huskfly *Rhagoletis completa*. Evolution. 38: 906-918.

Bryant, E.H.; van Dijk, H.; van Delden, W. 1981. Genetic bottleneck in *Musca autumnalis*. Evolution. 35: 872-881.

Collins, M.; Crawley, M.J.; McGavin, G.C. 1983. Survivorship of the sexual and agamic generations of Andricus quercuscalicis on *Quercus cerris* and *Quercus robur*. Ecological Entomology. 8: 133-138.

Cornell, H.V.; Hawkins, B.A. 1993. Accumulation of native parasitoid species on introduced herbivores: a comparison of "hosts as natives" and "hosts as invaders". American Naturalist. 141: 847-865.

Csóka, G. 1997. Plant galls. Budapest, Hungary: Agroinform Publishing.

Eber, S.; Brandl, R. 1997. Genetic differentiation of the tephritid fly *Urophora cardui* in Europe as evidence for its biogeographical history. Molecular Ecology. 6: 651-660.

Florence, L.Z.; Johnson, P.C.; Coster, J.E. 1982. Behavioural and genetic diversity during dispersal: analysis of a polymorphic esterase locus in southern pine beetle, *Dendroctonus frontalis*. Environmental Entomology. 1: 1014-1018.

Godfray, H.J.C.; Agassiz, D.J.L.; Nash, D.R.; Lawton, J.H. 1995. The recruitment of parasitoid species to two invading herbivores. Journal of Animal Ecology. 64: 393-402.

Gu, H.; Danthanarayana, W. 1992. Quantitative genetic analysis of dispersal in (*Epiphyas postvittana*) 1. Genetic variation in flight capacity. Heredity. 68: 53-60.

Hails, R.S. 1989. Host size and sex allocation of parasitoids in a gall forming community. Oecologia. 81: 28-32.

Hails, R.S.; Crawley, M.J. 1991. The population dynamics of an alien insect: *Andricus quercuscalicis* (Hymenoptera: Cynipidae). Journal of Animal Ecology. 60: 545-562.

Hewitt, G.M. 1996. Some genetic consequences of ice ages, and their role in divergence and speciation. Biological Journal of the Linnean Society. 58: 247-276.

Houard, C. 1912. The animal galls of northern Africa. Annales de la Societe Entomologique de France. 81: 1-45.

Jalas, J.; Suominen, J. 1987. Atlas Florae Europeae. Distribution of vascular plants in Europe. vol. 2. Cambridge, MA: Cambridge University Press.

Nei, M.; Roychoudhury, A.K. 1974. Sampling variances of heterozygosity and genetic distance. Genetics. 6: 379-390.

Nieves-Aldrey, J.L. 1987. Estado actual del conocimiento de la subfamilia Cynipidae (Hymenoptera, Parasitica, Cynipidae) en la Península Ibérica. Eos. 63: 179-195.

Parker, E.D.; Selander, R.K.; Hudson, R.O.; Lester, L.J. 1977. Genetic diversity in colonizing parthenogenetic cockroaches. Evolution. 1: 836-842.

Plantard, O.; Solignac, M. 1998. *Wolbachia*-induced thelytoky in cynipids. In: Csóka, Gyuri; Mattson, William J.; Stone, Graham N.; Price, Peter W., eds. The biology of gall-inducing arthropods: proceedings of the international symposium; 1997 August 14-19; Mátraküred, Hungary. Gen. Tech. Rep. NC-199. St. Paul, MN: U.S. Department of Agriculture, Forest Service, North Central Research Station: 111-121.

Richardson, B.J.; Baverstock, P.R.; Adams, M. 1986. Allozyme electrophoresis: a handbook for animal systematics and population studies. New York: Academic Press.

Schönrogge, K.; Stone, G.N.; Crawley, M.J. 1994a. Spatial and temporal variation in guild structure: parasitoids and inquilines of *Andricus quercuscalicis* (Hymenoptera: Cynipidae) in its native and alien ranges. Oikos. 72: 51-60.

Schönrogge, K.; Walker, P.; Crawley, M.J. 1994b. The distribution and abundance of alien, host-alternating *Andricus* species (Hymenoptera: Cynipidae) on oaks in Ireland. Proceedings of the Royal Irish Academy. 94: 265-274.

Schönrogge, K.; Stone, G.N.; Crawley, M.J. 1996a. Alien herbivores and native parasitoids: rapid development of guild structure in an invading gall wasp, *Andricus quercuscalicis* (Hymenoptera: Cynipidae). Ecological Entomology. 21: 71-80.

Schönrogge, K.; Stone, G.N.; Crawley, M.J. 1996b. Abundance patterns and species richness of the parasitoids and inquilines of the alien gall former *Andricus quercuscalicis* Burgsdorf (Hymenoptera: Cynipidae). Oikos. 77: 507-518.

Seitz, A.; Komma, M. 1984. Genetic polymorphism and its ecological background in tephritid populations. In: Woehrmann, K.; Loeschke, V., eds. Population biology and evolution. Heidelberg: Springer-Verlag: 143-158.

Sternlicht, M. 1968. The oak galls of Israel (*Quercus calliprinos* and *Quercus ithaburensis*). Contribution of the Volcani Institute of Agricultural Research Series 1198 E.

Stone G.N.; Sunnucks, P.J. 1993. The population genetics of an invasion through a patchy environment: the cynipid gallwasp *Andricus quercuscalicis*. Molecular Ecology. 2: 251-268.

Stone, G.N.; Schönrogge, K.; Crawley, M.J.; Fraser, S. 1995. Geographic variation in the parasitoid community associated with an invading gallwasp, *Andricus quercuscalicis* (Hymenoptera: Cynipidae). Oecologia. 104: 207-217.

Sunnucks, P.; Stone, G.N.; Schönrogge, K. 1994. The biogeography and population genetics of the invading gallwasp *Andricus quercuscalicis*. In: Williams, M.; Leach, C., eds. Plant galls: organisms, interactions, populations. Systematics Association Special Volume 49. Oxford: Clarendon Press: 351-368.

Sunnucks, P.J.; Stone, G.N. 1996. Genetic structure of invading insects and the case of the knopper gallwasp. In: Floyd, R.B.; Sheppard, A.W.; De Barro, P.J., eds. Frontiers of population ecology. Collingwood, Australia: CSIRO Publishing: 485-496.

Swofford, D.L.; Selander, R.B. 1981. BIOSYS 1: a FORTRAN program for the comprehensive analysis of data in population genetics and systematics. Journal of Heredity. 72: 281-283.

Williamson, M. 1996. Biological invasions. London: Chapman and Hall.

TESTS OF HYPOTHESES REGARDING HYBRID RESISTANCE TO INSECTS IN THE *QUERCUS COCCOLOBIFOLIA* X *Q. VIMINEA* SPECIES COMPLEX

William J. Boecklen and Richard Spellenberg

Department of Biology, New Mexico State University, Las Cruces, New Mexico 88003 USA

Abstract.—We examine patterns of gall-former and leaf-miner densities in the *Quercus coccolobifolia* x *Q. viminea* species complex to test several recent theories regarding the relative susceptibility of hybrid and parental host plants. Three species of leaf-miners and four species of gall-formers exhibited variation in their response to hybrid hosts. An hypothesis of hybrid intermediacy was strongly supported by four species. Densities of the cynipid, *Neuroterus* sp., supported an hypothesis of hybrid susceptibility, while those of the leaf-miner, *Nepticula* sp., supported the dominance hypothesis. The null hypothesis of no differences between host taxa was supported by one species; no species supported an hypothesis of hybrid resistance. At the community level, densities of gall-formers were most consistent with the dominance hypothesis, while those of leaf-miners gave strong support to an hypothesis of hybrid intermediacy and partial support to the dominance hypothesis.

INTRODUCTION

The study of hybrid zones recently has grown from a backwater of systematics to one of the most dynamic areas in evolutionary biology and ecology. Hybrid zones are centers of genetic variation (Endler 1977, Harrison 1990), and so provide an ideal arena in which to investigate evolutionary mechanisms thought to underlie population dynamics and species interactions. For this reason, ecologists have begun to study hybrid zones, motivated by interests in the effects of host hybridization on the ecology and evolutionary dynamics of associated organisms (Boecklen and Larson 1994, Boecklen and Spellenberg 1990, Dupont and Crivelli 1988, Floate *et al.* 1993, Fritz *et al.* 1994, Gaylord *et al.* 1996, LeBrun *et al.* 1992, Preszler and Boecklen 1994, Whitham 1989) and in the potential role of hybrid zones as centers of biodiversity (Martinsen and Whitham 1994, Whitham *et al.* 1994, Whitham and Maschinski 1996).

Ecologists interested in plant-herbivore interactions, in particular, have exhibited much recent interest in hybrid zones (reviewed by Strauss 1994). A small collection of studies have investigated the response of free-feeding and endophagous insects, epiphytic and endophytic fungi, as well as other types of herbivores to host plant hybridization. Interestingly, the response of herbivores to host plant hybridization is variable. When compared to parental hosts, hybrid hosts may support higher densities of herbivores (Boecklen and Larson 1994; Drake 1981; Floate *et al.* 1993; Fritz *et al.* 1994; Paige *et al.* 1990; Whitham 1989; Whitham *et al.* 1991, 1994), lower densities (Boecklen and Larson 1994, Boecklen and Spellenberg 1990), intermediate densities (Aguilar and Boecklen 1991, Drake 1981, Fritz *et al.* 1994, Gange 1995, Gaylord *et al.* 1996, Graham *et al.* 1995, Hanhimaki *et al.* 1994, Manley and Fowler 1969, Preszler and Boecklen 1994, Wu *et al.* 1996) or statistically equivalent densities (Boecklen and Larson 1994, Boecklen and Spellenberg 1990, Fritz *et al.* 1994).

The central question facing plant-herbivore ecologists working in host hybrid zones is determining which pattern of herbivore loads in host hybrid zones is the most prevalent in nature. The resolution of this question will require not only a larger compendium of cases studies, but may require a revision of the theoretical basis underlying studies of plant-herbivore interactions in host-plant hybrid zones. Variation in the responses of herbivores to hybrid hosts has led to a proliferation of hypotheses describing the relative susceptibility of parental and hybrid hosts to herbivores (table 1). Several of these hypotheses have been generated on an *ad hoc* basis, resulting in a set of hypotheses that are not

Table 1.— *Hypotheses and predicted patterns regarding herbivore loads in host hybrid zones.*

Hypothesis	Predicted Pattern	Source
Null	No differences among host taxa	Boecklen and Spellenberg (1990)
Hybrid Susceptibility	Loads greatest on hybrids	Boecklen and Spellenberg (1990)
		Fritz *et al.* (1994)
Hybrids-as-Sinks	Loads greater on hybrids	Whitham (1989)
Phenotypic Affinity	Loads greatest on hybrids most similar to parental with highest load	Morrow *et al.* (1994)
Hybrid Intermediacy	Loads intermediate on hybrids	Boecklen and Spellenberg (1990)
Additive	Loads intermediate on hybrids	Fritz *et al.* (1994)
Hybrid Resistance	Loads lowest on hybrids	Boecklen and Spellenberg (1990)
		Fritz *et al.* (1994)
Dominance	Loads greatest or lowest on one parent; same on all other host taxa	Fritz *et al.* (1994)
Threshold	Loads lowest on one parent; same on all other host taxa	Moorehead *et al.* (1993)

mutually exclusive. In fact, several of the hypotheses represent restatements of previous ones or special cases of others. For example, the Phenotypic Affinity Hypothesis (Morrow *et al.* 1994) can be viewed as a special case of the Hybrids-As-Sinks Hypothesis (Whitham 1989), and the Threshold Hypothesis (Moorehead *et al.* 1993) as a subset of the Dominance Hypothesis (Fritz *et al.* 1994). As it now stands, a given pattern of herbivore load can support several of the hypotheses simultaneously. Arguably, hypothesis generation has far outpaced our empirical base and has resulted in a theoretical system that limits broad-scale comparisons between studies.

Another potential problem with the current theoretical system is its reliance on discrete categories of host taxa and the assumption that host individuals can be identified as parental taxa or assigned to a number of finely divided backcross categories (hereafter, BC-1, BC-2, etc.) unambiguously (fig. 1). Usually, classification of host individuals is based on host morphology (e.g., Aguilar and Boecklen 1991, Boecklen and Larson 1994, Boecklen and Spellenberg 1990, Manley and Fowler 1969, Morrow *et al.* 1994, Whitham *et al.* 1994) or, increasingly, on the basis of molecular genetic data (e.g., Fritz *et al.* 1994, Howard *et al.* 1997, Paige *et al.* 1990, Paige and Capman 1993). It is generally accepted that morphological characters are too variable to reliably distinguish among parents and advanced backcross categories (but see Floate *et al.* 1992, 1994). Without accurate classification, hypotheses predicated on the responses of herbivores to finely divided taxonomic categories are not operational (sensu Brady 1979) and are, therefore, unfalsifiable in natural systems.

Even when molecular genetic data are used, classification errors are likely. Boecklen and Howard (1997) developed statistical models that relate the probability of

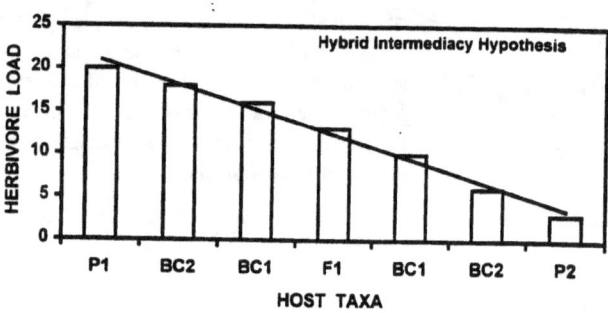
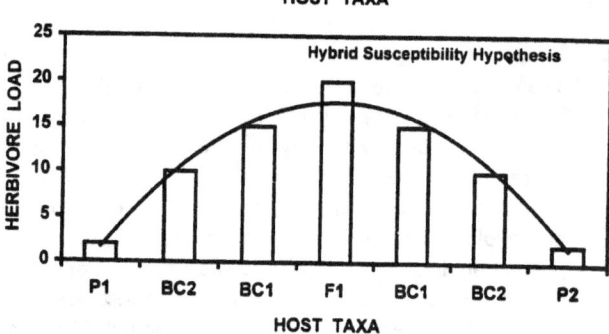
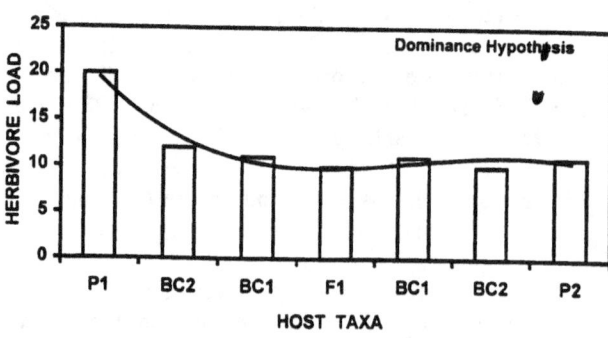

Figure 1.–*The Hybrid Intermediacy, Hybrid Susceptibility, and Dominance Hypotheses based on finely-divided taxonomic categories of host plants and approximations of these hypotheses using linear, quadratic, and cubic relationships, respectively.*

classification errors for various backcross categories to the number of diagnostic molecular markers examined. In general, large numbers of independent, diagnostic molecular markers are required to provide a reliable classification. For example, the probability that a BC-3 individual will be classified as a pure species is approximately 0.20 when 12 markers are examined, and does not drop below 0.10 until approximately 20 markers are examined (fig. 2). Error probability do not drop below 0.10 for BC-4 individuals until 30 markers are examined, and not until more than 70 markers are examined for BC-5 individuals. Boecklen and Howard (1997) conclude that the number of markers required to distinguish advanced backcrosses (BC-4 and BC-5) from pure species with low error is simply too large to be practical for most systems. In addition, the assignment of individuals to precise backcross categories is not practical, not only on the basis of the numbers of markers required, but also on the basis of heredity. Potential BC-1 genotypes represent all possible genotypes between F1 and pure species inclusive. This is also true of potential BC-2 genotypes, potential BC-3 genotypes, etc. Thus, the evolutionary history of an individual cannot be decided unambiguously on the basis of genotype alone.

An alternative to the use of discrete, taxonomic host-plant categories is to ordinate host individuals along a genetic or morphological continuum (see Aguilar and Boecklen 1991, Boecklen and Larson 1994, Christensen et al. 1995, Gaylord et al. 1996, Manley and Fowler 1969). The response of herbivores to host genotype can then be assessed along this continuum. Linear and non-linear models can be constructed (and tested) that capture the essential biological features of hypotheses predicated on discrete categories. For example, the Hybrid Intermediacy Hypothesis can be represented by a linear relationship, the Hybrid Susceptibility Hypothesis by a quadratic, and the Dominance Hypothesis by a third-order polynomial (fig. 2). In fact, all of the hypotheses presented in table 1 can be distinguished from each other by the order (linear, quadratic, etc.) of the best-fitting model and the signs of the regression coefficients. A major advantage of this approach is that the set of hypotheses can be evaluated in an hierarchical manner using well established statistical methods.

Here, we examine patterns of abundance for leaf-mining moths and gall-forming wasps (Hymenoptera: Cynipidae) on parental and hybrid hosts in the *Quercus coccolobifolia x Q. viminea* species complex. Specifically, we model the response of herbivore densities along a taxonomic continuum constructed on the basis of host plant morphology. We then construct linear, quadratic, and higher-order models in order to test the set of hypotheses described in table 1. We contribute to the compendium of case studies describing the responses of herbivores to host plant hybridization, and we present a preliminary analytical framework for testing hypotheses predicated on a taxonomic continuum.

METHODS

Study Sites and Organisms

We conducted this study in the Sierra Madre Ocidental, Chihuahua, Mexico. The region averages approximately 2000 meters elevation, with a few mountains and ridges reaching 3000 meters. The madrean vegetation represents a warm-temperate woodland (Brown et al. 1995) containing various species of conifers and evergreen dicot trees, including *Pinus*, *Juniperus*, *Cupressus*, *Quercus*, and *Arbutus*. The Sierra Madre Occidental is well suited to a broad-scale investigation of herbivory in

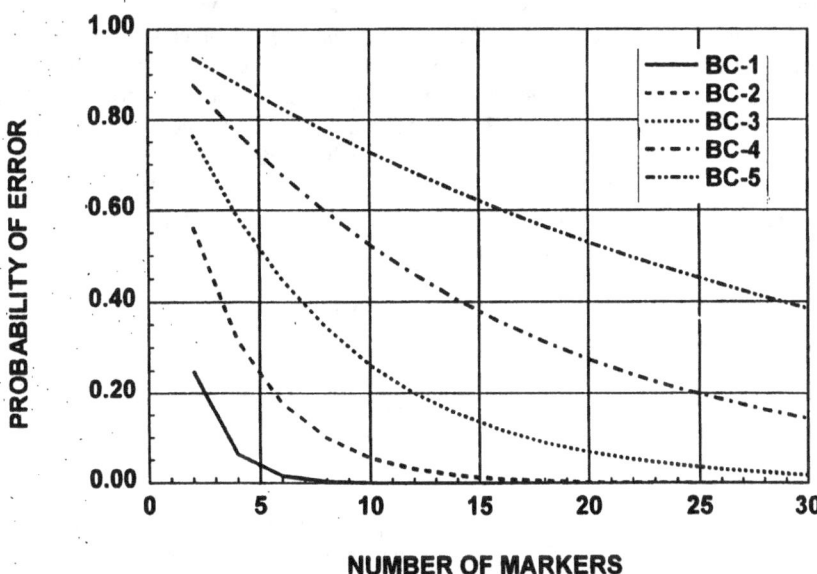

Figure 2.—Probability that an individual belonging to various backcross (BC) categories will be classified erroneously as a pure species as a function of the number of diagnostic genetic markers examined (after Boecklen and Howard 1997).

oak hybrid zones, as it contains approximately 50 species of oaks roughly divided between two sections commonly referred to as white (sect. Quercus) and black oaks (sect. Lobatae). Hybridization within a section is common, widely known and well documented (Trelease 1924; Muller 1952; Tucker 1959, 1961; Cottam et al. 1982; Spellenberg 1995). Oak species tend to be distributed in elevational bands in the Sierra. Canyons dissect the center of the Sierra, and the west slope contains precipitous cliffs and deep barrancas. Topography and micro- and macro-environmental gradients provide numerous situations where oak species are brought into contact and hybridization occurs.

The hybrid combination examined here is the *Quercus coccolobifolia x Q. viminea* species complex. Hybrid progeny in the complex have been described as *Q. knoblochii* (Muller 1942). The complex supports a diverse community of leaf-mining moths (Lepidoptera: Gracilariidae and Nepticulidae) and gall-forming wasps (Hymenoptera: Cynipidae). Owing to poor taxonomic coverage, most of the leaf-miners and cynipids cannot be identified at the species level. Fortunately, mine and gall morphologies are species-specific and quite distinct. We restrict our analysis to three most common species of leaf miners (*Nepticula* sp., *Phyllonorycter* sp., and *Tisheria* sp.) and to the three most common species of cynipids (*Andricus* sp., *Callirhytis* sp., and *Neuroterus* sp.). We also examine patterns of abundance for an unidentified homopteran galler (Homopteran sp.).

Quercus coccolobifola (Trel.) occurs throughout much of the mountainous regions of northwestern Mexico and is a conspicuous component of the pine-oak woodlands at moderate elevations. Trees are evergreen and are approximately 3-6 meters tall. Leaves are usually entire, 7-15 centimeters long, broadly ovate or orbicular, and often cordate at the base. The adaxial surface is moderately lustrous at maturity, while the abaxial surface is dull and is somewhat densely covered with golden glandular hairs.

Quercus viminea (Trel.) occurs primarily on the western slope of the Sierra Madre Occidental in arid open woodlands. Individuals are shrub-like and grow to approximately 3 meters. Leaves are lanceolate, 4-10 centimeters long, and glabrous. There are typically a few pronounced marginal teeth in the distal portion of leaves that bear long flexible seta at the tip.

Sampling Design and Statistical Analyses

We haphazardly sampled a total of 116 trees from five sites in west-central Chihuahua (see Appendix). The five sites consisted of two allopatric populations of *Quercus coccolobifolia*, an allopatric population of *Q. viminea*, and two hybrid zones that contained parental and hybrid trees. From each tree we haphazardly selected 10 leaves for morphological analysis. From each leaf, we measured area (AREA), perimeter (PER), blade length (BL), blade width (BW), distance from the tip to the widest point (TWD), petiole length (PL), blade width at a point 1/3 distal from the base (BWL), blade width at a point 1/3 proximal to the tip (BWU), number of major secondary veins on the right side of the leaf (VEINS), and number of teeth (TEETH). All variables were transformed to natural logarithms. In order to quantify variation between trees in leaf morphology and to create a hybrid index, we used canonical discriminant function analysis using individual trees as groups. We used scores along the first canonical axis to position individual trees along a morphological (taxonomic) continuum.

We censused 110 of the 117 trees used for the morphological analysis for herbivores. From each tree we haphazardly selected 300 leaves, on average, scored them for leaf-miners and gallers, and calculated densities (individuals/300 leaves) for the three species of leaf-miners and four species of gallers described above. To examine community-level responses of herbivores to hybrid hosts, we summed densities over all species of leaf-miners (Total Mines) and over all species of gall-formers (Total Galls).

We tested the set of hypotheses regarding hybrid resistance to herbivores as follows. We constructed linear, quadratic, and cubic regression models of herbivore densities against scores along the first canonical discriminant axis representing leaf morphology. We judged the lack of a significant relationship to be consistent with the Null Hypothesis. We considered a significant linear relationship (no significant higher-order terms) as support for the Hybrid Intermediacy Hypothesis. A significant quadratic relationship that was concave supported the Hybrid Resistance Hypothesis; whereas a convex quadratic relationship was consistent with the Hybrid Susceptibility Hypothesis. The Phenotypic Affinity Hypothesis was supported by a convex quadratic relationship with the maximum displaced from the average value along the discriminant axis. A significant cubic relationship was deemed consistent with the Dominance Hypothesis or Threshold Hypothesis.

RESULTS

Canonical discriminant function analyses identified two highly significant axes that collectively explained approximately 90 percent of the variation between trees in leaf morphology (table 2). The first axis explained roughly 84 percent of the variation and chiefly corresponded to increasing leaf width at a point 1/3 proximal to the tip (BWU), increasing leaf area (AREA), and

Table 2.—*Canonical discriminant function analysis of leaf morphology in the* Quercus coccolobifolia x Q. viminea *species complex. Values corresponding to morphological values are standardized discriminant function coefficients (CV-1 and CV-2).*

Variable	CV-1	CV-2
AREA	0.470	-0.355
PER	-0.137	4.302
BL	-1.515	-3.283
BW	0.468	0.942
PL	0.150	-0.161
TWD	-0.029	-0.075
BWL	0.168	-0.929
BWU	0.937	-0.805
VEINS	-0.013	-0.028
TEETH	-0.183	-0.705
Constant	-6.052	0.720
Eigenvalue	48.468	2.775
Percent variation	84.254	5.526

decreasing blade length (BL). The second discriminant axis primarily represented increasing leaf perimeter (PER) relative to decreasing blade length (BL). The morphological space contained three clusters of tree, representing *Quercus viminea, Q. coccolobifolia,* and hybrids (fig. 3). The first axis, in particular, served largely to ordinate trees from the lanceolate leaves of *Quercus viminea* to the broadly obovate leaves of *Q. coccolobifolia.*

We recorded a total of eight species of leaf-miners and nine species of gall-formers from the *Quercus viminea* x *Q. coccolobifolia* species complex. Herbivore abundances were generally low; leaf-miner densities averaged 5.38 mines per 300 leaves, while galler densities averaged 12.91 galls per 300 leaves. The homopteran galler was the most abundant species, averaging 3.86 galls per 300 leaves. *Neuroterus* sp. was the most abundant cynipid species and averaged 0.34 galls per 300 leaves. Average densities of *Andricus* sp. and *Callirhytis* sp. were 0.08 and 0.17 galls per 300 leaves, respectively. Of the leaf-miners, *Nepticula* sp. was the most common, averaging 2.77 miners per 300 leaves. *Tischeria* sp. was the least abundant leaf-miner with a mean density of 0.75 mines per 300 leaves. Densities of *Phyllonorycter* sp. averaged 1.07 mines per 300 leaves.

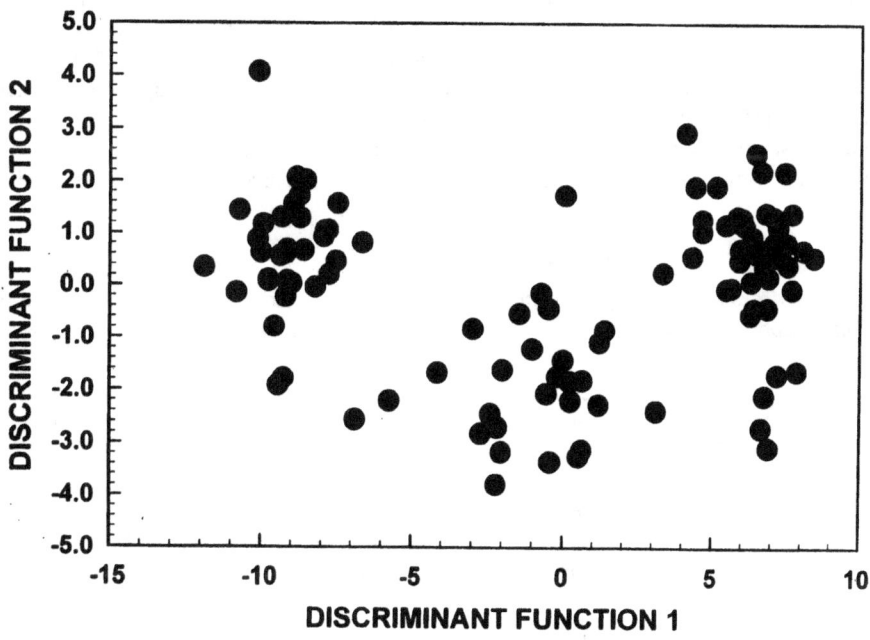

Figure 3.—*Canonical discriminant function analysis of between-tree variation in leaf morphology for host individuals in the* Quercus viminea x Q. coccolobifolia *species complex.*

All species except the cynipid, *Callirhytis* sp., exhibited statistically significant relationships between herbivore densities and scores along the morphological continuum defined by the first discriminant axis (table 3). In all, four hypotheses found full or partial support. Densities of *Callirhytis* sp. were most consistent with the Null Hypothesis. Four species strongly supported the Hybrid Intermediacy Hypothesis (*Phyllonorycter* sp., *Tischeria* sp., *Andricus* sp., and the homopteran galler), as these species exhibited significant linear relationships that were not significantly improved by including higher-order terms. Densities of the homopteran galler, *Phyllonorycter* sp., and *Andricus* sp. exhibited an increasing gradient from *Quercus viminea* to *Q. coccolobifolia*. In contrast, densities of *Tischeria* sp. exhibited the opposite gradient. The Hybrid Susceptibility Hypothesis was supported by *Neuroterus* sp. since the linear model was significantly improved by including a quadratic term that produced a convex relationship. Densities of the leaf-miner, *Nepticula* sp., were most consistent with the Dominance Hypothesis, as the inclusion of a cubic term made a significant contribution in explaining variation in mine density. In general, densities of *Nepticula* sp. were highest at the *Q. viminea* end of the continuum, but were roughly equivalent over the range of the continuum representing the hybrids and *Q. coccolobifolia*.

At the community level, densities of gall formers (Total Galls) supported the Dominance Hypothesis. Inclusion of a cubic term made a highly significant contribution in explaining variation in galler densities along the morphological continuum (table 3). In general, gall densities were highest at the *Quercus viminea* end of the continuum, but decreased rapidly and leveled-off over the range of the continuum representing the hybrids and *Q. coccolobifolia*. Densities of leaf-miners (Total Mines) were most consistent with the Hybrid Intermediacy Hypothesis, but also provided weak support for the Dominance Hypothesis, as a cubic term made a marginally significant ($p = 0.075$) contribution in explaining variation in miner density. Miner densities were highest at the *Q. viminea* end of the morphological continuum.

DISCUSSION

Patterns of herbivore loads in the *Quercus viminea* x *Q. coccolobifolia* species complex were most consistent with the Hybrid Intermediacy Hypothesis. Four of seven species examined supported the hypothesis, as did total leaf-miner densities. These results largely agree with those found in the *Q. grisea* x *Q. gambelii* species complex, where hybrid intermediacy was the most prevalent pattern for cynipids, leaf miners, and endophytic fungi (Aguilar and Boecklen 1992, Boecklen and Larson 1994, Preszler and Boecklen 1994, Gaylord et al. 1996). These results contrast with those found in the *Q. depressipes* x *Q. rugosa* and *Q. coccolobifolia* x *Q. emoryi* species complexes, where the dominant patterns was one of hybrid resistance (Boecklen and Spellenberg 1990). Interestingly, none of the species examined here supported the Hybrid Resistance Hypothesis.

It is clear that the development of a general and predictive theory of herbivory in hybrid zones awaits the resolution of two major questions. The first is "Which pattern of herbivore loads on host hybrids zones is the most prevalent in nature?" The second is, "What are the factors that determine which pattern is most likely to obtain in a given host hybrid zone?" The resolution of the first question requires a larger compendium of case studies than now exists. It also requires, as we have argued here, a change in the theoretical system underlying studies of herbivory in host-plant hybrid zones. We

Table 3.—*Response of herbivore densities along a morphological continuum created by canonical discriminant function analysis. Values represent significance levels associated with linear, quadratic, and cubic terms in regression models.*

Species	Linear	Quadratic	Cubic	Hypothesis Supported
Nepticula sp.	< 0.001	0.073	0.017	Dominance
Phyllonorycter sp.	< 0.001	0.062	0.686	Hybrid Intermediacy
Tischeria sp.	< 0.001	0.858	0.756	Hybrid Intermediacy
Total Mines	0.008	0.096	0.075	Hybrid Intermediacy Dominance
Andricus	0.017	0.339	0.512	Hybrid Intermediacy
Callirhytis sp.	0.092	0.617	0.611	Null
Neuroterus sp.	0.128	0.036	0.590	Hybrid Susceptibility
Homopteran Galler	< 0.001	0.612	0.788	Hybrid Intermediacy
Total Galls	0.051	0.857	0.009	Dominance

encourage hybrid zone workers to abandon the typological approach, namely the assignment of individuals to finely divided taxonomic categories, that has dominated the literature of late (but see Manley and Fowler 1969, Boecklen and Larson 1994, Christensen et al. 1995, Gaylord et al. 1996). The assignment of individuals to precise backcross categories is not practical and hypotheses based on such a classification may not be falsifiable in natural systems.

The resolution of the second question is a bit more challenging. First, a broad-scale investigation is required—one that considers hybrid combinations as the experimental units. Second, the set of potential factors is large. Boecklen and Spellenberg (1990) outlined several factors that may be important in determining patterns of herbivore loads in host hybrid zones. These included the structure (physical and genetic) of the hybrid zone itself, taxonomic relationships of the parental hosts, morphological relationships of the parental hosts, the relative importance of host defense versus herbivore ovipositional cues, and the autecologies of the herbivores. More recently, Preszler and Boecklen (1994) have demonstrated an important role for tri-trophic level interactions, while Gaylord et al. (1996) document important interactions involving host plants, endophytic fungi, herbivores, and natural enemies.

The problem of confounding factors probably is best avoided by using replicate herbivore assemblages within a set of closely related hosts that exhibit multiple and varied (with respect to a given factor under investigation) hybrid combinations. Such an experimental system would permit a rigorous and systematic examination of herbivory in hybrid zones without the confounding effects of comparing patterns derived from herbivores and hosts with widely disparate ecologies and phylogenetic histories. It is for exactly these reasons that we have been examining herbivore loads within oak syngameons of the southwestern United States and northern Mexico. In fact, oaks may represent the ideal experimental system in which to investigate patterns of herbivory in host hybrid zones. First, hybridization within a section is quite common, and well documented. Second, hybrid combinations within the syngameons exhibit variation in the structure (physical and genetic) of the hybrid zones, in the morphological and taxonomic relationships of the species involved, and in the number of species involved. This variation can be exploited by comparing patterns of herbivory in subsets of hybrid combinations that differ with respect to these factors. The large number of hybrid combinations ensures a rigorous test (via meta-analysis) of potential factors that may influence patterns of herbivory in hybrid zones. We hope that the results reported here contribute to this end.

ACKNOWLEDGMENTS

We thank R. Preszler for comments and suggestions that significantly improved the manuscript. C. Milner helped with the morphological analyses. C. Craddock, P. Gregory, and E. Ulasek helped with the field collections, and J. Thomason and A. Woolford helped with the herbivore censuses. This work was supported by NSF Grants DEB 9208109 and DEB 9632154.

LITERATURE CITED

Aguilar, J.M.; Boecklen, W.J. 1991. Patterns of herbivory in the *Quercus grisea* X *Quercus gambelii* species complex. Oikos. 64: 498-504.

Boecklen, W.J.; Howard, D.J. 1997. Genetic analysis of hybrid zones: numbers of markers and power of resolution. Ecology. (In press).

Boecklen, W.J.; Larson, K.C. 1994. Gall-forming wasps (Hymenoptera: Cynipidae) in an oak hybrid zone: testing hypotheses about hybrid susceptibility to herbivores. In: Price, P.W.; Mattson, W.J.; Baranchikov, Y.N., eds. The ecology and evolution of gall-forming insects. Gen. Tech. Rep. NC-174. Saint Paul, MN: U.S. Department of Agriculture, Forest Service, North Central Forest Experiment Station: 110-120.

Boecklen, W.J.; Spellenberg, R. 1990. Structure of herbivore communities in two oak (*Quercus* spp.) hybrid zones. Oecologia. 85: 92-100.

Brady, R.H. 1979. Natural selection and the criteria by which a theory is judged. Systematic Zoology. 28: 600-621.

Brown, D.E.; Reichenbacher, F.; Franson, S.E. 1995. A classification system and map of the biotic communities of North America. In: DeBano, L.F.; Ffolliet, P.F.; Ortega-Rubio, A.; Gottfried, G.J.; Hamre, R.H.; Edminster, C.B., tech. coords. Biodiversity and management of the madrean archipelago: the sky islands of southwestern United States and northwestern Mexico. Gen. Tech. Rep. RM-GTR-264. Fort Collins, CO: U.S. Department of Agriculture, Forest Service, Rocky Mountain Forest and Range Experiment Station: 109-125.

Cottam, W.P.; Tucker, J.M.; Santamour, F.S., Jr. 1982. Oak hybridization at the University of Utah. Publ. 1. Salt Lake City, UT: State Arboretum of Utah, University of Utah Press.

Christensen, K.M.; Whitham, T.G.; Keim, P. 1995. Herbivory and tree mortality across a pinyon pine hybrid zone. Oecologia. 101: 29-36.

Drake, D.W. 1981. Reproductive success of two *Eucalyptus* hybrid populations. II. Comparison of predispersal seed parameters. Australian Journal of Botany. 29: 37-48.

Dupont, F.; Crivelli, A.J. 1988. Do parasites confer a disadvantage to hybrids? Oecologia. 75: 587-592.

Endler, J.A. 1977. Geographic variation, speciation, and clines. Princeton, NJ: Princeton University Press.

Floate, K.D.; Kearsley, M.J.C.; Whitham, T.G. 1993. Elevated herbivory in plant hybrid zones: *Chrysomela confluens*, *Populus* and phenological sinks. Ecology. 74: 2056-2065.

Floate, K.D.; Whitham, T.G.; Keim, P. 1994. Morphological versus genetic markers in classifying hybrid plants. Evolution. 48: 929-930.

Floate, K.D.; Martinsen, G.D.; Whitham, T.G. 1997. Cottonwood hybrid zones as centers of abundance for gall aphids in western North America: importance of relative habitat size. Journal of Animal Ecology. 66: 179-188.

Fritz, R.S.; Nichols-Orians, C.M.; Brunsfeld, S.J. 1994. Interspecific hybridization of plants and resistance to herbivores: hypotheses, genetics, and variable responses in a diverse herbivore community. Oecologia. 97: 106-117.

Gange, A.C. 1995. Aphid performance in an alder (*Alnus*) hybrid zones. Ecology. 76: 2074-2083.

Gaylord, E.S.; Preszler, R.W.; Boecklen, W.J. 1996. Interactions between host plants, endophytic fungi, and a phytophagous insect in an oak (*Quercus grisea* x *Quercus gambelii*) hybrid zone. Oecologia. 105: 336-342.

Graham, J.H.; Freeman, D.C.; McArthur, E.D. 1995. Narrow hybrid zone between two subspecies of big sagebrush (*Artemisia tridentata*: Asteraceae). II. Selection gradients and hybrid fitness. American Journal of Botany. 82: 709-716.

HanhimNki, S.; Senn, J.; Haukioja, E. 1994. Performance of insect herbivores on hybridizing trees: the case of the subarctic birches. Journal of Animal Ecology. 63: 163-175.

Harrison, R.G. 1990. Hybrid zones: windows on the evolutionary process. Oxford Survey of Evolutionary Biology. 7: 69-128.

Howard, D.J.; Preszler, R.W.; Williams, J.; Fenchel, S.; Boecklen, W.J. 1997. How discrete are oak species? Insights from a hybrid zone between *Quercus grisea* and *Q. gambelii*. Evolution. 51: 747-755.

LeBrun, N.; Renaud, F.; Berrebi, P.; Lambert, A. 1992. Hybrid zones and host-parasite relationships: effects on the evolution of parasite specificity. Evolution. 46: 56-61.

Manley, A.M.; Fowler, D.P. 1969. Spruce budworm defoliation in relation to introgression in red and black spruce. Forest Science. 15: 365-366.

Martinsen, G.D.; Whitham, T.G. 1994. More birds nest in hybrid cottonwood trees. Wilson Bulletin. 106: 474-481.

Moorehead, J.R.; Taper, M.L.; Case, T.J. 1993. Utilization of hybrid oak hosts by a monophagous gall wasp: how little host character is sufficient? Oecologia. 95: 385-392.

Morrow, P.A.; Whitham, T.G.; Potts, B.M.; Ladiges, P.; Ashton, D.H.; Williams, J.B. 1994. Gall-forming insects concentrate on hybrid phenotypes of *Eucalyptus*. In: Price, P.W.; Mattson, W.J.; Baranchikov, Y.N., eds. The ecology and evolution of gall-forming insects. Gen. Tech. Rep. NC-174. Saint Paul, MN: U.S. Department of Agriculture, Forest Service, North Central Forest Experiment Station: 121-134.

Muller, C.H. 1942. Notes on the American flora, chiefly Mexican. American Midland Naturalist. 27: 470-490.

Muller, C.H. 1952. Ecological control of hybridization in *Quercus*: a factor in the mechanism of evolution. Evolution. 6: 147-161.

Paige, K.N.; Capman, W.C. 1993. The effects of host-plant genotype, hybridization, and environment on gall-aphid attack and survival in cottonwood: the importance of genetic studies and the utility of RFLP's. Evolution. 47: 36-45.

Paige, K.N.; Keim, P.; Whitham, T.G.; Lark, K.G. 1990. The use of restriction length polymorphisms to study the ecology and evolutionary biology of ant-aphid interactions. In: Campbell, R.K.; Eikenbary, R.D., eds. Mechanisms of aphid-plant genotype interactions. Amsterdam: Elsevier: 69-87.

Preszler, R.W.; Boecklen, W.J. 1994. A three-trophic-level analysis of the effects of plant hybridization on a leaf mining moth. Oecologia. 100: 66-73.

Strauss, S.Y. 1994. Level of herbivory and parasitism in host hybrid zones. Trends in Ecology and Evolution. 9: 209-214.

Spellenberg, R. 1995. On the hybrid nature of *Quercus basaseachicensis* (Fagaceae, sect. *Quercus*). Sida. 16(3): 427-437.

Trelease, W. 1924. The American oaks. Memoirs of the National Academy of Science (USA). 20: 1-255 (reprinted 1969 by Cramer; Stechert-Hafner Service Agency, New York).

Tucker, J.M. 1959. A review of hybridization in North American oaks. Proceedings of the 9th International Congress II: 404-405.

Tucker, J.M. 1961. Studies in the *Quercus undulata* species complex I: a preliminary statement. American Journal of Botany. 48: 202-208.

Whitham, T.G. 1989. Plant hybrid zones as sinks for pests. Science. 244: 1490-1493.

Whitham, T.G.; Maschinski, J. 1996. Current hybrid policy and the importance of hybrid plants in conservation. In: Maschinski, J.; Hammond, D.H.; Holter, L., eds. Southwestern rare and endangered plants: proceedings of the 2d conference; 1995 September 11-14; Flagstaff, AZ. RM-GTR-283. Fort Collins, CO: U.S. Department of Agriculture, Forest Service, Rocky Mountain Forest and Range Experiment Station: 103-112.

Whitham, T.G.; Morrow, P.A.; Potts, B.M. 1991. Conservation of hybrid plants. Science. 254: 779-780.

Whitham, T.G.; Morrow, P.A.; Potts, B.M. 1994. Plant hybrid zones as centers of biodiversity: the herbivore community of two endemic Tasmanian eucalypts. Oecologia. 97: 481-490.

Wu, H.X.; Yeng, C.C.; Muir, J.A. 1996. Effect of geographic variation and jack pine introgression on disease and insect resistance in lodgepole pine. Canadian Journal of Forest Research. 26: 711-726.

APPENDIX

Specimens of *Quercus* deposited at New Mexico State University, Universidad Nacional Autónoma de México, and elsewhere that document this study. All collections are from Chihuahua, México.

Q. coccolobifolia Trel. Mcpio. Guerrero, 13 km E of Tomochic on road to La Junta, 28°23'N, 107°47'W, 2120 m, 23 Sep 1991, *Spellenberg, Boecklen, Gregory 10905*; Mcpio. Ocampo, 6.4 km W of Ocampo at km 33, road to Moris, 3 km S of hwy, 28°11'N, 108°22'W, 2070-2200 m, 26 Sep 1991, *Spellenberg, Boecklen, Gregory 10908;* same locality, etc., ca 100 m to 1 km S of hwy, *10912*; 7 air km S of Huajumar, 28°07'N, 108°17'W, elev. 1820 m, 24 Oct 1993, *Spellenberg, Boecklen, Ulaszek, Craddock 11925.*

Q. coccolobifolia x Q. viminea. Mcpio. Ocampo, 6.4 km W of Ocampo at km 33, road to Moris, 28°11'N, 108°22'W, 2070-2200 m, 26 Sep 1991, *Spellenberg, Boecklen, Gregory 10914*; 7 air km S of Huajumar, 28°07'N, 108°17'W, elev. 1820 m, 24 Oct 1993, *Spellenberg, Boecklen, Ulaszek, Craddock 11923*

Q. viminea Trel. Mcpio. Ocampo, 6.4 km W of Ocampo at km 33, road to Moris, 28°11'N, 108°22'W, 2070-2200 m, 26 Sep 1991, *Spellenberg, Boecklen, Gregory 10913*; 13 km W of Ocampo at ca. KM 48, on road to Moris, *26 Sep 1991, Spellenberg, Boecklen, Gregory 10916*; 7 air km S of Huajumar, 28°07'N, 108°17'W, elev. 1820 m, 24 Oct 1993, *Spellenberg, Boecklen, Ulaszek, Craddock 11924.*

GENETIC AND ENVIRONMENTAL CONTRIBUTIONS TO VARIATION IN THE RESISTANCE OF *PICEA ABIES* TO THE GALL-FORMING ADELGID, *ADELGES ABIETIS* (HOMOPTERA: ADELGIDAE)

William J. Mattson[1], Jean Levieux[2], and Dominique Piou[3]

[1] North Central Forest Exper. Station, USDA Forest Service, Pesticide Res. Center, Michigan State University, East Lansing, Mi. 48824
[2] Department of Biology, University of Orleans, Orleans, France
[3] Barres National Arboretum, 45290 Nogent sur Vernisson, France

Abstract.—Norway spruce, *Picea abies*, varies widely among individuals in its susceptibility to the common gall-forming aphid, *Adelges abietis* (Homoptera: Adelgidae), an Old World species which now has a circumtemperate, circumboreal distribution. We estimated the genetic and environmental contributions to this variation in a polyclonal spruce plantation at the Barres National Aboretum in the southern Paris basin of France. Aphids infested 98 percent of the clones, but infestation levels were highly variable, ranging from an average of less than 1 to 95 galls/branch/ramet/clone. Neither altitude nor forest region of clonal origin made a significant contribution to variation in tree resistance. Instead, most of the variation was due to individual tree genetics because broad sense heritability of resistance against *A. abietis* was estimated to be 0.86. There was no consistent relationship between tree infestation by *A. abietis*, and another common, co-occurring adelgid, *A. laricis*.

INTRODUCTION

The spruce gall adelgid, *Adelges abietis*, is one of about 52 species of primitive, aphid-like insects (Homoptera: Adelgidae) which attack conifers in the family Pinaceae (Carter 1971, Ghosh 1983). *A. abietis* is probably the best known and most widely distributed of these species. It has a nearly circumboreal and circumtemperate distribution on several species of spruce, *Picea* spp., on which it completes its entire life cycle and forms characteristic pineapple-shaped galls (Rose and Lindquist 1985). North American spruces are recent hosts of *A. abietis* (since the 19th century); European and west Asian spruces, such as Norway spruce, *Picea abies*, are believed to be its ancestral hosts (Patch 1909).

As is typical for "intimate" plant-feeders such as aphids, scales, and gall-forming insects (Fritz and Price 1988, Mattson et al. 1988), there is abundant, unequivocal evidence showing that individual trees vary widely in their susceptibility to *A. abietis* (Ewert 1967, Thalenhorst 1972, Tjia and Houston 1975, Eidmann and Eriksson 1978, Rohfritsch 1981). That this variation is strongly affected by the genetics of the tree can be inferred from the fact that (a) progeny from different mother x father crosses exhibit different levels of infestation (Eidmann and Eriksson 1978), (b) the distribution of insect galls per tree in plantations is not random and is highly consistent from year to year (Ewert 1967, Tjia and Houston 1975, Eidmann and Eriksson 1978, Mattson et al. 1994), (c) putative resistant and susceptible trees differ in their reactions to aphid stylets inserted into their bud cortex tissues (Thalenhorst 1972; Rohfritsch 1981, 1988), and (d) and in their foliar phenolic profiles (Tjia and Houston 1975).

Because so very little is known about the genetic bases of woody plant resistance to insects (Fritz and Simms 1992), we proposed to quantify the extent to which such tree-to-tree variation in aphid infestations is due to genetic or environmental factors. To make such estimates we measured the broad sense heritability of tree resistance (H^2), i.e., the proportion of the total phenotypic variation (Vp) (where Vp = Vg + Ve) in a trait that can be attributed to total genetic variation (Vg) in the trait: H^2 = Vg/Vp (Falconer 1989). We hypothesized that broad sense heritability of resistance to *Adelges* spp. in Norway spruce is significantly greater than zero.

METHODS

To estimate the genetic and environmental components of variation in resistance, we used a 0.5-ha, 35 year-old (in 1991), polyclonal plantation of *Picea abies* at the Barres National Arboretum in the southern Paris basin, about 20 km west of Montargis, France. Plantation design consisted of 192 clones, each in a single, 5 or 10 tree plot, at 3 x 2 m spacing. For our purposes, this was not a perfect design because clones were not replicated accross the plantation and thus clone "effects" could be confounded with microenvironmental "effects" that might vary across the small plantation. Nevertheless, we felt that an investigation would be worthwhile because polyclonal plantations of spruce are rare, infested ones are even rarer, and the site was physically small and very uniform. To minimize distances and environmental differences between clones, we picked the 52 closest clones in the plantation meeting our selection criteria: (1) each clone must have at least four surviving ramets to allow a substantive measure of within clone variation, and (2) clones must represent a spectrum of altitudinal (400-1900 m) and/or geographic origins (Vosges, Jura, and Alps forests) in France.

Broad sense heritability of tree resistance can be estimated by substracting variation within clones from variation among clones to arrive at Vg, thereby allowing one to estimate the ratio Vg/Vp. In practice, this is done by using the variance components, so that:

$$H_c^2 = Sc^2/[Sc^2 + Se^2]$$

where Sc represents the variance component due to clones, and Se represents the variance component due to the error term of the ANOVA (Fritz and Price 1988).

Gall infestation per ramet (tree) was estimated as follows: We selected the first two whorls above ground (usually at 2m because the lower crown branches had been pruned) having nearly complete, or whole branches. In each of these whorls, we then selected the longest branch for counting all galls. Because galls of *Adelges laricis* were also present in small numbers on many of the clones, we also counted them to verify if there were any potential aphid x aphid interactions in tree resistance. These two galls are easy to distinguish because one is always spherical with short, vestigial needles emanating from its sides and at the terminus of a given year's shoot, and the other (*A. abietis*) is pineapple-shaped with long needles emanating from its sides, and always at the base of a given year's growth (Novak *et al.* 1976, Gaumont 1978, Rose and Lindquist 1985). Counts were recorded by branch, then, for each of the two gall aphids. Our gall counts represented a ramet's cumulative susceptibility over the life span of its two lower crown branches, i.e., during the past 10-13 years, because galls are persistent. Finally, we measured tree diameter as a potential covariate. Earlier observation of *A. abietis* infestations in the United States showed that infestations on susceptible trees generally increase with tree size because larger trees have more growing points for infestation (Mattson *et al.* 1994). This variate, if significant, would test for differences among clones after adjustment for differences in size. It also would remove variation among ramets within clones due to size caused by micro-environmental effects, and thereby lower the error variance.

We analyzed the data (mean no. galls/branch/tree) for each species of adelgid using a nested ANOVA, (ramets within clones, and clones within forest regions) with tree diameter as a covariate, after square root (x + 0.5) transformation. To test for differences among altitudes of origin, we plotted and regressed galls counts per ramet per clone against altitude of clonal origin. To test for interactions between the two aphid species, we plotted and regressed *A. laricis* gall counts per ramet against those of *A. abietis*, and also prepared 2x2 contingency tables showing the simultaneous presence or absence of the two gall aphids on each ramet (Pielou 1969). We estimated the variance components for heritability calculations using SAS Proc VarComp procedures, using a completely randomized ANOVA without covariates.

Finally, we plotted the spatial relationships between variously infested clones using Cartography II software for microcomputers to examine neighbor effects on resistance.

RESULTS AND DISCUSSION

Clonal Differences In Infestation

Adelges abietis

As expected, analysis of variance clearly showed highly significant differences (P < 0.001) in resistance among the various clones of Norway spruce (table 1). *Adelges abietis* infestation levels ranged from 0.13 to 95.2/branch/ramet/clone, the grand mean was 22.9 galls/branch. Only 1 of 52 clones was completely free of *A. abietis* galls, but six others averaged less than 1 gall/branch. In other words, about 13 percent were practically immune. The frequency distribution of clones by infestation classes weakly resembles a dome-shaped curve, but skewed right, implying that most trees (ca. 75 percent) have lightly-to-moderate susceptiblity, and the rest are split evenly between either negligible or very high susceptibility (fig. 1). The covariate, tree diameter,

Table 1.—*Analyses of variance testing the hypothesis that neither clones nor different populations of clonal origin (region) differ in their resistance to* Adelges abietis *and* Adelges laricis.

Source of variation	Degrees Frdm	Mean square	F ratio
Adelges abietis			
Regions	2	4.40	0.21
Clones	49	22.13	27.91**
Covariate (diam)	1	8.03	10.13**
Error (ramets)	155	0.79	
Adelges laricis			
Regions	2	1.84	1.85
Clones	49	1.03	9.25**
Covariate (diam)	1	0.29	2.62
Error (ramets)	155	0.11	

** $P < 0001$

Table 2.—*Contingency table breakdown of the number of sampled ramets having gall infestations of either* A. abietis, A. laricis, *or both.*

A. abietis present	A. laricis present		Grand totals
	yes	no	
yes	33	155	188
no	4	16	20
Grand totals	37	171	

was also significant, as expected; thereby confirming the pattern that larger trees tend to support more galls per branch. Variance due to diameter effect was twice the variance due to regions of forest origin.

To further demonstrate the strong between-clone effects relative to within-clone effects, while removing potentially confounding environmental effects, we arbitrarily formed 26 spatial "blocks" centered on 26 clones but also including 2 immediately adjacent clone, and compared differences in *A. abietis* infestations between (a) two members of each of the 26 clones (i.e., 1 vs 2, and 3 vs 4) and (b) these two members (1 & 3) and the means of the two adjacent clones. This gave a randomized block design with 2 treatments (differences within and differences among clones). The 26 clones and the two adjacent ones were treated as a block of homogeneity because members within the central clones were growing as close to the adjacent clone members as to one another. Analysis of variance showed that was a significant treatment effect: within-clone differences were on average less than half that of between-clone differences (9.86 vs. 22.14 galls/branch), and that "block" had a small but significant effect (table 2). There was no block x genotype interaction. At each locale, within-clone differences and between-clone differences consistently followed the overall average differences. We rejected, therefore, the null hypothesis that differences among ramets of the same clone are on average equal to differences between members of different clones when compared under nearly identical growing circumstances.

Adelges laricis

Just as for *A. abietis*, analysis of variance showed striking differences in susceptibility of clones to *A. laricis* (table 2). Infestation levels ranged from 0.13 to 15.6 galls/branch/ramet/clone, and the grand mean was 0.10/branch. Most of the 52 clones (34) had no *A. laricis* galls. Hence, the frequency distribution of infestation classes resembled a hyperbolic distribution, in striking contrast to that of *A. abietis* (fig. 1). The covariate, tree diameter, was not significant-because of the large number of uninfested trees.

Clonal Resistance and Place of Clonal Origin

Analysis of variance showed no apparent relationship between resistance of clones to either *A. abietis* or *A. laricis* and region of clonal origin (Alps, Jura, and Vosges Mountains) (table 1).

Likewise, plotting mean numbers of galls/ramet/clone against the altitudinal origin of the clonal material (fig. 2) revealed that altitude per se had little or no relationship with the resistance of the clones to either *A. abietis* or *A. laricis*.

Although larger tests need to be done, we suspect that the resistance of trees against these two adelgids is distributed within and among populations of spruce in France independent of geography and altitude. This may not necessarily be true in other countries, or for other species of adelgids (Loyttyniemi 1971, Canavera and DiGennaro 1979, Stephan 1987). For example, in North America, the general susceptibility of white spruce, *Picea glauca*, to *A. abietis* is apparently highly dependent on its population of origin. The least susceptible populations come from north central North America (e.g., northern Minnesota, Manitoba, w. Ontario) whereas the most susceptible come from eastern North America (i.e., east of 85° w. longitude) or from outlier populations in British Columbia and South Dakota (Canavera and DiGennaro 1979, Mattson et al. 1994).

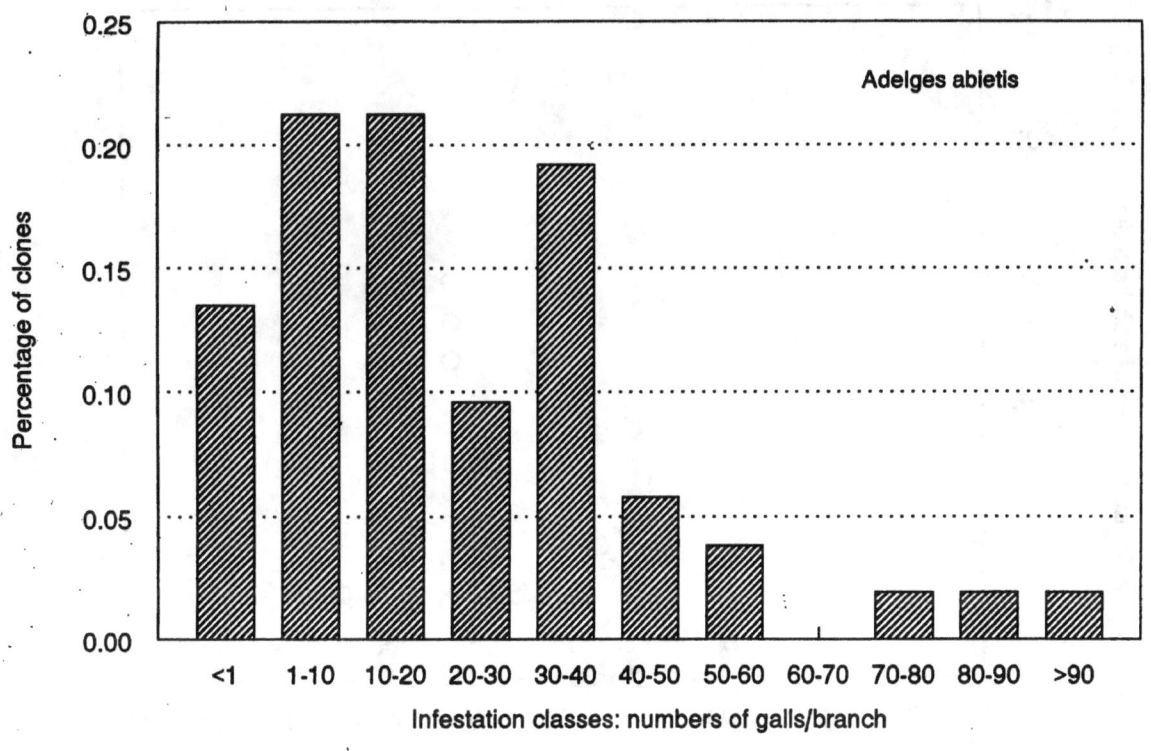

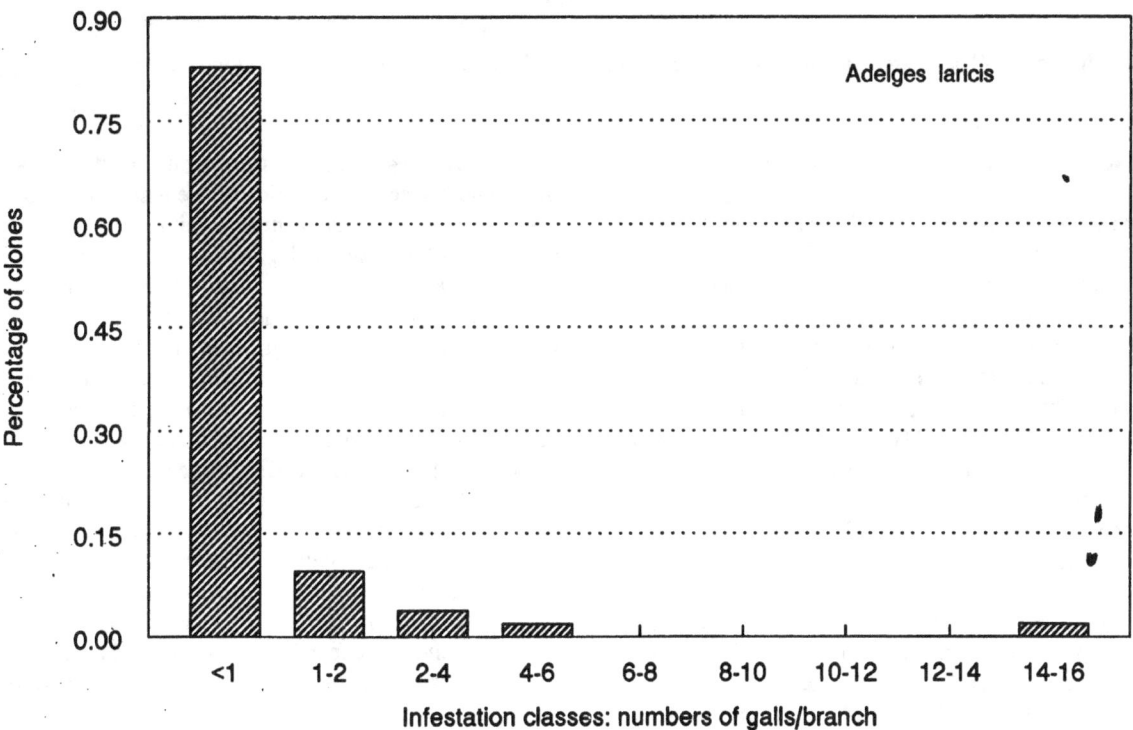

Figure 1.—*Frequency distributions (percent) of clones in different infestation classes for* A. abietis *(upper panel) and* A. laricis *(lower panel).*

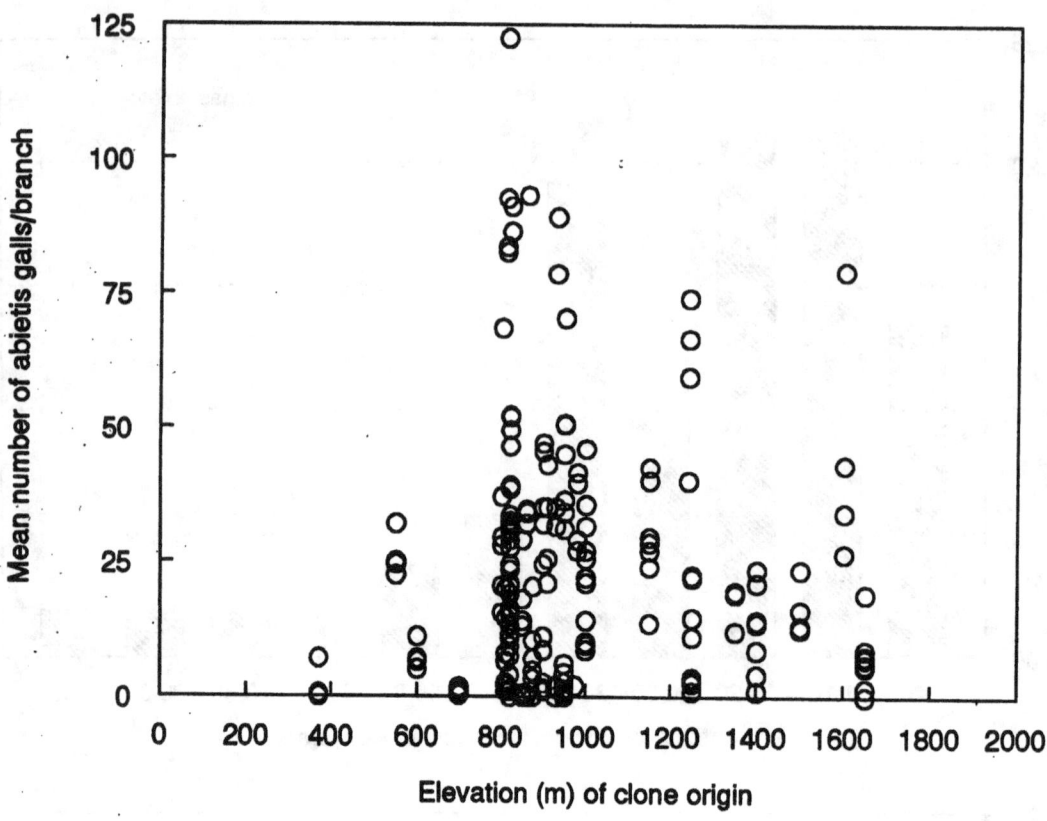

Figure 2.—*Plot of mean numbers of* A. abietis *galls/branch/ramet/clone vs altitude of clonal orgin.*

Spatial Distribution of Gall Damage in the Stand

If trees are truly resistant to gall formation, then having highly susceptible neighbors nearby should not change their susceptibility status. This is evident from the location of the resistant ("zero" class) trees in figs. 3a and 3b, for *A. abietis*, and *A. laricis*, respectively. It confirms the universal observations of putative resistant and susceptible trees growing intermingled (Ewert 1967, Thalenhorst 1972, Tjia and Houston 1975, Mattson *et al.* 1994). On the other hand, if many susceptible trees are growing close to one another, this spatial condition may enhance the probability of gall aphid population increase in such trees. For example, Eidmann and Eriksson (1978) have found that *A. abietis* gall numbers/tree increase linearly with the number of galls already present per susceptible host tree, just as Cranshaw (1989) found for *A. cooleyi* galls on *P. pungens*. Likewise, Butcher and Haynes (1960) reported that *A. abietis*-infested trees tend to be reinfested.

Relationship between *A. abietis* and *A. laricis*

Plotting mean numbers of *A. laricis* vs. *A. abietis* galls per ramet revealed no apparent, consistent relationship (fig. 4). One very interesting clone in the upper right corner of figure 4 was simultaneously the most susceptible to both species of adelgids. If we ignore this single case, the data imply that plant susceptibility to one species is not consistently linked with susceptibility to another. Likewise, contingency table analysis (table 3) showed no significant association between the two species. The same percentage of *abietis*-infested and non-*abietis*-infested trees were colonized by *A. laricis* (i.e., 18 vs 20 percent), and vice versa for *A. abietis* colonization of *laricis*-infested and non-*laricis*-infested trees (89 vs 90 percent). Trees may be highly susceptible to both adelgids, to neither adelgid, or to one but not the other as, Rohfritsch (1981, 1988) has already shown, based on the cellular evidence of how different plants react to the stylets of the two species of aphids. A plant resistant to *A. abietis* can be highly susceptible to *A. laricis* in spite of the fact that the two adelgids occupy almost the same plant tissue—their overwintering stages are separated by only a few mm. At least once, we have seen the same shoot supporting the growth of both types of adelgids—their galls fused together. However, the *A. laricis* data should be regarded as preliminary because its populations were small in our study area. Measuring the two aphids again in the near future could provide another more convincing test of their apparent lack of association.

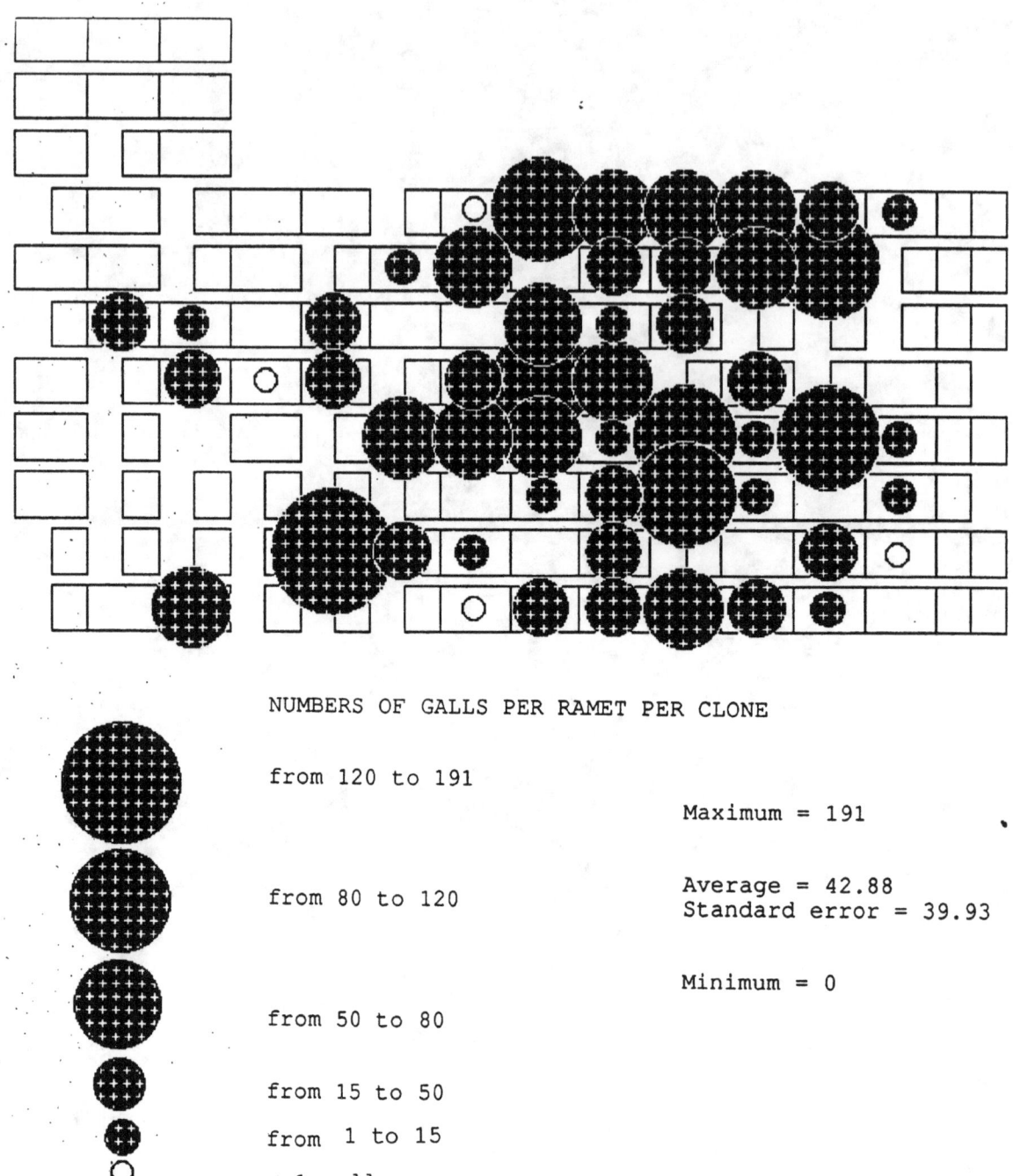

Figures 3a & 3b.—*Spatial relationships among all clones and those variously infested by* A. abietis *and* A. laricis. *Rectangles represent the 192 different clones, the size of each reflects the number of surviving ramets in each clone. Circles imposed upon certain rectangles reflect the mean infestation level/ramet (per 2 branches) of sampled clones, as indicated by the keys.*

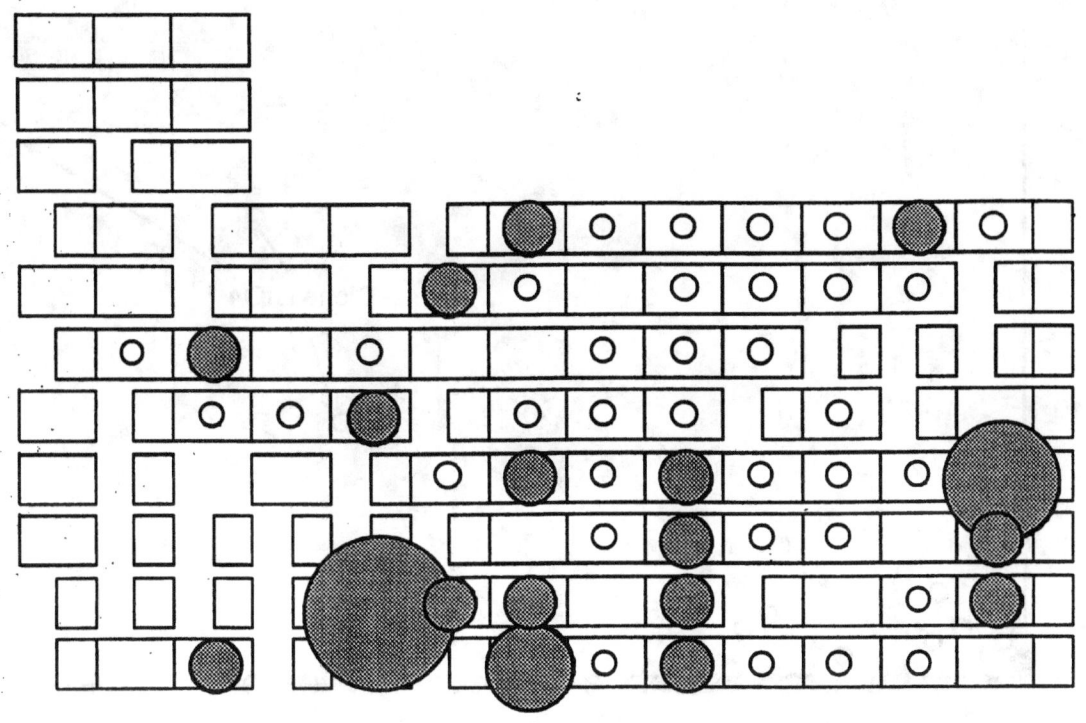

NUMBER OF GALLS PER RAMET PER CLONE

from 15 to 31   Maximum = 31

from 6 to 15   Average = 1.37
                     Standard error = 4.39

from 3 to 6   Minimum = 0

from 1 to 3

< 1 galls

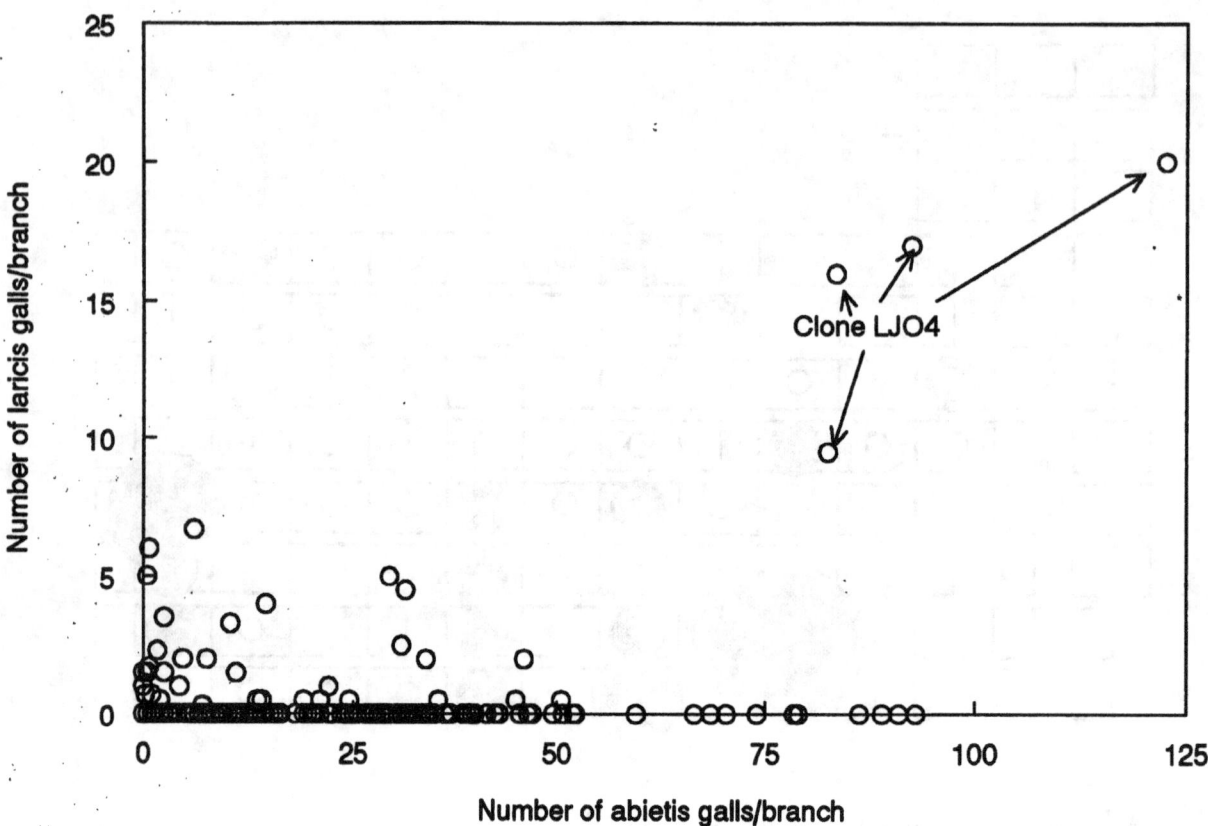

Figure 4.—*Plot of mean number of* A. laricis *galls per branch/ramet versus mean number of* A. abietis *galls per branch/ramet. All four ramets of clone LJ04 are identified because the clone was unique in having the highest populations of both species of adelgids.*

Table 3.—*Analysis of variance testing the hypothesis that within-clone and adjacent (same block) between-clone differences in gall infestations by* A. abietis *at 26 different blocks (locales within the plantation) are not different.*

Source of variation	Degrees Frdm	Mean square	F ratio
Blocks	25	5.82	1.90*
Within/Between clone differences	1	54.74	17.90**
B x WBCD	25	3.83	1.25
Error	52	3.06	

* $P < 0.02$
** $P < 0.001$

Estimating Heritability of Resistance

If we assume that clone effects as measured here are truly unbiased and minimially confounded with environmental effects, then we can make estimates of heritability in the broad sense, i.e., the proportion of individual phenotypic variation which is due to genetics.

To do this, we used a completely randomized ANOVA with clones being the only class variable and ramets nested therein. Broad sense heritability of resistance (H^2) against *A. abietis* and *A. laricis* is estimated at 0.86 (5.19/6.02) and 0.69 (0.25/0.36), respectively. Such high values are rather astonishing because they imply that less than 20-30 percent of the phenotypic variation in tree resistance to these aphids is due to environmental factors. That is, resistance/susceptibility is very strongly and tightly controlled by genetic mechanisms.

At this point we do not know whether the resistance mechanisms are oligogenic or polygenic in nature, nor whether they are primarily quantitatively or qualitatively expressed. The frequency distribution of infestation classes for *A. abietis* suggests that resistance is probably oligogenic, and of mixed inheritance (partly quantitative and partly qualitative) because of the relatively high frequency of "zero" classes (ca. 13 percent), and the clear deviation from the bell-shaped distribution that more often characterizes polygenic, quantitative inheritance. This pattern also agrees with our general observations on the distribution of the trait in other populations of Norway spruce at the INRA experimental forests (Forêt Domaniale d'Amance) near Nancy, France (unpubl. data) and in populations of *Picea glauca* in northern Great Lakes States, where about 30 and 20 percent, respectively, of the individuals were apparently immune to *A. abietis* (Mattson et al.1994). Our data for *A. laricis* are clearly too limited to make any general conclusions.

ACKNOWLEDGMENTS

This study was supported in part by grants from Institut National de la Recherche Agronomique, and USDA Office of International Cooperation and Development. We gratefully acknowledge the cooperation of Mr. Steinmetz who established and kindly permitted us to use the spruce clonal plantation reported on in this study. We thank Mr. B.A. Birr and Dr. R.K. Lawrence for their help with data analysis, and figure preparation. Likewise, we thank Drs. R.W. Fritz, R.K. Lawrence, and P.W. Price for reviewing an earlier draft of the manuscript.

LITERATURE CITED

Butcher, J.W.; Haynes, L.H. 1960. Chemical control of the eastern spruce gall aphid with observations on host preference and population increase. Journal of the Economic Entomological Society of America. 53: 970-982.

Canavera, D.S.; DiGennaro, J. 1979. Characteristics of eastern spruce gall aphid attack among 24 white spruce seed sources in central Maine. Proceedings of the Northeastern Forest Tree Improvement Conference. 26: 96-101.

Carter, C.I. 1971. Conifer woolly aphids (Adelgidae) in Britain. For. Commission Bull. no. 42. London. 51 p.

Cranshaw, W.S. 1989. Patterns of gall formation by the cooley spruce gall adelgid on Colorado blue spruce. Journal of Arboriculture. 15: 277-280.

Eidmann, H.H.; Eriksson, M. 1978. Unterschiede im befall der fichtengallenlaus, *Sacchiphantes abietis* L. an fichtenkreuzungen. Anzeiger fuer Schaedlingsk unDe. Pflanzenschutz Umweltschutz. 51: 177-183.

Ewert, J.-P. 1967. Untersuchungen uber die dispersion der fichtengallenlaus, *Sacchiphantes* (Chermes) *abietis* (L.) auf genwohnlichen kulturen, einzelstammabsaaten und klonen ihrer wirtspflanze. Journal of Applied Entomology. 59: 272-291.

Falconer, D.S. 1989. Introduction to quantitative genetics. New York: John Wiley. 438 p.

Fritz, R.S.; Price, P.W. 1988. Genetic variation among plants and insect community structure: willows and sawflies. Ecology. 69: 845-856.

Fritz, R.S.; Simms, E.L., eds. 1992. Plant resistance to herbivores and pathogens: ecology, evolution, and genetics. Chicago: University of Chicago Press. 590 p.

Gaumont, R. 1978. Tableaux pratiques de determination des principales formes de chermesides (Adelgides) de France. Revue Forestiere Francaise. 30(1): 21-36.

Ghosh, A.K. 1983. A review of the family Adelgidae from the Indian subregion (Homoptera: Aphidoidea). Oriental Insects. 17: 2-29.

Loyttyniemi, K. 1971. On the occurrence of *Physokermes* Targ. species (Hom. Lecaniidae) and *Sacchiphantes abietis* L. (Hom. Adelgidae) on various local races of *Picea abies* in Finland. Annales Entomologici Fennici. 37: 60-64.

Mattson, W.J.; Lawrence, R.A.; Haack, R.K.; Herms, D.A.; Charles, P.-J. 1988. Defensive strategies of woody plants against different insect feeding guilds in relation to plant ecological strategies and intimacy of association with insects. In: Mattson, W.J.; Levieux, J.; Bernard-Dagan, C., eds. Mechanisms of woody plant defenses against insects: search for pattern. New York: Springer-Verlag: 3-38.

Mattson, W.J.; Birr, B.A.; Lawrence, R.K. 1994. Variation in the susceptibility of North American white spruce populations to the gall-forming adelgid, *Adelges abietis* (Homoptera: Adelgidae). In: Price, P.W.; Mattson, W.J.; Baranchikov, Y.N., eds. The ecology and evolution of gall-forming insects. Gen. Tech. Rep. NC-174. St. Paul, MN: U.S. Department of Agriculture, Forest Service, North Central Forest Experiment Station: 135-147.

Novak, V.; Hrozinka, F.; Stary, B. 1976. Atlas of insects harmful to forest trees. Vol. I. New York: Elsevier Publ. Co. 125 p.

Patch, E.M. 1909. Chermes of Maine conifers. Maine Agricultural Experiment Station. Bulletin. 173: 277-308.

Pielou, E.C. 1969. An introduction to mathematical ecology. New York: John Wiley and Sons. 286 p.

Rohfritsch, O. 1981. A 'defense' mechanism of *Picea excelsa* L. against the gall former, *Chermes abietis* L. (Homoptera, Adelgidae). Zeitschrift fuer Angewandte Entomologie. 92: 18-26.

Rohfritsch, O. 1988. A resistance response of *Picea excelsa* to the aphid, *Adelges abietis* (Homoptera: Aphidoidea). In: Mattson, W.J.; Levieux, J.; Bernard-Dagan, C., eds. Mechanisms of woody plant defenses against insects: search for pattern. New York: Springer-Verlag: 253-266.

Rose, A.H.; Lindquist, O.H. 1985. Insects of eastern spruces, fir and hemlock. For. Tech. Rep. 23. Canada: Canadian Forest Service. 159 p.

Stephan, B.R. 1987. Differences in the resistance of Douglas fir provenances to the woolly aphid, *Gilletteella cooleyi*. Silvae Genetica. 36: 76-79

Thalenhorst, W. 1972. Zur frage der resistenz der fichte gegen die Gallenlaus *Sacchiphantes abietis* (L.). Journal of Applied Entomology. 71: 225-249.

Tjia, B.; Houston, D.B. 1975. Phenolic constituents of Norway spruce resistant or susceptible to the eastern spruce gall aphid. Forest Science. 21: 180-184.

POSTER ABSTRACTS

RECOLONIZATION OF NORTHERN EUROPE BY GALL WASPS REPLICATE ANALYSES OF THE GENETIC CONSEQUENCES OF RANGE EXPANSION USING ALLOZYME ELECTROPHORESIS OF FOUR GALL WASPS IN THE GENUS *ANDRICUS* (*HYMENOPTERA:CYNIPIDAE*)

Rachel Atkinson and Graham Stone

Department of Zoology Oxford University, South Parks Road, Oxford OX1 3PS, UK

Andricus quercuscalicis, A.kollari, A. lignicola, and *A. corruptrix* are strict cyclic parthenogens; the spring sexual generation emerges from galls on *Quercus cerris* and the autumn sexual generation emerges from galls on *Q. robur* and *Q. petraea*.

Q. robur and *Q. petraea* have been re-invading northern Europe from the Balkans since the Pleistocene (Huntley and Birks 1983). Extensive anthropogenic planting of *Q. cerris* outside its native range for the last 200-300 years has resulted in range expansion of these four cynipids. Their northerly spread has been documented in detail: *Andricus quercuscalicis* reached the UK in 1950's and *A. lignicola* and *A. corruptrix* were first recorded in the 1970's. *A.kollari* galls were brought to England from the eastern Mediterranean in the 19th century to provide tannin for the dye industry. The species has also spread across Europe from its native range.

The genetic consequences of range expansion in the four species was examined using allozyme electrophoresis. Populations were sampled in the species' native range (Hungary) and across western Europe to the United Kingdom (N = 1090 for *A. quercuscalicis* N = 638 for *A.kollari*, N = 482 for *A. lignicola* and N = 141 for *A. corruptrix*). Data was analysed using GENEPOP (Raymond and Rousset 1995) and BIOSYS-1 (Swofford and Selander 1981).

Preliminary results suggested that populations in the native range had a higher heterozygosity and larger number of alleles (including private alleles) than populations at the northernmost edge of the expanded range. The pattern was found in all four species.

More extensive data sets for *A. quercuscalicis* indicated a progressive loss of alleles with distance from Hungary (Stone and Sunnucks 1993). While the pattern of isolation-by-distance was also found in *A. kollari*, populations in the UK shared rare alleles with populations from Spain, Italy and the Balkans that were absent from the rest of Continental Europe. There were also some alleles found on the continent but not in the UK.

The replicated northerly decline in heterozygosity and allele number may be explained by serial mild founder effects. The result is the gradual loss of alleles present at low frequencies with distance from the native range. The presence of alleles in *A. kollari* shared between the UK and Southern Europe may be explained by gall importation to England from populations with high genetic variability in the eastern Mediterranean.

While allozyme electrophoresis is a cheap, rapid and relatively east method of analyzing population structure, selective neutrality of the markers is controversial and allelic diversity per locus is often limited. An alternative genetic marker is the microsatellite. These short lengths of non-coding DNA are composed of di-tri and tetra nucleotide repeats. The large variation in repeat number between individuals can be used to ask more detailed questions about population structure than allozymes permit. We are searching for microsatellites at present using an enriched library and wish to use then in a comparative study with the allozyme work and as a tool to investigate patterns of egg-laying in cynipids.

REFERENCES

Huntley, B.H.; Birks, J.B. 1983. An atlas of past and present pollen maps for Europe: 0-13000 years ago. Cambridge University Press, Cambridge.

Raymond, M.; Rousset, F. 1995. GENEPOP (version 1.2): a population genetics software for exact tests and ecumenicism. Journal of Heredity. 86: 248-249

Stone, G.N.; Sunnucks, P. 1993. Genetic consequences of a invasion through a patchy environment - the cynipid gallwasp *Andricus quercuscalicis* (Hymenoptera:Cynipidae). Molecular Ecology. 2: 251-268

Swofford, D.L.; Selander, R.B. 1981. BIOSYS 1: a FORTRAN program for the comprehensive analysis of data in population genetics and systematics. Journal of Heredity. 72: 281-283.

INSECT-INDUCED GALLS IN THE GUANDAUSHI FOREST ECOSYSTEM OF CENTRAL TAIWAN

Sue-Yen Yang[1], Jeng-Tze Yang[2] and Ming-Yih Chen[1]

[1] Department of Botany, National Chung-Hsing University, Taichung, TAIWAN
[2] Department of Entomology, National Chung-Hsing University, Taichung, TAIWAN

The insect-induced galls in the understory of four forest types in the long-term ecological research (LTER) site in the Guandaudshi ecosystem were investigated from September 1994 to March 1996. The four forest types were virgin hardwood forest (L1), artificial plantation (L2), secondary hardwood forest, and burned forest. Environmental factors (temperature, light intensity and relative humidity) were also investigated using data loggers.

Forty-five morphologically different types of insect-induced galls were found to be induced on plants belonging to 16 families and 27 species. Differences in the host plants galled betwen forest types did not correlate with differences in vegetation composition or environmental factors among different plots. However, variations in numbers of the galls induced on a given host plant by specific gall makers, such as the different developmental patterns seen in *Syziginm buxifolium* and the special distribution types in *Machilus thunbergii*, were probably due to differences in altitude between sites. Based on internal histology and external morphology of galls, we classify host plants on the basis of their gall faunas. We also try to select bioindicators which may reflect environmental changes in the ecosystem.

CECIDOMYIID-INDUCED GALLS OF *MACHILUS THUNBERGII* HAY. IN THE GUANDAUSHI FOREST ECOSYSTEM OF CENTRAL TAIWAN

Sue-Yen Yang[1], Jeng-Tze Yang[2] and Ming-Yih Chen[1]

[1] Department of Botany, National Chung-Hsing University, Taichung, TAIWAN
[2] Department of Entomology, National Chung-Hsing University, Taichung, TAIWAN

Three kinds of galls were found on *Machilus thunbergii* Hay. in the Guandaushi forest ecosystem, central Taiwan. Two galls (the swan-shaped gall and the club-shaped gall) are induced on leaves, and one (the spindle-shaped gall) is induced on twigs. All these galls are induced by cecidomyiid gall midges. Affected parts, galling insects, gall size and the general structures of the galls were studied and were described. The frequency of gall occurrence at different altitudes, average incidence, and the composition of plant communities were investigated and analyzed. The gall-making strategy and percentage inquiline emergence were also discussed.

CHALCID PARASITOIDS IN OAK CYNIPID GALLS (HYMENOPTERA: CYNIPIDAE) IN THE CARPATHIAN BASIN

Csaba Thuróczy [1], George Melika [1] and György Csóka [2]

[1] Savaria Museum, Kisfaludy S. u. 9, Szombathely, 9701 HUNGARY
[2] Forest Research Institute, 3232 Mátrafüred, P.O.Box 2, HUNGARY

About 140 species of oak cynipids (Hymenoptera, Cynipidae) are known from Europe, around 100 species from the Carpathian Basin. Chalcid parasitoids were reared from 59 species of oak cynipid galls. Approximately 95 species of chalcid wasps are known as parasitoids in oak cynipid galls in Europe. A total of 53 species of chalcid parasitoids (Hymenoptera: Chalcidoidea) were reared from oak cynipid galls by the authors; 5 Eurytomidae, 11 Torymidae, 2 Ormyridae, 15 Pteromalidae, 5 Eupelmidae, and 15 Eulophidae.

In the Carpathian Basin we include Hungary, The Czech Republic, Slovakia and the Transcarpathian region in western Ukraine. Parasitoid species are listed under host galls and generations. Only those parasitoid species reared by the authors are listed. A parasitoid listed is not necessarily one that has developed on the gall inducing cynipid, it could have developed on a cynipid inquiline (*Cynipidae, Synergini*) or as a hyperparasitoid of the gall inducer or inquiline.

GALL MIDGE FAUNAS (DIPTERA: CECIDOMYIIDAE) OF FOUR MEDITERRANEAN ISLANDS

Marcela Skuhravá [1] and Vacláv Skuhravy [2]

[1] Czech Zoological Society, CZ-128 00 Praha 2, Vinicná 7, CZECH REPUBLIC
[2] Institute of Entomology, Czech Academy of Sciences, CZ-370 05 Ceske Budejovice, Bransiková 21, CZECH REPUBLIC

The gall midge fauna of the Mediterranean area is very little known. To date, only 12 gall midge species were known to occur in Sardinia, 9 species in Cyprus, 2 species in Crete and none at all in Mallorca. During our investigations of gall midge faunas in the Mediterranean in 1996 and 1997 we found 21 gall midge species in Cyprus, 23 species in Mallorca, 35 species in Sardinia and 39 species in Crete. The majority of the gall midge fauna of these four islands are species whose larvae gall host plant species are Mediterranean plants. European and Euro-Siberian gall midge species were less common. Mediterranean gall midge species gall leaf buds, leaves, flower buds and fruits of host shrubs and trees. In contrast to European species, whose larvae usually leave their galls to hibernate or pupate in the soil, larvae of Mediterranean gall midges usually hibernate and pupate inside their galls. This is probably an adaptation allowing avoidance of unfavorable environmental conditions during the Mediterranean summer, in which there is little rainfall and daily temperature maxima may sometimes exceed 40°C.

Our investigations, together with data on the gall midges of Sicily and Corsica gathered by earlier authors, make it possible to evaluate the gall midge fauna of the whole Mediterranean area from a zoogeographical point of view.

REVIEW OF GIRAUD'S TYPES OF THE SPECIES OF *SYNERGUS* HARTIG, 1840 (HYMENOPTERA: CYNIPIDAE)

Juli Pujade i Villar and Palmira Ros-Farré

University of Barcelona. Department of Animal Biology. 645, Diagonal Ave. E-08028-Barcelona, SPAIN

Joseph-Etienne Giraud (1808-1877) was one of the greatest cynipidologists of the last century. His studies, though few in number (Giraud 1859, 1866, 1868a, b, 1871; Laboulbene 1877), greatly increased the numbers of known cynipid gall-forming species. Part of the material studied by this author was compiled in an unpublished manuscript deposited in the MNHN in Paris. Houard (1911) copied Giraud's manuscript. In this study a total number of 11 new *Synergus* species were described: *S. apertus, S. cerridis, S. clavatus, S. conformis, S. consobrinus, S. diaphanus, S. hartigi, S. inflatus, S. longiventris, S. subterraneus,* and *S. vesiculosus*.

The only one of these species that has been examined is *Synergus apertus*, which was synonymised with *Saphonecrus undulatus* by Pujade & Nieves-Aldrey (1990). In this work the typical series of the remaining *Synergus* species described by Giraud are studied, as well as all the specimens of this inquiline genus which were collected by this author and deposited in the MNHN.

The study of the *Synergus* species type-material is especially interesting in the Palaearctic zone because a total number of 31 of the 59 described species are considered dubious.

GALL-INDUCER AND GALL-INQUILINE CYNIPIDS (HYMENOPTERA: CYNIPIDAE) COLLECTED IN ANDORRA WITH A MALAISE TRAP

Palmira Ros-Farré and Juli Pujade i Villar

University of Barcelona. Department of Animal Biology. 645, Diagonal Ave. E-08028-Barcelona.

From August 1992 to December 1993, a Malaise trap was installed (Pujade 1997a) in an extreme evergreen oak grove located at an altitude of 1,050m in the district of Santa Coloma (Andorra). More than 450 cynipid specimens were captured (Segade *et al*. 1997), which were identified in this study. Species belonging to the tribes Rhoditini, Cynipini and Synergini were collected. The total absence of Aylacini captures is surprising.

Plagiotrochus amenti was collected in the study locality although the nearest individuals of its host, *Q. suber*, are 50 km away from the place where the Malaise trap was installed; the origin of this capture is discussed. The inquiline species: *Saphonecrus lusitanicus* and *S. barbotini*, typical of galls detected in *Q. ilex* dominated the number of captures of this biological group. Regarding the gall-forming species, the abundance of *Plagiotrochus britaniae* and *Neuroterus aprilinus* was notable.

The temporal dynamics of the species represented by a larger number of specimens is studied. The usefulness and efficiency of the Malaise trap in the study of the Cynipidae communities is discussed, comparing these results with those obtained through active sampling methods based on gall collecting (Pujade 1994a, b, 1996, 1997b, c).

GALL MIDGE-PARASITOID INTERACTIONS IN A PATCHY LANDSCAPE

Olof Widenfalk

Swedish University of Agricultural Sciences, P.O.Box 7044 S-750 07 Uppsala, SWEDEN

The population structure of a monophagous gall midge and its two parasitoids on a patchily distributed herb were studied at the landscape scale (200 km^2).

The gall midge was abundant in a majority of patches and its occurrence increased with patch size, whereas densities actually decreased as patches became larger. No effects of isolation were found.

The attack rates and the relative role of the two parasitoids varied much over the landscape. This variation was affected by patch size, surrounding environment, differences in spatial density dependence and competitiveness.

INDUCED RESISTANCE IN WILLOW AGAINST A GALL-FORMER: ACTIVE DEFENSE OR LACK OF RESISTANCE?

Peter Saarikoski, Solveig Höglund and Stig Larsson

Department of Entomology, Swedish University of Agricultural Sciences, Box 7044, 75007 Uppsala, SWEDEN

The aim of this project is to investigate the mechanisms behind resistance in *Salix viminalis* against a gall-forming insect, *Dasineura marginemtorquens* (Diptera: Cecidomyiidae). Earlier research has documented great genetic variation in resistance among willow genotypes. The resistance is manifested through high larval mortality. Patterns of the distribution of galls on resistant and susceptible willows in field situations strongly indicate the presence of systemic induced resistance. If so, this would be one of the first examples of intraspecific genetic variation in induced resistance against insects. We propose to test this hypothesis by means of a series of field and laboratory inoculation experiments. If the induction hypothesis is supported in these experiments, then another series of experiments will be designed to seek molecular mechanisms behind resistance. Two alternative explanations are proposed in order to explain the resistance: (1) The *defence explanation* predicts that substances toxic to the larva are produced as a result of the induction. (2) The *recognition explanation* suggests that the resistance acts by inhibiting gall formation whereby young larvae will starve to death. In order to gain more insight into the process of resistance, we are screening for differentially expressed genes in resistant and susceptible willow genotypes. In this project we are using a novel system of RNA fingerprinting termed cDNA-AFLP.

EXPRESSION OF GALL MIDGE RESISTANCE IN A WILLOW GENOTYPE: EFFECTS OF SHADING

Solveig Höglund and Stig Larsson

Swedish University of Agricultural Sciences, Department of Entomology, Box 7044, 750 07 Uppsala, SWEDEN

There exists great genetic variation in resistance among genotypes of *Salix viminalis* against the gallmidge *Dasineura marginemtorquens*. The resistance is manifested through high mortality of newly hatched larvae. Ovipositing females do not discriminate against resistant genotypes. There is, however, variation in survival within the same resistant clone over time. This could be due to environmental effects. Differences in larval survival were examined on a resistant genotype growing in two different environments, sun and shade. Although the two environments resulted in plants with different leaf morphologies (toughness, specific weight) there was little difference in larval survival.

BROAD OVERVIEW OF A GALL-SYSTEM: FROM PLANT ANATOMY TO ECOLOGICAL INTERACTIONS

Milton Mendonca Jr. [1]*, Helena Piccoli Romanowski[1] and Jane Elizabeth Kraus[2]

[1] Curso de Pos-Graduacao em Biologia Animal, Instituto de Biociencias, UFRGS; Av. Paulo Gama, s/n:, Porto Alegre, CEP 90040-090, RS, BRAZIL
[2] Dept. de Botanica, Instituto de Biociencias, USP; Caixa Postal 11461, Sao Paulo, CEP 05422-970, SP, Brazil.
*current address: Dept. of Biology, Imperial College at Silwood Park, Ascot, Berks, SL5 7PY, UK
*current address: Dept. of Biology, Imperial College at Silwood Park, Ascot, Berks, SL5 7PY, United Kingdom.

This work aims to give a broad overview of a biological system centered on a gall and its gall-maker, considering ecological relationships between component parts. The gall is induced by *Eugeniamyia dispar* (Cecidomyiidae), a newly described species, on young leaves and stems of *Eugenia uniflora* (Myrtaceae), a native southern Brazilian shrub. This insect is multivoltine, with up to nine generations a year. One parasitoid (*Rileya* sp., Eurytomidae) and three predatory ant species (*Pseudomyrmex* sp., *Pheidole* sp. and an unidentified species) attacked the gall-maker larvae.

Data resulted from: (i) 2 years of weekly field sampling, (ii) laboratory rearing, and (iii) anatomical studies of the gall (hand-cut and microtome sections for light microscopy; also scanning electronic microscopy).

E. uniflora leaves show a delayed-greening pattern, making it possible to assign them an approximate age according to their color. New leaves are produced all year round except for a short period (about 1.5 months) of "quiescence" corresponding to the winter. Quiescence can start earlier for shaded plants (late summer). There is a consistent burst of leaf production in late winter among plants. During the rest of the year, three lower peaks of leaf abundance can be distinguished. There is no synchronization between gall-maker oviposition dates and leaf production peaks, nor any relation between new leaves' abundance and number of shoots galled per generation. However, plant leaf phenology determines the availability of induction sites, so limiting the occurrence of gall-makers (and parasitoids) in space/time, for the insects are absent during the "quiescent" period.

The plant also hampers larval gall induction by a localized hypersensitive response. Dark spots on the induction sites result from tissue necrosis after accumulation of phenolic substances and development of a cicatrization tissue. Smaller plants showed a more pronounced hypersensitive reaction (ANOVA, $F = 4.456$, $p = 0.012$, $n = 5$), but no differences in mortality rates were found among leaves with different gall densities (regression analysis, $F = 0.0192$, $p = 0.891$). The host plant response to this insect parasite is qualitative; and does not seem to occur in a quantitative scale.

E. dispar galls are unilocular, white, round swellings formed by hypertrophy rather than hyperplasy. They are usually 3.5 to 4 cm diameter. Column-shaped cortex cells (mesophyll) are responsible for gall volume, shape and a spongy consistency. There is a core region (chamber ceiling) acting as a nutritive tissue. High quantities of starch are found in all cells. No schlerenchyma is found in the gall, and the larva thus gets no protection from gall hardness. Gall anatomy traits probably provide the gall-maker with protection from desiccation (cortex cells are liquid-rich), parasitoid oviposition (mature galls are thick) and from chewing herbivore attack (phenolic substances present on cortex cells, *Phocides polybius* (Lep.: Hesperiidae) caterpillars avoid the galls when eating the leaves). They may also offer an energy source for the parasitoid larvae and ants inside the gall through sugary (starch derived) exudates from the nutritive tissue.

Gall aggregation on leaves seems to influence parasitoid but not ant attack. *Rileya* sp. attack galls primarily on higher gall density leaves in each generation, but ants seemed to attack galls irrespective of their densities on leaves or shoots.

The gall inducing habit adds another component to the community, a compartment in which energy is perhaps extracted and exchanged in a more efficient way than in normal plant tissues. One part of the community is interacting tightly with another creating a different microhabitat, or an ecosystem subset, where ecological processes can happen differently from the rest of the community.

TWO SPECIES OF EURYTOMID GALL FORMERS WITH POTENTIAL AS BIOLOGICAL CONTROL AGENTS OF STRAWBERRY GUAVA, *PSIDIUM CATTLEIANUM*, IN HAWAII

Charles Wikler[1]* and Jose Henrique Pedrosa-Macedo[1]

[1] Universidade Federal do Paran, C. P. G. E. F. Rua Bom Jesus, 650 -80.035-010 Curitiba-PR, Brazil
* Current Address: International Institute of Biological Control, Silwood Park, Buckhurst Road, Ascot, Berkshire - SL5 7TA, UK e-mail: c.wikler@cabi.org

Psidium cattleianum (Myrtaceae), known as strawberry guava, is a fruit plant originally from Brazil. It was introduced to Hawaii in the last century where it has become a serious weed in all the islands. Collections of host-specific natural enemies were made within its native range in the state of Parana, Brazil. Among the insects, two eurytomids showed potential as biocontrol agents. An unidentified species of *Eurytoma* (Hymenoptera: Eurytomidae) causes a gall in the stems, limiting its growth and reducing the flowering and fruit production of strawberry guava. The gall is formed after the oviposition of the insect in young stems and 1 year after adult emergence, the gall dries out and in the following year the stem dies. A species of *Sycophila* spp. makes its oviposition in the seeds of strawberry guava enclosing them in a hard woody gall. Seeds failed to germinate in the fruits where this gall was found. The main biological and ecological aspects of both gall formers were studied under laboratory and field conditions to investigate the possibility of their use as specific biological control agent for strawberry guava in Hawaii.

Csóka, Gyuri; Mattson, William J.; Stone, Graham N.; Price, Peter W., eds. 1998. **The biology of gall-inducing arthropods.** Gen. Tech. Rep. NC-199. St. Paul, MN: U.S. Department of Agriculture, Forest Service, North Central Research Station. 329 p.

    This proceedings explores many facets of the ever intriguing and enigmatic relationships between plants and their gall-forming herbivores. The research reported herein ranges from studies on classical biology and systematics of galling to molecular phylogeny, population genetics, and ecological and evolutionary theory. Human kind has much to learn and gain from understanding the fine details of how plants and their gallers interact.

KEY WORDS: Biogeography, biodiversity, phylogeny, evolution, resistance, plant-herbivore interactions, Cynipidae, Cecidomyiidae, and Tenthredinidae.

Our job at the North Central Forest Experiment Station is discovering and creating new knowledge and technology in the field of natural resources and conveying this information to the people who can use it. As a new generation of forests emerges in our region, managers are confronted with two unique challenges: (1) Dealing with the great diversity in composition, quality, and ownership of the forests, and (2) Reconciling the conflicting demands of the people who use them. Helping the forest manager meet these challenges while protecting the environment is what research at North Central is all about.

www.ingramcontent.com/pod-product-compliance
Lightning Source LLC
Chambersburg PA
CBHW081234180526
45171CB00005B/420